Tri-ang HORNBY

The Story of Rovex
Volume 2 1965-1971

PAT HAMMOND

New Cavendish Books

DEDICATION

This book is dedicated to my wife and family
without whose patience and understanding
this volume would not have been completed.

**First edition published in Great Britain
by New Cavendish Books – 1998
Reprinted 2003**

Volume 3 by Pat hammond covering the story of the system from 1972-1996,
under the `Hornby Railways´ name, will be published in 2004.

The publishers thank Rovex Ltd and subsequently
Hornby Hobbies Ltd for the use of the name
Tri-ang Hornby and the reproduction of certain pictorial and text material.

Please note that all items illustrated in this book are no longer produced
and any enquiries should be addressed
for the `authors attention´ at the publishers address.

Design & typesetting – John B Cooper
Editorial & production director – Narisa Chakra

Printed and bound in Thailand by Amarin Printing & Publishing Co., Ltd.
under the supervision of River Books, Bangkok.

New Cavendish Books
3 Denbigh Road, London W11 2SJ
Email: sales@cavbooks.demon.co.uk
www.newcavendishbooks.co.uk
ISBN 1 872727 58 1

Contents 3

As with **Volume 1**, the success of this book is dependant upon the work of many people. I have again been astounded by the freedom with which people have given their time to conduct research on my behalf and to read and correct drafts from the manuscript.

Once again, pride of place must go to Richard Lines who has been especially generous with his time which I understand is even more restricted now that he is in retirement doing all those things that for years he planned to do when the time came. Most of the records of the company, the reasons why things were done and who said what to whom, have come from his memories of a lifetime in toy making. He has seen many changes in the industry, not always for the better, but he is rightly proud of the achievements of the company and has been one of its principal advocates of quality.

Richard, and later John Cains and Tom Carfrae in whose capable hands he left me, provided me with access to factory records and the archives both of which have been important sources of information.

IN BRITAIN
I am also indebted to many people who have allowed me to study their collections and photograph models for the book. These include:
Laurence Moger, Tony Blackman, Stuart Bean, Chris Greenwood and Matthew Petzold.
Specialist information was received for the following topics:
Tri-ang Big Big – Tim Curd, Dave Moss and John Foreman
Tri-ang Wrenn – Stuart Bean
Electrical Equipment – Malcolm Pugh
Model-Land – James Day
Minix – Brian Salter, Robert Newson, David French and James Day
Minic Motorway – John Rogers and James Day
Tri-ang for France – Malcolm Pugh
Crown Railways – Peter Hilton and Tony Stanford

Arkitex Station – Martin Jarrett
Arkitex – Peter Gurd and James Day
Rovex Tri-ang Liquidation – John Rogers
Minic TMNR – Malcolm Pugh
New Zealand Models – Matthew Petzold
Battle Space – Nigel Swain
Thanks also go to Alistair Rolfe, who through his early study and listings of Tri-ang models while at Platform 2 (the spiritual home of the Tri-ang Hornby Collector's Club of the 1970s) set the foundation stone of these books. He has provided copious notes from his archives.

Further information was received from:
Matthew Petzold, Bob Palmer, Chris Wyman, Jack Holmes, Robert Hampton, Edward Hunter, Tony Le Grys, Peter Randell, Andy Wakeford, Bob Leggitt, Rolande Allen (who has an eye for the unusual), Tony Brown, Bernard Taylor, Clive Washbourne and Neil Bowsher.

Models have been supplied by:
Bill Bourne, Peter Corley, John Ridley, Charlie Pickup, Michael Greenwood, Rolande Allen, Russel Foster, Richard Doyle-Davidson, A P A Wykes, Adrian Barnard, Malcolm Pugh, Geoff Killick and Matthew Petzold.
My special thanks go to the late Gary Cordier who worked for a number of years as a production engineer at Margate and who has been an excellent source of information both for this volume and the next. He had immeasurable patience as he tried to explain to me the finer design details of Tri-ang Hornby locomotive mechanisms and the problems that occasionally arose when things went wrong. He will be much missed both as a friend and helper.
I also acknowledge the British Transport Commission who were the source of some of the photographs of railway scenes and Hornby Hobbies for the studio pictures and other material.

IN AUSTRALIA
My principal contact in Australia this time has been Dennis Johnson who has an uncanny knack of turning up Australian sets not previously recorded. Several of the photographs of Australian sets were supplied by him. Further information and pictures were received from: Gary Pola, Teddy Schaefer, Peter Yaxley, David Neale (including pictures of the former Australian factories), Peter Duckmanton and the late Howard Lever. Bronte Watt provided useful historical notes that included the disposal of the Australian and New Zealand factories in 1971.

IN CANADA
I am particularly indebted to Gary McKinnon who has researched the Canadian scene for me and provided extensive information used in the book. In this I know that he has been helped by some of Canada's leading Tri-ang collectors including David Cochrane, Chris Clarke and Peter Zimmerman. Much of the American information also came from Gary McKinnon. My thanks to you all.

IN NEW ZEALAND
Henry Deer has continued to channel information to me and I am particularly grateful to Paul Gourley of Auckland who checked all New Zealand references for me and provided additional literature, several of the photographs used in this volume and information about strange local practices.

IN HOLLAND
Michael Peters advised me on the marketing of Tri-ang Hornby in Holland.

My thanks to you all. Without your help this would have been a far less interesting book.

FOREWORD

When I started selling model railways in the early 1980s, serious collecting largely meant Hornby O gauge and even Hornby-Dublo had a limited following. Tri-ang collectors were few and far between and usually viewed with something nearer pity than amused tolerance! It was obvious to me, however, that the huge scale of Tri-ang production must have resulted in many variations and oddities.

Delighted to discover my original Tri-ang No.1 catalogue, I soon found that original literature was easily accumulated. One thing led to another and although not originally a Tri-ang Railways boy, I soon became enamoured of the oddities of the system, with patient and friendly guidance from Pat Hammond and other knowledgeable collectors.

In many ways the publication of **Volume 1** in this series really established Tri-ang Railways as a 'mainstream' collectable. Certainly, a Tri-ang item at any toy auction today commands brisk bidding and achieves a respectable price. This does highlight the intrinsic appeal of the toy which is not only an acceptable scale model but also a realistic working miniature.

The Tri-ang Railways period ended with the acquisition of the ailing business of Meccano Ltd, makers of Hornby-Dublo. Then a Dublo owner myself, I well remember the shock on hearing of its takeover by the makers of the rival system but, in practice, the Dublo boys did very well as shops cleared their old stock at bargain prices. The new 'merged' Tri-ang Hornby system had little connection with the former Hornby-Dublo, however, the purists who look askance at 'Margate' Hornby should reflect that the name 'Hornby' has been synonymous with the best of British model railways for well over 70 years.

The name 'Tri-ang Hornby' always looked a little uncomfortable (I felt that it should be 'Tri-ang-Hornby') but the 1965-71 period quickly proved to be the zenith of the railway system and its associated products such as Minic Motorway, Model-Land and Arkitex. Although development of track, rolling stock and lineside accessories was not overlooked, the wide range of locomotive power had always been the envy of other systems and this range was to be increased by the introduction of company liveries at the end of the 1960s. From Stephenson's 'Rocket' to the Midland Pullman and Polly to the E3001, there was a locomotive for every age and pocket and competitive pricing was always a key consideration in the development and marketing of this range.

This, indeed was the most enduring feature of the Tri-ang Hornby system and its successor, Hornby Railways. Today's youngsters can still buy wagons new or used at pocket money prices and locomotives for only a little more. Let us hope this accessibility for the novice will continue to outweigh the pressure of the collecting market where many other collectables have become too expensive for the beginner's pocket.

Pat Hammond's comprehensive treatment of this important and popular period of Rovex production will resolve many queries for the collector and historian alike; the completion of the series by the third volume will be eagerly awaited.

Peter Gurd
NOVEMBER 1997

PREFACE

Among Tri-ang Collectors, the 1960s have long been a period of special interest. In the 1950s Tri-ang Railways were rather toy-like. The coaches were short, the wagons strange colours and the pride of the fleet in 1959 was still an under-size Princess Class locomotive which had first made its appearance in 1950.

My childhood lay in the 1950s and consequently, this was the period in the history of the company that initially interested me most of all. Anyone who does not feel the magic of these early toy-like trains, may well have a greater affinity for the models of the next decade; one in which many of today's collectors had their childhood. I missed out on the Tri-ang Hornby years as they were that time, which comes to every young man, when other attractions seemed more important.

Once I had settled down and my interest was re-awakened in the late 60s and early 70s, I started to collect a wide range of British makes of model railways. Initially I developed good collections of Hornby-Dublo and Trix but, after a visit to see Albert Chaplin's collection in Essex, my prime interest became those more obscure makes that appeared after the Second World War and survived only a short time before fizzling out. Names like Gaiety, Ever-Ready, Rowell, KMR, Acro and Nucro, Kirdon and Scalemaster had a special attraction for me for a while.

It is difficult to say when Tri-ang started the big fight back into my affections but I think it was when I joined the Tri-ang Hornby Collector's Club. Interest in collecting vintage toys started in the 1960s but exploded into activity in the 1970s. This was when the main railway collectors clubs sprang up and I joined them all. The publishing industry had been slow to react to the demand for information about vintage toys and so collectors came together to share their knowledge and swap their spares and this was the origin of Swap-meets.

The Tri-ang Hornby Collector's Club was principally run by John Shimwell from his shop, Platform 2, in Wimbledon and with a band of fellow enthusiasts, including Alistair Rolfe who has helped with this book, he set about establishing lists of Tri-ang's locomotives, coaches, wagons and sets. For the first time, the finer detail differences between chassis, couplings and wheels were recorded. I became fascinated by the immense variety to be found in the Tri-ang range, a richness that set it apart from any of the other makes, and I have tried to pass that enthusiasm on to readers in **Volume 1**.

When the club eventually folded, my interest remained and I extended the classifications started by the club, throughout the Tri-ang Railways, Tri-ang Hornby and Hornby Railways ranges. I continued to research the subject and collect the model variations but, with the release of Michael Foster's excellent book on Hornby-Dublo, I felt that I wanted

to do the same for the products of Rovex. I quickly obtained the agreement of New Cavendish Books and the six years that it took to produce the first volume was a labour of love.

With the publication of **Volume 1** out of the way, I was able to turn my attention to the Tri-ang Hornby years. Those lost years of my youth. I had assumed that, as **Volume 1** had covered 14 years of considerable growth and change, **Volume 2**, in covering only seven years when individual models varied little, would be only half the size. Far from being a subdued period it has proved to be a most fascinating one. The range of products made was much greater than I had imagined from looking in the catalogues of the period and there were so many side alleys up which the story led me and many fascinating previously untold tales waiting to be told. In consequence of this, it has been difficult to keep this volume to a reasonable size.

INTRODUCTION

This is the second of three volumes planned which together will span the period 1950 to the present day, whenever that is. The trilogy deals with the Rovex company, and its successor Hornby Hobbies Ltd, and the country's most successful model railway system which was developed by them over half a century. In this respect it would be well to clarify from the outset that the products sold in the shops today as Hornby Railways are the direct descendants of Tri-ang Railways and not of Hornby-Dublo with which they are sometimes confused.

The system's roots can be traced back to 1950 when a small company, Rovex Plastics Ltd started manufacturing a train set for Marks and Spencers. Short of money, the company sought an investor and found Lines Bros. who were destined to become Britain's largest toy company and whose products were sold worldwide under the name of Tri-ang.

Lines renamed the model railway Tri-ang Railways and Rovex Plastics Ltd was renamed Rovex Scale Models Ltd as, in those days, 'plastic' implied an inferior product. They moved production to a purpose-built factory at Margate in Kent; the home that it still occupies today.

Thanks to Lines Bros. experience in toy making, Tri-ang Railways quickly became the market leader and in 1964 Lines Bros. Ltd took over their principal rival in the UK, Meccano Ltd, who made the Hornby-Dublo model railway system.

Production of Hornby-Dublo had ceased by then but in order to placate former Dublo customers, Rovex renamed their own model railway system 'Tri-ang Hornby'. At the time, the process was described as a merging of the two great systems but in reality, with the exception of absorbing a few Hornby-Dublo building kits into their range for a short period, Tri-ang Railways continued little changed under the new name. The Hornby-Dublo surplus stock and tools were sold off to a subsidiary

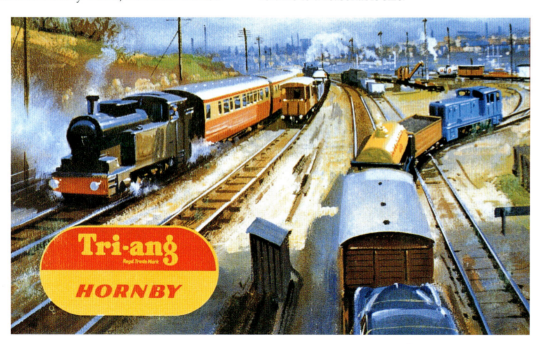

who re-launched the system first as 'Tri-ang Wrenn' and then as 'Wrenn Railways'.

This volume describes how Rovex Scale Models Ltd developed Tri-ang Hornby between 1965 and 1971. These were lean and hungry years during a recession in the model railway industry and the book looks at how Rovex weathered the storm and emerged a strong and healthy company despite the collapse of its parent company in 1971.

In January 1972, the model railway system was again renamed and **Volume 3 - *Hornby Railways*** will tell how the system developed under its new name.

LAYOUT OF INFORMATION

The task of deciding where in the book a particular item should appear is a very difficult one. Rovex were constantly changing their mind when designing the price list. Should buffer stops, for example, belong with 'Track' or with '**Buildings**' and what should be included under the title of '**Sundries**'? While Rovex may have changed their minds over the years, for the sake of this series of books, we have kept to a single pattern throughout.

This means that the reader or researcher who is familiar with using **Volume 1** will quickly be able to trace what they are looking for in **Volume 2**. The same chapter headings have been used and the main chapters have been broken down into the following three sections:

Introduction

This provides a chronological account of the changes that occurred to the range over the period covered by this book as though the reader were looking through the annual catalogues one by one. Major changes in style such as the adoption of pre-nationalisation liveries at the end of the 1960s or the change from brown to bright red station buildings are dealt with here.

The Introduction to each of the major chapters provides a summary of the range and is probably the most suitable part for those desiring only a nostalgic read.

Check Lists

The check lists used in **Volume 1** proved so popular that this method of providing information has been extended in this volume. Following each Introduction, a list of the models dealt with in the chapter provides a quick source of basic information. In all cases the 'R' number and dates of availability are provided and in some cases the original price (earliest known Tri-ang Hornby price) and the minimum quantity available to retailers when ordering, is also given.

The latter is of interest as it is a guide to how some of the smaller items were packaged. This number varied over the years and so the earliest information for the Tri-ang Hornby period is given. Towards the end of the period retailers were required to buy larger quantities than they had at the start; e.g. three signal boxes instead of one. An entry such as '2x6' suggests that they were sold to retailers in dozens but were packed in boxes of six.

In other cases, additional detail such as type of wheels, couplings and chassis is provided. In each case consideration has been given to what is most likely to interest collectors, subject to the space available.

Directories

The last part of the chapter contains the bulk of the information and describes each model in detail. The notes may include details on the prototype, the reason for its choice, the date it first appeared, the number made and the variations that may be found. Where models have already been dealt with in **Volume 1**, information given in this volume is likely to be limited to numbers and variations relevant to the Tri-ang Hornby years.

Index

Much time and effort went into the 12 page index in **Volume 1** as it was felt that this was an important starting point in a search for information. The same attention has been given to the index for **Volume 2**.

DATES

Although every effort has been made to inform the reader of the dates when items were available it is extremely difficult to be definite about these. Before work on this book was started a decision had to be made as to whether the dates of availability of a model should be those when it was listed in the price list or illustrated in the catalogue (assuming that it actually appeared in the catalogue), the dates they were known to arrive in the shops or when they were still available in the stores at Margate, or the years they were actually manufactured.

The easiest solution for me as the writer would be to stick to published dates but these can be misleading. Instead, I decided to try to be more accurate and to apply greater realism to the information provided here in the hope that it may be of more value to the reader. I have therefore, wherever possible, quoted the years that the model was in production. Thus it will be found that many of the dates do not correspond with those normally assumed.

While I believe that they provide a better guide than price lists, etc., I cannot claim that they are indisputable. At the end of the day knowledge is based on available information plus a fair amount of guesswork. In this case, much of the knowledge comes from the census sheets kept by the company. These were similar to an annual stock take of what passed through the stores in a given year but anomalies frequently occur for which an explanation has not been found.

Now enjoy the book.

THE COMPANY

BACKGROUND

As we have seen, Rovex Plastics Ltd had been founded in 1946 and acquired by Lines Bros. Ltd in 1951. Its model railway system was re-launched as Tri-ang Railways in 1952 and the company was renamed Rovex Scale Models Ltd in 1953. Production was moved to a purpose-built factory at Margate in 1954 and, by the end of the decade, Tri-ang Railways had become the market leader.

Up until 1964, Rovex Scale Models Ltd had been a subsidiary of Lines Bros. Ltd, a company which manufactured Tri-ang toys in its own factories in Birmingham, Merton, and Merthyr Tydfil. In 1964 its manufacturing function was transferred to a new company set up for the purpose and called Tri-ang Toys Ltd and Lines Bros. Ltd became a holding company for the group.

The Lines Dynasty

Lines Bros. Ltd had been formed after the First World War by the three brothers – William John Lines, Arthur Edwin Lines and Walter Lines. They were later joined by their fourth brother George. W.J. Lines gave up executive duties in 1945 but remained a director until his death in 1962. Walter Lines was Chairman and Managing Director and Arthur Lines was Deputy Chairman and Joint Managing Director at the time he retired in May 1961. The latter was replaced by Walter Moray Lines, son of Walter Lines, whom he succeeded as Chairman when Walter Lines became President for life.

With the disappearance of the founding brothers in the early 60s, four sons, Walter Moray Lines, William Graeme Lines, Arthur John David Lines and Hugh Richard Lines, became joint Managing Directors of the vast toy empire. Moray Lines had decided that he did not wish to run everything himself as his father had done and so he divided up the group into a number of units, each with a degree of autonomy.

Richard Lines, the son of Arthur Lines, was responsible for Rovex Industries Ltd, while the factories at Belfast, Merton, Merthyr Tydfil and Birmingham became the charge of his brother, John Lines. Graeme Lines was responsible for Marketing and European Sales. While the divisions were to have a degree of independence they all had their headquarters at the Merton factory. Rovex Industries was in the old IMA building and it was from here that sales were controlled.

Rivals

The mid 1960s were a strange time. The model railway market was in decline, production of Hornby-Dublo had ceased and there was the question of what to do with the tools. To add to the confusion, the owners of British Trix were looking for a buyer and approached Lines Bros., while Lone Star were wanting Lines to buy their N gauge Trebl-o-lectric system from them.

MECCANO LTD

Lines Bros. Ltd had bought Meccano Ltd in February 1964 for their name. They had no plans to amalgamate the railway systems, indeed, when Richard Lines visited the Binns Road factory soon after the takeover he found the whole of the train department closed down and large stocks of unsold products. He asked Joe Fallmann, the Managing Director at Meccano Ltd, to supply him with a list of the previous year's sales. This was the first time they discovered how bad things were at Liverpool.

Meccano had damaged themselves in their conversion of Hornby-Dublo to two rail and a major contributing factor had been the complicated trackwork that resulted from the adoption of live frogs instead of self-isolating points. The chief draughtsman had been a scale railway modeller and the resulting track was too fragile for children to use. By the time they rectified the points problem with the introduction of the Simplex points the battle had been lost. Too many people had abandoned the system for Tri-ang and other makes. Meccano's loyalty to its three-rail customers had also been damaging. It is hard enough to profitably manufacture one system let alone two.

An account by a former employee in the marketing department at Binns Road describes the company as being out of touch with its customers. When Meccano eventually changed to plastic moulding for its wagons it had to produce a higher quality model than anyone else. The search for quality without the essentials of common-sense economy, was expensive and the new tooling swallowed up vast sums of money. Products were also issued too fast for the public to keep pace with them and so they sold in smaller numbers than had been envisaged.

The Early Days of Success

The problem had been personalities. In 1910 Frank Hornby took on Jones and Beardsley. Beardsley

The former Richmond factory c.1994.

The Meccano factory showing Frank Hornby with a group of guests.

The Margate factory in the late 1960s.

Hornby-Dublo literature.

looked after production and Jones dealt with sales. The team worked extremely well.

George Lines, who was a trained engineer, went ever year to the Leipzig Fair with Beardsley and they were great friends. Both were always on the lookout for ways of streamlining production. Meccano marketing was excellent. Wherever you went in the world there was a Meccano price list in the language and currency of the country.

Death of Frank Hornby

On the death of Hornby in 1936, his son Roland became Chairman of the company and George Jones became Managing Director. For a while the firm continued to prosper. Jones understood the market and one of his successes was the introduction of Hornby-Dublo. The War brought plenty of work but afterwards things became difficult. When Jones died, Beardsley took control but as a production man it seems that he did not have the same flair as Jones for marketing.

Trains could not be made in those early post-war years and so the company concentrated on Dinky Toys and small Meccano sets. In the meantime other people were coming into the trains market and Meccano Ltd did not do enough to counter the challenge. They were slow to change to BR livery, and with no one there to say "I want it next month", there was a lack of urgency which meant the market lead was slowly lost.

The Slippery Slope

Meccano carried on making Dinky Toys just as they had always done despite competition from Matchbox and Corgi. Meanwhile, Lego established a lead in the construction kit market and again Meccano failed to respond adequately to the changes taking place and to recognise the importance of plastic as the material of the present let alone the future.

Dale, another production man took over after Beardsley but he did not stay for more than a couple of years. It seems that Roland Hornby's influence in the organisation was limited and that he was happy to leave control of the company to the Managing Director.

At toy fairs Meccano staff would visit the Rovex stand to see what the competition were doing, but they gave the impression that they saw Hornby-Dublo as being on a different and higher plane to Tri-ang which was just a toy. The fact that Tri-ang was improving and expanding so fast that it was destroying their business seemed to be unimportant. The company's strength was felt to reside in its past reputation irrespective of whether the public still wanted its products. It was this complacency in the late 1950s which did much to create the problems of the early 1960s.

Meccano brought in an accountant called Tattersall who was new to the toy business. He looked around for new things for Meccano to market but none were successful. Too late Dinkys were fitted with windows.

Too Little Too Late

Too late they brought in an Austrian, Joe Fallmann, who was very good and saw the potential of Dinky Toys based on film and TV series. He later worked well with Lines, but the old guard continued to worry more about devaluing Dinky than about selling products.

Plastic Meccano, which was eventually developed at Canterbury by Moray Lines in a unit attached to Minic, would have been unthinkable at the time when it could have helped save the company, but this and a colour change to the main range, when introduced by Lines, helped to revitalised Meccano. Dinky toys were also up dated under Lines management.

At the time of the takeover, the manufacturing costs at Meccano were found to be very high and the company over staffed. There were development staff but little development was taking place. A few weeks later they had all gone with little effect on the company. Lines production facilities were far better than Meccano's and Lines set about revitalising the company.

AMALGAMATION

What Plans?

In an interview for *Model Railway Constructor* published in February 1966, Richard Lines was asked how much of the original Hornby-Dublo system Rovex intended to continue. In reply he said:

"When we acquired Meccano Ltd we immediately sought the reason why the excellent Hornby-Dublo railway system had sold less well in recent years. Amongst our conclusions were the factors of price and a relatively elaborate electrical system. We found that the reason for the price difficulty was largely due to the use of the more traditional materials and methods of manufacture and, consequently, the tooling was generally not as up-to-date and economic to use as comparable tooling on Tri-ang Railways. On the other hand, there were a number of models which were quite different from any in the Tri-ang system.

"Adding together these factors we came to the conclusion that it was desirable to amalgamate the systems and convert certain Hornby-Dublo items so that they could be fully integrated with Tri-ang models. Reference to our 1966 Tri-ang Hornby catalogue will show which items are maintained as Tri-ang Hornby, but in addition to this there are still a large number of original Hornby-Dublo models available from various model railway stockists and we anticipate that this position will remain for at least another year. We will be introducing other new locomotives and it is intended that these will have a 'Hornby' influence."

At this time there were still 30 Hornby-Dublo wagons available and Rovex produced a price list for these and other surplus stock. It was also stated at the time that Rovex intended to undertake servicing of Hornby-Dublo models at Margate.

The Future of Tri-ang TT

On the question of continuing TT production, Richard Lines said at the start of 1966:

"We have certainly had some problems with TT. Because of its small size, it is no easier to make than OO and consequently the price differential has not been as great as perhaps the external appearance would lead one to expect. However, we certainly intend to keep the basic TT items in production for an unlimited period but we are no longer selling sets.

We have also decided to withdraw some of the less popular items, again because the quantities required became uneconomical to justify production."

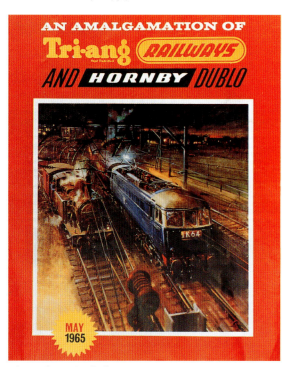

The amalgamation leaflet.

Nine months later, responding to a writer who complained that Tri-ang were not interested when he asked them to produce more locomotives in TT scale, H.W. Hendon, who was a Director of Rovex Scale Models Ltd, said :

"Quite wrong! Tri-ang are extremely interested in both gauges (TT and N). We believe, in fact, that we introduced TT gauge too soon – perhaps we have learnt a lesson and perhaps we shall introduce N gauge at the right time".

The Amalgamation Leaflet

The eight-page full colour leaflet published by Lines Bros. in May 1965 and called 'An Amalgamation of Tri-ang Railways and Hornby-Dublo', gave the impression that it would be just that – an amalgamation.

The introduction told us:

"Following the incorporation of Meccano Limited, the manufacturers of Hornby-Dublo trains, into the Lines Bros. Group (Tri-ang), a careful analysis was undertaken to study the advantages and disadvantages of maintaining production of two entirely separate Model railway Systems. The evidence showed clearly that there was excessive duplication of products in all fields and that since both systems operated on the 2-rail 12 volt DC standard, there were no fundamental obstacles to amalgamation.

"The two railways will, therefore, be progressively brought together under the name Tri-ang Hornby and will use Tri-ang Super 4 Track and Tri-ang Couplings.

The sting, however, was in the final paragraph:

"Existing owners of Hornby-Dublo will continue to be able to purchase Hornby-Dublo components while stocks last and can then go on to Tri-ang Hornby track by means of a special converter rail. A converter wagon with mixed couplings is also now available, so that no Hornby-Dublo system will become obsolete. It must be recognised, however, that running trains with mixed couplings does not permit full remote uncoupling operations."

This clearly indicated that it was not intended to make any more Hornby-Dublo.

Converter Track and Wagon

The leaflet contained pictures of the converter track and wagon, the latter being the 7-plank open wagon already in the Tri-ang Railways range but shown with a Tri-ang tension-lock coupling at one end and a Hornby-Dublo coupling at the other. This was to be supplied free with each of five Hornby-Dublo locos that were to be available in the Tri-ang Hornby range. The locos were to be the West Country, Co-Bo, N2 tank, R1 tank and the diesel shunter. These were all given 'R' numbers as were a range of station and lineside buildings. The Hornby-Dublo EMU and E3002 were also illustrated but it was pointed out that these two and other models in the Hornby-Dublo range, not illustrated, could not be converted to the tension-lock couplings. The leaflet also announced that a passenger converter vehicle would be available once the stock of Hornby-Dublo coaches ran out.

It was also explained that Hornby-Dublo 2-rail locos and rolling stock operated satisfactorily on Super 4 track and that most modern Tri-ang Railways locomotives and rolling stock would operate on Hornby-Dublo 2-rail track.

A report of the amalgamation in the May issue of *Model Railway Constructor* emphasised the advantage of amalgamating the two station systems and at the end of the day this was the area that saw the greatest integration, albeit for a very limited period. The article said:

"Hornby and Tri-ang station and lineside architecture have always been very different and the bringing together of the two into one system means that the model railway enthusiast can now build up a comprehensive layout unequalled anywhere in the world and yet use only Tri-ang Hornby products."

In fact many of us were doing this long before the 'amalgamation'.

Models initially selected for inclusion in the Tri-ang Hornby range.

The article went on to say:

"Tri-ang Hornby will be marketed by Rovex Scale Models Ltd. There are several Hornby-Dublo items which it has not been found possible to integrate into Tri-ang Hornby and these will continue to be available under the name Hornby-Dublo. The servicing of Hornby-Dublo items will continue to be undertaken by Meccano Limited, Binns Road, Liverpool.

"The activities of the Hornby Railway Company in helping users to get the best value from their trains will be superseded by the model railways Operators section of the Tri-ang Club. This Club issues a Bulletin at regular intervals and details of membership are published in the monthly Tri-ang Magazine."

Supplementary Australian 1965 price list included old Hornby-Dublo stock.

June 1965 Hornby-Dublo price list.

West Country loco and Co-Bo in the 1966 Tri-ang Hornby catalogue.

Public Reaction

The news of the amalgamation was not well received by the public and it led to a stream of letters in the model railway press expressing concern on a range of issues from the standardisation on Super 4 track to the superiority of Hornby locos. The general feeling was that whatever happened, the trade never listened to the modellers. The people at Rovex must have heard that many times before and since and no doubt felt frustrated that modellers did not understand the trade, perhaps feeling that the fate of Meccano Ltd should have taught modellers something about the business of toy making and the importance of the right commercial decisions.

Reacting to the continuous stream of criticism in the model railway press Hendon wrote:

"At times one gets the feeling that we go our own way oblivious to the wishes of the general public. This I assure you is not the case... Sir, we are very interested in the requirements of the public, but can not afford to let our hearts rule our heads."

To a gentleman who was complaining about the amalgamation and wanted Tri-ang to make more of the former Hornby-Dublo coaches he said:

"Let me repeat that stocks of Hornby-Dublo models are still available and if he has any difficulty in obtaining those he needs, we will certainly be in a position to supply him through his regular retailer."

Today, the thought that Rovex might give up the manufacture of its highly successful Tri-ang Railways range in order to take over the production of the Hornby-Dublo system seems laughable but in 1965 it seemed a possibility simply because we were unaware of the huge stocks of unsold Hornby-Dublo and the reasons for Meccano's failure.

Mr W.A. Eaglesham, Liaison Officer for NMRA British Region, writing to the *Model Railway Constructor*, made the point:

"... had many of those modellers who are mourning the loss of Hornby bought this range when it was available the takeover would not have taken place."

I suspect that this was too simplistic. They probably had bought them but there were just not enough modellers to keep the system in production. The real money came from the sale of model railways for use by children and here the price was all important. Meccano tried to recapture this market too late.

Five years later, the following letter appeared in the *Model Railway Constructor*:

"Sir: It is almost five years since I wrote the Editor of MRC deploring the apparent annihilation of Hornby-Dublo by its (then) new overlord, Tri-ang.

"It would be fair, therefore, that following the issue of the new (1970) Tri-ang Hornby catalogue, I should be among the first, to publicly congratulate Messrs. Lines Bros. upon a superb production and a most comprehensive collection of OO gauge model railway products. Here, we are offered, in one excellent publication, most of the better items from both Tri-ang and Hornby improved and rationalised in matters of track and couplings.

"The many enthusiasts, who have been urging such a policy upon Messrs. Tri-ang will, I feel sure, join me in applauding this new approach to the seventies."

THE DISPOSAL OF HORNBY-DUBLO TOOLS

Huge Unsold Stocks

These were an international problem which Richard Lines had to sort out – what was going on in Australia, how could Canadian dealership be rationalised etc. – at the same time as sorting out what to do about Hornby-Dublo.

The initial thought was that the Hornby-Dublo production line would start up again once the huge stocks at Binns Road had been disposed of. It was decided that there was to be no big sell-off as flooding the market with cheap Hornby-Dublo would damage Tri-ang's sales. It was also found that retailers were holding large stocks that they could not sell, thus a major sell-off would not produce more orders and could cause them considerable harm.

Rovex therefore published a retail price list and continued to sell off the surplus stock over a number of years. Both Tri-ang and Trix were experiencing reduced demand for their products and it was felt that if Hornby-Dublo was off the market there would be one less competitor.

One major purchaser was Hattons of Liverpool who announced a massive acquisition in August 1966 and were still trying to clear the stock 30 years later.

During the months following the Meccano take-over, Graeme Lines was sent in to study the company. He got on well with Joe Fallmann who proved to be very Meccano Ltd minded although he had been there only a short time.

Trix Interest

By 1966, Trix had gone through a major overhaul which had cost British Celanese, who had an 82% holding in Trix, large sums in writing off obsolete stocks, tools etc. The company's turnover in 1965/66 had been £55,000 and they had sustained a £20,000 trading loss. They predicted a doubling of their turnover in 1966/67 including £5000 in exports to Australia and South Africa. Their break-even point was £80,000.

At this time they were concentrating on kits of locos, coaches and wagons sold through 400 retailers. British Trix employed 16 on production and had a sales staff of three. Two of their senior staff had come to them from Meccano Ltd.

British Celanese were reluctant to spend money on new tools and decided to see whether they could acquire some of the Hornby-Dublo tools.

While on a visit to the 1966 Nuremberg Toy Fair, Richard Lines was in conversation with Mr Rosza, who managed British Trix Ltd, when the subject of the old tools came up.

Trix catalogue cover.

Initially the company was interested in leasing the tools to make the following models for sale in finished form and as kits:

2224 2-8-0 8F
2221 4-6-0 Cardiff Castle
4644 Hopper Wagon
4658 Prestwin Silo Wagon
4626 Presflo Bulk Cement Wagon
4665 Saxa Salt Wagon
4625 Grain Wagon
2900 Re-railer

Hiring was thought by Rovex to be out of the question and in March, Meccano were asked to prepare prices for sale of tools to make these models.

By now it was assumed that the Hornby-Dublo tools, if not sold, would be a complete write-off. If a buyer were to be found a reasonable price was required as they would be used in competition with Tri-ang Hornby. Meccano confirmed that all the tools were still available including those for the manufacture of the motor.

Cornard Model Co. Interest

At this time Cornard Model Co. of Sudbury (Suffolk) had also approached Meccano Ltd about the possibility of buying the Hornby-Dublo tools. If they were to be split it could lead to complications as the motors were shared by models.

As it happened Mr Little of Cornard Models decided not to buy the tools but did clear the stocks of completed super-detail wagons left in the factory. As Rovex were already selling surplus Hornby-Dublo stock they had bought from Meccano they insisted that any remaining stocks sold to other customers were sold at the same price they were paying and that the purchasers were required to sell them at not less than Rovex's prices.

Lines Bros is Invited to Buy Trix

The tools Trix wanted were offered to them at the end of March at a quarter of their original purchase price and Trix came back with a counter offer in the third week of April.

While this was going on Richard Morris of British Celanese Ltd had opened discussions with Graeme Lines of Lines Bros. about the possibility of British Trix Ltd being bought by the Lines Group.

Mr Morris was chairman of a number of British Celanese companies including British Trix and British Lego both of which were housed in buildings on a large industrial complex built by British Celanese on a wartime ammunition factory at Wrexham. Trix occupied one of the old ammunition buildings and Lego another.

British Celanese wanted to get out of model mak-

ing and suggested that initially Lines Bros. took a minority holding in Trix. Lines would supply the Hornby-Dublo tools and when the company became profitable Lines should take over the company or acquire a majority share holding.

Wrenn Enter the Field

While this was happening, as a result of some considerable pressure from the public to reintroduce Hornby-Dublo locomotives, Rovex were considering low-volume manufacture of two or three of the better models through their subsidiary G&R Wrenn Ltd. Indeed, just one week after Richard Lines visited Trix at Wrexham to consider their proposal he was due to meet George Wrenn at Liverpool to look at the Hornby-Dublo tools. But with their existing contacts for die-casting and their painting facilities Trix looked a good alternative for the reintroduction of some Hornby-Dublo models.

Underlying all this there was the feeling that Rovex Industries business might become too stretched and divided on policy if it were responsible for making two different classes of trains at different ends of the country. If Trix were to be bought it would be sensible to link it to Meccano Ltd at Liverpool because of the closeness of the two factories but the feeling was that, having got Meccano

> "I feel that the right people to make use of Meccano's old tools **must** be the Model Division, Rovex preferably, but Wrenn if you like."
>
> "I feel it should be under a name _other_ than Tri-ang, and perhaps this indicates Wrenn as a possibility. I would be strongly against selling _any_ old Meccano H.O. tools to Trix under any circumstances. This would simply aid a possible competitor."

An extract from a memo dated 3 June 1966 from Moray Lines to Richard Lines following an approach from British Trix asking to be able to buy the former Hornby-Dublo tools.

out of train production, it was foolish to go back in again. It was better to keep the train production in the Model Division centred on Margate and, as Wrenn was linked to the Model Division, it should have the tools.

It is known that Walter Moray Lines, the Chairman of Lines Bros. Ltd, was concerned about the vacuum left in the market by the close down of the Hornby-Dublo production line and felt that it was important that Rovex developed a first-class range for the 'model' trade alongside the toy range. Wrenn presented this possibility and Wrenn should be given the opportunity, rather than Trix, to make use of the tools.

And Trix?

On 16th June 1966 Lines Bros. turned down the British Celanese offer to sell them British Trix. In the same letter interest was expressed by Lines Bros. in acquiring British Lego and this led to a meeting at Margate and discussions about co-operation in the field of model railways. Included was an offer by Rovex to supply motors and couplings; an offer that Trix took up. This episode led to the Whisky wagon

> "I believe that all the problems you may have with your Company exist in a similar manner with ours, e.g. cost of new tools, cost of selling, etc. and with the present economic situation one cannot help being reluctant to embark on further enterprises unless these are virtually guaranteed of success and quickly.
>
> "Thus, there could really be no question of our taking an interest in British Trix. On the other hand, we would be very happy to quote for the manufacture of components, such as, electric motors, couplings, etc. if you feel that this source of supply might be of benefit to you."

An extract from a letter dated 17 October 1966 from Richard Lines to Dick Morris of British Trix Ltd, who, after having their request to purchase tools turned down, invited Tri-ang to take over their company.

fiasco, a story which is told elsewhere in this book.

Thus passed the opportunity for all three of the big names of post-war train production in Britain to come together.

Production under the name of Trix would survive in Britain for a few more years but in the 1970s and 1980s the model railway/toy train scene would be dominated by Hornby Railways (the new name for Tri-ang Hornby), chased hard by Lima and two newcomers, Airfix and Mainline, who concentrated on the model trade. Wrenn Railways were also to remain in production for 26 years using the ex-Hornby-Dublo tools to good effect – but that's another story.

In passing, it is worth noting that Dick Morris of British Celanese also approached German Trix about a possible takeover. This was more successful. Guenter Kurz of Trix Modellbahnen joined the board of the new company Thernglade Ltd and the product was renamed Trix Trains. German Trix was doing particularly well at this time. Financial backing from the Von Bohlem element of the Krupp family had taken the company into second place in the German toy train market displacing Fleischmann. Märklin, who in 1966/67 had a turnover of over £6m, still lead the field. There was also a rumour at the time that Lesney might be interested in Trix.

A full history of British-made Trix may be found in Tony Matthewman's excellent book *The History of Trix HO/OO Model Railways in Britain* which is published by New Cavendish Books in this series.

ROVEX INDUSTRIES LTD

Rovex became the core of the Models Division, as it came to be known, and sometime between April and September 1967, a new holding company was formed called Rovex Industries Ltd. The division, despite its apparent large size, made up only about 10% of Lines Bros. Ltd. It included Rovex Scale Models Ltd at Margate, Minic Ltd at Canterbury,

International Model Aircraft Ltd at Merton, Spot-On Ltd at Belfast and, later, Minimodels Ltd at Havant. Meccano Ltd was not included.

As we have seen, the new company had as its subsidiary G&R Wrenn Ltd of Basildon, a company previously fully owned by George Wrenn, and still managed by him, and which seemed to become a clearing house for products abandoned by other members of the group.

Gradually the companies within the model division lost their independence and became outposts of Rovex Industries Ltd.

Pedigree Dolls Ltd, who made the very popular Sindy series and was based at Merton, was transferred to Canterbury in 1967 and thus became part of Rovex Industries.

GLOBAL EMPIRE

UK Companies

By now Rovex was just one of 40 companies, world wide, owned by Lines Bros. Ltd. In addition, there were a further four associated companies. 22 of the owned companies were in the UK.

The list below shows Minimodels Ltd as an independent

Company	Premises	Equity capital held
Tri-ang Toys Ltd	Merton, Birmingham and Merthyr Tydfil	100
Pedigree Dolls Ltd	Merton	100
Pedigree Soft Toys Ltd	Belfast	100
Rovex Industries Ltd	Merton, Margate, Canterbury and Belfast	100
G&R Wrenn Ltd	Basildon	66.67
Select Nursery Goods Ltd	Leamington Spa and Paisley	100
Northland (S.N.G.) Ltd	Merton	100
Westline (S.N.G.) Ltd	Market Harborough	100
Lines Bros. (Ireland) Ltd	Belfast	100
Lines Bros. (Richmond) Ltd	Richmond (Surrey)	100
Shuresta (A. Mirecki) Ltd	Coventry	100
Good-Wood Toys (Lavant) Ltd	Lavant (Sussex)	100
J. Schowanek Ltd	Lavant (Sussex)	100
Minimodels Ltd	Havant and Langstone (Hampshire) and Hayes (Middlesex)	100
Scale Figures Ltd (Subbuteo)	Tunbridge Wells	100
The Medway Tool Company Ltd	Tunbridge Wells	51
Meccano Ltd	Liverpool	100
A.A. Hales Ltd	Potters Bar (Hertfordshire)	66.67
Raphael Lipkin Ltd	Lambeth	100
Arrow Games Ltd	Waltham Abbey	79
Hamley Brothers Ltd	London	57.34
Youngsters Ltd	Merton	100

pendant company within the group but, as we have seen, this and Pedigree Dolls Ltd were soon to be absorbed by Rovex Industries Ltd.

When one considers that Tri-ang Hornby was just one product of one of four factories run by just one of these subsidiary companies, you begin to get things into perspective. Of course, the companies differed in size and importance and Rovex Industries Ltd was one of the most important and productive in the group. Another was Meccano Ltd, the makers of Dinky Toys, who, as we have seen, had been acquired in 1964 after ceasing production of the rival Hornby-Dublo model railway system; the tools later being sold to Rovex subsidiary – G&R Wrenn Ltd.

Australia
The next largest geographical concentration of companies was in Australia where Lines had bought into the country's leading toy company, Cyclops Toys, in the late 1950s. By 1966 they owned the Australian companies shown at right:

Of these subsidiaries, Moldex Ltd was the most important, from the point of view of our story, as this was where plastic-moulded toys for the Australian market were made.

Mainland Europe
There were also six companies on mainland Europe:

The first of these was the result of an amalgamation between the French Meccano company and the Lines Bros. French subsidiary with its main manufacturing base now at Calais in the former Tri-ang factory.

Other Companies
Besides the above there were five more companies owned by Lines Bros:

Both Lines Bros. (South Africa) (Pty) Ltd and Lines Bros. (NZ) Ltd had made Tri-ang Railways, the latter continuing until 1971.

Associated Companies
The following were associated companies at that time:
De Luxe Topper Corporation, Elizabeth (New Jersey, USA)
Exin-Lines SA Barcelona, Spain
Regal Trading Co. (Pty) Ltd. Johannesburg, South Africa
Daymond Industries Pty Ltd, Revesby, NSW, Australia

Renaming of Companies
1968 saw the renaming of some of the companies to link them more closely with their products. At this time, all of the Australian companies were made subsidiaries of the newly renamed Cyclops Tri-ang

(Aust) Ltd. Within this group, Moldex Ltd was renamed Tri-ang Moldex Ltd and the same year ceased manufacture of locomotives and rolling stock, importing them instead from Rovex.

Across the water in Auckland the New Zealand manufacturer of Tri-ang Railways was renamed Tri-ang Pedigree (NZ) Ltd in 1969 and back at home the makers of Scalextric became Minimodels Tri-ang Ltd.

ROVEX TRI-ANG LTD

More name changes within the Lines Bros. Group occurred in 1969 including that of Rovex Industries Ltd which became Rovex Tri-ang Ltd. The Directors

Company	Premises	Equity capital held
Cyclops & Lines Bros. (Aust) Ltd	Sydney, Brisbane and Adelaide	100
Moldex Ltd	Melbourne	100
H.W. Rice Pty Ltd	Sydney	100
Segals Pty Ltd	Fitzroy (Victoria)	100
Toy Traders (Pty) Ltd	Sydney	100
Joy Toys (Pty) Ltd	Richmond (Victoria)	100
Steelcraft Baby Carriages Pty Ltd	Sunshine (Victoria)	51

Meccano Tri-ang Lines Freres SA	Paris, Calais and Nogent	99.96
Lines Bros. (Holland) NV	Amsterdam	100
Lines Bros. (Belgium) SA	Brussels	100
Lines Bros. (Deutschland) GmbH	Frankfurt	100
Lines Bros. (Italiana) SpA	Milan	100
J. Schowanek, GmbH	Piding (W.Germany)	75

Lines Bros. (Dublin) Ltd	Dublin	55
Lines Bros. (NZ) Ltd	Auckland	86.11
Lines Bros. (Canada) Ltd	Waterloo and Sutton (PQ) and Toronto	90.63
Lines Bros. (South Africa) (Pty) Ltd	Durban, Johannesburg, Bulawayo and Cape Town	100
Bunsen Properties (Pty) Ltd	Johannesburg	100

of Rovex Tri-ang Ltd were H.R. Lines, W.M. Lines, R.E. Wadsworth, H.W. Henden, K. Edey, H. Gordon and O.D. Kutluk and the registered address was Westwood, Margate. It was about this time that Richard Wadsworth became Managing Director of Rovex Tri-ang Ltd having joined the firm in the early 1960s as Marketing Manager.

Other notable changes were:
Meccano Tri-ang Ltd (formerly Meccano Ltd), Tri-ang Pedigree Ltd (formerly Tri-ang Toys Ltd) and Lines Bros. (South Africa) (Pty) Ltd became Tri-ang Pedigree (South Africa) (Pty) Ltd.

LIQUIDATION

First Tremors
In July 1970, it was registered that Lines Bros. Ltd had a net loss of £121,000. Shares, that were 21/- earlier that year, fell to 7/3 after this news. A major shake-up followed and the board of management was re-organised.

The Editor of *Model Railway Constructor*, writing in the March 1971 edition of his magazine noted at the Toy Fair in Tri-ang House that year:

"The vast area of Tri-ang House was well filled with toys and models of every description from the large Tri-ang Empire, although we thought that some of the subsidiary companies were not showing quite the large range they have done previously and it was noted that the wholesale firm of A.A. Hales was not represented. Fortunately there seems to be no recession on the model railway side and we were pleased to see a display of 'Big-Big Trains' (gauge O) that dispelled the rumour that this particular line was to be 'axed' following the recent financial difficulties."

The Announcement
The announcement that Lines Bros. Ltd had failed came on Thursday 19th August 1971 following a withdrawal, the previous Monday, of a £5 million rescue bid by the cigarette company, Gallahers. Lines Bros. Ltd had debts of about £17.5m.

Managers were briefed the following day and in turn the rest of the staff were told. Shock must have struck the 8,500 people around the globe that worked for this giant of the toy world. No doubt most had felt themselves safe in such a large organisation and could not have contemplated it crumbling around them.

The Cause
The rot had started far away from the heart of the organisation. In the 1950s the group saw its future in overseas expansion but in order to get around trade barriers thrown up by many countries after the war, Lines Bros. established production and marketing bases overseas. Many of these catered for much smaller local markets than Lines were used to in the UK. This meant that production had to be kept to an artificially low level. As barriers were lowered, customers in these overseas countries could buy a better and cheaper product from outside. Ironically Lines' British factories became principal competitors for their overseas subsidiaries.

Instead of disposing of their unprofitable overseas subsidiaries in the 1960s, Lines Bros. subsidised them and in so doing weakened their own UK base. The recession in the toy market during the second half of the 1960s and an onslaught on the British toy market by American manufacturers made matters worse, until eventually the centre fell in and Lines Bros. Ltd went into liquidation.

A Confident Future
In the trading subsidiaries such as Rovex Tri-ang Ltd there was a degree of confidence that they would survive. Harold Henden, the General Manager at Rovex, was quoted as saying:

"We have a large order book and deliveries are being made from our warehouse in the normal way. I am confident that nothing will happen to people's jobs at present.

Cutting from the Chatham Evening Post *20 September 1971.*

A LAST-MINUTE bid to save the Kent-based Tri-ang and Meccano toy firm Lines Brothers has failed.

The life-line was expected to be thrown by the giant American toys combine, Marx, through its British offshoot Dunbee-Combex-Marx.

But Dunbee Combex directors have issued a statement saying it was clear the cash needed to save the Lines was much "in excess of that originally hoped for".

"We do not feel it would be justified for our company to raise its offer to the level required," said the directors.

Only two weeks ago shareholders gave Lines a reprieve after a dramatic intervention by director Mr. Moray Lines at a meeting to put the group into voluntary liquidation.

Dunbee Combex's original £3m. offer for Lines was rejected.

There is only one other company still known to be interested in taking over Lines: the American foods giant, General Foods.

Shareholders are due to meet tomorrow week to ratify the Lines' board's recommendation to voluntarily wind up the company, which employs 1,000 at its Margate toy factory.

Nor do we contemplate any colossal redundancies in the future."

The Works Manager at Canterbury, Geoffrey Greenaway, showed similar confidence saying that they had many more orders than last year and wanted more people to help with production.

The Strength to Survive
While in 1964 it had largely been the loss of sales of Hornby-Dublo that had brought Meccano Ltd to its knees, in 1971 it was failures of other parts of the empire, and particularly overseas, that brought about the collapse of Lines Bros. Indeed, it was the Tri-ang Railways system and Rovex Ltd as a whole that offered most to a future investor. These were the plum that the receiver had to dispose of and he hived down the company and created a new one

Factory's future settled

The uncertain future of Canterbury's Rovex Tri-ang factory was settled this week with the news that the company is to be bought by another giant toy-making concern as part of a £2,260,000 cash deal.

There was apprehension over the survival of the Market Way factory and its counterpart at Margate when the parent company, Lines Brothers, went into liquidation in September.

Since then Rovex, makers of Tri-ang Hornby model trains, Scalextric cars and Pedigree and Sindy dolls, has been operating at a profit. Increased turnover and profitability is expected this year.

Now Dunbee - Combex - Marx, one time contender in the bidding for Lines Brothers, has stepped in by offering to take over the total assets of Rovex, valued net at £3,350,000.

The transaction, made through the Lines Brothers liquidators, takes in the freehold factories and land at Canterbury and Margate, valued together at £741,000

Mr. Richard Beecham, managing director of Dunbee-Combex-Marx, said on Wednesday that the Rovex range of toys was "entirely complementary" to those produced by his companies.

Dunbee, which will become "one of the biggest toymakers in Europe" through the deal, will pay £1,009,000 on completion and a further £1,250,000 at the end of two years, free of interest.

Cutting from the Kentish Gazette *14 January 1972.*

under the name Pocketmoney Toys Ltd to allow it to carry on trading while a purchaser was found.

The collapse of Meccano in 1964 had been a sad time for Hornby-Dublo as, at that time, few in the industry saw much future for a model railway system that had been out marketed. As time was to show, its days as a complete model railway system had gone but a market remained for some of those models whose cost could be kept to the minimum by batch production and low investment in new product development.

By contrast, the situation was quite different at Rovex in 1972 where the Tri-ang Hornby model railway system was very much alive and kicking and the receivers allowed the development of new products to go on uninterrupted.

Purchase of Rovex Tri-ang Ltd

In January 1972, it was announced that Dunbee-Combex-Marx would buy Rovex Tri-ang Ltd for £2,260,000. This would include the factories at Margate (Rovex) and Canterbury (Pedigree Soft Toys) valued at £741,000. They could not, however, be persuaded to buy any of the other companies in the Group.

Richard Beecham, Managing Director of DCM claimed that 92.5% of the Rovex toy range was entirely complementary to those produced by his companies and that the takeover would make DCM one of the largest toy makers in Europe. Rovex would add between £4.5m and £5m turnover to the £3.5m already enjoyed by DMC.

The financial arrangement was that DCM would pay £1,009,000 on completion and a further £1,250,000 at the end of two years, free of interest. In exchange they received assets valued at £3.35m including trademarks, £1m in cash and £900,000 of debtors.

Rovex Ltd

Meccano Tri-ang Ltd had already been sold to Airfix

Ltd and Tri-ang Pedigree Ltd, together with the Tri-ang name, was sold to the Barclays Security Group. Rovex Tri-ang Ltd was the last of the three large British-based subsidiaries of Lines Bros. Ltd to be disposed of. The sale of the subsidiaries so far had raised about £11m.

The Lines trademarks are thought to have raised about £2.5m on their own. As the model railway system could no longer be sold under the name Tri-ang, from 1 January 1972, it became 'Hornby Railways' and from the same date the manufacturing company was renamed Rovex Ltd.

Its range of products including Big Big Train, Frog Kits, Hornby Railways, Pedigree Playtime, Pedigree Soft Toys and Dolls, Scalex Boats, Scalextric, Jump Jockey, Pennybrix and Sindy. The company also owned a number of dormant trademarks including Hornby-Dublo and Minic.

But what happened to G&R Wrenn Ltd? With the Hornby-Dublo tools secured and now providing the company with a steady income, the collapse of Lines Bros. provided George Wrenn with the opportunity to cast himself loose and return to being an independent manufacturer once more and this he did, buying out Lines' share of his company. The result of this together with the development of Hornby Railways under DCM, and later Hornby Hobbies, will be dealt with in **Volume 3**.

PRODUCTS MADE BY ROVEX

Such were the changes that occured to the Rovex company over the years that it is sometimes difficult to keep up with the range of products they made.

The following list has therefore been prepared to help clarify this but is not claimed to be comprehensive. It should be noted that the dates quoted refer only to the name of the company and not the period during which the toys were made. Some of the brands listed had a very short life.

Rovex Plastics Ltd *1946-53*
Rovex and Tri-ang Railways.

Rovex Scale Models Ltd *1953-67*
Tri-ang Railways (including TT), Model-Land, Minix, Tri-ang Lionel, Hornby-Dublo (disposal of surplus stocks) and Tri-ang Hornby.

Rovex Industries Ltd *1967-69*
Tri-ang Hornby, Miniville, Model-Land, Frog, Big Big, Minix, Minic, Car Play, (Arkitex), Tri-ang Wrenn, Scalextric, Jump Jockey, Tri-ang Science, Pedigree Dolls, Pedigree Soft Toys and Sindy.

Rovex Tri-ang Ltd *1969-71*
Tri-ang Hornby, Tri-ang Minic and Car Play, Tri-ang Model-Land, Tri-ang Wrenn, Clockwork Toy Trains, Miniville Trains, Frog, Scalextric, Jump Jockey, Big Big, Tri-ang Science, Pedigree Dolls, Pedigree Soft Toys and Sindy.

Pocketmoney Toys Ltd *1971*
The name used while the company continued to trade while in the hands of the Receiver.

Rovex Ltd *1972-80*
Hornby Railways, Hornby Minitrix, Big Big Train, Frog, Scalextric, Scalex Boats, Hornby Stamps, Minic Ships, Shipwright, Hornby Steam, Pedigree Dolls, Pedigree Soft Toys, Pedigree Playtime and Sindy.

Hornby Hobbies Ltd *1980-96*
Hornby Railways, Hornby Steam, 3-D-S, Scalextric, Hornby Minitrix, Playtrains, Thomas & Friends, Hammant & Morgan Power Units, Minimites, Flower Fairies, Magical Mist Fairies, Pound Puppies, BooBoos, Baby Hush-a-bye, The Twins, Fireman Sam, Super Heros, Tom & Jerry, Were Bears & Terror Teds, Rag Doll, Quints, Konami, My Pretty Ballerina, Guard Dogs, Gro Toys, L.A. Gear, Yawnies, Karate Kid, Kool Kats, Puppy/Kitty Care,

Thunder Force, Gremlins 2, My Pal 2. (Many of the above were made under licence during the 1980s and early 1990s.)

Hornby Hobbies Ltd. *1996-*
Hornby Railways, Scalextric.

Undated cutting from the Financial Times.

BIDS AND DEALS

Dunbee-Combex buying Rovex Tri-ang

BY NICHOLAS LESLIE

Dunbee-Combex-Marx is to acquire for nearly £2.26m. cash the business of Rovex Tri-ang, which makes the well-known ranges of Hornby Electric trains, Scalextric model racing cars and Pedigree dolls and soft toys. This represents the sale of the only remaining major U.K. subsidiary of **Lines Brothers**, parent company of the toy group, which was voted into liquidation last September.

In addition, DCM has purchased from the joint liquidators of Lines the factories and land at Margate and Canterbury occupied by Rovex for £741,000 cash, payable on completion.

Mr Richard Beecham, joint managing director of DCM, pointed out yesterday that 92.5% of the Rovex merchandise was complementary to DCM products, which left duplication on only 7.5% of the Rovex range which was "absolutely unimportant."

Rovex had made profits from September last and would add around £4.5m to £5m turnover to the Combex and Marx toys sales of some £3.5m. DCM had had a "very healthy year" in 1971, said Mr Beecham, and he was quite optimistic for the future of the enlarged group.

Of the £2.26m consideration, just over £1m. will be paid by DCM on completion, with the balance payable after two years free of interest. Assets being acquired total £3.35m. and include trademarks, £1m. in cash and some £900,000 of debtors.

From the proceeds of sales negotiated so far of Lines subsidiaries, which amount to over £11m. and relate to Tri-ang Pedigree, Meccano Tri-ang and other UK and overseas offshoots, a picture is emerging of the total of funds which may eventually be available for Lines' creditors.

After claims have been met from each

subsidiary's creditors, there is a potential £5m. available at present. Of this, about £2.5m. relates to the sale of trademarks (the majority of which were held by Lines) and certain freehold properties. The balance is accounted for by Lines' investments in its U.K. and overseas subsidiaries.

Since creditors of Lines at August 1⁰, 1971, amounted to £17.55m., there is on a pro rata basis getting on for 30p in the £ available for meeting creditors' claims. This is, however, an over-simplification of the matter since there are legal points yet to be cleared on whether the banks' claims on the subsidiaries rank ahead of certain loans by the parent company. Nonetheless, with the major sales now having been negotiated or completed, it is crystal clear that Lines' shareholders have no hope of any return on their investment.

TRI-ANG HORNBY

Let us turn our attention now to the Tri-ang Hornby model railway system which was to dominate the market for seven years. What were the principal developments during that time?

A Brave New World

The loss of John Hefford had been a blow to the company. He had been an ideas man and without him Rovex were preciously short of good ideas. Rovex entered the new era with the model railway market in recession. Balanced against this was the fact that its principal British rival, Hornby-Dublo,

was now out of the frame and the other rival, Trix, was struggling to re-establish after a drastic attempt to modernise itself. This left Tri-ang Hornby with a weak market but little competition.

There was, however, competition abroad but as yet the models made there were rarely of British prototypes. This was partly because Britain had adopted a scale all of its own and partly because British locomotives were peculiar to Britain and did not cross international boundaries or appear in the liveries of other European countries. It would be the 1970s before foreign manufacturers would take the bold step of manufacturing models exclusively for the British market. The respite did much to save Tri-ang Hornby.

Another potential threat was from the smaller gauges. Trebl-o-lectric, made by Lone Star, was too crude to have much appeal but by the late 1960s this was all to change with three attractive N gauge systems on the market and a growing range of support equipment from smaller manufacturers. Back in 1965, however, that threat had not yet appeared.

1965-1967

End of the Old

The 1965 catalogue, prepared in the autumn of 1964, carried the Tri-ang Railways name and featured a Class 31 crossing a viaduct below which steamed

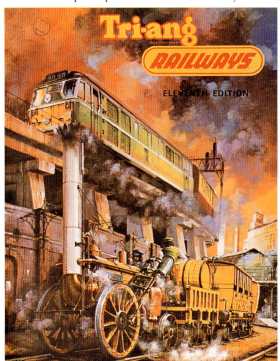

1965 Tri-ang Railways *catalogue.*

Picture from the 1965 catalogue showing the integration of Tri-ang product ranges.

Rocket with a train of coaches. This edition contained the end of the Tri-ang Railways range rather than the beginning of the Tri-ang Hornby models.

Two new train sets of particular note were the much sought-after RS61 Old Smoky and, a firm favourite, the RS62 Car-a-belle. The new locomotive was to be an English Electric Type 3 Co-Co which in fact did not arrive until early the following year. Besides these there was very little else that was new. However, more models had disappeared as part of a rationalisation process.

For Canada, the year was marked by the first Canadian-liveried models; all in Canadian National colours. There was also evidence of a new flurry of activity in Model-Land with new batches of about 15 of the range being made. These included, for the first time, two of the figure sets. The King Size scatter series also appeared in the shops. Another new arrival was the revised edition of the Super 4 track plan book which was released during the year.

The back page of the catalogue showed four Tri-ang logos at the head of the page: Tri-ang Railways, Minic Motorways, Arkitex and Model-Land. The picture below this showed how the four ranges could work together to create an exciting integrated layout. Later we will see how the four were more formally linked together, Model-Land production having already been moved to the Margate factory.

New Locomotives

The advertising at the start of 1966 depicted four locomotives that had been fitted with Synchrosmoke. These were the Jinty, B12, Britannia and the Deeley Class 3F. By April magazine advertisements concentrated on the R753 Bo-Bo Class E3001 and the R751 Co-Co English Electric Type 3.

The 1966 catalogue was the first to show the new trademark and it revealed two new locomotives planned which would strengthen the Western Region stable. These were Albert Hall and the Hymek. Important new train sets were the 'Inter-

Announcing the forthcoming Battle Space models.

city Express' made up of a green Co-Co with a train of Pullmans and 'The Midlander' containing the maroon Deeley and two BR maroon coaches. From the Hornby-Dublo range, Barnstaple and the Co-Bo were illustrated on the locomotive pages. The first blue coaches had also appeared. The first Tri-ang Hornby wagon was the cement wagon which was initially released in grey. A selection of modern shops and offices was added to the Model-Land range.

Battle Space

The most noticeable new face was Battle Space which was presented as two new sets and eleven items of rolling stock. Previously this had appeared as a small selection of military wagons later in NATO green livery. Now they were in khaki with a totally fictitious logo.

To accompany many of the models there was a set of Battle Space Commandos. The prize for the strangest model of this series must surely go to the propeller-driven Battle Space Car, which sped round the track under propeller-power. This was a very fast model and carried a long spike on its front. The last time I saw one demonstrated, it was on a friend's layout. He liked to view his trains from track level. My friend left hospital a week later – a far wiser man! I have heard of a space car that embedded itself in a wall like a dart in a dartboard. They were later fitted with rubber spikes.

Lineside buildings were looking particularly impressive in the 1966 catalogue with one page devoted to the Arkitex Ultra Modern Station kit and another displaying an aerial view of a large complex of Hornby-Dublo stations and engine sheds. This picture, however, showed no Hornby-Dublo locomotives or rolling stock.

Big Big

1966 saw two side events of note. The first was the moving of Frog aircraft kit production from IMA at Merton to the Rovex factory at Margate where it was to become an important product for a number of years. The other was the launch of the Tri-ang Big Big large scale toy railway system for children. This was based on O gauge and intended for garden use. While in many ways very toy-like, the mouldings of some of the models were exceptionally good and found their way onto the layouts of serious O gauge modellers. It was made at Margate by Rovex and is described in some detail in a section of its own at the back of the this volume.

Rovex Industries Ltd

In 1967 Rovex Industries Ltd came into being as an umbrella company for all the Lines Bros. model-making companies. It gained a new Export Sales Manager by the name of Roy Kibby formerly of Peco. His task was to sell Tri-ang Hornby, Minic and Big-Big trains abroad but also to look after the wholesale side of other products.

During the year, Minic Motorway was moved into the Margate factory from Canterbury thus physically bringing together the motorway system with the model railway. While Spot-On became part of Rovex Industries, production at Belfast ceased during the year and preparations were started to clear the factory. Rovex Industries headquarters at this time were in the Merton offices.

The M7 Tank

The March edition of *Model Railway News* gave an account of a visit to the London Trade Fair and particularly the display at Tri-ang House in the Edgeware Road. A model of the proposed Tri-ang Hornby M7 tank was running on a circuit and the new blue and grey coaches were on display either as ready to run or as assembly packs. The new

Freightliner wagon was on display with the van carrier.

The catalogue featured the new M7 tank with its firebox glow and opening smokebox door as its frontispiece and it was the subject of the Cuneo picture on the front cover. The catalogue extolled the virtues of the system with its Magnadhesion, Synchrosmoke, foam underlay, clip-fit accessories, lighting systems and automation uncoupling of rolling stock. For former Hornby-Dublo operators, the horse box converter wagon was now available.

Wrenn OO and N

Also on display at the Tri-ang House trade fair, early in the year, were the first Wrenn N gauge sets and an N gauge layout. A working model of the proposed AL6 E3001 was not available but the mock-up was on show together with the blue and grey Mk1 coaches, the mineral wagon and standard BR goods brake van. These resulted from a deal struck by Lines Bros. under which Lima N gauge would be sold in the UK through G&R Wrenn and Lima would have a licence to make Big-Big.

Tri-ang Hornby was also on display at the Model Railway Club's Easter Exhibition in the New Horticultural Hall. Here there was an 8' x 4' combined railway and roadway showing all the latest in locomotives, rolling stock and cars. In addition there was a small airport with the latest Frog aircraft. A second layout 5' x 2'6" displayed the new Wrenn N Micromodels. Boxed sets and Tri-ang Big-Big trains were also exhibited.

Wrenn released their first locos during the year. These were made from the former Hornby-Dublo tools which they had purchased from Meccano Ltd. Back at the Margate factory, in response to requests from the trade, there was production of small batches of a few of the TT models. This was also the year that Rovex started manufacturing models in American liveries for sale to an American import company called American Train & Track (ATT).

Model or toy?

The Big Big Train. With an 'O' gauge loco 14" long—authentic in detail and scale. It's battery powered and it runs on a tough polypropylene track. So it's safe, simple and very much an out-door type. The track is pliable and takes the ups and downs of a garden layout in its stride. And rain can't harm it. A fine model by any standards. But also a marvellous toy for very young children.

Trip switch gives auto-reverse. Clip this special trip switch on one side of the track and it stops the loco. Clip it on the other side, it puts the loco into reverse. This means the layout need not join up in a circle or an oval but can be laid end to end.

The Big Big Train comes in 2 sets. Loco, 4 trucks, 2 switches and 18 ft track for 99.6. Or Loco, 2 trucks and 12 ft circular track 69.6. Extra track, locos, trucks and switches are available.

THE BIG BIG TRAIN

ROVEX SCALE MODELS LTD · WESTWOOD · MARGATE · KENT

The launch of Big Big.

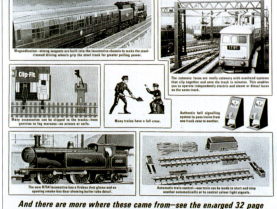

Look at these extra reasons for choosing Tri-ang Hornby

Magnadhesion—strong magnets are built into the locomotive chassis to make the steel-rimmed driving wheels grip the steel track for greater pulling power.

The catenary locos are really authentic with overhead systems that clip together and onto the track in minutes. This enables you to operate independently electric and steam or diesel locos on the same track.

Clip-Fit Many accessories can be clipped to the tracks—from gantries to fog warnings—no screws or nails.

Many trains have a full crew.

Authentic bell signalling system to pass trains from one track zone to another.

The new R254 locomotive has a firebox that glows and an opening smoke-box door showing boiler tube detail.

Automatic train control—one train can be made to start and stop another automatically or to control colour light signals.

And there are more where these came from—see the enlarged 32 page edition of the Tri-ang Hornby catalogue out now!

Tri-ang HORNBY

MADE IN BRITAIN BY ROVEX INDUSTRIES LIMITED, MARGATE, KENT

Magazine advert promoting the advantages of Tri-ang Hornby.

A leaflet promotes the Miniville series.

A NEW CLOCKWORK TRAIN WITH IT'S OWN PUT TOGETHER TOWN AND CARS!

MINIVILLE TRAIN & STATION

A bright little clockwork train with three goods trucks, an oval of track and Miniville Station. Boxed complete & ready to run.

MINIVILLE

Four Houses and Buildings moulded in bright colours. Completely finished parts ready to stick together plus a beautiful little car with windows and detailed chassis to assemble. Packed separately each with a car and a coloured card base for the building.

above, Manor House W20   below, Bahama Villa W21   above, Ideal Homestead W22   below, Super Service Station W23

Made in England

INTRODUCING THE WRENN 2-8-0 8F FREIGHT . . .
The second of three fine new WRENN "OO/HO" locomotives

Available in limited supply

PRICE £5-19-6

Don't miss The WRENN 2-8-0 8F FREIGHT and two more wonderful WRENN 'OO/HO' locomotives—the WRENN 2-6-4 B.R. Tank, price 5 gns. and the WRENN "Cardiff Castle", price 5½ gns.

Wrenn Locomotives are engineered to the highest standard of quality with finely detailed die castings and powerful D.C. motors. Each is equipped with a choice of couplings, TRI-ANG as fitted, with an exchange clip on HORNBY-DUBLO type included.

Fully guaranteed with a first class after sales service, it is anticipated that WRENN OO/HO Locomotives will become the most sought after locomotives on the market. Ask your retailer for details.

WRENN G. & R. WRENN LTD., Bowlers Croft, BASILDON, Essex

An advertisement for Wrenn.

1968-1969

The Complete Transport System
In 1968 the catalogue carried a triple name – 'Tri-ang Hornby Minic'. Minic Motorway production got fully underway at Margate and was included at the back of the catalogue. The occasion was marked with the release of a new Moto-rail set which, in June and July that year, was to be the feature of the monthly advertisement in the model railway press under the slogan:

"More than a Railway – A Complete Transport System."

The advert went on to say:
"Most 00 gauge railways can become complete transport systems almost overnight with all the fascinating problems of control and organisation of a road and rail system.

The basic units your railway needs

Introducing containers.

The Moto-rail set celebrated the transfer of Minic Motorway production to Margate.

A Minic motorway layout is in scale. Links in with 00 gauge railways. It can have a Motorail terminal where Minic cars drive onto rail transporters, and Freightliner Depots where Freightliner trains pull in alongside Minic lorries which deliver their containers to factories on the Minic motorway. It can have the latest level crossings, where Minic cars have to be fitted into train timings. Operating Tri-ang Hornby Minic rail and motorway vehicles as one complete transport system gives your railway a new dimension it can get in no other way."

Rovex Industries were making the most of the rationalisation that had brought together the model ranges under one roof.

Miniville
In order to get their model ranges into the small corner shops which were unable to commit themselves to the larger ranges, Rovex Industries set up a Wholesale Department. Their task was to establish a range of cheap and toylike models to be sold under the Miniville label.

Rather than produce new models, the subjects were selected from the existing model ranges. Thus Frog aircraft, Model-Land buildings, Minix cars, and Tri-ang Hornby models found themselves packaged as Miniville with their true identity removed. Not all were sold under the Miniville name. The Wholesale Department also sold train sets simply marked 'Tri-ang'.

Blue Liveries
In the Tri-ang Hornby Range, the new 20" curves were introduced to provide a third radius and the Flying Scotsman was planned. Sets were given a new look with window-type boxes or fully illustrated lids. The diesel fleet turned electric blue over night and the Mark 2 coaches were featured for the first time.

Trix grain wagons in a Tri-ang catalogue took a lot of people by surprise but this story is told in detail, later, in the chapter on 'Wagons'. In the Model-Land

Frog kits, now made at Margate, were promoted along side Tri-ang Hornby and Minic Motorway.

range the factory buildings were released and Minix Playpacks arrived despite the first Minix deletions occurring. The third edition of the track plan book arrived together with another publication called the *Tri-ang Hornby Book of Trains*. This was based on the *Hornby Book of Trains* produced by Meccano Ltd before the war to promote their O gauge railways.

This was also the year that the company returned to the Nuremberg Toy Fair and Roy Kibby of the Export Department was able to report orders for 100,000 Tri-ang Hornby train sets. These were to go to Australia, South Africa, Canada, the USA and a number of other countries including Iceland (which had no railways)!

Overseas
Canadian variations increased and more models were turned out in American liveries for ATT. During the year, production in Australia ceased except for track and packaging. In the summer, Richard Lines visited the Lines Factory in Natal, South Africa. He was asked if he could find a buyer for the redundant tools used to make the South African Tri-ang Railways sets in the 1950s. A possible market was India. Spot-On production started in New Zealand.

Big Big was further expanded and Wrenn took over dispatch of the remaining stocks of TT announcing that they were bringing out sets which, although appearing to be reissues were in fact boxed-up surplus stock. Wrenn also reintroduced some of the former Hornby-Dublo wagons in private owner liveries and the horn unit arrived.

Cut Backs
The toy industry was by now in a spiral of decline and Lines Bros. were pulling in their horns. Expenditure cuts were necessary and strict limitations were placed on capital investment which meant few new tools could be bought. This led to greater concentration on the old tools and the production of existing models in new liveries.

Under consideration at that time was a series of airfield buildings for the Frog range of model aircraft kits made from the former Hornby-Dublo station and the Tri-ang TT signal box which looked like a control tower, but the idea was abandoned.

Rail Blue Era Arrives
The big news was the expansion of container traffic with five new wagons and a contained depot crane. The diesel and electric fleet was 'repainted' in Rail blue and the Pullman DMU adopted the grey and blue livery like the prototype. Carriage lighting was now available and Series 5 track was being planned. More of this later in the chapter on '**Track**'. The Cartic car carrier arrived in 1969 as did the china clay tank and the large hopper wagon. The former Hornby-Dublo terminus station buildings were now shown in maroon with cream girders and grey platforms so that they matched the rest of the Tri-ang Hornby station series.

Rovex Tri-ang Ltd (as the company was now called) used a can motor for the first time in one of its locomotives. This was a new small diesel designed for the beginner's market and based on a Swedish design. This proved to be the thin end of the wedge.

For the first time Wrenn locomotives were advertised in the Tri-ang Hornby catalogue. A page was devoted to them and five were illustrated under the name 'Tri-ang Hornby by Wrenn'. Despite this the 'Tri-ang Wrenn' logo came into use during the year. Frog aircraft kits also got a plug on the back page of the catalogue.

Up until now, Canadian buyers had had to be content with Canadian National livery but in 1969, Canadian Pacific livery was added. Amro in America imported Tri-ang Hornby models to make up their own sets in presentation cases aimed at a supposed market of ex-patriots which failed to materialise. In India S. Kumar Ltd ordered the tools to make the former Primary Series saddle tank.

1970-1971

1970 Customer Survey
1970 was an important year in the history of the model railway system, seeing the beginning of considerable changes to make Tri-ang Hornby more attractive to customers. It also saw the introduction of System 6 track and work on a revolutionary new locomotive.

From the start of 1970, Moray Lines had a research project undertaken to establish how the company's products could be improved to meet customer's needs. Representatives were asked to complete returns and a list of the ideas was assembled. This was the year that Rovex also carried out a customer survey. A questionnaire was supplied in each copy of the catalogue which purchasers were asked to fill in. By April over 7,000 had been returned, some from a considerable distance such as the Philippines, Czechoslovakia and Poland. By the end of the year, replies totalled 30,000 and 50% of them came from people over the age of 18. One was sent in by an American Senator!

Principal Suggestions
Based on the responses received Rovex would plan their future. Suggestions included the following:
- Scale wheels.
- Better fitting fishplates for the new System 6 track.
- Seats for LNER coaches.
- Stronger clips for the Freightliner crane.
- Modification to the motor bogie housing on the R357 and R351 locomotives.
- More business-like power units.
- Return to BR green for diesels.
- More remote control assessories.
- More elaborate track work.
- A Continental type level crossing with lifting barriers and flashing lights.
- Heavier weights in rolling stock to make them look more authentic when moving.
- Motorising the Big Big 0-6-0 steam loco.

Pre-Grouping Locomotives and Coaches–
by **Tri-ang HORNBY**

The 1970 edition of the Tri-ang Hornby catalogue contains full details of the new range of pre-grouping locomotives and coaches. Unfortunately, space limits us to showing you only two locomotives and two coaches from the range, but *every* one is up to the usual high-precision engineered standards of Tri-ang Hornby—from the superb body detailing to the 12-volt D.C. motors with self-aligning sintered metal bearings. And, in addition to authentic liveries many models have a choice of alternative nameplates. They are all available at your local model shop—now. Make sure you see them!

R.258G 4-6-2 PRINCESS ROYAL CLASS LOCOMOTIVE "PRINCESS ELIZABETH" with crew and tender. Decorated in authentic L.M.S. livery with a choice of 3 alternative names included. **77/-**

R.745 A superb new item of rolling stock. I..N.E.R. FULL THIRD COACH. Also **R.746** L.N.E.R. BRAKE THIRD COACH. **16/-**

R.743 G.W.R. COMPOSITE COACH with seating. Also **R.744** G.W.R. BRAKE THIRD CLASS COACH with seating. **18/6**

R.868 0-4-4 M7 CLASS TANK LOCOMOTIVE in authentic S.R. markings complete with crew. Has firebox glow and opening smoke box door to show boiler tube detail. **65/-**

CLERESTORY COACHES
You keep asking for them— so here they are. We have re-introduced the famous Tri-ang Hornby Clerestory Coaches— available now.

R.332 G.W.R. Third Class Clerestory Coach and **R.333** G.W.R. Brake Third Class Clerestory Coach. **16/-**

FINE SCALE WHEELS
For modellers with fine scale trackage (by the way, have you seen System 6 yet?) a pack of fine scale wheels to convert your Tri-ang Hornby coaches. At only **2/6d.** Why pay more?
Ask for:
R700 Pack of 8 wheel/4 axle assemblies to convert one Tri-ang Hornby coach to fine scale.

All prices quoted are recommended retail prices.

Tri-ang HORNBY *The model railway system designed for the 70's*

ROVEX TRI-ANG LIMITED, WESTWOOD, MARGATE, KENT

It seems that steam locomotives were preferred to diesels by a ratio of 5:1 and pre-nationalisation liveries were preferred to British Railways by 2:1.

Pre-Nationalisation Liveries

In 1970 the catalogue, which had reverted to landscape format with an improved quality of printing, showed a strong swing towards the adoption of pre-nationalisation railway liveries. This was clearly a decision taken in the light of the success reaped by the model of Flying Scotsman introduced in 1968. The results of the public opinion survey would see this trend extended further the following year.

A new feature in the 1970 catalogue was a symbol which told the reader the average current consumption in milliamperes for each model.

The most exciting news was the plan to introduce a model of the streamlined Princess Coronation Class and the adoption of pre-nationalisation liveries for many of the existing steam models, coaches and wagons. Alternative names for some of the larger locomotives were to be available and a 100 ton tank wagon and a ferry van were also planned.

Going Fine Scale

As we have seen, 1970 was the year that saw the introduction of the long-awaited Series 5 track which had been renamed 'System 6'. It had scale sleepers moulded in black plastic and engraved for a wood-grain finish. The new system included, for the first time, flexible track and this was offered in a choice of steel or nickel-silver. To go with the new track, there was a System 6 plan book. There were also to be a new station building and footbridge.

In addition to introducing scale track, Rovex made 1970 the year of the scale wheel. These had already been fitted to the MkII coaches which were released at the end of 1969 but the model railway press had been asked not to draw attention to this point at the time for fear that Rovex would be deluged with orders for the wheels before they were

ready. Early in 1970 the coach wheels were available in packs of four axles for 2/6 a pack. They were plastic wheels on metal pin-point axles – a fine scale version of the existing pin-point wheels in use on carriages and wagons.

The *Model Railway Constructor* described them as: *"Very good and to an even finer scale than the old Hornby-Dublo coach wheels"*

It went on to say:

"The introduction of these will stop many criticisms of Tri-ang wheels, and the running over scale points … will now be trouble free. This is yet another sign that Tri-ang's are becoming more scale minded."

Changes and Moves

1970 was the year that saw the end of the American importer ATT and the disposal of American-held stock to Model Power and British-held stock offered to regular retailers at home and abroad.

A lot of interest in Flying Scotsman was generated during the year with the visit of the Prototype to America. In consequence the Tri-ang Hornby model of Flying Scotsman in LNER livery became a best seller.

In February the decision was taken to cease development of the Minic Motorway system and wind it up. It had never been a good earner but its association with the model railway system was probably not just a one-way benefit. There is little doubt that the model railway system was more attractive for having a motorway equivalent.

This decision coincided with the move of Scalextric production to Margate and the closure of the Minimodels factory in Havant. This move was to have the most far-reaching significance providing the company with two strong products in the years ahead.

The Minix range of cars, which had always been made at Margate, was to cease to be a model range in its own right after 1970 but remained in production as loads for Tri-ang Hornby wagons.

Finally, the Series 3 track tools were possibly shipped out to India to be put to work there in 1970. Another associated company, this time in Spain, launched the Ibertren model railway system which seems to have been completely independent of its distant cousin – Tri-ang Hornby.

Looking to the Future

With the approach of 1971, which, unbeknown to those involved, was to be the last year of Tri-ang Hornby, it is interesting to note what changes were taking place in the thinking at Margate.

We know, for example, that there was commitment to the new System 6 track and they were also looking at the possibility of 2' radius points to see how they could be fitted into the geometry of the existing track system.

Work was progressing with a revolutionary new locomotive which would lead to major changes throughout the range of model locos made by the company. There was also an expansion of the range of pre-nationalisation liveries for the existing models. Some of the engines were to be given nickelled wheel tyres. Steam sound tenders were being developed and a longer term study of couplings was to be instigated.

Besides Evening Star, new models proposed included a low price coal wagon, large breakdown crane and coach and several other things, some of which arrived in 1971 while others slipped into the Hornby Railways period. There were also a number of items that never saw the light of day such as a long-bodied clockwork loco (R884), a Dubonnet wagon (R674) and a new caboose for the US and Canadian market (R482). A battery box disguised as a station was also proposed but did not arrive until 1993!

Other models planned at this time took some years to materialise. There was to be an LNER fish van using the R14 body which arrived in 1972 but with a new body of a Hull & Barnsley van. A girder bridge that could break down to be used for either a high or low level system and as either single or double track eventually arrived in 1973 as R657. A track bed (R656) was also proposed but did not materialise. The Caledonian Single in LMS livery was also proposed but had to wait until 1983 before becoming a reality.

A Sign of Things to Come

Once again the new Tri-ang Hornby range was on show at Tri-ang House in Edgware Road, London, during late January 1971, for inspection by the trade. The model railway press reacted well to the new catalogue with special praise for the Cuneo picture that adorned the front cover and, incidentally, the cover of this volume. This time, for those searching for the Cuneo mouse in the picture, Rovex provided the answer on page 4.

The most exciting new model was Evening Star which was the subject of Cuneo's picture and which, we were assured, would be up to European standards; a completely new departure for Tri-ang Hornby. This was a clear sign to some people of Rovex's intention to make scale models.

The model was displayed at the Tri-ang House trade toy fair well secured inside a clear plastic showcase. The maroon and gold Coronation, noticeably with nickel-tyred wheels (another sign of the path to be taken by Rovex in the years ahead), also came in for praise. It was noted that a gloss finish was now being applied to these early locomotive liveries.

Chuff-Chuff

Customers were also introduced to 'exhaust steam sound'. In their continuing search for ways of making their models more attractive to children, Rovex had come up with a very cheap way of adding a 'chuff-chuff' noise to the motion of its engines. A description of how this worked will be found in the chapter on 'Components'.

The disadvantage was that it was suitable only for tender engines with engine-mounted motors and so it could not be fitted to the tank engines or the new Evening Star with its revolutionary X800 motor bogie tender. In 1971, it would be fitted to only four locomotives.

1971 was the year that the Caledonian coaches were introduced in other liveries and while today we would wince at this idea, then, there was delight that the coaches would allow the proprietary modeller to take a step further back in time. Certainly, without close study the models in most cases were plausible copies of coaches run by the Southern Railway (ex-LSWR and SECR) and LMS (ex-CR) in the dim and distant past. The wagon range was strengthened with several additions including the coke wagon and the promised giant breakdown crane and matching workmen's coach. The year saw the fitting of fine scale wheels to more rolling stock and white-rimmed wheels to period coaches. Fine scale wheels for freight rolling stock in both disc and spoked form were also available to buy for converting stock.

Some excellent artwork made the train sets look very attractive in 1971, the square boxes of some being a notable feature. These returned to lift-off lids after a year when end-flap boxes were tried as an economy measure, but had proved unpopular.

In Canada the Canadian Pacific livery was updated to 'CP Rail' and the product name finally changed to Tri-ang Hornby.

To go with a new range of station sets for 1971, Rovex brought out their own version of the over-all station roof which was much cheaper to manufacture than the ex-Hornby-Dublo one.

The End of the Beginning

This brought to an end the fruits of amalgamation of the two systems and provides a fitting point to leave this review of the Tri-ang Hornby years. Two models with a long history that were dropped from the

catalogue in 1971 were the black Jinty and the Utility Van. Both were to return much improved in years to come but their loss at this time helped to emphasise the end of an era. The following year the name of the system would change to 'Hornby Railways' and the products available in the shops under this name will be reviewed in **Volume 3**.

The parent company went into liquidation in August.

As the year closed, stocks of packaging ran out and plain white or brown boxes were used for some models with labels printed in the factory stuck on the ends to identify the contents. Tri-ang Hornby's last locomotive was not to be Evening Star but a much smaller model. At the end of 1971, while the company was trading under the name of Pocketmoney Toys Ltd, it released a GWR Pannier Tank (R51S). This continued to be made in very large quantities in the years ahead rivalling the success of the Jinty which it initially replaced.

A small item, almost lost on page 439 of the December 1971 edition of *Model Railway Constructor*, will have meant little to most people at the time. It was about the Model Railway Society of Ireland awarding Honorary Membership to Mr and Mrs Cyril Fry, the builders and operators of the 0 gauge Irish International Railway and Tramway System. Those who have read my **Volume 1** may remember that it was Cyril Fry whose synchronised smoke for railway engines spurred on Rovex to invent their own Synchrosmoke.

NEW ZEALAND

The New Zealand outpost of the Lines empire could claim to be one of the oldest overseas companies in the group having been acquired at the end of the Second World War. Although its output was low, catering as it did only for a small home market, it continued to produce Tri-ang Hornby models after the other overseas subsidiaries had ceased production and were importing most of their requirements from the UK.

New Name but Old Barriers

Partly due to the low output and therefore the infrequent need to reprint packaging, the models continued to be sold in Tri-ang Railways boxes long after the name change in the UK. Even when the change did come in 1968, it involved sticking a printed label on the old boxes. Interestingly this coincided with a short spell during which boxes were stamped on the inside of an end flap with the date of packaging.

The New Zealand company retained the name Lines Bros. (NZ) Ltd until 1968 but the 1969 illustrated price list showed it changed to Tri-ang Pedigree (NZ) Ltd. While the old boxes may have remained in use, the product was renamed 'Tri-ang

Hornby' in the 1966 illustrated price list. Despite this some models continued to carry the road name 'Tri-ang Railways' to the end.

While South Africa and Australia relaxed their import controls and thus allowed the import of more ready-made Tri-ang Hornby, the case was very different in New Zealand where, even as late as 1968, two thirds of the value of a train set had to be manufactured domestically. Thus the New Zealand company continued to manufacture most of their requirements until the end of production in 1971 or 1972.

The search for new products – The L1

The company was looking for new products to make and towards the end of 1967 it was suggested at Margate that the 4-4-0 L1 was no longer required and might go to New Zealand. Harold Hendon was

asked to prepare a list of tools required and this was sent to John Paul at Lines Bros. (NZ) Ltd.

It was clear that some of the tools such as those for the bogies and wheels were not available as they were in use on other models. The driving wheels, for example were used on the new M7 tank engine. New Zealand were offered the available tools for £100 presumably with the idea that, where a tool was not available, the parts would be supplied ready-made for assembly at Auckland.

John Paul must have thought the project unviable as the tools were not sent.

Visit by Richard Lines

Instead, in July 1968, Richard Lines went out to New Zealand to advise them on their range. He found that the company was making eight different electric train sets but were selling only 2,000 electric

So slow was the turnover of boxes that old stocks had to be used up with Tri-ang Hornby labels suitably affixed.

Sets in the 1969 New Zealand illustrated price list.

train sets, and 2-3,000 clockwork sets, per year. His recommendations included the dropping of the least popular sets, the introduction of new cheaper sets, improved packaging and simplification of set codes.

A breakdown of the company's sales was produced at the time and this makes interesting reading. The exchange rate at the time the list was compiled (May 1968) was 2.1492 NZ$ to the £.

The company was introduced to the new cheap electric 0-4-0 tank, R852, and the integral wagons that were being tooled up that summer. The latter could be supplied to New Zealand bulk packed for 8d each and the loco for 6/8 each.

Other Might Have Beens

New Zealand were keen to get their hands on more redundant tools and it was suggested that they might have the Hornby-Dublo tools for the R5020 Goods Depot, R5030 Island Platform and R5086 Platform Extension. Other tools discussed at this time for possible shipping to New Zealand were the Series 3 incline and high level piers and the R43 signal. The latter would be available when Rovex produced their new signals.

It was also suggested that the parts for operating accessories such as the R752 Turbo Car, R216 Rocket Launcher, R128 Helicopter Car and R127 Crane Truck should be sent out for assembly in New Zealand but that brighter colours should be used for these.

Recommendations

The management at Lines Bros. (NZ) Ltd were also asked to look at the bubble packs used for sale of points and diamond crossings to see if they would be suitable for Series 3. It was further proposed that there should be Series 3 versions of the three track packs made in the UK for Super 4.

Other recommendations included:
1. The Jinty to have Synchrosmoke and Magnadhesion.

2. Points to be sold as hand operated only but with a conversion pack available.
3. Reduction in the tightness of eyelets used to fix bogies and couplings.
4. Better mixing of plastic colours to achieve the correct opacity (a lot of New Zealand wagons are almost translucent).
5. Roofs on rolling stock and station buildings should be glued in place.
6. Improvements to the quality of heat printing.
7. White lining on platforms to be printed rather than hand painted.
8. Reduce flash on track mouldings.
9. Series 5 (System 6) track to replace Series 3, eventually. (In fact they listed System 6 track in their price list from 1970 and never sold Super 4.)

Finally it was recommended that the New Zealand company should undertake harder selling of beneficial features of Tri-ang Hornby in their literature. Features such as the pulling power of locomotives, smoke and Magnadhesion, more detail of power units, and the long-lasting fun to be got from a model, were all mentioned.

Final days

In 1970 the company made a tax-paid profit of $61,000 (about $120,000 before tax).

By mid 1971, with Lines Bros. Ltd in difficulty, Tri-ang Pedigree (NZ) Ltd became 84% owned by the Australian company Cyclops Tri-ang Ltd and, between them, the two companies were budgeting for a A$1.5 million pre-tax profit in 1971.

At the time it was intended that Tri-ang Pedigree (NZ) Ltd would become an independent Australian-owned unit but, by the end of the year, Cyclops Tri-ang Ltd had been bought by Tube Investments Ltd and, with it, went the New Zealand subsidiary. TI already owned a company (Radiation New Zealand Ltd) in Dunedin which made Champion and New World kitchen stoves.

For a while Tri-ang Pedigree (NZ) Ltd continued to trade under the Tri-ang Pedigree name, although the 1972 North Island Illustrated Price List showed the product as simply 'Hornby'. The range was no different from what had been illustrated and listed in previous editions. In the final days it seems that models were leaving the factory that did not match the standard specification as stocks of materials and parts became used up. The 1972 North Island price list still carried the name Tri-ang Pedigree (NZ) Ltd but the 1973 North Island list was titled 'Cyclops by Tri-ang Pedigree' and gave no address. It listed much more than model railways.

The company eventually disappeared in 1978 but it seems that between 1972 and 1978 it imported Hornby for local consumption and continued to assemble System 6 sets unique to New Zealand, using all imported parts.

1968 New Zealand Price List.

AUSTRALIA

Agency Role – Background

Moldex Ltd, in Melbourne, were the manufacturers of Tri-ang Hornby in Australia and were themselves a subsidiary of Cyclops & Lines Bros. (AUST) Ltd of Sydney. Manufacture of the model railway system in Australia had begun in the late 1950s and continued until 1968.

The Australian company also became the selling agents for all Tri-ang Hornby Trains in Australia and were responsible for 'operating the market'. This meant establishing retail selling prices, booking indent orders for Rovex from wholesalers and master retailers, running service centres, advertising and informing Rovex of the activities of their competitors. For these functions Moldex received a 10% commission on all goods shipped by Rovex.

Moldex were free to choose the media for advertising but Rovex required that they emphasised the long-lasting fun of model railways as opposed to many other throw-away toys. They were to draw attention to the exclusive features of Tri-ang and had to be aware of the qualities of competitive products.

Acting as Rovex's service agents could have its moments. One unhappy incident was the failure of 825 clockwork mechanisms fitted in R660 and R854 locomotives which Moldex had to replace with new ones sent out from Margate. They paid 1/- each for the replacements plus the cost of transport and had to absorb the cost of their labour. They were responsible for repairing not only items sold in Australia through themselves but also models imported by other wholesalers.

Australian Supplement and Price List

The Australian catalogue supplement, that was slipped into the standard Tri-ang Hornby catalogue, was designed and printed by Moldex. They also produced the Australian Trade and Retail Price Lists which had to be in circulation by 1st February each year. To achieve this Rovex sent out photostats of the standard catalogue artwork by mid September together with approximate prices. Moldex would then select the train sets and solo items they considered from past experience would sell in Australia and would justify packaging by Moldex. They would then ask Rovex to quote their best prices for supplying these items bulk packed.

By the beginning of December Rovex would supply the proof for their Export Price List from which Moldex could work out the prices at which they would sell the items they would package in their factory. These were the ones that went in the Australian Trade and Retail Price List, the draft of which was with the printers by Christmas. After 1968 the practice of showing the origin of models with an 'A' (Moldex) or 'E' (Margate) in the price list was to be dropped.

Wholesaling

Moldex were themselves wholesalers, ordering from Rovex, holding stocks and supplying goods to other wholesalers who had run short of stock. On their own indents Moldex received 10% commission and a further 10% wholesale discount.

It is interesting to note the basis for costing models imported from the UK. This was done as follows :

		£
FOB England		100.00
Less 10% Commission		10.00
	=	90.00
		A$
Conversion £ to A$(x2.16)		194.40
Plus Duty at 17.5%		34.02
	=	228.42
Plus Freight at 5% of gross FOB		10.80
Plus Landing at 5% of gross FOB		10.80
Plus Confirming House Charges at 2.5%		5.40
	=	255.42
Plus Wholesale Margin of 50%		127.71
therefore Trade Price	=	383.13
Plus Sales Tax at 15%		57.47
	=	440.60
Plus Retailer's Margin of 50%		220.30
therefore Retail Price	=	660.90

Rovex agreed not to supply Australian importers with items that were already available from Moldex.

*The Moldex Ltd factory in Melbourne, c. 1995 (a picture of the factory when it was making Tri-ang Railways will be found on page 13 of **Volume 1**).*

The Cyclops Toys factory in Sydney as it was c.1960 and in the mid 1990s.

This, however caused trouble as larger wholesalers could obtain better deals by importing large quantities direct from Rovex. In order to overcome this problem it was agreed to equalise these margins.

1968 Changes – End of Manufacturing

By July 1968 Moldex was experiencing serious difficulties. They were undertaking the assembly of

The cover of the 1966 Australian catalogue supplement.

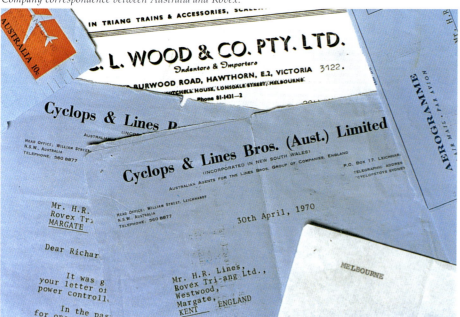

points and some rolling stock and locomotives from parts sent out from Margate. Points made in the UK were retailing at $1.69 each while those being assembled by Moldex retailed at $1.95. As if this was not bad enough, Lima points were selling in Australia for only $1.40 each.

This problem was not just confined to points. Similar discrepancies occurred on many other items including ordinary track. However, as the track was used in the assembly of sets by Moldex it was not thought to be so serious.

To make matters worse, wholesalers were complaining about the quality of Australian track. It was found that the sleeper web was not holding together properly and samples of the web clips used in the Margate factory were sent out to Moldex.

An agreement was reached in mid-1968 that the track and power unit production would continue

but that the assembly of other models would cease. The assembly of sets would continue at Moldex using home-made track but locos and rolling stock would be imported from Rovex. Stocks of parts could be used up or passed over to the New Zealand factory, or sent back to the UK. Surplus stock at that time included factory components, finished goods (including Minic) and some service spares.

While most sets to be assembled by Moldex would be copies of Rovex ones, they had the licence to make additional sets of their own using their knowledge of what sold well in Australia. This was only partially successful. Moldex complained of delays in delivery of parts and Rovex complained about the high degree of non-standard parts being ordered and the problem of supplying them.

Initially they were also allowed to box up for solo sale any bought in models not used in the sets. This

concession was withdrawn soon after but not before a number of Rovex-finished models had been boxed up in Moldex Tri-ang Hornby boxes and sold.

System 6

To cut the cost of track production and to compete better with Lima, it was agreed to send out tools for the production of the elements of the new System 6 track, or Series 5 as it was still referred to at that time, sufficient for the continued production of sets. It was planned that this would be introduced simultaneously with Rovex and that Super 4 would continue in production for a while. Samples of packaging for System 6 track were sent out to Moldex in February 1970.

As an economy measure, Moldex was advised to pack its popular items of track in boxes of 24 and not 12 and was asked to use double straights rather than two single straights wherever possible. They were also advised to follow Rovex's example of minimising the use of stock numbers by combining items into sets, like the R663 points remote control pack, instead of selling the components individually.

Packaging

Under the 1968 agreement Moldex had been allowed to continue to package goods for the Australian market but Rovex had laid down specific rules to be observed in the use of packaging:

1. Where a set was a copy of a Rovex set it must have the same number as the set copied. If contents were varied a new number had to be used.

2. Rovex would send samples of any new packaging so that Moldex could keep theirs looking the same but greater licence in box design would in future be allowed for Moldex's own sets.

3. Moldex could vary the packaging material but the finished article should have an equivalent look of quality. In particular attention should be given to the covering of bare chipboard edges with red paper, sharper edges on vacuum forming and vacuum-formed inserts to sit properly on the bottom of the box. There was a general agreement against the use of hot wire cut expanded polystyrene slabs but if they had to be used Rovex would supply aluminium castings for use as masters.

4. Track layouts and accessories were not to be illustrated on the cover of the box as this could lead to obsolescence when items were changed.

5. Solo track to be packed in lidded boxes.

6. Train sets to be clearly divided into two categories: Tri-ang Hornby Model Railways and Tri-ang Hornby Toy Trains.

7. Wording on boxes was to follow, more closely, that on Rovex boxes.

Master Wholesalers

Under the 1968 proposals Moldex were to become 'Master Wholesaler' stocking the complete range of Rovex's products while other wholesalers had only selected items. Interstate offices of the company while not being able to retain the complete range would hold stocks of basic items only.

As we have seen, independent importers were always a problem for Moldex and there was sometimes jealousy over availability of clearance stock at favourable prices. Major wholesalers included C.L. Wood of Melbourne, H.W. Rice (another member of the Lines group) of Sydney, Southern Models of Adelaide and Toys & Hobbies of Brisbane. A large wholesaler might order as many as 400 of each set they were importing in a given year. This was all business not going through Moldex.

Final Days – New Name, New Owner

With the reshaping of the Lines Group in 1969 the Australian parent company became Cyclops Tri-ang (Aust) Ltd and Moldex became Tri-ang Moldex Ltd. At that time Cyclops had a total of nine subsidiaries in Australia.

As we have seen, production of locos and rolling stock had ceased in 1968 but track production carried on for a while. Production at Moldex had completely ceased, however, by the end of 1970 and correspondence in January 1971 referred to the need to clear the company's stocks of Tri-ang Hornby products before orders from customers were referred to Margate.

In mid 1971, Cyclops Tri-ang (Aust) Ltd acquired 84% of the stock of Tri-ang Pedigree (NZ) Ltd which became a subsidiary of the Australian company in the same way that Tri-ang Moldex (Aust) Ltd already was. At the time it was intended that Cyclops Tri-ang (Aust) Ltd would became an independent Australian-owned unit but, by the end of the year it had been bought by Tube Investments (Australian Holdings) Pty, a subsidiary of the UK company Tube Investments Ltd. Tri-ang Moldex was wound up in 1976.

CANADA

Getting it Right – In the Beginning

From its foundation in 1947 until the mid 1950s, Lines Bros. (Canada) Ltd had not been profitable. The Transcontinental range was devised specifically to give the Canadian company a wider range of toys to market but the model railway market had been a very difficult one to capture due to the strong competition from the large American toy companies across the border.

It had been thought that because trains had been successful in the UK they would be popular in Canada. Sadly they were produced too quickly and without sufficient attention to what the Canadian market demanded. As a result the Transcontinental products of the mid 1950s lacked realism.

By the late 1950s the standard of model had improved but they still carried the Tri-ang Railways brand name instead of authentic liveries. The argument for this was that this made it easier to sell them world wide. It was not until 1965 that the first Canadian liveries arrived. Initially the choice was limited to Canadian National but in 1969 a range of Canadian Pacific models was added. Eventually, in 1971, following the updating of the real Canadian Pacific Railroad, a third series, CP Rail, was added.

Scope for Further Development

Today, there are a good number of Canadian collectors of Tri-ang Hornby and the help of some of them in compiling this book has added much interest to the subject.

It has been suggested, for example, that Lines Bros. could have better exploited their informal association with Lionel. This might have avoided the limitations of the system which prevented it

The Lines Bros. (Canada) Ltd factory in Montreal.

R139CN in order to distinguish it from the Westwood Pickles version. Despite this, the system worked well until it was decided to introduce an updated Canadian Pacific livery – CP Rail. At this point it was decided to abandon suffixes and adopt four figure numbers. To do this an '0' was added to the end of the original 'R' number for the private owner liveried model, Canadian National models had a 1 added, Canadian Pacific a 2 and CP Rail a 3. Thus the Pulpwood car was numbered as follows :

R235 The standard Transcontinental model available world wide.
R2350 International Logging Company
R2351 Canadian National livery
R2352 Canadian Pacific livery
R2353 CP Rail livery

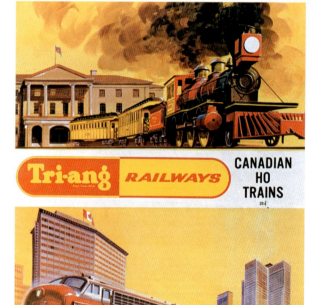

from becoming Canada's market leader but there was an informal agreement between the two companies not to tread on each other's toes.

The single biggest problem in selling Tri-ang Railways/Tri-ang Hornby in Canada was its incompatibility with products made to NMRA standards. Beginners who started with Tri-ang inevitably became exposed to scale products which in turn led to crude attempts at modification or the disposal of Tri-ang in favour of scale HO products. Despite this, the robust character of the Tri-ang Transcontinental models did endear them to many people who later, as adults, let their children play with their much-prized sets.

Numbering of Canadian Models

As we have seen, the first Canadian livery to appear was that of Canadian National (CN) in 1965. That year the Canadian price list contained three locomotives in CN livery, three passenger cars and six freight cars.

These reliveried models were given a CN suffix to their original 'R' numbers. Thus the Canadian National version of the R115 caboose became R115CN. This meant that when the Canadian Pacific liveries came along they could be given a CP suffix. Unfortunately CN did not always mean that a model carried the Canadian National livery. The Heinz pickle car, for example was numbered

1968 Canadian Tri-ang Railways *catalogue.*

Final Days – Name Change

Lines Bros. (Canada) Ltd did not wish to change the brand name in 1965 when Tri-ang Railways became 'Tri-ang Hornby' as Hornby continued to have other meanings in Canada, as we shall see. The logo, however, was revised to resemble that of the Tri-ang Hornby one. The name in Canada eventually changed to 'Tri-ang Hornby' in 1971.

Closure of the Montreal Factory

The financial problems of the Canadian company continued. There were problems with changes in import duties and adverse movements in the value of the Canadian dollar, necessitating drastic measures in the mid 1960s. At about the time of the takeover of Meccano Ltd, Lines Bros. (Canada) Ltd decided to sell their Montreal factory and move the company to smaller premises in Waterloo, Quebec.

Prior to this, a warehouse at the Montreal factory had provided storage accommodation for Canadian-produced Tri-ang products (which were sold under the 'Thistle' brand name) and goods imported from the UK factories in the Lines Group. With the disposal of the factory this warehouse was no longer available and the opportunity was taken to combine Tri-ang storage with Meccano Ltd at Brown's Line in Toronto.

Roland G Hornby Ltd

No.95 Brown's Line was a very modern building and seemed suitable, although Richard Lines remembers it as a claustrophobic place as it had no opening windows in the offices. It had been set up as an independent company by Meccano (Canada) Ltd under the name of Roland G. Hornby Ltd for the sole purpose of wholesaling in order to comply with Canadian taxation law. Now it took over wholesaling of Tri-ang as well.

One Canadian collector remembers the Brown's Line building with its stocks of Dinky Toys, Meccano Building Sets, Thistle Baby Carriages and Tri-ang Railways. He also remembers Hong Kong-made toys in the store in the late 1960s and Frog kits in 1970 and 1971.

The Canadian-themed models made at Margate were normally made in the first half of the year and so the Canadian company had to get its orders in by mid April. While there had been some market

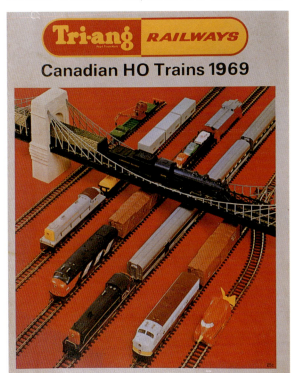

1969 Canadian Tri-ang Railways *catalogue.*

1970 Canadian HO Trains catalogue.

1970 Canadian Tri-ang Railways *catalogue.*

1973 Canadian Hornby Railways *catalogue.*

growth during the late 1960s, Canadian sales were eventually undermined by Lima.

After Brown's Line closed down, the bulk of their stock was acquired by a chain known as Playtime Stores who were selling the stock up until the time they themselves went into liquidation in 1979.

Louis Marx

With the break-up of the Lines Group, Louis Marx Industries (Canada) Ltd bought Lines Bros. interests in Canada. They initially decided to continue the

Lines Bros. Inc. 1966 Tri-ang Hornby price list.

range and even ordered a number of new models from Rovex. During 1972 and 1973 these were sold in the new Hornby Railways packaging. The company must have had second thoughts, for after 1973, no more Canadian catalogues were produced and it is assumed that no further models were ordered from Rovex.

Marx sold off their surplus stocks of Tri-ang/Hornby models locally but around 1980 a large stock of Canadian freight cars and other models was uncovered in Canada and initially offered for sale locally. Canadian collectors were able to buy models very cheaply at that time and what did not sell was shipped to the UK where it was sold by the Zodiac chain of toy shops.

USA

Lines Bros. Inc. had been the American subsidiary of Lines Bros. Ltd occupying a sales office at 200 Fifth Avenue until about 1966. Lines also had an 11% share in a company called De Luxe Topper Corporation of Elizabeth, NJ.

AMERICAN TRAIN & TRACK CORP.

Mike Tager

Mike Tager worked as sales agent for Atlas, the American model railway company. With the closure of Lines Bros. Inc. in the mid 1960s, Tager left Atlas to become the Tri-ang Hornby agent in New York taking an office in the same building, on Fifth Avenue, formerly occupied by Lines Bros.

At about this time he set up American Train & Track Corp. (ATT) at the Fifth Avenue address. ATT did not manufacture models but bought in special liveried models from Rovex and other manufacturers, including German kit makers, and packaged them for the American market. Using Atlas track, he made up sets at a plant in Cedarburg, Wisconsin.

Mike Tager also imported plastic aircraft kits,

from various manufacturers around the world including Frog kits from Rovex, and repackaged them under another name of his own – Universal Powermaster Corporation (UPC). His period of operation in the case of both model railways and aircraft kits seems to have been 1967-1970.

It is interesting to note that while Margate's American range was made only for ATT, the models were displayed at the Trade Show at Tri-ang House in both 1968 and 1969. This suggests that Rovex were interested in finding other buyers for them.

The models ATT bought from Rovex included the second series TC passenger cars, Old Time coach, two freight cars, Budd RDCs, stations, bridges, Rocket and its coach and some starter set models. Part of the deal was that Rovex would fit NMRA couplings to the models before dispatch.

Passenger Cars

The Tri-ang Hornby TC passenger cars were in four liveries and offered by ATT solo, in presentation packs of one livery and in train sets where they were matched with an FT power unit and dummy unit, made by a different manufacturer but in the correct livery. The four sets (one for each railroad company) containing these passenger cars were numbered 107-110. The coach liveries were:

> Pennsylvania
> Burlington
> Santa Fe
> Baltimore & Ohio

The old time coach was offered in two liveries:

> Pullman
> Central Pacific

Freight Cars

The two models chosen were the pulp wood car and the liner train with three 20' containers. The pulp wood car was available in the following three liveries:

> Northern Pacific
> Southern
> Chesapeake & Ohio

May 1967 ATT price list showing Stephenson's Rocket set and 1967/68 ATT catalogue.

Far right:
ATT set showing Tri-ang 2nd series TC passenger cars on the box lid and the 0-4-0 continental tank inside.

ATT Rocket set.

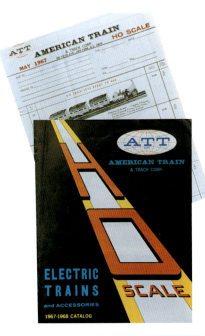

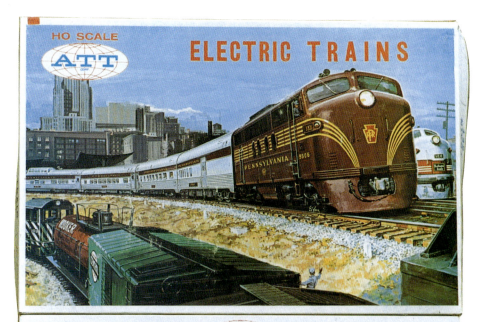

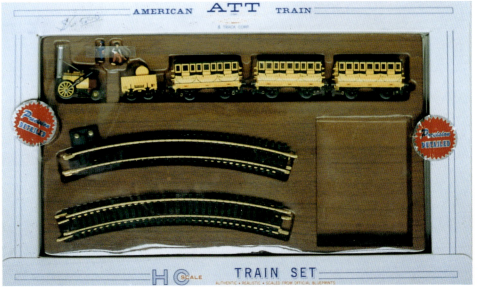

The liner train containers were finished in silver but printed with the following two American road names:

Santa Fe
The Milwaukee Road

Budd RDCs

The Budd RDCs were added to the American range in 1968 and produced in the following four liveries:

Northern Pacific
Santa Fe
Chesapeake & Ohio
Reading Lines

The models were available as both powered and non-powered versions which were sold separately or together in a presentation pack with a circle of Atlas track. These presentation sets were numbered 1545-1548.

Models in ATT boxes.

Train Sets

Besides the four sets containing the TC passenger cars in American liveries and their presentation packs, Tri-ang Hornby models were used in a number of other sets. The most obvious was set No.100 Stephenson's Rocket which contained the loco, with its driver and fireman, together with three of its coaches and a circle of Atlas track.

Three of the old time coaches in Central Pacific livery also appeared in set No.106 with a Yard Bird loco and tender of unknown made.

The R852 electric 0-4-0 Continental tank loco for starter sets was used in the No.1020 train set with three freight cars from a different manufacturer. ATT also produced their own 'Miniville' sets. The one I have seen was based on the W1 set made by Rovex and using all Rovex parts except the track which was from Atlas.

Buildings

The Tri-ang bridges appealed to ATT partly because of their low price but also the range of structures modelled. They therefore ordered the Victorian Suspension bridge, the river bridge and the viaduct.

In the station line, many of the components for the second series stations were bought and assembled into three station sets which were numbered from 610-612.

610 was called the Standard City Station and consisted of two straight platform sections, platform with underpass, two ramps, the steps unit, ticket office, seat unit and platform accessories.

611 had a straight platform unit, two platforms with under passes, two ramps, the island waiting room with a canopy on top, seat unit and platform accessories.

612 was a curved station made up of three large curved platform units with canopies, two ramps, the island waiting room, seat unit and platform accessories.

Packaging

ATT boxes came in 3 different types :
1. Sleeved box with item rubber stamped on the end and a yellow plastic tray insert.
2. Shrink wrapped box also with a plastic insert.
3. Box with a lift-off lid and an orange card insert.

The train sets and station sets came in heavily built boxes with lift-off lids. Some had illustrations of the contents and other available accessories on the lids while others were window boxes. Presentation packs were made up of solo-packaged models in their ATT boxes, shrink wrapped into a slab with a specially printed header card.

MODEL POWER

Acquisition of Stock

Sales were not as good as expected and Mike Tager closed ATT in 1970 and sold his remaining stock. The Tri-ang Hornby and models of other makes were bought by PTI Products of 180 Smith Street, Farmingdale, New York, who repackaged them in blue boxes and sold them under the Model Power trademark. It seems that Model Power started to box Tri-ang Hornby after the Summer of 1971.

What did PMI stand for? Was there a connection with Polk's Model Craft Hobbies of 346 Bergan Ave., Jersey City, NJ, who had a hobby department store at 314 Fifth Avenue, New York.

Model Power boxes listed the following 8 items:
Pulpwood car Northern Pacific (#471)
Pulpwood car Southern (#472)
Pulpwood car L & N (#473)
Container wagon (unspecified)
Coach (unspecified)
Combine (baggage/dormitory?)(unspecified)
Observation car (unspecified)
Shifter and tender (not Tri-ang Hornby)
It is quite possible that this list is not complete.

Remnant ATT stock from the Margate store boxed for sale in Australia and New Zealand.

Ex-ATT stock boxed for sale by Model Power.

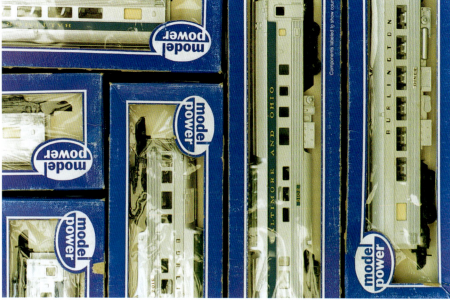

Remnant Stock at Margate

Some of the stock manufactured for ATT but not sent to the States ended up in Australia packaged in Tri-ang Hornby boxes. The boxes carried the 'R' number of the contents rubber stamped on an otherwise blank end flap panel. It is assumed that these were amongst the stores residue disposed of by Rovex during 1970.

In February 1970 retailers in the UK were being offered 16 different coaches, freight cars and Budd RDCs in US liveries with NMRA couplings which were described as 'frustrated stock'. By the March letter to stockists the range available was down to 9.

Other Sets Supplied to the USA

The census of production kept at Margate showed that a few other special sets were ordered by unnamed customers in the USA during 1970. These were the RS33 special set of which 648 were dispatched and the RS605D Set of which 48 were sent.

AMRO LTD

Exclusive Authorised Distributor

In 1969 Hugh and Mike Stephen's (believed to have been father and son) ran a company called American Rovex, or Amro for short, selling British outline models, made by Rovex, to ex-patriots. The company was based in Cedarburg, Wisconsin. As we have seen, ATT also had a plant at Cedarburg, Wisconsin 53012. Presumably this was the same plant but what was the connection?

Amro sold the Tri-ang Hornby range by mail order in the USA and advertised the complete Tri-ang Hornby British range with the exception of starter sets, Minic Motorway and Tri-ang Wrenn. The company's publicity material claimed that they were the 'Exclusive Authorised Distributor and Service Station for the United States' for Tri-ang Hornby.

Flying Scotsman Exhibition

The timing of the exercise seemed to be good with the pending arrival from the UK of the real engine, the double-tendered Flying Scotsman and its exhibition train consisting of 3 Pullman cars, 4 Gresley full brake coaches painted to match the Pullmans and the ex-Devon Belle observation coach. Amro tried to capitalise on this by offering what they called an 'executive' ET4 set containing Flying Scotsman and Pullman cars. They also offered spare R856 tenders for the Scotsman with extra '4472' number sets to allow for converting these to the water-carrying version that travelled with the loco. The total additional cost was $3.95.

The Margate factory census sheets provide no evidence of Amro having been supplied with bulk packed parts for its set. Did Amro receive complete sets (possibly the RS605D of 1970) or individual boxed items and re-box them? There is also no evidence of the company receiving any other locos bulk

Polk's Hobby Dept. advertisement disposing of remnant stocks of ATT models.

Locomotives

BRITANNIA CLASS Heavy Pacific, No. R259S is painted in British National Railways dark green with red trim, yellow numbers and BNR insignia on the tender. Also smoke unit with capsule of smoke oil and engine crew. Plenty of power, was used in heavy passenger service. Price, ready-to-run, $18.95

FLYING SCOTSMAN, 4-6-2 locomotive of international fame is offered in original L.N.E.R. light green with red, white and black trim, gold and red lettering. One of England's great steam locomotives. Price, ready-to-run, $18.95

THE PRINCESS ROYAL, 4-6-2 another powerful Pacific used in fast passenger service. This engine comes in maroon with yellow and black trim, British Railways herald on tender. Price, ready-to-run, $18.95

ALBERT HALL, 4-6-0 Express locomotive, often used on fast passenger runs. Smooth, quiet running, engine is painted in dark British Railways green, red trim and brass capped smoke stack. Price, ready-to-run, $15.95

FLYING SCOTSMAN, a beautifully illustrated, full text book about the locomotive, its owners, the history and events surrounding its many years of service and final preservation and rebuilding. The 64 pages will hold you spellbound — and give you a complete reference for this world famous locomotive. The book is authored by the owner of the Flying Scotsman, Alan Pegler, together with Cecil J. Allen and Trevor Bailer. Price, $3.50, postpaid.

AMRO leaflet.

Executive Set
FLYING SCOTSMAN

L.N.E.R.

UNIQUE, COMPLETE HO train set consisting of a Flying Scotsman locomotive and five Pullman cars, station, 47 sections of track, four switches and a power pack plus a Track Plans Book. Truly a complete model railroad in a beautiful simulated wood carrying case. Here is a gift item unsurpassed — even if you are giving it to yourself. Ideal as a Corporate gift from an executive to a special customer or account that is hard to please. Limited edition. Price, complete, postpaid, $69.95

The Flying Scotsman Executive Train Set contains track to create this unusual, double loop model railroad layout — or you may create your own. The Triang/Hornby line of model railway locomotives and rolling stock is one of the most extensive offered. Consult the Triang/Hornby catalogue included in every Flying Scotsman Executive Train Set.

The Flying Scotsman BUILT FOR THE EXECUTIVE ENGINEER

packed and so they may have received them ready boxed. In this case they would be included in the total export figures on the factory census sheets and could not be easily distinguished.

Amro even had its own travelling display in one of the exhibition coaches attached to the Flying Scotsman train while it was on tour. The display was in a wall-mounted glass cabinet which contained mainly steam locomotives in pre-nationalisation liveries. It was sited close to a Tri-ang Hornby layout which was a feature of the exhibition.

Expansion and Disposal

Early in 1971 Amro began to expand the range of makes they stocked to include Fleischmann, Rowa and Liliput. They continued to sell Hornby Railways after the collapse of Lines Bros. and the takeover of Rovex by Dunbee Combex Marx, and by mid 1972 they were selling Wrenn Railways which apparently had not been possible while Wrenn was part of the Tri-ang Group.

In December 1972, the company was sold to Paul Wagner who owned Paul's Model R.R. Shop in New Oxford, Pennsylvania. Amro Ltd was moved to this address and continued to be the official importer of Hornby Railways until the summer of 1976 when, protesting at what it considered to be unreasonable price demands, it gave up this role. Throughout the period that it handled products from Margate it had an excellent reputation for its speed of service. Amro Ltd was still operating in the mid 1990s from Paul's Model R.R. Shop in New Oxford, Pennsyvania.

THE TRI-ANG HORNBY EXHIBITION LAYOUT

We mentioned above that the 'Flying Scotsman' on tour in the USA in 1970 contained in one of the coaches a Tri-ang Hornby layout.

The following is an extract from the 'Flying Scotsman Express', a news sheet, published in August 1970 by Green Bay, Wisconsin, which described the exhibit:

"No exhibition train is complete without a model railway. A Tri-ang Hornby layout with representative British diesel, electric and steam engines (including the 'Flying Scotsman') can be seen. The layout is complete with sidings, stations, signal boxes and signals. A new train set, featuring 'Flying Scotsman' with Pullmans, train station, power unit and other accessories is also featured, and similar sets can be ordered on the train from the manufacturers. Other British trains shown here in HO gauge are the LMS Princess Pacific, the Great Western Hall 4-6-0, the LNER (misspelled LMER) B12 4-6-0, and the BR 'Britannia' Pacific. Also displayed are the English Electric Type 3 Diesel, the British Type 2 Diesel, the Southern 0-4-4T and the Southern 'Battle of Britain' Pacific."

Gary McKinnon of Barrie, Ontario, Canada, remembers his visit to the 'Flying Scotsman' when the train was at the Canadian National Exhibition in Toronto in the last week of August. He describes the layout as being approximately 4' x 6' and encased in a perspex cover. The layout plan was fairly simple but with numerous disconnected sidings for the displaying of models. At the time he saw it, the layout was not in operation. Perhaps, surprisingly, the 'Flying Scotsman' displayed was the BR version.

The rest of the models and accessories were all out of the 1970 catalogue, but not all from the British edition. A number of CN and CPRail items had found their way onto the layout thanks to the proprietor of George's, one of Toronto's leading model shops and a major service dealer for Tri-ang Hornby.

INDIA

EARLY CONTACTS

RI Sets

New evidence suggests that two sets may have been made in the UK, in the late 50s or early 60s, specially for India as there is reference in correspondence with an Indian company to an RI.A and an RI.B set (the 'I' presumably standing for 'India'). The RI.A set contained the R56 Baltic tank and R230 primary set coaches and the RI.B consisted of the Baltic tank with wagons.

1963 Correspondence

As early as 1963 there was thought of selling tools to an associated company in India. We do not have the correspondence that explains how this came about but we do have a list, dated 8th May 1963, which shows the models the Indian contact was interested in making. At Margate the availability of the tools was checked and the Costing Department sent a list of tools needed, together with samples of the models, to the Group's headquarters at Merton.

The following is the list:

R50/R30 'Princess Victoria' and Tender
The tools for the chassis were available except those for the pony truck wheels and the front bogie. The tender tools and the loco body mould were also unavailable as the loco was still in production.

0-4-0 C/W Top Tank Loco – There was no complete locomotive available to send as a sample and the tools were still required.

R28 Mainline Brake 2nd – The body mould was available but cannot have been sent to India as it was later to go out to New Zealand.

R29 Mainline Composite – This body tool was still required for the red and yellow coach for starter sets.

R224 Restaurant Car – The body, roof and interior moulds were available as twere the window -printing tool for the maroon and cream version of the model, the table-top spray mask and the printing die. The bogie and wheel moulds were unavailable. The tools can not have gone to India as they were later sent to New Zealand.

R230 Primary Coach – Details of these tools had already been supplied and so none were given here.

R10 Open Wagon – Tools unavailable.

R11 Goods Van – Tools unavailable.

R12 Petrol Tank Wagon – No sample was available. Tools unavailable.

R13 Open Wagon with Coal Load Tools unavailable.

R14 Fish Van – Tools unavailable.

R15 Milk Tank Wagon – Tools unavailable.

R16 Brake Van – Chassis tool only available.

R110 Bogie Bolster Wagon – Only one sample could be found. Tools unavailable.

R217 Primary Open Truck – Body mould only available.

R218 Primary Van – Body and roof moulds available but not wheels.

R60 Ticket Office – All moulds and spraymask available.

R61 Signal Box – All moulds and spraymask available.

R62 Waiting Room – Only one sample could be found. All moulds and spraymask available.

R63 Central Platform Unit – Mould and spraymask available.

R64L Left Hand Platform End – Mould and spraymask available.

R64R Right Hand Platform End – Mould and spraymask available.

R65 Platform Ramp – Mould and spraymask available.

R66 Porter's Room – All moulds and spraymask available.

R67 Approach Steps – Mould available.

R75 Water Tower – Both moulds and spraymask available.

R76 Engine Shed – (no sample was available)

R175 Single Track Level Crossing – Gate mould available. The tool for the body of the crossing would have had to be converted to take Super 4 track which would have been a big job.

Super 4 Track – A six-impression mould which made two straights, two double curves and the mouldings for the power connector clip was available.

No further correspondence has survived from this period and it thought that the matter was allowed to drop. It has, however, been suggested that the correspondent was a Mr Kumar.

KUMAR & LINES BROS.

S.Kumar & Co. (Pvt) Ltd

Three years later, in 1966, Lines Bros. were corresponding with Mr Kumar of S.Kumar & Co. (Pvt) Ltd of Calcutta on a proposal that, in exchange for shares in his company, Lines Bros. would supply tools for the manufacture of Tri-ang products in India. At the time, Indian importers were finding it hard to obtain licences to buy in ready-made goods from the UK but the import of tools for the purpose of manufacturing within India was encouraged.

S.Kumar & Co. (Pvt) Ltd was associated with Aurora (who had a number of companies in India) but their note paper, which was decorated with just Tri-ang product names, claimed 'A Collaboration of Kumar & Lines Bros.' Despite this, the company was not listed amongst the associates in the Lines Bros. annual reports of the time.

Kumar had his head office at 15 Shakespeare Sarani, Calcutta 16 and factories at 63/64 Foreshore Road, Shibpur, Howrah, Calcutta and at 13/7 Mathura Road, Faridabad, Delhi. The correspondence from August 1966 to March 1970 has been made available to me and forms the basis of this research.

Earliest Correspondence

The earliest piece of correspondence is an internal memorandum from the Lines Bros. head office at Merton, a copy of which was sent to Kumar & Co. This contained the following list of models the tools for which Kumar were interested in buying:

R252 0-4-0 Steeple Cab Locomotive – No further reference was made to this model in subsequent correspondence.

R56 Baltic Tank – to be supplied without valve gear.

R24 1st Series Transcontinental Passenger Car

R25 1st Series Transcontinental Observation Car

R130 1st Series Transcontinental Baggage Car

R324 1st Series Transcontinental Dining Car

R217 Primary Open Truck

R218 Primary Van

R60 Ticket Office

R61 Signal Box

R62 Waiting Room

R63 Central Platform Unit

R65 Platform Short Ramp

R66 Porter's Room

R67 Approach Steps

R68 Station Name Boards

R73 Island Platform Canopy Set

R74 High Level Pier

R76 Engine Shed

R79 Inclined Piers

R89 Sidewalls

R90 Curved Sidewalls

R187 Pedestrian Crossing Set

Series 3 Track –

R190 Straight Track

R191 Quarter Straight Track

R192 Eighth Straight Track

R193 Curved Track

R194 Half Curved Track

R195 Large Radius Curve – Not available.

R197 Power Connecting Clip – It was felt unlikely that this would be available

R198 Uncoupling Track

R290 Diamond Crossing

R291 Left Hand Point

R292 Right Hand Point

1967

In the 1967 capital expenditure budget, Lines Bros. had included a sum of £1,450 to cover the cost of a pierce-and-blank tool, motor-pole piece, laminated stacking tool and a progression tool for coupling bars. It was intended that this would release the existing tools for India so that Kumar could make his own motors and couplings but the money was later deleted from the budget as it was agreed that

Cable : 'TRADISCORP'

Phones : 44-2764
44-995.
44-932
$ Line

S. KUMAR & CO., (Pvt.) LTD.

Head Office :
15, SHAKESPEARE SARANI,
CALCUTTA-16.

a collaboration of
KUMAR & LINES BROS

Factories :
Calcutta : 63/64, FORESHORE ROAD,
SHIBPUR, HOWRAH,
(Phone : 67-2047 & 67-2782)
Delhi : 13/7, MATHURA ROAD,
FARIDABAD.

Our Ref.   SK:b   Your Ref.   Dated. October 29,   1969.

DINKY:
   Toys

Mr. J. B. Hartley,
Lines Bros. Ltd.,
Morden Road,
Merton,
London SW 19

[handwritten: Mr H R Lines — If you really desire could I have a copy please? MRM]

Dear Mr. Hartley,

TRIANG :
   Trains
   Toys

        Could we have the following details
from Rovex?

PEDIGREE :
   Dolls
   Prams

i)   What machine capacity is required for their
      plastic moulds?

ii)   Is it possible to get a drawing die for the
      rails?

iii)   Details of rail-cropping m/c.

iv)   Details of motor RPM testing fixtures.

v)   Samples of components used in the locos, car-
      riages and accessories.

vi)   Samples of accessories in current production
      (so that our moulds can be made for producing
      the same designs) with their drawings.

FROG :
   Engines
   Model Kits

vii)   Details of further items of rolling stock they
      may have discontinued.

viii) A few sets of components of the electric motor
      will be helpful in getting our assembly line
      going.

ix)   Design of diecast chassis instead of pressed
      chassis -- as per their current practice - for
      R-153 & R-56 Locos.

MECCANO :

      With thanks,

              Yours sincerely,

              (S. Kumar)

[handwritten margin notes: L.R.; see vol; all items of tool; layout drawings; L.R.; L.R.; P.T.O.]

Associate Companies :
Aurora Model Mfg. Co. Pvt. Ltd.
India's Hobby Centre Pvt. Ltd.
Aurora Teaching Aids Mfg. Co. Pvt. Ltd.
Aurora Pram & Cycle Co.

Showrooms :
1A, Russell Street, Calcutta-16
1/155, Mount Road, Madras-2
Opp : Plaza Cinema, New Delhi (42121)
53B, Queen's Road, Bombay-2

---

any tools that could not be supplied by Rovex would be made in India. In the meantime, Hendon, the General Manager at Margate, was asked to prepare drawings of the parts for which Kumar would have to have new tools made.

1969

Things then appear to have gone quiet until October 1969 when Lines Bros. received an order from S. Kumar & Co. (Pvt) Ltd for the tools to make the following models:

   R153 Saddle Tank
   R217 Primary Open Wagon
   R218 Primary Van
   R193 Series 3 Small Radius Curve
   R197 Series 3 Power Clip
   X04 Motor

The Indian company also confirmed its continued interest in the Baltic Tank, the Series 3 straight and points, the 1st Series TC coaches and the Primary coach. They also asked for samples of current trackside accessories with a view to having the tools made for them locally.

By the end of October, enquiries from India lead onto details of machine capacity required for the Rovex moulds, drawing dies and cropping tools for rail production and the equipment to test electric motors.

There was also a request for samples of the components used in the assembly of locos, carriages and accessories for use in obtaining quotations from the many small companies in India that could take on

Letter from S.Kumar & Co., (Pvt.) Ltd of Calcutta with whom Rovex were discussing sale of tools for setting up a manufacturing plant in India in the 1960s. Note the 'a collaboration of KUMAR & LINES BROS'. At this particular stage it was the tools for the R153 saddle tank, the R217 and R218 Primary wagons, the R193 Series 3 curve and the R197 power clip that were being requested.

A Crown Railways set. Whilst this set is of more recent manufacture, Crown Railways show strong Tri-ang Hornby similarities.

Boxed Crown Railways wagons.

this work. In those days Indian businessmen were not accustomed to giving quotations with only a drawing of the component to look at. They were used to being handed a sample and asked to give a quote for a fixed number of copies. As Mr Kumar put it, "their visualising powers are not fully developed."

The South African Baltic Tank

In the summer of 1968, Richard Lines had gone out to South Africa and, while there, discovered that Tri-ang Pedigree (South Africa) Pty Ltd still had the tools which had been sent out to them in 1955 for the production of the Baltic Tank, two short passenger cars and three freight cars and were wanting to find

a buyer for them. In December 1969, the thought occurred that these might be suitable for India and Lines Bros. wrote to the South African company in an attempt to establish what was available and what price was being sought for them.

In February 1970, Kumar received a letter from Mr L.W. Reed of Lines Bros. who was dealing with the sale of the tools to India, saying:

"I have placed an order as per the enclosed list (missing), with Rovex Tri-ang, Margate, for all the tools which they have available for the train set which you intend to manufacture in India. These tools should be in transit to you very shortly.

"In addition, I am enclosing a catalogue from Lines Bros. (South Africa), who have available for disposal com-

plete sets of tools for R26 (TC Short Coach), R115 (Caboose), R27 (TC Short Vista Dome), R114 (Short Box Car) and R116 (Gondola) rolling stock, together with the housing for the RT42 control unit. If these tools are of any interest to you, would you please inform me as soon as possible..."

It is interesting to note that Lines did not offer the South African Baltic tank, possibly because they wanted to dispose of the tools of the Margate version.

Series 3 Track

In March, Kumar wrote asking for the Super 4 track tools instead of Series 3 which he had been offered.

At this point the surviving correspondence dried up and we are left to guess the outcome of this episode. Were the tools finally sent and was the offer of the South African tools taken up? Did Kumar get Series 3 or Super 4 track?

Part of the answer may lie in the tool census carried out at Margate in 1982. This showed no sign of the tools for Series 3 track but all the Super 4 tools were still there. The tools for the original series of TC passenger cars were missing but the Baltic tank was there. Perhaps India had the South African tools for this after all, or may be none.

It should be noted that Kumar's interest in tools from Lines Bros. was not limited to the trains but included Pedigree dolls and Arkitex as well.

Crown Railways

If you are visiting India today, look out for electric trains sold as 'Crown Railways' or later as 'Model Railways' by Electronic Toys (India) of Bombay. See how many similarities you can find with Tri-ang Railways and Tri-ang Hornby models and design.

FRANCE

MECCANO (FRANCE) LTD

Background

When Meccano Ltd joined the Lines Group of companies in 1964 it brought with it its French subsidiary. Meccano (France) Ltd had been established in 1912 and in 1930 had moved to premises at Bobigny near Paris. Here it occupied a factory with 4,000 m² of floor space. Reorganisation of the company in 1951 left it with 14,000 m² and one of the largest toy manufacturing plants in France. At the time of the Meccano take over by Lines, Meccano (France) Ltd employed 600 staff of which 400 were women.

It was not until 1960 that Meccano (France) Ltd started making its Hornby Acho train system which was a close cousin of the Hornby-Dublo, made by the parent company at its factory in Liverpool.

Calais

Lines Bros. Ltd, as we saw in **Volume 1**, had chosen to build a factory at Calais to establish a production base within the European Common Market of which Britain at that time was not a member. The Calais factory was built in the style of those at Margate (Rovex Ltd), Canterbury (Minic Ltd) and Havant (Minimodels Ltd) and had 10,000 m² of floor space and sat on a site of 60,000 m² which allowed for future expansion.

Calais had been chosen as it was close to Kent and moulds could be transferred between factories with the minimum of cost. The factory provided employment for 150 people and its principal product was Scalextric for the Continental market although it did produce a good range of Frog plastic aircraft kits which initially were sold under the Tri-ang name so as not to offend the French.

The company also had a pram manufacturing plant at Nogent-sur-Seine.

The amalgamation of Lines and Meccano interests in France was carried out under the control of Meccano (France) Ltd and its managing director M. Marcel Chanu. The main products of the enlarged French subsidiary were Scalextric, Dinky Toys, Hornby Acho, Play Doh (made under an American licence), Tri-ang (Frog) Aircraft Kits, Meccano and Tri-ang Toys. Meccano (France) Ltd moved into the Calais factory and made this their production base. From 1965, with a staff of 800, they traded under the name Meccano Tri-ang.

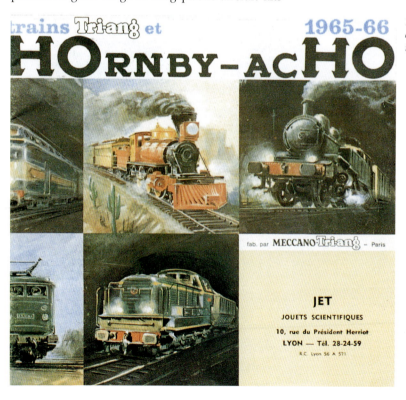

1965-66 Hornby-Acho catalogue showed already the Tri-ang influence.

Tri-ang Hornby Products

As the Lines Group's foot in mainland Europe it was required to widen its range to incorporate products from other Lines factories that were thought likely to sell, particularly in France. One result of this was that while Meccano's French Hornby Acho model railway system would remain in production its catalogues would include models from the Tri-ang Hornby system.

Where they could get away with it Rovex left their own couplings in place but where it would be necessary for an item to connect with a French one, Acho couplings supplied by Meccano Tri-ang were provided in the box. The Rocket set and Pullman DMU were supplied with only the tension-lock Tri-ang couplings but the wagons came with the French design of coupling as well. These models were given an 'F' suffix to their 'R' number; thus R128 became R128F.

The B12 had the Acho coupling at the rear and a Tri-ang coupling at the front. It was sold in a standard Tri-ang Hornby DPL2 size window box but on the end flaps there was a 65 x 7 mm blue sticker with the text 'LOCO. VAPEUR 4.6.0 REF. 150S'. It is strange that '4.6.0' was used as, under the French classification, it should have been '2.3.0'.

The wagons were sent over in standard Tri-ang Hornby window boxes marked 'REF R216' etc.

The End

This arrangement was not very satisfactory. Few Tri-ang Hornby products were sold in France. Hornby Acho itself faced competition from other European companies such as Jouef and Lima. As in the case of Hornby-Dublo it suffered from too little standardisation of parts which resulted in higher prices in the shops. Sadly the Rovex technique for capturing the market had not been learnt and production of Hornby Acho ceased in 1973.

Hornby-Acho couplings fitted to Tri-ang Hornby models.

Hornby Acho catalogue extract.

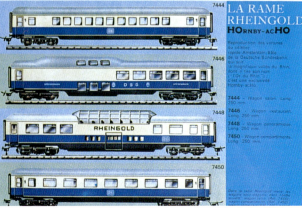

INTRODUCTION

A lot of the secret to the success of Tri-ang Hornby lay in decisions made at the development stage. Rovex provided a unique balance of quality and economy which allowed them to produce the models people not only wanted to buy but could afford. The latter depended on models sharing parts wherever possible without seriously compromising realism. In this chapter we will look at how a model was developed and made and what interchangeable components went into producing it.

MODEL DEVELOPMENT

Choosing the Models

Early each year the Development Department, in consultation with the sales side of the company, would decide what products they wanted in the catalogue the following year. The list was by no means stable, as many of the ideas required investigation before a final decision on production could be taken, and would undergo many changes in the forthcoming months. Changes would be based on information fed back to the company from the men in the front line, the salesmen, who went from dealer to dealer listening to suggestions and complaints, and also from many years of experience of supplying the toy industry.

In order to see how the company determined what it wanted to make and ensure that the catalogue was correct and delivered on time it is useful to look at the production process over the period of a year.

Draft Programme

By 1970, the company was holding planning meetings to determine manufacturing policy. These started in February and were organised by the Development Department at Margate.

At this early stage a draft programme of new products, and proposed deletions, was prepared based on information being received from the salesmen. This was called the Programme and the first draft was circulated in mid-February. It proposed the models to be deleted from the catalogue and those that might be added. Comments on the proposals were then invited.

There were four Production Periods each year which were known as:

PP1 = January to March
PP2 = April to June
PP3 = July to September
PP4 = October to December

As this series of planning meetings continued, instructions went out concerning the restriction of production of those models that were likely to be deleted at the end of the current year. It was important to avoid having stocks in the store once a model was no longer in the catalogue as these would be very difficult to dispose of. It was also particularly important that large stocks of customised boxes were not left over as this represented wastage that affected profits. Frequently the order was given to continue manufacturing until the boxes ran out or to include in the catalogue one more year in order to clear stocks of a model.

At this early stage the cost of tooling and factory costs had to be calculated in order to determine the actual cost of producing each model and therefore the likely retail price. This would be an important factor when deciding what to make.

Drawings and Patterns

Once it had been decided what new models were to be made, a member of staff would be dispatched to British Rail or the National Railway Museum at York to obtain drawings. If possible a preserved example would be found and a team sent to photograph it from all angles. Occasionally the plans could not be obtained from official sources and then the company had to turn to modellers' drawings of the type found regularly in the model railway press.

From the photographs, drawings were prepared in the drawing office at Margate. These showed the model from all angles and included cross-sections to show the thickness of the moulding to be made. The drawings would be kept for many years and be updated as modifications were made to the model; these modifications were dated and today these dates provide us with a useful record of changes to the model.

With models that had complicated curves, such as the Coronation Class loco, the drawings were not sufficient for the toolmaker and so a wooden pattern was made from which the measurements could be taken. This was a large scale solid model showing the external shape that had to be recreated. I was fortunate some years ago in obtaining the pattern for the Swedish style 0-4-0 diesel. This was made from a block of ash which I understand was the usual wood to use for this purpose.

Pre-production Models

Next, a proving model was produced from plasticard in order to test the drawings These were hard to make and often an existing chassis was adapted to take the new body. These proving models were painted and detailed on one side so that it was possible to see how the finished model would look. Many of these have survived and are sought by collectors.

Where only livery changes were proposed, an existing model would be painted up and detailed by hand to show what the finished model would look like. These generally needed to be available by mid March. If adopted it was common practice to hand paint a second example and sometimes more. One would be for the catalogue picture and another for the toy fair. This explains why more than one pre-production model of a wagon sometimes turn up. To prepare mock-ups, the company had their own system for making transfers from artwork and pre-production models were often decorated with these.

TOOLING

Tool Costs

Once the proving model and the tooling costs were approved the order was placed for the tools to be made.

Tools were expensive and ways of avoiding or reducing these costs were constantly being sought. This often included using existing parts or modifying existing tools. A good example of this was the Wild West 0-4-0 loco for the RB6 set of 1971. To make this locomotive, the tool for the body of the top tank 0-4-0T was taken out of the store and altered. The cowcatcher and smokestack of 'Davy Crockett' were used and the tender was made from an integral low-sided wagon and the top of the Crockett tender.

The cost of new tools varied according to the complexity of the model to be made. A useful comparison is provided by the following estimates for the tools of a number of models planned for 1971:

R65	Platform Section	£350
R781	Coke Wagon	£800
R739	Large Breakdown Crane	£4,500
R861	'Evening Star'	£17,000

These figures may not seem large today thanks to the effect of inflation but the cost had to be spread over the first 50,000 models made from the tools. This, for example, increased the wholesale cost of the coke wagon from 3/5d to 3/9d. By contrast, the existing open wagon in the new SR livery had a wholesale price of 2/8d.

Toolmaking

Tools were made in the tool room at Margate when time allowed but during busy periods the work was put out to outside companies. This was particularly so in the early days when the young and growing company was under considerable pressure for new products.

Die-casting was not done at Margate in the early years. Rovex sent out drawings to a tool manufacturer inviting them to quote for making and supplying a tool or alternatively for making the tool and supplying the castings in the numbers required. In the first case Rovex might have had to pay, say £1,000, for the tool and in the second case they may have paid £400 and 7d per casting. Rovex tended to go for the latter, if the opportunity existed, so that they owned the tool and reserved the right to take possession of it on payment of the balance of £600, which they never did. It was always the die-caster's responsibility to retool when things went wrong.

The tool makers model of the 0-4-0 Swedish diesel shunter together with a sample of the model to give it scale.

This also allowed a replacement tool to be an improvement on the original.

The replacement of a tool by the die-caster was sometimes done without being noticed by Rovex and this could account for variations in the wagon chassis listed on pages 74-77 of **Volume 1**. Two dies used at the same time may have been the result of inviting quotes from two companies and accepting both in order to meet the high demand for the item.

Toolmakers

One of the earliest toolmakers used by Rovex was TAL Developments Ltd of Enfield, who made the tool for the first Tri-ang wagon chassis. They were a good firm and were to have a long association with Rovex. They also made the early loco weights in mazak.

Universal Tools of Richmond were the company's first tool supplier and as we saw in **Volume 1**, were responsible for early moulds for the track. They also made the moulds for the Princess loco and coaches. When the track was modified to meet the new standards required for Tri-ang Railways, the tools were sent back to Universal for modification.

The sintered wheels used on early locomotives all came from Rigby's of Cleckheaton who had been presented with drawings of what was required and had come back with suggestions for several modifications, Murry's Die Cast Supplies made the mazak wheels for the early locos.

Pressure from the public for new models was so great in the early years that Rovex constantly had to find new toolmakers to meet their growing needs. This was often difficult. Some of those they used were very small companies. One of these was Mr Law who worked in a shed in Streatham and made the coupling tools.

When Rovex moved to Margate they looked for local toolmakers but could not find many. One they found was PDS who were quite well known. They also used the Riverhead Tool Company (RTC) of Sevenoaks in Kent.

The pre-production model of 'Albert Hall' showing how it has been detailed on one side only.

As time went on, the more expensive toolmakers were dropped when cheaper ones were found. One that came into use that way was Die Casting Tool & Engineering (DCTE).

The original tools for the Series 3 track were made by Perfect Die Sinkers (PDS). They also tooled the Dock shunter and Deeley 3F. Another firm that was asked to produce Series 3 moulds was Plastalls Ltd (PL3).

John McClark (JMC) of Southfields in Wimbledon made excellent tools including those for the original station buildings. They hardened their tools and they always worked first time they were put on the machine. Everyone in the factory loved them but they were expensive compared to the others used. The very successful Super 4 moulds were made by John McClark.

Testing Tools

When the body tool was ready it would be tested on an injection-moulding machine. A number of shots would be taken and checked for the ease with which they left the mould, the cleanness of the moulding and the way in which it would fit the chassis.

To test some tools, especially those for locomotive bodies, a material called crystal was used for a few shots. Crystal was completely clear which meant that if the body when fitted to the chassis was sticking, the engineer could see where the fault lay. Probably a dozen shots would have been taken each time a new locomotive body was being fitted. These were normally broken up after use but occasionally examples escaped from the factory.

How a moulding tool behaved once it was on the machine could determine when a model was finally

Monthly newsletter to retailers.

ROVEX Tri·ang Regd. Trade Mark

Rovex Tri·ang Limited
Westwood
Margate
Kent
England

Telephone
0843 22294
Telegrams
Rovextois Margate

Your ref   Our ref   HRL/EH   Date   January, 1971

To All Stockists

TRI-ANG HORNBY BULLETIN NO.1 - JANUARY, 1971

WHEELS

The result of the analysis of the Questionnaire Forms which we included in our 1970 Catalogues has proved very useful. Far and away the most requests for improvement related to wheel standards, the broad wheels which we have always used being generally considered ungainly, although suitable for the younger users.

We have, therefore, set ourselves to effect a gradual changeover to a finer scale wheel profile.

As will be seen from the 1971 Catalogue, the 2-10-0 Evening Star Locomotive will be fitted with new type wheels from the outset. Other Locomotives with nickelled tyres will have finer scale bogie and pony truck wheels but the profile of the driving wheels will remain unchanged for the present.

On Coaches, we have fitted finer scale wheels to the Inter-City models with interior lights for some two years and for 1971 the 16 Famous Company Coaches will all have finer scale wheels with white rims. These white rims set off the Coaches very well and also serve to identify the finer scale wheels. A few coaches, mostly Pullmans, were put out in the Autumn of 1970 with white rims on old type broad wheels for test purposes. B.R. Coaches will change to finer scale wheels during 1971, but without white rims.

New 4 wheel wagons and those with Famous Company Markings are also due to be introduced with finer scale wheels early in 1971.

Other models will follow gradually. There are some items which are made in comparatively small quantities - principally tenders - and these will be the last to be changed.

FINER SCALE WHEEL PACKS

We have sold some millions of units of rolling stock over the years with the broad type wheel. It, therefore, seems likely that sales of Finer Scale Wheel Packs can be very considerable. It should be noted that it is only practicable for the 'average' owner to change the wheels on rolling stock which has closed axle boxes and provision for needle point bearings. Earlier pattern rolling stock where the ends of the axles can be seen from the outside are not suitable for conversion, but such items were mostly phased out in 1963.

P.T.O.

Directors H.R.Lines W.M.Lines R.E.Wadsworth H.W.Henden K.Edey H.Gordon O.D.Kutluk

released. A tool that played up had to go back to the tool room to be worked on, making it very difficult to advise retailers when to expect the new models.

Having perfected the tools, a few sample models would be made up for testing to ensure that the design would not produce problems after purchase. This sometimes lead to modifications and unavoidable delays. Samples were also required for demonstrating with at toy fairs or at Tri-ang House. These were proof runs. At least four dozen of each new set were made for customer sampling.

PRODUCTION

Programming
Once the model was cleared of any defects it could go into production. However, not everything could be made at once and so the factory worked to a

No, not a ghost but a locomotive body moulded in crystal for testing its fit on a chassis during the development stage.

schedule. Thus, the new models were released at intervals over the year with sets normally timed for release at Christmas.

Despite good planning things could go wrong causing delays to models. Frequently the factors were external such as a delay in the delivery of transfers or fishplates. Sometimes there were delays in the development stage when it was found that what seemed like a good idea did not work as well as it should when the model was track tested and so modifications had to be made. Rovex would not market a model until it performed perfectly well.

The process was complicated by the sheer number of items in the catalogue. In 1970, for example, there were 242 models in the range and by March that year, ten of these were behind schedule. Retailers were kept up to date with revised release dates through a monthly newsletter issued by the factory.

Moulding

The moulding tool was made of a number of blocks which were pushed together by hydraulics. The plastic was forced into the machine in granule form. It did not flow and so had to be heated to a semi-liquid state to be forced under pressure into the mould. The plastic was abrasive and wore out tools.

Moulding tools were heavy because of the high pressure under which they are used and, when not in use, were stored in a warehouse. All were numbered and stored with jigs etc. Die-casting tools were a lot lighter.

In later years, batch production became popular in the model railway industry. Five years stock of a model would be made at one go and then the tools were cleaned up and stored away until next time.

Such was the demand for rolling-stock wheels that the wheel-making machine in the factory was going non-stop.

Plastics

Rovex received their plastic from a single supplier with whom they had a contract. This unfortunately tied them to orders of at least two tons of any colour and that became a problem when they were asked to supply small batches of models in a particular livery. One such example was the Australian trains in Victorian Railway blue. This particular colour was not required for anything else.

The material used for the model mouldings was known as thermoplastic as it was shaped by means of heat. Normally it was stable to 80°C but Rovex sometimes had complaints from retailers that models were warping in shop windows. Their advice was to keep the models out of the sun and to avoid displaying them in window cartons. Around 1970 they were experimenting with more heat-stable materials.

Mouldings that could not be used were ground up and the material used for moulding tunnels where the colour of the plastic did not matter.

Diesel cab windows and other clear parts were moulded in crystal; the same material used for testing tools. Although brittle, it was cheap and was also used for cleaning machines when changing to different coloured plastics. Being clear, any impurities in the moulding would show up and it could take as many as 50 shots to clean the machine. Some mouldings in crystal have escaped the factory and may turn up from time to time.

Metal Parts

Metal parts were die-cast or stamped. Couplings were made in the factory and assembled by a girl. Eyelets were used for fixing couplings and bogies to models as they were cheaper than rivets. Valve gear was also stamped out in the factory and zinc plated. Zinc was used from the late 60s to mid 70s having replaced nickel plating as it was cheaper but was later dropped when people became worried that it was poisonous. It was replaced by nickel plating.

Screws and whistles, etc, were made by outside firms and bought in.

FINISHING

Painting

The plastic could provide only one of the colours of the finished model and the rest of the colour had to be applied to the surface of the plastic. Fine detail was applied as transfers, printed on by hot foil or painted on by hand. By 1974, but outside the period covered by this volume, the company had acquired tampo-printing machines which made transfers and hot-foil printing virtually obsolete and greatly improved the appearance of models. The lining, including boiler bands, pipes, hand rails, was done with a mapping pen. The model was held in the hand and the whole boiler band was done in one stroke; a task which took a lot of practice.

For larger areas, paint sprays were used and for spraying on a colour such as black for the locomotive smoke box and foot plate, a mask (or stencil) was required. The mask was a metal jig which was brought into place with levers before the paint was applied and had then to be quickly released to get the moulding out without delay. These could not be made until the first mouldings, on which it would be used, had been produced. The mask had to fit snugly to avoid overspray.

Lacquering

Lacquer was hand sprayed onto loco bodies by a girl who held the moulding in a gloved hand. It was used to give the model a gloss finish. It had been used in the 1950s and was reintroduced in 1970 when a gloss finish was once again required.

There were problems in getting this right and some early 'Coronation' and 'Lord of the Isles' locos did not look as good as later ones due to a treacle effect. This problem was cured after experimenting with various materials. Besides giving the model a glossy look, the lacquer had the benefit of protecting the painted and printed detail from damage.

'Chroming' Plant

For some models a shiny 'chromed' plastic part was required but this effect was produced with aluminium and not chromium. The part to be 'chromed' was placed in a plating chamber which looked like a decompression chamber used by deep sea divers. A block of aluminium was placed inside to be bombarded with atoms. It was important that the recipient of the plating was clean and in particular free of oil or else the plating peeled off.

'Chrome'-plated Locomotives

Occasionally one can come across a Tri-ang Hornby or Hornby Railways locomotive in a shiny 'chromed' finish. These were retirement presents for members of the staff leaving the factory.

The retiring person would normally receive a chrome-plated 'Flying Scotsman', with detailing (although they can be found without any), mounted on a piece of track on a wooden plinth. If they did not want a 'Flying Scotsman' they could select another model off the production line, but it had to be something in production at the time. The loco line was asked to supply the chassis. The following are some of the chromed models that have been found:

LNER Flying Scotsman
BR Flying Scotsman
Coronation
Hall
Evening Star
Black 5
Jinty
TT Merchant Navy and coach

The company no longer does its own plating as this is now done outside under contract.

Lettering

Until 1976 all the lettering and numbers on the models were designed in the factory by Norman Lyson.

Chromed Flying Scotsman and Coronation.

NUMBERING OF PARTS

Rovex used a sequence of 'S', 'X' and 'R' numbers to identify parts, assemblies and complete models. The 'S' number came from the original drawings. Thus the drawings made in 1960 for the NSWR Suburban Electric 3-car unit were numbered as follows:

NO.	PART
S.5804C	body common to all three cars
S.5805C	roof of R.451
S.5806C	roof of R.450/452
S.5812	power switch on R.450
S.5813A	dummy switch on R.452
S.5814B	bogie plate
S.5815B	seat units
S.5816B	underframe
S.5817C	bogie frame - non motorised
S.5820	bogie frame - trailer car

The suffix 'A', 'B' and 'C' denoted the size of the plan; 'C' being the largest size.

MOTORS AND MOTOR BOGIES

Introduction

The X04, X05 (XT60) and X500 electric motors, which were developed for Tri-ang Railways locomotives, are described in some detail in the 'Components' chapter in **Volume 1** and they remained in use during the Tri-ang Hornby period. Indeed, they were once again the subject of the monthly Tri-ang Hornby advertisement in the model railway press in April 1968. The X800 used in 'Evening Star' and the can motor adopted for some starter set locos, however, were used for the first time during the period covered by this volume and are therefore described in greater detail here. All but the can motors were made by Rovex at the Margate factory.

In **Volume 1** we also saw how the motor bogies were developed by John Hefford from earlier Zenith designs. The Mk I motor bogie incorporated an X04 motor with a spur drive to a parallel lay shaft on

which two worms were fitted. This was replaced by a newly developed Mk II motor with twin worms on the armature shaft and a block magnet with sintered iron pole pieces. A further development was the Mk III motor bogie which used phosphor bronze stampings to conduct electricity from the wheels to the brushes.

Having discovered how to make a power bogie with an armature shaft and worms, Rovex were able to develop additional models, based on diesel and electric prototypes, relatively easily, irrespective of whether the power bogie had four or six wheels. The so-called 'six-wheeled' Mk VII motor bogie (in fact a four-wheel assembly with two dummy centre wheels) arrived with the EM2 electric locomotive in 1961.

The Tri-ang Hornby period saw adaptations of the Mk II and Mk III assemblies for new models.

In the early days the motor bogies were identified by their 'Mark' numbers but later the 'X' number became a more popular term of reference. For continuity with **Volume 1**, both numbering systems are provided here.

While the day of the clockwork model railway had passed, clockwork power was still a means of introducing beginners to the Tri-ang Hornby system through cheap and easy-to-use starter sets. Thus the only clockwork motor now made at Margate powered a cheap 0-4-0 chassis which was manufactured in its hundreds of thousands and for which no spares were available.

MOTOR CHECK LIST

The following is a list of the electric and clockwork motors and motor bogies used in model locomotives made at Margate during the Tri-ang Hornby years. Examples of locomotives that carried each motor or motor bogies are given.

MOTOR	ELECTRIC MOTORS	
Mark IXT60	TT steam locomotives	
Mark IV	X04	most steam locomotives
Mark V	X05	R354, R553
	X500	R651
	X800	R861
	Can	R853, R858
	ELECTRIC BOGIES	
Mark IIB	X3121	R157, R555
Mark IIC	X3122	R159, R253
Mark IIG	X3169	R753
Mark IIIA	X239	R352
Mark IIIB	X3170	R758
Mark VIIA	X213	R351
Mark VIIB	X337	R357, R751
	CLOCKWORK MOTOR	
Mark III	all clockwork models	

ELECTRIC MOTORS

X04

This was one of the most successful miniature motors ever made. Until the early 1970s it was used in almost every electric steam locomotive made by Rovex. It used a current of 0.4 of an amp, had a reduction ratio of 20:1 and ran at 20,000 rpm.

X04T

This was the same as the X04 but it had a single-start worm instead of a two-start and had been developed for use in the R45 electric turntable (the reason for the 'T' suffix). It also had a front-end fixing screw as well as one at the rear. While the X04T motor had been specified for the 0-4-0 locomotive chassis, from January 1961 it was replaced by the X04. Despite this, the X04T remained available until 1969.

XT60

This motor was developed for TT locomotives but also remained available on the spares order form

Tri-ang Hornby motors.

until 1969 as it was still used in the turntable. It was referred to as the Mk I motor presumably because it was the Mark I TT motor.

X05

The X05 motor was similar to the XT60 but had a twin-start worm. It was slimmer than the X04 and was used in the 'Lord of the Isles' and Caledonian Single. It consumed a current of 0.3 of an amp, had a reduction ratio of 26:1 and ran at 15,000 rpm.

X500

This was the motor developed for the 'Rocket'. It used 0.1 of an amp current and had a reduction ratio of 33:1. The motor turned at 12,000 rpm and had a two-start worm.

X800

For the model of 'Evening Star', a tender-mounted motor was required. Rovex had acquired the technology of the Ringfield motor when Lines Bros. took over Meccano Ltd in 1964 and there was much speculation at the time as to how they would make use of it in their own range of locomotives. As the 1960s passed, nothing seemed to be happening. Eventually, in 1970, there was talk of a revolutionary new motor being designed at Margate for an exciting new locomotive and naturally speculation pre-

dicted that it would be a version of the former Hornby-Dublo motor. In actual fact, the design that emerged was almost an exact replica of a motor used in Fleischmann models as early as 1962. The Hornby-Dublo motor had been a unit that was cumbersome and expensive to make, while the Fleischmann motor was more suited to tender-driven locomotives.

Whereas previous motors used in Tri-ang models had been free-standing assemblies that slotted into the chassis of the locomotive, this was not so with the X800. A plastic tender chassis was fitted to a six-wheeled trolley/motor block around which the motor was assembled and over which sat a casting which filled the rest of the inside of the tender body giving it weight. The plastic-moulded tender body slipped over the top of the weight and motor trolley to engage the tender chassis.

The motor had a disc commutator which, with the

armature, revolved inside a large cylindrical magnet mounted in the centre of the tender frame. As the motor was thus at right angles to the traditional placing, no worm was required and, instead, motion was transferred to the wheels through spur gears held in place by a circlip. The carbon brushes were held against the commutator with spiral springs in a similar way to those on the earlier Hornby-Dublo locomotives.

Neoprene tyres were fitted to all six tender wheels to give better traction and this made it unnecessary to have Magnadhesion. Power pickup was through the locomotive wheels (see details of the 2-10-0 chassis).

Can Motors

The Mabuchi FA-15 Motor, which was imported from Japan, first made its appearance in the 'Swedish' 0-4-0 diesel used in starter sets in the late

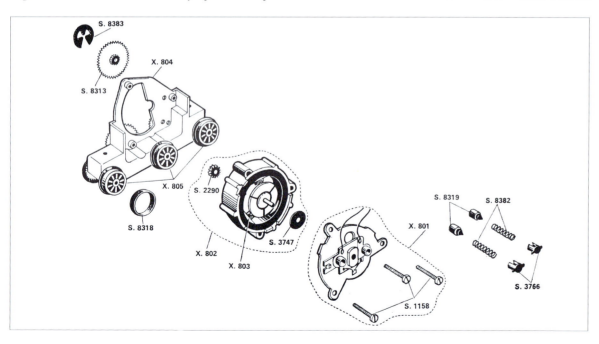

A can motor.

X800 motor chassis.

1960s. Compared with later can motors, it was quite a large unit. It had a specially-designed plastic chassis block to hold it at the right angle so that the pinion wheel on the end of the motor shaft engaged a contrate wheel on the axle of one of the pairs of wheels. The pre-production model of the motor block has survived.

The use of imported can motors enabled Rovex to keep the cost of these electric starter set locomotives to a very low level.

Service Sheets
Drawings of the above motors may be found on the following service sheets:

MOTOR	SERVICE SHEETS *
X04	1 (Nov 55 revised May 63), 51 (Apr 68)
X04/T	10 (Jul 56), 19 (Jan 60)
XT60	21 (Jan 60)
X05	32 (Aug 62)
X500	41 (Dec 64), 52 (Apr 68)
X800	78b (Sep 71)

5-pole Motors
Rovex did not change to 5-pole motors because they are complicated to wire. It is interesting to note that a lot of continental companies have now dropped them and are skew winding motors instead.

MOTOR BOGIES

Mark IIB (X3121)
This was a four-wheel unit used for the Metropolitan-Cammell DMU and the Pullman DMU. Models made in the Tri-ang Hornby period had a modification to the bogie-frame casting which gave it a loop at the back end. The phosphor bronze wipers gave way to spring wire pickups in the early 1960s.

* NB. Dates shown sfter service sheet numbers refer to the modification of the model where known or the publication of the sheet, if not.

Mark IIC (X3122)
This was used in most of the Transcontinental diesel locomotives. It was similar to the X3121 but had a larger frame and wheels. It continued to be used for the Dock Shunter and the Double-Ended Diesel and, for the Canadian market, in the A unit and the Bo-Bo Switcher.

Mark IIG (X3169)
The X3169 was the first motor bogie to be developed after the change to Tri-ang Hornby. It was required for the E3000 and instead of basing it on the newer X239 (Mk III), it was an adaptation of the older X3121 (Mk II) design. This was probably because the latter had already been designed for use with overhead power supply and was used in the R257 TC Electric Locomotive. It had a standard 3-pole Tri-ang motor with worm-and-gear drive onto both axles. The gear ratio was 13:1 compared with 11:1 on the Hornby-Dublo version which had the 3-pole Ringfield motor.

A track test by *Model Railway Constructor* in June 1966 revealed that the E3000 chassis weighed 6 oz. The wheel diameter was 17 mm making them about 3 scale inches larger than prototype. They provided a 37 mm (9' 3") wheel base which was 1.5' too short. Back-to-back measurement was 13.5 mm.

Mark IIIA (X239)
The Mark III was developed for the Budd RDC but as it went on to be developed for another model, we have given it the 'A' suffix. It was very similar to the Mk VII (X213) and shared many parts but was shorter in order to represent a four-wheel bogie.

Mark IIIB (X3170)
When the model of the Hymek diesel hydraulic locomotive was being developed in 1965 thought was given to a suitable motor bogie for it. As the Budd RDC had the most advanced four-wheeled unit this was naturally chosen and given a suitable

bogie frame with Commonwealth pattern side plates. These were as fitted on the prototype, although the 9' 3" (37 mm) wheel base was some 15" too short in scale length. The wheels, which had a diameter of 13 mm, were not knurled like earlier ones used with this bogie frame but they were the normal Tri-ang type, fitted with traction magnets. The gear ratio was 13:1 and in its chassis it weighed 6.75 oz. The wheel back-to-back measurement was 13.5 mm.

Mark VIIA (X213)
The motor bogie was designed for the Class 2700 'Electra' and remained exclusive to this locomotive. It had a die-cast frame which incorporated the bogie side plates and a double-ended armature shaft with a worm at each end. It had two small traction magnets mounted between the wheels to provide Magnadhesion. The power pickups were copper stampings which served both as wipers to collect electricity from the wheels and as power conductors to the motor thus eliminating the need for wiring.

The worms drove nylon cogs mounted on each outer axle. The centre axle was missing.

Mark VIIB (X337)
The X337, or Mark VIIB, introduced in December 1962, was identical to the X213 but had a different frame and was used for the model of the A1A-A1A (Class 31) locomotive. A minor difference in the assembly of the motor was that instead of having two washers above the insulating washer on top of the top housing (the plastic moulding that formed the outer shell of the motor), the Mk VIIB had a single tab washer to which the capacitor was soldered. By September 1969 the Mk VIIA motor bogie had been similarly adapted.

In 1965, work was progressing on the model of the Co-Co (Class 37) and a motor bogie was needed. The similarity of the prototype's bogie to that of the Brush Type 2 suggested that this should be adopted without alteration. The result was covered in a

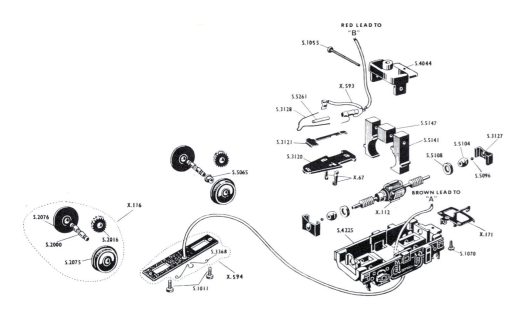

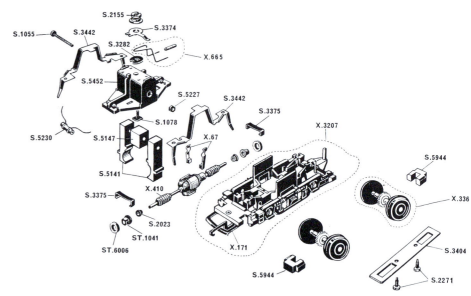

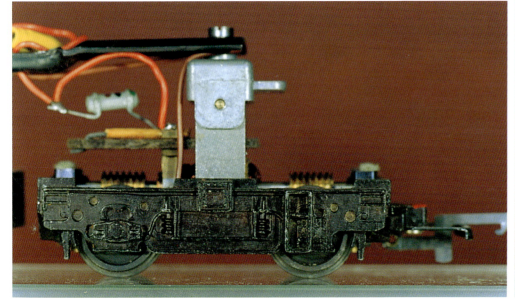

MkIIG (X3169) motor bogie used for the E3000.

MkIIIB (X3170) motor bogie developed from the Budd bogie for the Hymek.

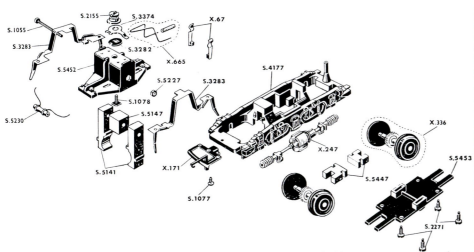

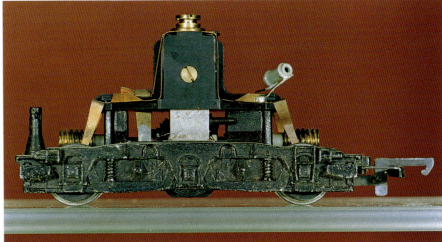

MkVIIB (X337) motor bogie used for the new English Electric Co-Co.

test track review conducted by *Model Railway Constructor* in May 1966. This told us that the motor bogie weighed 4.75 oz and the gear ratio was 13:1. It had 13 mm diameter wheels at 28 mm centres. This meant that the wheels were 6 scale inches too small but only 3 scale inches too far apart. The wheel back-to-back measurement was 13.5 mm.

The bogie lasted until 1976 when it was replaced by one of the new clip-in six-wheel motor assemblies with a Ringfield motor.

Service Sheets

The above motor bogies are detailed on the following service sheets:

BOGIE		SERVICE SHEETS
Mark IIB	X3121	
Mark IIC	X3122	
Mark IIG	X3169	
Mark IIIA	X239	31 (Aug 62)
Mark IIIB	X3170	
Mark VIIA	X213	30 (Aug 62) 64 (Apr 68)
Mark VIIB	X337	37 (Aug 63) 45 (Dec 66)
		66&69 (Apr 68)

CLOCKWORK MOTORS

0-4-0 Motor

This was a very cheap clockwork chassis designed for use in locomotives to be used in starter sets. Further details may be found in **Volume 1**. It is interesting to compare one of these motors made in the early 1960s with one made thirty years later. The design, but for very minor details, was unchanged.

R884 Long-Bodied Clockwork Locomotive

Early in 1970 it was suggested that there might be a market for a larger clockwork 0-6-0 locomotive and the idea of producing a long-bodied clockwork engine with a reversing gear and trackside triggered brake was being investigated. Fitted with a chuff-chuff engine sound device, it was hoped to use it in sets in 1972. It was to accommodate the existing clockwork mechanism but it did not come to anything.

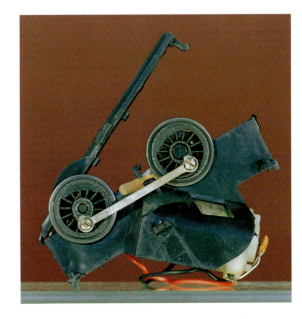

New 0-4-0 chassis (the factory model from Margate).

DUMMY DIESEL AND ELECTRIC MOTOR BOGIES

As only one of the bogies on a model diesel or electric locomotive was motorised, the other bogie was a plastic dummy, the frame of which was specially moulded to match the motorised one. The following dummy motor bogies were used on models. Further details about them may be obtained from the service sheets listed. Examples from the Tri-ang Railways period are included here as they were not listed in **Volume 1**.

4 WHEEL BOGIES		SERVICE SHEETS
X286	R352 Budd RDC	31 (Aug 62)
X302	R157/R158 DMU	7a (Jan 60)
X303	R225 EMU	7a (Jan 60)
X387	R55/R57/R58 A & B Units	5a (Jan 60)
	R155 Switcher	
	R159/R250, Double-Ended Diesel	
	R257 TC Electric	
X387	R155 Switcher R55 A Unit	54 (Apr 68)
	R159 Double-Ended Diesel	
X564	R753 E3000	47 (Dec 66)
		70 (Apr 68)
X620	R758 Hymek	75 (Apr 68)
6 WHEEL BOGIES		
X214	R351 EM2	30 (Aug 62)
		64 (Apr 68)
X348	R357 A1A-A1A R751 Co-Co	37 (Aug 63)
X348	R751 Co-Co	45 (Dec 66)
		69 (Apr 68)
X348	R357 A1A-A1A	66 (Apr 68)

STEAM LOCOMOTIVE CHASSIS

Introduction

This section carries on the story of the development of the various Tri-ang chassis described in **Volume 1**, but also includes details of the service sheets for all Tri-ang model locomotives made from 1952 until the start of 1972.

0-2-2 Chassis

Only the 'Rocket' used this chassis and the X500 motor it carried was designed specially for this model. It remained available until 1969 but from 1st March 1966 the model was not fitted with a smoke unit. Instead, a resistor was fitted into the funnel with its longest lead soldered to the brush clip and the shortest lead soldered to the left-hand capacitor clip.

MODEL	SERVICE SHEETS
R651s 'Rocket'	40 (Dec 64)
R651 'Rocket'	68 (Apr 66)

4-2-2 Chassis

The Caledonian Single was dropped after 1965 and only very few Dean Single locomotive models were made in 1965 and 1966. A batch of 1,000 were produced in 1967. The Dean was reintroduced in 1970 and the Caley in 1971 but there is no evidence that these later models had any chassis variations. This assumption is supported by the fact that the service sheet issued in 1962 was not subsequently updated.

MODEL	SERVICE SHEETS
R354/R354s 'Lord of the Isles'	32 (Aug 62)

0-4-0 Chassis

This chassis dates from 1959. Between then and 1968 the main changes were the abandoning of the collector insulator, the change to spring collectors, fluted coupling rods and the fitting of see-through wheels with steel tyres. All these improvements had been introduced prior to the start of the Tri-ang Hornby period. The pre-production chassis block was kept by the designer when he left the company and is now in a private collection.

In 1968 work was proceeding on a revolutionary new 0-4-0 chassis designed to further cut the cost of locomotive production and thus reduce the price of starter electric train sets. The chassis consisted of a single plastic moulding that was hinged. The largest part formed a cradle, for a large can motor, and had notches on the bottom side into which the two axles fitted. The hinged section formed the bottom plate which swung round trapping the axles in position. The worm on the end of the motor engaged a gear wheel on the leading axle and the weight of the chassis was principally provided by the motor. The chassis was held inside the locomotive body by two pegs that snapped into holes in the body's sides. The wipers were copper strips and the wheels and coupling rods were from 'Nellie'.

MODEL	SERVICE SHEETS
R252/R254 Steeple Cabs	19 (Jan 60)
R252 Steeple Cab	62 (Apr 68)
R355/R359 'Nellie' etc	
R559 Diesel Shunter	

0-4-4 Chassis

The chassis was used for only the M7 tank and was first made in 1967 and last made in 1988.

It had 24 mm diameter driving wheels and a 33 mm + 37 mm + 24 mm wheel base. Their back-to-back measurement was 14 mm. The chassis weighed 5 oz and was fitted with a TV interference suppressor. It was fitted with an X04 motor, mounted low and driving the leading axle through worm and gear. The gear ratio was 20:1 and the maximum scale speed running light was 200 mph. The chassis was fitted with Magnadhesion and, initially, firebox glow which was dropped after 1970.

The bogie, attached at the rear of the model had 12 mm diameter wheels and was held down onto the track by a phosphor bronze spring.

MODEL	SERVICE SHEETS
R754 M7	74 (Apr 68)

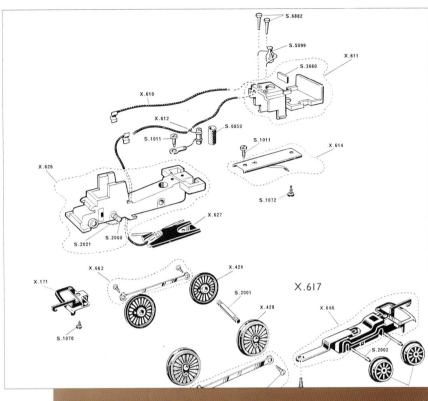

X.617

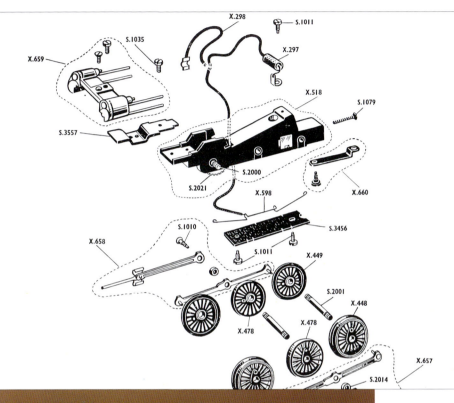

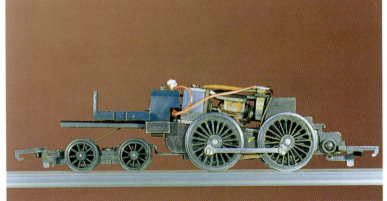

0-4-4 chassis and X617 bogie.

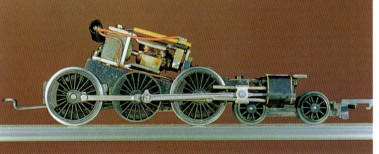

Hall chassis.

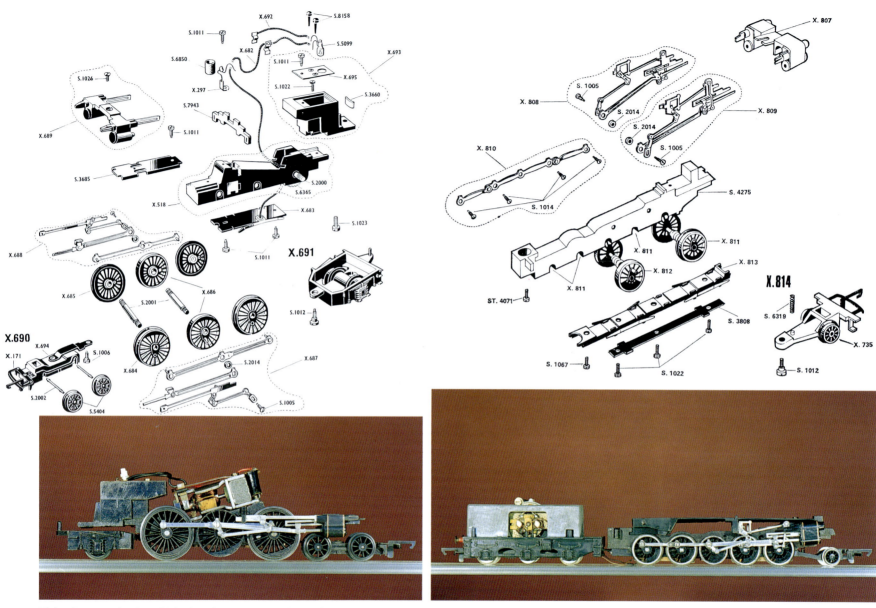

X.692
S.8158
S.1011
X.682
S.5099
X.693
S.6850
X.695
S.1011
X.297
S.1022
S.1026
S.3660
X.689
S.1011
S.3685
S.2000
S.6365
X.518
X.683
S.1023
S.1011
X.691
X.688
S.1012
X.685
S.2001
X.686
X.690
X.694
S.1006
X.171
X.684
S.2014
X.687
S.2002
S.5404
S.1005

X.807
S.1005
X.808
S.2014
X.809
S.2014
X.810
S.1005
S.1014
S.4275
X.811
X.811
X.811
X.812
X.813
X.814
X.811
S.3808
S.6319
ST.4071
X.735
S.1067
S.1022
S.1012

Flying Scotsman chassis, X690 bogie and X691 pony truck as produced after June 1969

Evening Star chassis and X814 pony truck.

4-4-0 Chassis

The chassis dates from 1960 when the Class L1 locomotive was added to the Tri-ang Railways range. It was the first Tri-ang chassis to have see-through driving wheels and these were the only see-through wheels to be cast in one piece as they predated Magnadhesion. By May 1961 these had also been fitted with steel tyres. Also by this date, a smoke generator had been fitted in the chimney in place of the body-fixing screw but Synchrosmoke was not fitted to this chassis. The chassis had spring collectors from the start which were held in place with a black Paxolin plate held by a single screw.

The chassis remained in production until 1967 but was reintroduced in 1972 when the L1 returned in Southern livery. In this form it had nickel-plated wheels.

MODEL	SERVICE SHEETS
R350 L1	29 (Jan 61)

0-6-0 Chassis

This was one of the most important and most widely used chassis, especially early on. It had started life in 1953 when the Jinty first appeared and has remained in production ever since. During this time it has undergone many changes but virtually no changes occurred during the Tri-ang Hornby years.

MODEL	SERVICE SHEETS
R52 Jinty	2 (Nov 55)
R152 Diesel Shunter	
R153 Saddle Tank	
R52 Jinty	2a (Jan 59)
R152 Diesel Shunter	2b (May 61)
R153 Saddle Tank	36 (Jan 63)
R251 Deeley	
R52s Jinty	2c (Jan 64)
R152 Diesel Shunter	58 (Apr 68)
R251s Deeley	

2-6-0 Chassis

This chassis, introduced in August 1962, was based on the 0-6-0 chassis but had the front end of the casting built up to take the smoke unit. The wheels were allocated the same 'S' number as those for the 0-6-0 chassis but each had a 'Y' suffix to indicate that they were yellow.

MODEL	SERVICE SHEET
R358 Crockett	36 (Jun 62)

2-6-2 Chassis

This chassis was produced for the R59 BR Standard tank in 1955 and found a second use in the Continental Prairie. The chassis never looked right as the wheels were too small but at the time it was first introduced, the choice had been the Princess or Jinty wheels and the latter won.

The wheels had an 18 mm diameter and the centre ones were flangeless. The coupled wheel centres were 33 mm and 34 mm; the front and rear pony truck wheel centres were 37 mm from those of the outer drivers. The chassis carried an X04 motor.

MODEL	SERVICE SHEETS
R59 Standard Tank	9 (Nov 55)
	9a (Jan 61)
	9a (Sept 62)

4-6-0 Chassis

The chassis was specially designed for the B12 loco which was first made in 1963 and is described in **Volume 1**. In 1966 it was also used on the Hall Class loco which came out that year.

It had the X04 motor driving the leading axle through worm and gear. The gear ratio was 1:20 and the chassis weighed 6oz. The driving wheels were 24 mm in diameter and the wheel base for the drivers was 29 mm. The centre wheels were flangeless and the back-to-back measurement of the drivers was 13.5 mm. Magnadhesion was provided on the rear wheels.

The bogie had 13 mm diameter wheels which were further apart on the Hall and so, while very similar in appearance, the bogies were different

lengths on the two classes of loco. The driving wheels on the two models were the same but the front chassis extensions were different to allow for attachment of the cylinder unit on the Hall.

No substantial change was made during the Tri-ang Hornby years.

MODEL	SERVICE SHEETS
R150 B12	38 (Jan 63) 38 (Jun 64)
R150s B12	67 (June 64)
R759 Hall	49 (Dec 66) 73 (Apr 68)

4-6-2 Chassis

The chassis weighed 10.5 oz and was powered by an XO4 motor with a gear ratio of 20:1.

The Battle of Britain model was track tested by the *Model Railway Constructor* in December 1966 and the following information was given about the 4-6-2 chassis:

The driving wheels had a wheel base of 27 mm + 26 mm and a diameter of 24 mm. As such they were just 2 scale inches under size. The leading bogie had 10 mm wheels, which were much too small, and a 24 mm wheel base. The pony truck had 12 mm wheels which were just about exactly correct. The centre driving wheels were flangeless.

MODEL	SERVICE SHEETS
R50 Black Princess	0a 0b 3 3a (Jan 61)
R53 Green Princess	4 (Nov 55)
R54 TC Pacific	
R53 Green Princess	4a (Jan 61) 59 (Apr 68)
R258 Maroon Princesses	
R54 TC Pacific	
R259 'Britannia'	28 (Jan 61) 63 (Apr 68)
R356/R356s Battle of Britain	33 (Aug 62)
R356s Battle of Britain	65 (Jun 64)

4-6-2 Chassis Mark 2

For the 'Flying Scotsman' a new adaptation of a reversed B12 chassis was designed. This consisted of a two-piece casting, one part forming the main chassis with holes for the driving axles and the other a bolt-on firebox. The latter was initially wired with a bulb to give a yellow glow when the model was running but from 1974 the electrical fittings for this were omitted.

The chassis was lengthened at the leading end with a plate screwed in place, which provided an anchor for the front end of the body. The chassis was driven by a standard XO4 motor mounted backwards so that it drove the rear pair of driving wheels. The reviewer in the *Model Railway Constructor* told us:

"The 20:1 ratio gear set consists of a brass worm and nylon gear wheel – the first time this combination has been used on a Tri-ang Hornby steam outline locomotive"

The wheel sizes and spacing were reasonably accurate for the 'Flying Scotsman' although the pony truck wheels were 8" too small. The wheel base of the bogie was 6" short and in contrast the distance between the rear wheels of the bogie and the leading drivers was 6" overscale. The rest were perfect and the wheels had the correct number of spokes. The centre drivers were flangeless and the back-to-back wheel measurement was 14 mm.

The fluted valve gear consisted of the usual metal stampings representing a rather simplified reproduction of the Walschearts valve gear. It was, the reviewer found, 'sturdy and functional in design'. Current collection was through phosphor bronze wipers contacting front and rear drivers on one side.

The weight of the chassis was 9.5 oz.

MODEL	SERVICE SHEETS
R850/R855 'Flying Scotsman'	76 (Jul 69)

2-10-0 Chassis

This chassis was unique to one model – 'Evening Star'.

The original mechanism Rovex designed for the loco had a clutch drive by Barry Stevenson intended to provide ultra slow running. It had a motor in the loco and worked well until taken up an incline when it was found that on a downhill slope you could not control it as it free wheeled.

They next tried the revolutionary step of putting the motor in the tender (R862) and using traction tyres on two of the tender wheels to improve grip. The result was so successful that other models in the range were later remodelled on this principle.

Tender Drive

The final model had all three axles driven through spur gearing from the motor and, with all six wheels fitted with neoprene tyres, the power of the locomotive on the track was greatly increased. The first models of 'Evening Star' therefore had fully-flanged tender wheels and a die-cast weight fitted over the whole of the motor giving the tender maximum weight.

These tenders were found to derail on curves and several were sent back to the works. To correct the problem the centre wheels were made flangeless. Years later the flanges were to return but without tyres. The tender was permanently connected to the engine and current collection was through the front and rear pair of loco wheels. As we have seen, contrary to popular belief the Ringfield motor used, described earlier in this chapter, was based on a design used by Fleischmann and not the Hornby-Dublo version.

Wheels

The locomotive wheels, were die-cast, individually hand turned on a lathe and nickel plated. They were free wheeling and the locomotive was pushed along by the motor in the tender. 'Evening Star' was also the first locomotive to use a new type of axle. This was thinner than that used on other models and therefore generated less friction. Sadly, the opportunity to bring all other models up to this new standard was not taken at that time. It was the only Hornby loco still made in the mid 1990s with engineered screws instead of self-tapping screws.

A new type of wheel bush was introduced for the 'Evening Star' model. This used acetyl resin instead of plastic and consequently lasted longer. During the period of change, in order to identify the acetyl resin bushes from the old plastic ones they were initially made in a blue-grey colour. Once the old stock had been used up and all locomotives had changed to acetyl resin (around 1977) the colour reverted to black. Locomotives with blue grey wheel bushes therefore date from this period.

MODEL	SERVICE SHEETS
R861 'Evening Star'	78a (Sep 71)

STEAM LOCOMOTIVE BOGIES AND PONY TRUCKS

In the early days models frequently shared bogies and pony trucks but as greater attention was given to detail, these parts were specially designed for each model. The following list gives the part number of each bogie or pony truck, the locomotive it was used with and the number and date of the service sheet that illustrated it. The list includes Tri-ang Railways models, as details of the bogies and pony trucks for these were not provided in **Volume 1**. The list does not include models made after 1971 as these will be found in **Volume 3**.

BOGIE	MODEL	SERVICE SHEETS
X33	R54 TC Pacific	5 (Nov 55)
	R56 TC Baltic Tank	
X34	R56 TC Baltic Tank	5 (Nov 55)
X35	R50 Black Princesses	5 (Nov 55)
	R53 Green Princesses	
X228	R350 L1	29 (May 61)
X228	R350 L1	53 (Apr 68)
	R553 Caledonian Single	
X258	R356 Battle of Britain	33 (Aug 62)
		65 (Apr 68)
X273	R354 'Lord of the Isles'	32 (Aug 62)
X333	R54 TC Pacific	5a (1/60) 5b (5/61)
	R56 TC Baltic Tank	
X333	R54 TC Pacific	53 (Apr 68)
X334	R56 TC Baltic Tank	5a (Jan 60)

BOGIE	MODEL	SERVICE SHEETS
X335	R50 Black Princess	5a (Jan 60)
	R53 Green Princess	
	R258 Maroon Princess	
X335	R50 Black Princess	5b (May 61)
	R53 Green Princess	
	R258 Maroon Princess	
	R259 'Britannia'	
X335	R53 Green Princess	53 (Apr 68)
	R259 'Britannia'	
X416	R150 B12	38 (Aug 63)
X455	R56 TC Baltic Tank	5b (May 61)
X566	R759 Hall	49 (Dec 66)
		53 (Apr 68)
X617	R754 M7	74 (Apr 68)
X690	R850, R855 'Flying Scotsman'	
		76 (Jun 69)

PONY TRUCKS		
X26	R50 Black Princess	5 (Nov 55)
	R53 Green Princess	
X31	R54 TC Pacific	5 (Nov 55)
X66	R59 2-6-2 Tank	5 (Nov 55)
X232	R259 'Britannia'	5b (May 61)
		53 (Apr 68)
X259	R356 Battle of Britain	33 (Aug 62)
		65 (Apr 68)
X365	R358 'Davy Crockett'	36 (Aug 63)
X366	R59 2-6-2 Tank	5a (Jan 60)
		5b (May 61)
X366	R59s 2-6-2 Tank	77 (Jun 70)
X453	R50 Black Princess	5b (May 61)
	R53 Green Princess	
	R54 TC Pacific	
	R258 Maroon Princess	
X453	R53 Green Princess	53 (Apr 68)
	R54 TC Pacific	
X691	R850, R855 'Flying Scotsman'	
		76 (Jun 69)
X814	R861 'Evening Star'	78a (Sep 71)

LOCOMOTIVE WHEELS

Mazak

Rovex locomotive models had started in 1950 with plastic wheels followed by sintered iron and eventually mazak. These wheels were die-cast, stamped out to clean them and were then put into a rumbling machine before going to the blacking shop. They were promatised to blacken them but this reduced conductivity and so they were buffed. They were made this way until the introduction of Magnadhesion which would not work with mazak wheels.

Wheel Tyres

To overcome the magnetism problem, locomotive wheels were fitted with steel tyres which were blacked by negridising. In the late 1960s the rims were nickel plated to brighten them up. At one stage the plating was done with zinc but zinc is rather porous and could not be nickel plated without first copper plating.

It was quite a difficult and expensive process made worse by the fact that the nickel plate would not take paint very well and even today personnel on the packing line have pots of paint with which to touch up the paint work when it chips. Now that the company does not use Magnadhesion the models do not need steel tyres.

Fine Scale Wheels

In February 1970 it was decided to introduce nickelled tyred wheels to the maroon Coronation, LNER 'Flying Scotsman' and 'Britannia' for 1971. The centre wheels, which did not have tyres, were nickel plated and then stove enamelled.

'Evening Star' had finer scale wheels from the start. All locomotives with nickelled tyres were fitted with finer scale bogie and pony truck wheels when they first appeared in 1971 but for a while the driving wheels remained coarse scale.

ADDING REALISM

Magnadhesion

This was invented by Rovex and first used in 1961. By inserting magnets in the locomotive chassis it was possible to magnetise the wheels which greatly improved grip. It was eventually abandoned in Hornby Railways time with the introduction of nickel-silver track.

Synchro-smoke

This had been invented by Rovex as a replacement for the Seuthe smoke system which was used in earlier locomotives and was expensive to buy. The Synchro-smoke unit carried the patent number '961630' These required approximately 400 milliamperes to work.

Fire Box Glow

This was first used in the M7 tank engine and later in the 'Flying Scotsman'. Not all of these models were so fitted as the feature was dropped when there was a need to reduce the production cost of the model. The effect was produced with a bulb suitably placed at the front of the cab.

Exhaust Sound

In their continuing search for ways of making their models more attractive to children, Rovex had come up with a very cheap way of adding a 'chuff-chuff' noise to the motion of its engines. At a meeting on 2nd February 1970, it was decided to introduce steam sound tenders to the following locomotives for the 1971 catalogue:

'Flying Scotsman'
'Albert Hall'
B12
'Princess Elizabeth'
'Winston Churchill'
'Britannia'
The device was added to the tender and consisted

of a projection of the rear tender axle which scraped against a 'sandpaper' strip with each revolution of the wheels. A hollow compartment in the tender to which the strip was attached acted as a sounding box and the effect was quite convincing. The disadvantage was that it was suitable only for tender engines with engine-mounted motors and so it could not be fitted to the tank engines or the new 'Evening Star'.

Originally the system was referred to as 'Choo-Choo' but in May 1970 this was changed to 'Chuff-Chuff'. It had been decided that the models fitted with Chuff-Chuff should have a 'C' suffix to their 'R' number but later it was felt that this might lead to some confusion as several other models already had this for other reasons. The decision was thus taken to use an 'N' (for noise) instead.

Steam Sound tenders.

COUPLINGS

Tension Lock
Since December 1958, the standard Rovex coupling was the tension-lock type which had been developed for TT stock. As it proved successful, a larger version was made for the 00 range.

The coupling (X171) went through a number of minor changes over the next few years which are described in **Volume 1**, but the version being used in May 1965, when the system officially became Tri-ang Hornby, was the type IIIc. This remained the only version in use until the late 1960s when it was modified again to form the type IIId. Although we know that the future of the tension-lock coupling was being further considered early in 1970, it was not until the late 1970s that the type IIIe re-profiled coupling hook was adopted for closer coupling and this will be covered in **Volume 3**.

Type IIIc – The type IIIc coupling came into use in 1962 and unlike earlier versions it used a brass eyelet as the hook's fulcrum instead of a black stud. Examples can be found with an extra hole at the rear. The coupling remained in use until the late 1960s and may sometimes be found on models made as late as 1969, even though the type IIId coupling had been introduced more than a year earlier. This was no doubt due to old stock being mixed with the new.

Type IIId – There had been some difficulty in getting rolling stock to couple properly to the plastic integral coupling bar at the rear of the starter set locomotives. The bar was moulded in a semi-flexible material which if packed badly, when sets were being put away, could be left bent. To solve this problem the metal coupling hook on the coupling was enlarged, not just for starter set rolling stock but for all models. This was the type IIId coupling which was first used in 1967.

Other Types of Coupling
Five other types of coupling were also fitted for special purposes during the Tri-ang Hornby years:

Hornby-Dublo – used at one end of the two converter wagons and possibly supplied by Meccano Limited.

Hornby Acho – supplied with stock sold through the Hornby Acho catalogue and probably made by Meccano France Ltd.

Lima – supplied with two locomotives sold through Lima who presumably made the couplings.

NMRA – made by Rovex and fitted to stock supplied to ATT but also offered in packets of six (R698) in the Canadian catalogue for those who wanted to convert their Tri-ang Hornby stock.

Hook & Eye – used on the large crane to attach the two small trucks.

ROLLING STOCK WHEELS

Coarse Wheels
The wheels of wagons and coaches had already undergone considerable change by the start of the Tri-ang Hornby era. By 1965 pinpoint axles were used on nearly all rolling stock but the wheels were still coarse. This was to remain the case until the end of the decade when fine scale wheels were introduced.

The only sleeved axle (pre-pinpoint) to survive through this period was the DS3 which had two holes in the disc and which carried mould numbers 1-12. These were used on the grain and Engineer's wagons which retained their metal chassis and so could not be fitted with pinpoint axles.

With one minor exception, coarse scale wheels did not appear with white rims. The exception occurred in August 1970 when Rovex experimented with applying white to the wheel rims of some carriages going through the factory at the time to see what the public's reaction would be. It seems that most of the coaches so treated were Pullman cars. The reaction was good enough for a policy change in 1971 which resulted in white rims for certain coaches and wagons. By the time this was introduced, however, the company had changed to fine scale wheels. It therefore seems likely that coaches with coarse wheels and white rims date from August 1970.

Fine Scale Wheels

Fine scale wheels were first introduced for the coaches made for Tager of the USA in order to comply with American standards. The first British coaches to receive them were the Inter-City Mk2 coaches fitted with interior lights which first appeared in the shops in 1969 but it did not become standard practice until 1971. There had not been an earlier change for fear that the increased costs, which had to be passed on to the purchaser, would damage sales.

The customer survey, held in 1970, revealed that finer scale wheels were by far the most popular improvement demanded by customers and it was this that spurred on the company to make the change. The message must have come through strongly in the early questionnaire returns as it was on 2nd February 1970 that the decision was made to immediately change to finer wheels for all coaches made at Margate. In actual fact it seems that these did not arrive until January 1971. A note to retailers that month suggests that the '16 Famous Company Coaches' with finer scale white rim wheels were just becoming available and that plain finer scale wheels would begin to appear on BR coaches as the year progressed. Care had been taken to find a plastic which did not form a deposit on the track.

As for wagons, all new 4-wheel models in the catalogue and all those in pre-nationalisation liveries were available with finer scale wheels early in 1971. Other models were gradually added to the list as time allowed. Models with special wheels which consequently were made in only small quantities were the last to be converted.

Mk 2 coach bogie.

The new 16.5 mm gauge wheels, which could be bought in packs of four sets for 2/6d, had a back-to-back measurement of 14 mm and a thickness of 3 mm. The flange thickness was 0.75 mm and the depth 1 mm. Despite their thinness, the wheels still did not strictly conform to the accepted BRMSB or NMRA standards. They had to be a compromise to allow the rolling stock to be used on the widest range of proprietary track in order to maximise sales.

In the summer of 1971, Rovex Tri-ang reported having difficulty with some coaches derailing over points. The fault was not the flanges but the back-to-back measurement of some wheels being 0.015" oversize. Retailers were advised by the factory to fit new wheels when this problem arose.

White Rim Wheels

As we have seen, as an experiment, a few coaches were put out in August 1970 with white rimmed wheels to test the market. These were mainly Pullman cars and the wheels would have been coarse scale.

White rims appeared on the 'Famous Company Coaches' at the time they changed to finer scale wheels and so any of these coaches found with the new scale wheels without white rims will almost certainly have had their wheels changed after leaving the factory. At the same time, BR coaches changed to non-white rimmed finer scale wheels.

WHEELS/AXLES USED ON TRI-ANG HORNBY
ROLLING STOCK

DS3 disc, 2 holes in disc, numbered 1-12. 1962-75
DP1 disc, coarse scale, inverted numbers 1-12.
1963-69
DP2 disc coarse scale, two holes, no numbers,
1965
DP3 disc, coarse scale, numbered 1-12. 1968-70
DP3a as DP3 but with white rims. 1970
DP4 disc, fine scale, '+-' in moulding,
numbered 1-4. 1970-72
DP5 as DP4 but with white rims. 1971-73
SP1 spoked, coarse scale. 1962-71
SP2 spoked, fine scale 1971-74
MP1 small, coarse scale, small boss,
numbered 1-12. 1963-70?
MP2 small, fine scale, large boss. 1971-74

NB. The full range of wheels used over the period covered by all three volumes appears on page 53 of **Volume 1**.

COACH BOGIES

Short Coach Bogie

This short bogie dates back to 1952, but by 1965 only one piece of coaching stock used it and that was the travelling post office (TPO). The version of the bogie in use at that time was the b2b which had closed axle boxes, pinpoint axles and had been reinforced with braces on the underside.

BR Mk1 Standard Bogies

This was the standard bogie of the series of 9"

coaches made from the mid-1950s. It went on to be used on the scale length Mk1 coaches of the 1960s and various other stock such as the GWR clerestories and the Caledonian coaches. The version in use in 1965 was the c2b which remained until 1972. This had pinpoint axles and closed axle boxes. It had narrow slots cut in its base for the wheels.

BR Mk2 Standard Bogies

This was a new bogie introduced in 1969 for the new Mk2 coaches. It remained in use until 1976 when it was replaced by a clip-on version. The bogie was

based on the B4 type developed by British Rail for use on all Mk2 prototype coaches. It was steel fabricated and fitted with coil springs. Coaches fitted with it were restricted to 100 mph or, with special maintenance, to 110 mph.

DMU Bogies

The DMU originally had standard Mk1 bogies but, by 1960 a special dummy motor bogie tool had been made. It was re-tooled in 1962 and the result was a slightly longer bogie – the d2a. This remained in use throughout the Tri-ang Hornby period.

Transcontinental Bogies

Very few Transcontinental coaches were made during the Tri-ang Hornby era, but those that were were fitted with the t3b trucks which had a 'P' marked on the moulding. This stood for 'pin-point' and referred to the type of wheels that needed to be fitted.

Check List

The following table lists the bogies and trucks used on coaching stock made at Margate between 1965 and 1971. The first column gives the collectors code for the bogie (full explanation on page 69 of **Volume 1**) and the second column shows the 'X' number the bogie carried.

COACH BOGIES USED ON TRI-ANG HORNBY ROLLING STOCK

b2b	X310(r)	Short bogie for TPO.	1961-72
c2b	X531	Used on most other stock.	1963-1971
d2a	X870	Special bogie for DMU.	1962-76?
e1	X680	B4 type used for Mk2 coaches.	1969-76
t3b	X522	Plastic trucks for series 2 cars.	1963-?

Australian and New Zealand Trucks

The Australian home-made metal truck seems to have remained in production until 1968 when the tool was probably sent to New Zealand.

A4a, A4b, A4c and A5 wagon chassis. Only the first two were used during the Tri-ang Hornby period.

WAGON CHASSIS AND BOGIES

Classification

All five of the basic wagon chassis and bogie types developed before 1965 remained in production and four new ones were added. These were as follows:

CHASSIS/BOGIES	EXAMPLE
Pre-1965	
A 9′ wheel base chassis	Open Wagon
B Four-wheeled bogie	Bogie Bolster Wagon
C 13′ wheel base chassis	Cattle Wagon
D Six-wheeled bogie	Trestrol Wagon
E Crane chassis	Mobile Crane
Post-1965	
F liner train bogie	Container Wagon
G 15′ wheel base chassis	'Whisky' Wagon
H 24′ wheel base chassis	Ferry Van
I Four-wheeled bogies	100T Tank Wagon

NB. The background to the classification used for wagon chassis and bogies will be found on pages 72-77 of **Volume 1**.

9′ Chassis

This was a direct descendant of the original Trackmaster chassis of 1950. It was now made of plastic and the version in use in 1965 was the A4a (the first plastic type). The only other type used during this period was the A4b which dated from 1966 and survived until 1972. It differed from its predecessor in having the name 'Tri-ang' erased from its base.

Four-Wheeled Wagon Bogies

Like the 9′ chassis, the four-wheeled bogie was made in plastic by 1965 and was the B3a. This remained in use until the end of the year when it was replaced by the B3b which can be immediately identified by its two dimples or recesses on the underside of each side-frame. The B3b remained in use for many years.

13′ Chassis

Once again, the first plastic version of the chassis was in use by 1965. This was the C3a which was replaced by the C3b sometime in 1966. It is very difficult to tell these two apart but the C3b had lost detail from around the brake handle. It remained in use until 1974.

Six-Wheeled Wagon Bogies

This had been made for the Trestrol wagon and consequently found itself on many Battle Space cars which were based on this wagon. Only the D1b version was used during the Tri-ang Hornby years.

Crane Chassis

This was made exclusively for the R127 mobile crane which remained in production throughout this period.

Liner Train Bogies

The liner trains were fitted with Ridemaster or Ride Control bogies and these were so characteristic of the rolling stock that Rovex had little choice but to tool up new bogies when they planned to introduce a range of container wagons. This was the type F.

15′ Chassis

The chassis was based on that of a bulk grain wagon of the mid 1960s of the type used to run grain from East Anglia to the Scottish distilleries. It was first used on the R214 BR large hopper wagon in 1969 and was the G1a.

24′ Chassis

The chassis was made specially for the ferry van and was not used on any other wagon. This is chassis type H.

Four-Wheeled Railtank Bogies

The bogies were typical of those found on some petroleum carrying railtanks built around 1967/68. This is wagon bogie type I.

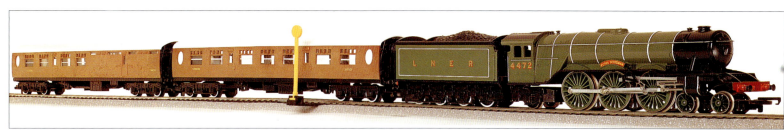

RS2E battery-powered mixed traffic set, 1971.

RS608 (1971) 'Flying Scotsman' sets.

RS606 'Express Goods' set, introducing Steam Sound, 1970/71.

RS607 'Local Passenger' set, 1971.

RS609 'Express Passenger' set, 1971.

INTRODUCTION

As we saw in **Volume 1**, the train set was an important tool to get new recruits to the Tri-ang Hornby system. Sets were manufactured ready for the Christmas trade and the aim was to dispose of all sets from the stores by Christmas. Any left by then would be difficult to sell. Having bought a set, it was hoped the public would return to the store to spend Christmas money on accessories.

In this chapter we look at how the train set developed during the Tri-ang Hornby years. In the final years of Tri-ang Railways we saw changes beginning to occur to the outward appearance of the set with a move away from standard box illustrations and towards box lids designed specifically for each set. These included some excellent artwork which is of collecting interest in its own right.

Once again sets were designed for different markets, from the young novice to the more discerning modeller. Sets continued to be assembled in Canada, Australia and New Zealand and each had their own characteristics and methods of manufacture. Making it easy to trace sets in a book of this kind is very difficult as different readers are working to different clues. The order in which sets have been dealt with in this chapter is described below, but for those with just a set code number to work with, a numerical list of sets is included in the Index under 'Sets'.

ORDER OF LISTING

In **Volume 1**, we were able to list sets in alphabetical and numerical order as newly-introduced sets followed a sequence. In **Volume 2**, this would have been very confusing as numbers for new series of sets were picked at random and, in some cases, earlier numbers have been reused. British sets during this period fell into one of the following categories:

A. Tri-ang Railways Sets Carried Over. These were sets originally carrying the Tri-ang Railways name which remained available after May 1965. In some cases their retention was to use up existing stock but in other cases batches were assembled in reprinted boxes (often with the same design of box). As these sets have already been covered in **Volume 1**, less space is given to them here.

B. Starter sets. These were sets specifically designed to be inexpensive to entice first-time buyers from amongst the 'not so well off' and those parents who wanted to establish their child's level of interest before spending money on one of the mainstream sets.

C. Mainstream sets. These are the ones that were usually featured in the catalogues and were later described as 'Model Trains' as opposed to 'Toy Trains'. They were more likely to be bought by parents convinced that their child would play with it and that it was likely to lead to an interest in railway modelling. The contents had to be more realistic as the child would not be satisfied with less.

D. Presentation sets. These were sets made up of a locomotive and train but no track or controls. They were useful for adding more stock to an existing layout. The principle was not widely used by Rovex but the Canadian company produced quite a few presentation packs for their home market.

E. Motorway Sets. Only a few joint railway/motorway sets were made, mostly before May 1965, and these were principally produced to help sell Minic Motorway on the back of Tri-ang Railways. Only one new set was made for general distribution in the Tri-ang Hornby period and this was to promote Minic Motorway after production had been moved to the Margate factory.

F. Canadian sets. Canada did not make any Tri-ang Hornby models but they did assemble their own sets which, unlike those made in other overseas factories, bore little resemblance to sets made in the UK as they had completely different packaging. All such sets will be marked as having been made in Canada

G. New Zealand sets. These were sets assembled at the New Zealand factory in Auckland for the local market. The packaging indicates that they were made in New Zealand and so, although they look like British sets, there should be no confusion.

H. Australian sets. Looking like British sets, these were clearly marked as having been made by Moldex Ltd. Some of these were copies of UK sets, with the same number, but were assembled in Australia with locally-made track and boxes.

I. French Sets. These were sets made in the UK but sold through Meccano France Ltd. They came with Hornby Acho track replacing the Super 4 and in some cases a set of Hornby Acho couplings was provided in the box to allow the purchaser to convert one item of stock. The boxes usually had a locally-printed label on the ends describing the contents in French. The track, couplings and labels are the means by which these sets may be identified.

Mail Order and Special Sets. Over the years Rovex made many special sets for large stores and mail order houses. These sometimes had different packaging and a different assortment of contents. Only one wagon had to be changed for the store to claim that the set was exclusive to them. Almost always an existing set was used so that special packaging did not have to be made. These special sets have been listed with their original contents wherever possible. Some of those listed among the mainstream sets may in fact have been starter sets.

CUTTING THE COST

Cheaper Locomotives
As we have seen, Rovex manufactured a large range of sets for beginners and produced them in great numbers. Many did not appear in the catalogues and records have not been kept of what was made and what each set contained.

The principal feature of these sets was the cheaply-made locomotive based on one of the following (*see* table below) seven 0-4-0 types which, in some cases, were available in clockwork and electric forms. The name by which they are referred to in this book is given in brackets, after each, together with the 'R' numbers of known variations:

As time went on, attempts at reducing further the cost of production of these models by leaving off parts such as transfers, buffers and coupling rods created new variations. There were also colour changes. Each change was given a different 'R' number. Usually a change of locomotive type or selection of wagons led to a renumbering of sets.

Integral Rolling Stock
Prior to 1969 inexpensive sets often contained unprinted versions of standard wagons. These carried a 'NP' (for non-printed) suffix to their 'R' number. In 1969, when the toy train sets were renumbered from 'RS1' or 'RB1' (those with plastic track), there was a need for a range of cheap-to-make rolling stock for these sets. What resulted were a coach and four basic wagons that could take a variety of loads.

These were cheap because, with the exception of the coach, the body and chassis were a single moulding. In the factory they were referred to as 'integral' wagons and this is the term I have used in the book. The vehicles each had four wheels. Three of the wagons had long wheel bases and the fourth was extra long to take the 30' container. The coach, while still referred to as 'integral', was made in two parts; the body was yellow and the chassis black or dark brown. One wagon was a tanker which seems to have always been moulded in yellow plastic and sometimes printed with a Shell symbol, but the flat and open wagons were made in various shades of blue, green, yellow or red and even black.

To add interest the integral flat wagon often carried a Minix car or van as a load and some sets were given a level crossing.

Integral wagons were to survive into the '90s although, over the years, the tools underwent a number of modifications.

The track, in all but the RB sets, was originally Super 4 and not the new System 6. This may have been seen as a way of disposing of the large stock of Super 4 left in the factory stores when the new track came into use, but the explanation given to retailers was that the larger track was easier for small children to join up. The track was supplied as either a circle or an oval.

Set Numbering
Besides set numbers changing when the contents were modified, to add further to the confusion, special versions of sets were produced for export. These may have had a different box or a different selection of contents. There were also particular variations put together for special customers. Each of these set variations was given a number of its own.

The result of all this is a very confused picture which is like a jigsaw puzzle with a number of the pieces missing. What I have tried to do here is provide some order to the scene and fill in some of the gaps with well-educated guesses (marked with a question mark).

Type		Referred to as	Models	Notes
1	Dock Shunter/Yard Switcher	'Dock'	R253, R353.	
2	Industrial Tank ('Nellie' & friends)	'Industrial'	R335 (R335B), R335R, R335Y, R335G, R359.	
3	North British type Diesel Shunter	'N British'	R557, R559, R654, R756.	probably not used after 1965.
4	North British Diesel Shunter & lump on top	'Barclay'	R858.	cab was enlarged to take a can motor.
5	Tank Engine with a box tank on top	'Top Tank'	R657, R659, R660, R660A, R660T, R873.	moulding was later used for the Wild West loco.
6	Continental Tank Engine	'Continental'	R852, R852A, R852CN, R852T, R854, R854T.	not to be confused with German type tank of early 1970s
7	Swedish type Diesel Shunter	'Swedish'	R853.	

CHECK LIST OF SETS

The description in the second column of this Check List represents that given by the company if shown with capital letters, eg. Action Set and the author's description if shown in small letters, eg. clockwork export set.

To simplify searches for information, the sets have been listed under the same headings used in the Directory of Sets that follows.

A. TRI-ANG RAILWAYS SETS CARRIED OVER

For ease of reference these have not been separated out here but if there is a possibility that they were issued in Tri-ang Hornby packaging before being withdrawn, they are included in the appropriate list below.

A list of Tri-ang Railways sets carried over will be found in the Directory of Sets which follows this check list.

B. STARTER SETS

2. Miniville Sets

Sets made in the UK for starters but to be sold through the Wholesale Department under the name 'Miniville'.

The name was dropped in the spring of 1971.

No.	Miniville Sets	Contents	Dates
W1	Clockwork Train Set	c/w Top Tank 0-4-0T, 3 wagons, station	68-70
W2	Clockwork Train Set	c/w Top Tank 0-4-0T, 2 wagons, station	68-70
W3	Clockwork Train Set	c/w Top Tank 0-4-0T?	68
W4	Clockwork Train Set	c/w Top Tank 0-4-0T, 2 wagons	69-70
W4W	Special for Waltons (Australia)	c/w Top Tank 0-4-0T, 3 wagons ?	69
W4W	Special for Waltons (Canada)	'M-T Express' c/w Top Tank 0-4-0T, 3 wagons	69
W10	Electric Train Set	Barclay 0-4-0DS, 3 wagons	69-71

1. Cheap Sets. Sets made in the UK for starters.

No.	Cheap Sets	Contents	Dates
RS12	clockwork export set	contents unknown	69-70
RS13	clockwork set for Sweden	contents unknown	69
RS13M	clockwork export set (Miniville?)	contents unknown	69
RS18	Action Set	c/w Top Tank 0-4-0T, 1 wagon	66-68
RS18E	export version of RS18	c/w Top Tank 0-4-0T, 1 wagon	66
RS19	Goods Set	c/w Top Tank 0-4-0T, 1 wagon	66-68
RS19E	export version of RS19	c/w Top Tank 0-4-0T, 1 wagon	66
RS19M	special order of RS19 for Myers	contents unknown	?
RS20	Goods Set – Battery Operated	Top Tank 0-4-0T, 2 wagons	66-68
RS20A	See Australian sets		
RS43	export clockwork goods set	c/w North British 0-4-0DS, 2 wagons	62-65
RS43E	export clockwork goods set	c/w North British 0-4-0DS, 2 wagons	66
RS49	Clockwork Passenger Set	c/w Top Tank 0-4-0T, coach	63-67
RS49E	export version of RS49	c/w Top Tank 0-4-0T, coach	65-67
RS70	Clockwork Goods Set	c/w Top Tank 0-4-0T, 2 wagons	64-67
RS70E	export version of RS70	c/w Top Tank 0-4-0T, 2 wagons	64-67
RS71	Junior Goods Train Set	North British 0-4-0DS, 2 wagons	64-65
RS71E	export version of RS71	North British 0-4-0DS, 2 wagons	64-65
RS74	Export Military Set	North British 0-4-0DS, 2 wagons	65
RS82	Clockwork Train Set for Canada	c/w Top Tank 0-4-0T, 1 wagon	67
RS85	Goods Set	c/w Top Tank 0-4-0T, 2 wagons, level crossing & cars	67
RS86	Goods Set	c/w Top Tank 0-4-0T, 2 wagons	68
RS86(W)	RS86 set for Woolworths	contents unknown	68
RS86A	Special batch of RS86 for Waltons	contents unknown	68
RS87	Goods Set	c/w Top Tank 0-4-0T, 3 wagons	68
RS87M	Special batch of RS87 for Myers	contents unknown	68
RS88	Goods Set	c/w Top Tank 0-4-0T, 4 wagons, crossing	68
RS88A	See Australian sets		
RS91	electric goods	Continental 0-4-0T, 3 wagons	68?
RS96	Woolworths clockwork set	C/W Continental 0-4-0T ?	68
RS97	Woolworths clockwork set	C/W Continental 0-4-0T ?	70
RS100	Large Clockwork Steam Freight Train	contents unknown	69-70

3. Toy Trains

Sets made in the UK for starters. From 1969 sets classified as 'Toy Trains' but with ordinary track were numbered from 'RS1' onwards. These sets normally had integral 'rolling stock'.

No.	Toy Train Sets	Contents	Dates
RS1	Clockwork Steam Freight Train	c/w Continental 0-4-0T, 2 wagons	69-71
RS1M	export version of RS1	contents unknown	69-70
RS1E	Battery Powered Steam Freight	Continental 0-4-0T, 2 wagons	69-71
RS1EN	See Australian sets		
RS2	Clockwork Steam Passenger Train	c/w Continental, 2 coaches	69-70
RS2M	export version of RS2	contents unknown	69
RS2E	Battery Powered Mixed Traffic Set	Swedish 0-4-0DS, 3 wagons, 1 coach	69-71
RS2EN	possibly for export to Australia	Swedish 0-4-0DS , 3 wagons, 1 coach ?	70
RS3	Large Clockwork Steam Freight Train	c/w Continental 0-4-0T, 4 wagons, crossing	69-71
RS3A	See Australian sets		
RS3W	RS3 for Woolworths	contents unknown	69-70
RS4	Electric Powered Steam Passenger Train	Continental 0-4-0T, 2 coaches, crossing	69
RS4A	See Australian sets		
RS4E	Battery Powered Steam Passenger Set	Continental 0-4-0T, 2 coaches, crossing	70-71
RS5	Diesel Freight Train Set – Battery Powered	Swedish 0-4-0DS, 3 wagons	69-70
RS5A	See Australian sets		
RS6	Large Diesel Freight Train Set – Battery Powered	Swedish 0-4-0DS, 4 wagons, crossing, 2 cars	69-70
RS6A	See Australian sets		
R852C	Go Electric conversion Set	Continental 0-4-0T, battery box	71-72

4. Starter Sets with plastic track

Sets made in the UK for starters which contain plastic track.

No.	Plastic Track Sets	Contents	Dates
RB1	Goods Set	c/w Continental 0-4-0T, 2 wagons	70
RB1M	Miniville version of RB1?	contents unknown	70
RB1MS	Special order RB1M?	contents unknown	70
RB2	Mixed Traffic Set	c/w Continental 0-4-0T, 1 coach, 2 wagons	70
RB2M	Miniville version of RB2?	contents unknown	?
RB2MS	Special order RB2M?	contents unknown	70
RB2S	Special version of RB2?	contents unknown	70
RB3	Large Goods Set	c/w Continental 0-4-0T, 4 wagons	70
RB3M	Miniville version of RB3?	contents unknown	70
RB3T	Special version of RB3?	contents unknown	71
RB4	Goods set	c/w Continental 0-4-0T, 2 wagons	71
RB4M	Clockwork Train Set	contents unknown	71
RB5	Clockwork Train Set with Station	c/w Continental 0-4-0T, 2 wagons, station, crossing	71
RB5M	Miniville version of RB5?	contents unknown	71
RB6	Wild West Train	c/w Wild West 0-4-0, coach, 4 Cowboys and Indians	71
RB7	See **Volume 3**		

C. MAINSTREAM SETS

Sets made in the UK for sale at home and abroad.

No.	Mainstream Sets	Contents	Dates
RS7	Local Passenger	0-6-0DS, utility van, coach	69
RS8	The Midlander	Deeley, 2 coaches	65-68
RS9	Inter-City Express	Class 37, 3 coaches	66-67
RS10	export set	Industrial 0-4-0T 'Nellie', 5 wagons	65-66

No.	Mainstream Sets	Contents	Dates
RS11	The Goods	Industrial 0-4-0T 'Polly', 3 wagons	65-67
RS11A	export set for Australia	contents unknown	?
RS16	Strike Force 10 Set	Battle Space Jinty, 2 wagons, figures	66-70
RS17	The Satellite Set	Battle Space 0-4-0DS, 2 wagons, figures	66-70
RS20A	See Australian sets		
RS24	The Pick-Up	Industrial 0-4-0T 'Nellie', 3 wagons	62-67
RS24A	The Pick-Up	Industrial 0-4-0T 'Polly', 3 wagons	68-69
RS24A	See Australian sets		
RS29	Holiday Express	B12, 2 coaches	63-66
RS33	special set for USA	contents unknown	?
RS34	See **Volume 1**		
RS34A	Blue Streak set for Australia	Double Ended Diesel, 4 passenger cars	65
RS34F	See French sets		
RS50F	The Defender see French sets	0-6-0DS, 3 wagons	65
RS51	Freightmaster	green Class 31, 7 wagons	65-67
RS51	Freightmaster	Electric blue Class 31, 7 wagons	68
RS51	Freightmaster	Rail blue Class 31, 7 wagons	69
RS51A	See Australian sets		
RS52	The Blue Pullman	Pullman DMU, dummy car, coach (blue, crest)	65-67
RS52	Diesel Pullman Train Set	Pullman DMU, dummy car, coach (blue, yellow front)	68
RS52	Diesel Pullman Train Set	Pullman DMU, dummy car, coach (grey, blue)	69
RS52F	See French sets		
RS52MG	special set for Myers	contents unknown	67
RS52MP	special set for Myers	contents unknown	67
RS61	Old Smoky	grimy Deeley, 2 coaches	65
RS62	Car-a-Belle	Jinty, 3 wagons	65-69
RS62A	RS62 for Australia	Jinty, 3 wagons ?	66
RS89	Rail Freight	0-6-0DS, 2 wagons	68-69
RS89G	Ideal Freight for Australia	0-6-0DS, 2 wagons ?	69
RS90	The Pullman Express	Hall, 3 coaches	68-69

No.	Mainstream Sets	Contents	Dates
RS92	Turbo Car Set	Battle Space Turbo Car, ?	not made
RS93	Blue Train for South Africa	contents unknown	68
RS94	freight train for South Africa	contents unknown	68
RS94M	special set for Myers	contents unknown	68
RS100	special set for Waltons	contents unknown	71
RS101	The streamliner Set for Australia	A Unit, 3 passenger cars	71
RS101A	See **Volume 3**		
RS102	The Express Freighter Set for Australia	TC (Aust) Pacific, 5 wagons	71
RS103	The Steam Freighter Set for Australia	Jinty, 4 wagons, freight cars	71
RS103A	special set for Waltons	contents unknown	71
RS104	The Huskey Freighter Set for Australia	Dock 0-4-0DS, 3 wagons	71
RS105	The Diesel Freighter Set for Australia	Double Ended Diesel, 5 wagons	71
RS106	special set for Myers	contents unknown	71
RS600	Junior Freightliner Set	0-6-0DS, 2 wagons	70
RS600A	special order of RS600?	contents unknown	70
RS601	Steam Freight Set	Jinty, 4 wagons	70
RS601A	special order of RS601?	contents unknown	70
RS602	Senior Freightliner Set	Hymek, 2 wagons, depot crane, container	70-71
RS602A	special order of RS602?	contents unknown	70
RS603	Local Diesel Set	DMU	70
RS604	Night Mail Set	Class 37, 3 coaches	70-71
RS604A	special order of RS604?	contents unknown	70
RS604B	special order of RS604?	contents unknown	70
RS605	Flying Scotsman Set	Scotsman, 3 coaches, crossing, signal	70
RS605A	Flying Scotsman Set	Scotsman, 3 coaches, crossing, signal	70
RS605B	special overseas order of RS605?	contents unknown	70
RS605C	special order of RS605?	contents unknown	70
RS605D	special US order of RS605?	contents unknown	70
RS606	Express Goods Set	B12, 4 wagons	70-71
RS607	Local Passenger Set	M7, 2 coaches	71
RS608	Flying Scotsman Set	Scotsman, 2 coaches	71
RS609	Express Passenger Set	Princess, 2 coaches	71
RS611	Special Export Local Freight Set	Industrial 0-4-0, 3 wagons	71
RS612	Local Freight Set	Industrial 0-4-0, 3 wagons	71
RS613	Steam Freight Set	Jinty, 4 wagons, crossing	71

No.	Mainstream Sets	Contents	Dates
RS614	Pick Up Goods	Industrial 0-4-0, 3 wagons	71
RS615	The Railway Children Set	Jinty, 2 coaches	71
RS616	See **Volume 3**		
RS624	Pick-Up Set	Industrial 0-4-0T, 3 wagons	70
RS651	Freightmaster	Class 31, 7 wagons	70-71
RS651A	special version of RS651	contents unknown	70
RS652	Diesel Pullman Set	Pullman DMU, dummy car, coach	70
RS801	special set for Kays	contents unknown	69
RS1051	export set	contents unknown	65

D. PRESENTATION SETS

Sets which do not include track and which were made in the UK.

No.	Presentation Sets	Contents	Dates
R346	Stephenson's Rocket	'Rocket', 1 coach	63-67
R346C	Stephenson's Rocket	'Rocket', 3 coaches	68-69
R346S(F)	See French sets		
R397	satellite set	0-4-0DS, 2 wagons	not made
R398	Battle Space set	BS Jinty, 2 wagons	not made
R399	promotional Shunting Set	contents unknown	69
R619	special set for Littlewoods	contents unknown	
R640	Lord of the Isles Presentation Pack	'Lord of the Isles', 1 coach	64-65
R640(F)	See French sets		
R641	Davy Crockett Presentation Pack	'Davy Crockett', 1 coach	65-68
R641S(F)	See French sets		
R644	Inter-City Train Pack	Electric blue 'E3001', 3 coaches	68-69
R644A	Inter-City Train Pack (with lighting)	Rail blue 'E3001', 3 coaches	69-70
R645	Freightliner Train Pack	Hymek, 3 wagons	67-70
R695	Flying Scotsman set	Scotsman, 3 coaches	not made
R702	special set for GUS	contents unknown	68-70
R703	special set for Littlewoods	contents unknown	68
R704	special set for IIE	contents unknown	69
R792	special set for an unknown customer.	contents unknown	71
R793	Flying Scotsman set	Scotsman	71
R795	special set for Kays	contents unknown	71

No.	Presentation Sets	Contents	Dates
R796	special set for Kays	contents unknown	71
R797	special set for IIE	contents unknown	70
R798	special set for IIE	contents unknown	70
R799	special set for IIE	contents unknown	70

E. MOTORWAY SETS

Sets containing a mixture of Tri-ang Hornby railways and Minic Motorway.

No.	Motorway Sets	Contents	Dates
RMD	Moto-Rail Set	Dock Shunter, 1 wagon, ramp, 1 car, siding	68-70
RMZ	special set for Littlewoods	contents unknown	68
RM?	special set for Kelloggs	B12, 4 wagons, motorway parts	67-68?

F. CANADIAN SETS

Sets assembled in the Tri-ang factory in Canada.

No.	Canadian Sets	Contents	Dates
CTS1	Steam Freighter Set	Pacific, 3 freight cars	65-66
CTS2	Le Champlain Diesel Passenger Set	A Unit, 3 passenger cars	65-66
CTS3	Diesel Freighter Set	Switcher, 3 freight cars	65-66
CTS4	Rocky Mountain Avalanche Train	Yard Switcher, ambulance car, 2 freight cars	65-66
CTS5	Works Train	Yard Switcher, 2 freight cars	65-66
CTS6	The Canadian Diesel Passenger Set	A Unit, 3 passenger cars	65-66
CTS662R	special set for Eatons	A Unit, 2 freight cars	66
CTS65??	special set for Eatons	contents unknown	65
TS671	Diesel Freight Set	Yard Switcher, 3 freight cars	67-68

No.	Canadian Sets	Contents	Dates
TS672	Cross Canada Diesel Freight Set	Switcher, 4 freight cars	67-68
TS673	Pacific Steam Freight Set	'Hiawatha', 5 freight cars	67-68
TS674	The Canadian CPR Diesel Passenger Set	A Unit, 4 passenger cars	67-68
TS675	Centennial Steam Passenger Set	'Davy Crockett', 2 passenger cars, caboose	67-68
TS67??	special set for Eatons	Switcher, 4 freight cars	67
TS676	Industrial Diesel Freight Set	A Unit, 3 freight cars	67
TS680	8 Car Set	contents unknown	68
TS681?	special set for Eatons	contents unknown	68
TS901	Cross Canada Diesel Freight Set	A Unit, 3 freight cars	69
TS902	Canadian Pacific Diesel Freight	Switcher, 4 freight cars	69
TS903	Steam Freight Set	Pacific, 5 freight cars	69
TS904	Canadian National Passenger Set	A Unit, 4 passenger cars	69
TS905	Canadian National Diesel Freight	A Unit, 4 freight cars	69
TS906	Steam Switcher Freight Set	'Chugga' 0-4-0, 3 integral wagons	69
TS9??	special set for Eatons	contents unknown	69
675	Davy Crockett Steam Set	'Davy Crockett', 2 passenger cars, caboose	71
900	special set for Eatons	contents unknown	70
913	Cross Canada Diesel Freight Set	A Unit, 3 freight cars	70
914	CP Rail Diesel Freight Set	A Unit, 4 freight cars	70-71
915	Steam Freight Set	Pacific, 5 freight cars	70
916	Canadian National Diesel Freight	Switcher, 4 freight cars	70-71
917	CP Rail Passenger Set	A Unit, 4 passenger cars	70
918	Steam Switcher Freight Set	'Chugga' 0-4-0, 3 integral wagons, caboose	70-71
919	International Container Express Set	Switcher, 4 container cars, caboose	70-71
926	Canadian National Diesel Freight Set	A Unit, 3 freight cars	71
927	Canadian Pacific Mixed Steam Freight Set	Pacific, 5 freight cars	71
928	Canadian National Passenger Train	A Unit, 4 passenger cars	71

No.	Canadian Sets	Contents	Dates
7301	Side-Tank Switcher Freight Set	Continental 0-4-0, 3 freight cars	73
7302	Diesel Switcher Freight Set	0-6-0DS, 3 freight cars	73
7303	Canadian National Short Haul Freight Set	A Unit, 3 freight cars	73
7304	CP Rail Diesel Freight Set	Switcher, 4 freight cars	73
7305	CP Rail Inter-City Express Set	A Unit, 4 passenger cars	73
7306	Pacific Steam Freight Set	Pacific, 5 freight cars	73

Canadian Presentation Sets

No.	Canadian Sets	Contents	Dates
CPS101	Passenger Assortment	A Unit, 3 passenger cars(#1)	65-66
CPS201	Freight Assortment	Switcher, 3 freight cars(#2)	65-66
CPS301	Freight Assortment	A Unit, 3 freight cars	65
CPS401	Freight Assortment	A Unit, 2 freight cars	65
#4	Freight Cars	3 freight cars	66
#5	Freight Cars	4 freight cars	66
PS167	Passenger Presentation Set	3 passenger cars	67-68
PS267	Works Train Presentation Set	4 freight cars	67-68
PS367	All-Action Presentation Set	4 freight cars	67-68
R367	All-Action Presentation Set	4 freight cars	69
R467	Canadian National Dayliner Diesel Train	Budd (3 car)	69-71
R468	Dockside Containerisation Switcher Set	Dock 0-4-0DS, 3 container cars	69
R469	Works Train Presentation Set	4 freight cars	69-71

G. NEW ZEALAND SETS

Sets assembled in New Zealand at the Tri-ang factory in Auckland.

No.	New Zealand Sets	Contents	Dates
NC1	clockwork goods set	c/w 0-4-0T Top Tank, 2 integral wagons	69-70
NC1	clockwork goods set	c/w 0-4-0T Continental, 2 integral wagons	71-72
NC3	clockwork goods set	c/w 0-4-0DS, 3 integral wagons	69-70

No.	New Zealand Sets	Contents	Dates
NC4	clockwork goods set	c/w 0-4-0T Top Tank, 3 integral wagons	69-70
NC4	clockwork goods set	c/w Continental 0-4-0T 3 integral wagons	71-72
NE1	electric goods set	Continental 0-4-0T, 3 integral wagons	69-73
NE2	Local Goods Set	Jinty, 3 wagons	69-73
NE3	Diesel Shunter with Siding	0-6-0DS, 3 wagons, point, buffer stop	69-73
NE4	Diesel Express Freight Set	Double Ended Diesel, 3 freight cars	69-72
NE5	Blue Streak Diesel Passenger Set	Double Ended Diesel, 3 passenger cars	69-71
NE5	Diesel Passenger Express	Double Ended Diesel, 3 coaches	72-?
NE6	Freightmaster Set	Hymek, 5 wagons/freight cars	69-73
NE7	Transcontinental Freight Express	'Hiawatha', 4 wagons/freight cars	69-73
NZ3DG	Diesel Express Freight Set	Double Ended Diesel, 4 wagons	58-68
NZ3RG	Local Goods Set	Jinty, 3 wagons	63-68
NZ50G	BR Express Goods Set	'Princess', 5 wagons	64-68
NZ50P	BR Express Passenger Set	'Princess', 3 coaches	62-68
NZC3	clockwork goods set	c/w 0-4-0DS, 3 wagons	66-68
NZC4	clockwork goods set	c/w Top Tank 0-4-0T, 3 wagons	66-68
NZDBS	Blue Streak Diesel Express Passenger Set	Double Ended Diesel, 4 passenger cars	64-68
NZDSH	Diesel Shunter Set	0-6-0DS, 3 wagons, point, buffer	68
NZHYF	Freightmaster Train Set	Hymek, 5 wagons	68
NZRC	Rail Car Set	EMU 2 car	64-67
NZST	Suburban Train Set	EMU 2 car, brake, utility	64-67
NZTCF	Transcontinental Freight Express	'Hiawatha', 4 freight cars	64-68

H. AUSTRALIAN SETS

Sets made or assembled in the Moldex factory in Melbourne.

No.	Australian Sets	Contents	Dates
MR94A	special set for Myers?	Jinty, 4 wagons	
R4A	Steam Freight Set	Jinty, 3 wagons	66-67
R4B	Countryman	'Princess Royal', 2 coaches	66-67
R4D	Trans-Australia Express	Double Ended Diesel, 3 coaches	62-67
R4DA	Transcontinental Express	Double Ended Diesel, 3 coaches	68-71
R4F	Diesel Freight Set	Double Ended Diesel, 5 wagons	65-67
R4FA	Diesel Freight Set	Double Ended Diesel, 5 wagons	68-71
R4N	Interstate Freight Set	'Hiawatha', 3 wagons	66-67
R4W	Southern Aurora	A Unit, 3 coaches	66-67
R4Y	Suburban Set	NSWR Suburban Electric (3 cars)	62-66
R4YA	Sydney Suburban Set	NSWR Suburban Electric (3 cars)	68
RS1EN	Battery Powered Steam Freight Set	Continental 0-4-0T, 2 wagons	70-71
RS3A	clockwork goods set	c/w Continental 0-4-0T, 4 wagons	69
RS4A	Battery Powered Passenger Set	Continental 0-4-0T, 2 coaches	69
RS4CA	Freight Set	'Hiawatha', 5 wagons	70
R4DA	Transcontinental Express	Double Ended Diesel, 3 coaches	68
RS4DA	Transcontinental Express	Double Ended Diesel, 3 coaches	69
RS4DA	Transcontinental Express	A Unit, 3 coaches	70-71
RS4E	See RS4A above		
RS4FA	Diesel Freight Set	Double Ended Diesel, 5 wagons	69-70
RS5A	Electric Powered Diesel Freight Train	Swedish 0-4-0DS, 3 wagons	69-70

No.	Australian Sets	Contents	Dates
RS6A	Large Electric Powered Diesel Freight Train	Swedish 0-4-0DS, 4 wagons	69-70
RS11A	special set for an unknown customer	Industrial 0-4-0T, 3 wagons	?
RS12	special set for Myers	contents unknown	70
RS18	clockwork goods set	c/w Top Tank 0-4-0T, 1 wagon	67
RS19	clockwork goods set	c/w Top Tank 0-4-0T, 1 wagon	67
RS20A	Freight Set	BR Jinty, 3 wagons	64-65
RS21A	See R4B		
RS24A	Pick Up Goods Set	Continental 0-4-0T, 3 wagons	68-71
RS24XA	Electric Train Set	Continental 0-4-0T, 3 wagons	69-71
RS38	Snow Rescue Set	Dock 0-4-0DS, 2 wagons, coach	66-67
RS43E	Goods Set	c/w 0-4-0DS, 2 wagons	66
RS49	Passenger Set	c/w Top Tank 0-4-0T, coach	67
RS51A	Freightmaster Set	Class 31, 7 wagons	68-69
RS61	Old Smokey	grimy Deeley, 2 coaches	66
RS62A	Car-a-Belle Set	Jinty, 3 wagons	66
RS70	Goods Set	c/w Top Tank 0-4-0T, 2 wagons	67
RS85	Goods Set	c/w Top Tank 0-4-0T, 2 wagons, crossing	67
RS88A	Clockwork Goods Set	c/w Top Tank 0-4-0T, 4 wagons	68

	Australian Presentation Sets		
R346	Stephenson's Rocket	'Rocket', coach	63-67
R640	Lord of the Isles	'Lord of the Isles', coach	66
R641	Davy Crockett	'Davy Crockett', coach	65-68

I. FRENCH SETS

Sets made in the UK to which Hornby Acho track was added (not in the case of the presentation sets) in the Calais factory and French labelling was added to the box.

No.	French Sets	Contents	Dates
RS34F	Le Transcontinental	Double Ended Diesel, 4 passenger cars	65-70
RS50F	Le Train Militaire	0-6-0DS, 3 wagons	67-69
RS52	Le Pullman Bleu	Pullman DMU, dummy car, coach	65-70
R346	Train Rocket de Stephenson	'Rocket', 1 coach	67-70
R346S	Train Rocket de Stephenson	'Rocket', 1 coach	65-66?
R640	Train XIX siecle Lord of the Isles	'Lord of the Isles', 1 coach	65-70?
R641	Train Western Davy Crockett	'Davy Crockett', 1 coach	65-66?
R641S	Train Western Davy Crockett	'Davy Crockett' with smoke, 1 coach	67-70

MYSTERY TRAIN SETS

Sets about which too little is known to allow them to be allocated to one of the categories above.

No.	Name or Description	Contents	Dates
R?	Lyons Maid Set		
RS1051	Export Set		

DIRECTORY OF SETS

A. TRI-ANG RAILWAYS SETS CARRIED OVER

A Change of Image
The following is a list of Tri-ang Railways sets which survived into the Tri-ang Hornby era and therefore probably appeared in the new packaging. Indeed, some of the sets such as the Pick-up Goods and the Freightmaster had a long life and are more commonly found in Tri-ang Hornby than Tri-ang Railways boxes.

Not only did the brand-name on the box change but, during the mid 1960s, we saw a change in the type of box construction. Out went the card in-fill and partitions and in came the moulded plastic former, and later still, the expanded polystyrene block. Many sets survived across these changes.

In Canada the sets continued to be sold under the Tri-ang Railways name until 1971 when for a very brief moment it changed to Tri-ang Hornby.

Manufactured at Margate
The following former Tri-ang Railways sets, manufactured at Margate, will be found in the sections indicated:

RS24 'Pick-up Goods' *see* Mainstream Train Sets
RS29 'The Holiday Express' *see* Mainstream Train Sets
RS43 'Clockwork Goods' *see* Starter Sets
RS49 'Clockwork Passenger' *see* Starter Sets
RS50 'The Defender' *see* Mainstream Train Sets
RS51 'The Freightmaster' *see* Mainstream Train sets
RS52 'The Blue Pullman' *see* Mainstream Train Sets
RS61 'Old Smoky' *see* Mainstream Train sets
RS62 'Car-a-Belle' *see* Mainstream Train Sets
RS70 'Clockwork Dockmaster' *see* Starter Sets
RS71 'Passenger Set' *see* Starter Sets
R346 'Stephenson's Rocket Train'
     see Presentation Sets

Sets carried over in the Australian and New Zealand ranges are described in those sections of this chapter.

B. STARTER SETS

Introduction
From the beginning of Tri-ang Railways there have always been inexpensive sets aimed at beginners. When cheap sets from France were imported by Playcraft in the late 1950s, Tri-ang responded vigorously with a wide range of even cheaper sets planned to under cut the imported ones. The battle raged on into the 1960s but it was not until 1969 that the company officially recognised that it was catering for two distinctly different markets and started to list their train sets under the 'Model' and 'Toy' headings. As we have seen earlier in this chapter, to keep sets inexpensive Rovex had to find cheaper ways of making locomotives, rolling stock and track and in this they were very successful.

Whereas the minimum order quantity on sets was normally three, beginners' sets, because of their low profit margin, had to be bought by retailers in packages of six or 12. Miniville sets were sold in packs of 24 or 36.

Four Types of Sets
During the period 1965-1971 starter sets fell into four categories:
1. Cheap sets – 1965-68. These had cheaper contents, e.g. clockwork engines and wagons without markings.
2. Miniville sets – 1968-71. Similar to the above but produced under a different brand-name.
3. Toy Train sets – 1969-71. This was the RS1-RS6 series with 'integral' rolling stock.
4. Plastic Track sets – 1970-71. These were similar to the above but had plastic track instead of the normal Super 4 and carried an 'RB' prefix.

1. Cheap Sets – 1965-68
As far as their code (RS) numbers were concerned, these jumped about according to what numbers were spare at the time and so they were lost amongst more conventional train sets. For our purpose they have been extracted from the products list and re-listed in numerical rather than date order. This is so that the reader might find it easier to trace a set by its number.

Variations are listed under each basic set.

RS12 Clockwork set – details unknown.
This appears to have been an export set which went out to Australia and Canada, possibly for mail order houses there. Australia received 3,000 in 1969 and Canada 144. In 1970 a batch of more than 3,000 went to Myers stores in Australia as a special order.

RS13 Clockwork set.
Not much is known about this set other than that 5,000 were made in 1969 which went to Sweden.

RS13M
This may have been a version of the set sold under the Miniville trademark. Only about 3,900 were made, all in 1969, which were exported to Australia.

RS18 Clockwork Top Tank 0-4-0 (R660), rocket launching vehicle & circle of Super 4.
Introduced in 1966 as a train-and-track pack, it survived for three years. The loco was black with red wheels. The packaging was cheap, consisting of a two-sided display box with a circle of track beneath a card former. This also meant that they did not take up much shop space. It is thought that about 28,000 were made, a figure which possibly includes export sets (see opposite), and it is known to have been available in Canada in 1966 and Australia in 1967 (when sets may have been assembled there). RS18 was dropped from the export list in 1968.

RS18E This was an export version of RS18 available during 1966 and looked the same as RS18. It was exported in cartons of 12 which measured 27x27x65 cms.

RS19 Clockwork Top Tank 0-4-0 (R660), bogie bolster, 3 Minix cars & circle of Super 4 track. RS19 was similar to RS18 but with a different wagon. About 29,000 were made and sold between 1966 and 1968 in similar train-and-track display packaging. The set was possibly also assembled in Australia in 1967.

RS19M This was a special set of which 840 were made for Myers of Australia. It is assumed that the 'M' suffix stood for 'Myers' and not 'Miniville'.

RS19E The export version of RS19 with a similar box to that for RS18E. The 1966 Canadian catalogue has an illustration of the set but the following year Canada received the similar RS82 instead.

RS20 Electric Top Tank 0-4-0 (R659), flat wagon with a Minix car and a short brake van, circle of Super 4 & RP40 battery connector.

The locomotive was black and the set was sold in the chunky train-and-track pack window boxes with a plain brown outer sleeve. Some 14,000 sets were

made and sold in 1966 and 1967, the vast majority of them selling in Britain between 1966 and 1968.

In the 1965 Australian catalogue, RS20 was illustrated as an Australian Jinty tank engine and three wagons. This appears to have been an example of incorrect labelling. The set shown was R4A but the RS20 set was also imported in packs of three by Australia where it retailed at $A11.55.

RS20A *See* H. Australian Sets

RS43 Clockwork Goods Set described in **Volume 1**.
A little over 1,000 were made in 1965 for export but it is not known if Tri-ang Hornby boxes were used.

RS43E An export set, illustrated in the 1965/66 Canadian catalogue and also assembled in Australia in 1966. The 0-4-0DS was red and had a green open wagon and brown short brake van. The Canadian catalogue described it as a 'Mechanical Freight Set'.

RS49 Clockwork Top Tank 0-4-0 (R660), R720 coach & circle of Super 4 track.
The set remained in production until the end of 1967 also selling in the train-and-track display boxes. During the last three years 86,000 were made and most sold in the UK. Early sets had the R657

loco and a red and yellow coach but the ones sold in Tri-ang Hornby boxes had the R660 locomotive and, towards the end, a plain red coach. The set was possibly assembled in Australia in 1967.

RS49E This was a special export version of the clockwork RS49 set available between 1965 and 1967. It was also illustrated in the 1965/66 Canadian catalogue where it was described as a 'Mechanical Passenger Set'.

RS70 Clockwork Top Tank 0-4-0 (R660), two wagons (R10 & R14) & circle of Super 4 track.
This Dockmaster set was described in **Volume 1** but further batches were made in 1965, 1966 and 1967 under the Tri-ang Hornby brand-name. During this period over 150,000 were sold. It was possibly assembled in Australia in 1967.

RS70E This was an export version of RS70 available between 1964 and 1967. During the last two years about 73,000 were made and 90% of them sold in the UK.

RS71 Electric North British 0-4-0DS (R654), open wagon, short brake van & circle of Super 4 track.
Described in **Volume 1**, the set continued in production in 1965 when about 7,000 were made but it is not known whether these were in Tri-ang Hornby boxes. Some advertisements showed the set with an unmarked red dock shunter (without a front coupling or buffers) instead of the blue diesel.

RS71E This is thought to have been a special export version of the above set. It appeared on the 1965 export trade list and was also listed in Australia.

RS20 battery operated goods set, 1966/68.

RS72, RS75, RS76, RS77 and **RS78**
These were made in 1965 and may have been sold in Tri-ang Hornby boxes. *See* **Volume 1**.

RS74 Clockwork North British 0-4-0DS, four rocket launch wagon, closed van & circle of Super 4 track.

The loco was black with red wheels and the wagons khaki. The van was a standard moulding in khaki plastic with a white roof but the other wagon was possibly unique to this set. It consisted of the short wagon chassis with the turret of the four rocket launching wagon mounted on it. The turret car-

ried no markings but there were four rockets packed in the box. The lid of the box carried an attractive picture which as far as I know was unique to this set.

The set was made in 1965 and sold in a Tri-ang Hornby box. 2,500 were made, all of which were sold abroad.

RS82 Clockwork Top Tank 0-4-0 (R660), bogie bolster with three Minix vans & circle of Super 4 track.

Only one batch of 150 sets was made and these went to Canada in 1967. In the Canadian catalogue illustration the Minix vans were shown as red, white and blue.

RS85 Clockwork Top Tank 0-4-0 (R660), 7-plank open wagon and short brake van & circle of Super 4 track, level crossing & two Minix cars.

Manufacture of this set was limited to 1967 when a little over 24,800 were made, 10% of these going overseas. This set may have also been assembled in Australia in 1967.

RS86 Clockwork Top Tank 0-4-0 (R660), mineral wagon, van & circle of Super 4 track.

The van was described as a 'cattle truck'. Just over 90,000 of these sets were made in 1968 almost half of them going abroad.

RS86(W) A further 12,000 of the sets were made for Woolworths that year and must have been slightly different to the main batch and may have carried a 'W' suffix.

RS86A 1968 also saw a batch of 2,000 special sets go to Walton's Stores in Australia. Another 60 of these passed through the Margate stores in 1970 and went out to Australia that year.

RS87 Clockwork Top Tank 0-4-0 (R660), three wagons (brick wagon, flat with a Minic car and a container wagon) & oval of Super 4 track.

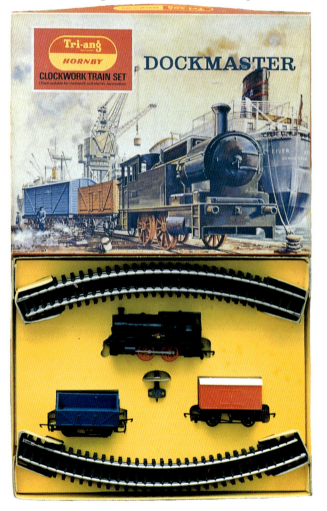

RS70 'Dockmaster' clockwork set, 1964/67.

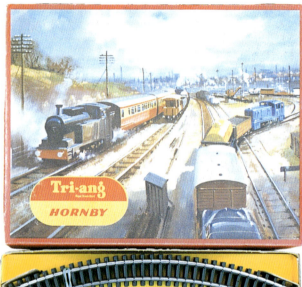
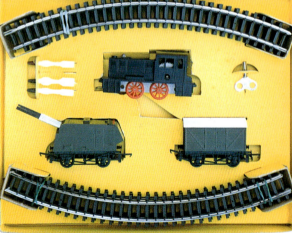

Rare RS74 export military starter set, 1965.

W1 Miniville set, 1968/70.

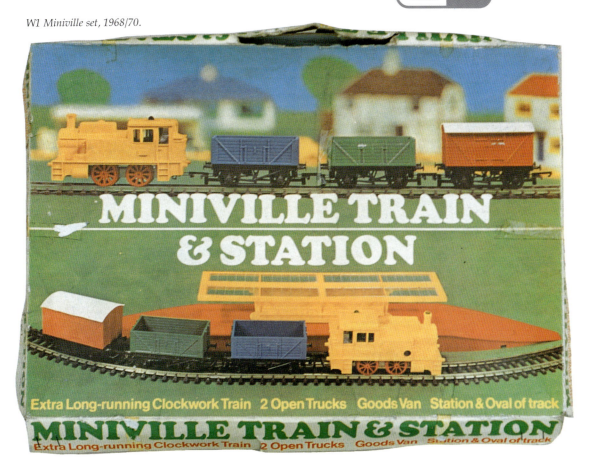

MINIVILLE TRAIN & STATION

Extra Long-running Clockwork Train   2 Open Trucks   Goods Van   Station & Oval of track

MINIVILLE TRAIN & STATION
Extra Long-running Clockwork Train   2 Open Trucks   Goods Van   Station & Oval of track

This set also dates from 1968 when 34,000 were made. One third of these went overseas and all but 30 of the remainder sold in the UK during the year.

RS87M This was a special version of the set made for Myers of Australia during 1968. 1,030 were dispatched.

RS88 Clockwork Top Tank 0-4-0 (R660), four wagons (bogie bolster with three Minix cars, tanker, open wagon and drop-side wagon), level crossing & oval of Super 4 track.

The colour of the loco in sets RS85 – RS88 is thought to have been dark green. A loco of this colour has certainly been found in an RS88 set. Nearly 82,000 sets were made in 1968 and all but 130 were sold that year; mostly in Britain. In Canada it sold for $10.

RS88A This variation was also made in 1968 and assembled in Australia. It contained the same loco, level crossing, track and assortment of wagons but all had been assembled in the Moldex factory. It sold for $A8.25. *See* H. Australian Sets.

RS91 Electric Continental Tank 0-4-0 (R852), three wagons and a circle of Super 4 track?

Not much is known about this set including whether it was made. We do know that in 1968 over 73,000 orders for it were received within the UK and a further 4,500 from abroad.

RS96 C/W Continental 0-4-0T? – other contents unknown.

This was described in the factory records as a special clockwork set made for Woolworths Australian stores and is thought to have had the Continental tank 0-4-0 (R854). In 1968 about 4,700 sets were made but in February 1970, 35 sets still remained at Margate and were disposed of in Britain during the year. The trade price was 10/2d.

RS97 Clockwork set – contents unknown.

4,700 of this special clockwork set were made for Woolworths of Australia and it was available in limited quantities in February 1970. It is thought that it may have contained the R854 locomotive.

RS100 Contents unknown but said to have been similar to the RS3 clockwork set (*see* Toy Train Sets).

The first batch were made in 1969 but not sold until 1970 when nearly 8,000 left the stores, a third of them going abroad. A further 3,000 of the sets were sold to Walton's of Australia in 1971.

2. Miniville Sets

Up until the late 1960s Tri-ang salesmen were confined to trading with registered dealers and the small corner shop had great difficulty in buying British model railways to sell. Lines Bros. decided to tap this potentially lucrative market and to do so adopted a new brand-name – 'Miniville'. The products sold under the Miniville name were drawn from various Tri-ang product ranges, manufactured by Rovex Industries Ltd, and often involved a combination of products from different lines. This was the time that the name 'Tri-ang' started to disappear from mouldings; note particularly the standard wagon chassis.

The 'W' prefix used for Miniville products stood for 'Wholesale' as the range was sold through the company's Wholesale Division. It appears that some of the RB (plastic track) sets were transferred to the Miniville range and, where these received a new box lid to denote this, they received an 'M' suffix added onto their 'RB' codes (see 4. Plastic Track Sets). Finally, in May 1971, it appears that the remnants of the 'Tri-ang' RB sets were handed over to the Wholesale Division to dispose of through their Miniville customers.

Few of the Miniville set boxes have survived. They were doomed from the start because, on the back of each, was a model of a tunnel to be cut out!

The models themselves were set in a pre-formed pale blue plastic tray contained in a fairly crude cardboard outer (on which was printed the cut-out tunnel) whose top was mainly a cellophane window.

The clockwork locomotive was normally the Top Tank 0-4-0 in yellow with red wheels and the electric loco was a blue Barclay 0-4-0. The wagons carried no printing and may have varied in colour.

Amongst the Miniville range were the following train sets:

W1 Clockwork Top Tank 0-4-0 (R660), three wagons (blue and green open wagons and a red van), island platform with TT shelter & circle of Super 4 track.

The strangest feature of this set was the island platform which was from the TT range but was fitted with 00 curved ramps! The platform was bright red and the shelter bright yellow. 18,000 of these sets were made in 1968 and another 18,000 in 1969. Sales ceased early in 1971. The set retailed initially at £1-19s-11d and sold wholesale in packs of 24.

W2 Clockwork Top Tank 0-4-0 (R660), two open wagons, platform with two TT seat units (blue) & circle of Super 4 track.

28,000 sets were made during 1968 and 1969, retailing initially at £1-9s-11d each, but it was withdrawn early in 1971.

W3 Clockwork (probably the R660) but contents unknown. Super 4 track.

1968 was the only year this set was made and 33,000 were produced, 90% of them selling during the year. The retail price was £1-3s-0d but they could be bought from the Wholesale Department at £7-9s-6d per dozen.

W4 Clockwork Top Tank 0-4-0 (R660), open wagon, van & circle of Super 4 track.

W10 Miniville sets, 1969/71.

RS4 electric starter set, 1969.

RS1E electric starter set, 1970.

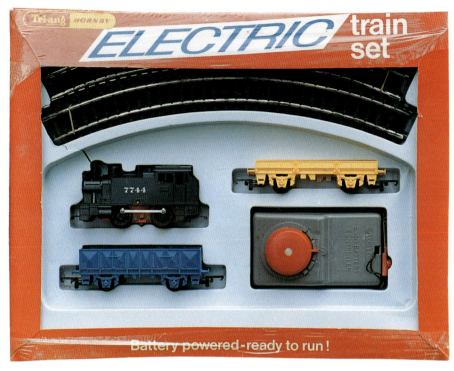

In 1969 54,000 sets were made and only 9,400 were left over to sell the following year.

W4W This seems to have been a special version of the set made for Walton's stores in Australia and Canada in 1969. 1,020 were sent to Canada and 2,100 to Australia. Evidence found in Canada suggests that the loco was a Rainbow blue clockwork 0-4-0 top tank with black wheels and no markings but the buffer stocks filled in. The freight cars are thought to have been the brick wagon, caboose and track-cleaning car, all unprinted but, in the case of the Canadian sets, carrying 'M-T EXPRESS' stickers applied on arrival at the Canadian factory. The brick wagon was black, the track-cleaning car orange-brown and the caboose, vermilion.

W10 R858 Barclay 0-4-0DS, flat wagon and Minix car, flat wagon and orange container, red brick wagon, yellow P40 battery top controller, circle of Super 4 track & power connector.

The loco was electric with a bright blue body and a red plastic chassis block. The publicity photo showed the set with a Dock Shunter which looked uncomfortable because the space in the plastic tray was designed to take the Barclay. 20,000 were made in 1969 and a further 6,900 in 1971. All were sold in the UK and a large proportion was sold to wholesalers. Its final price was £3.

The Miniville series also included a number of kits from the Model-Land range which are described in the 'Model-Land' feature in the chapter on 'Related Series', towards the end of this volume.

3. Toy Train Sets

In 1969 the hotchpotch of beginner's sets was brought together under the heading Toy Train Sets and given a fresh set of numbers – RS1-RS6. They were also provided with the new integral wagons developed especially for the sets and the new Continental and Swedish styled 0-4-0 locomotives.

In 1970 it was decided to leave this range of sets unchanged as there were large stocks of unsold sets or surplus packaging. Instruction was given that no new packaging was to be ordered unless sales could be completed by the end of November 1971.

The sets were sold in batches of 12 while other sets were mainly sold in threes.

RS1 C/W Continental 0-4-0 (R854), two keys, two integral wagons (low-sided and high-sided) & circle of Super 4 track.

Described as a 'Clockwork Steam Freight', it was introduced in 1969 and almost a quarter of a million were made over a period of three years; a third of them going abroad. The catalogue illustrations showed the tank engine as red, the low-sided wagon green and the high-sided wagon red, but the Canadian catalogue showed the high-sided as blue. The set originally retailed at £1-2s-11d (£1.15) but the price rose sharply to £1.95 in just two years.

RS1M This was a special export version sent out to Sweden, Denmark and Lebanon, and possibly other places. Some 8,300 of these were made, the last of them being sold in the UK in March 1970.

RS1E Electric Continental 0-4-0 (R852), two integral wagons (high-sided & low-sided), circle of Super 4 track, battery control box & power clip.

Here we had a 'Steam Freight' battery-operated version of RS1 with a blue R852 tank engine and a battery box provided. Although 700 of the sets were assembled in 1969 it was not available until the following year. During 1970 and 1971 5,000 of these sets sold in the UK and a further 3,000 went abroad. The set was withdrawn the following year. The example in my own collection has a black loco numbered '7744' with a red chassis and can motor. The high-sided wagon is blue and the low-sided wagon yellow.

RS1EN A set sold in Australia during 1970 and 1971 which was possibly the same as RS1E but locally assembled and boxed. It sold alongside RS1E but for A$0.50 less.

RS2 C/W Continental 0-4-0 (R854), two keys, two integral coaches & circle of Super 4 track.

The 'Clockwork Steam Passenger Train' was also

introduced in 1969, priced £1-7s-11d. It was available for only two years in the UK lists but longer overseas. The locomotive was red and the coaches yellow. About 60,000 were made of which about a third were sold abroad. The set may have been withdrawn at the start of 1971 despite being featured in the catalogue for that year.

RS2M 4,000 of this special export variation are known to have been sold in Denmark in 1969.

RS2E Electric Swedish-style Diesel 0-4-0 (R853), four integral vehicles (low-sided wagon, tanker, high-sided wagon and coach), a circle of Super 4 track & a battery box.

This was a completely different set to RS2. Called the 'Mixed Traffic', it was first manufactured in 1969 but not sold until the following year when about 1,700 were delivered in the UK and 1,150 overseas. In 1971, only some 1,000 odd were sold in Britain, 144 of them going for mail order, and a further 1,070 sold abroad.

RS2EN Available in Australia in 1970 and possibly assembled there. It was replaced by RS2E the following year.

RS3 C/W Continental 0-4-0 (R854), two keys, four integral wagons (low-sided wagon & Minix car, low-sided wagon & Minix van, high-sided wagon, tanker), oval of Super 4 track and a level crossing.

The 'Large Clockwork Steam Freight Train' was another 1969 introduction. 83,600 were made the first year of which only 12,500 went abroad. A further 1,000 were made specially for Woolworths although it is not known how these differed from the normal issue sets. These seem to have sold well as it appears that a further 10,300 were produced for that customer the following year; at least, that quantity of a set coded RS3W left Margate during 1970. The same year also saw a further 42,600 of the nor-

mal set sold in Britain and 6,000 abroad. Another 47,000 RS3 sets sold in 1971 of which we know that 20,000 went for mail order. The set is similar to RS100.

RS3W *See* above.

RS3A This was to be found in the Australian price list in 1969 and is listed under H. Australian Sets below.

RS4 Electric Continental 0-4-0 (R852), two integral coaches, a circle of Super 4 track, an RP43 battery box, power clip & level crossing.

The loco was bright blue with a red chassis/motor block and without any markings, or it was black and numbered 7744. The coaches were the standard yellow and the battery box was grey. The contents were dropped into a pale blue vacuum-formed plastic tray.

16,000 of the 'Electric Powered Steam Passenger Train' set were made in 1969 and over 10,000 of these sold although it was not available until the summer. The following year the set was re-coded RS4E to conform with other battery sets but of the 13,000 available that year only 2,600 sold. A further 6,100 were sold in 1971, 1,400 of them going for mail order. About a quarter of the sets ended up overseas. It was withdrawn at the end of the year.

RS4E *See* above.

RS4A Assembled in Melbourne in 1969 but later re-coded RS4E for sale along side imported sets of the same coding but costing the purchaser A$0.20 less! *See* H. Australian Sets.

RS5 Electric Swedish Diesel 0-4-0 (R853), three integral wagons (low-sided & Minix van, high-sided wagon, tanker), circle of Super 4 track, an RP43 battery box & power clip.

Called a 'Diesel Freight Train', the locomotive was shown as blue and the wagons green, red and yellow in the catalogue illustration. It arrived halfway through 1969 when 15,000 were made of which about 10,000 sold that year, mainly in Britain. The remainder of the stock sold in the UK between January and May 1970.

RS5A An Australian-assembled version of the same set which was available locally in 1969 and 1970. *See* H. Australian Sets.

RS6 Electric Swedish Diesel 0-4-0 (R853), four integral wagons (low-sided & Minix car, low-sided & Minix car or van, high-sided, tanker), oval of Super

R852C 'Go Electric' conversion set, 1971.

4 track, RP43 battery box, power clip & level crossing.

The 'Large Diesel Freight Train' was illustrated with a blue locomotive and wagons of green, blue, red and yellow respectively but the only example I have seen had a red loco with a black roof and carrying the number 4718. The two low-sided wagons (blue and yellow) each had a Minix car, the high-sided wagon was red and the tank was yellow without printing. The level-crossing base was grey but the battery box was yellow. The box tray was a pale blue and vacuum formed.

15,000 were made in 1969 and it was in the shops by the summer. 6,000 of these did not sell until May 1970. The retail price was £3-7s-6d.

RS6A Again this was an Australian-assembled version of the RS6 available in 1969 and 1970. *See* H. Australian Sets.

R852C Electric Continental 0-4-0 (R852), battery box controller.

This was a special conversion set allowing one to convert a clockwork set into an electric one. It arrived in March 1971 and about 1,250 were sold in Britain and a further 500 abroad. The retail price was £1.65. The only examples I have seen contained a black locomotive marked '7744' in white. It remained available in 1972 although there had been a proposal to replace it with an R355 0-4-0T and an RP40 battery box.

4. Plastic Track Sets

This was a further attempt at reducing the cost of starter sets and appealing to an ever-younger clientele. The sets, introduced in 1970, were made for the Wholesale Department for sale to small retailers rather than model shops. Each carried the prefix 'RB' and almost all were sold abroad.

Most sets contained the 'Continental' 0-4-0 loco which was SR green with red wheels. They all had plastic track which was grey until November 1970 when it changed to black. At this time the locomotive changed to red in some sets. The wagons were the 'integral' type in their usual bright toy-like colours. The coach was also the yellow four-wheeled integral type.

The most important feature of the RB sets, to collectors today, was the four special container wagons that they spawned. A long integral flat wagon was used as a carrier for a standard container moulding which carried printed stickers on their sides. The containers (especially the Coca-Cola one) are now much sought after by collectors.

A number of the standard sets were produced with a different lid to the box, possibly indicating that they were 'Miniville' rather than 'Tri-ang' sets. These were given an 'M' suffix.

For 1970 the RB1, RB2 and RB3 were made. It was intended to replace these in 1971 and so instructions were given to get rid of all made-up sets and packaging by 10th December 1970. RB4 and RB5 were the 1971 replacements.

RB1 Clockwork Continental 0-4-0 (R854), R710NP integral high-sided wagon, R712NP integral low-sided wagon & circle of grey R412 track.

In 1970 34,000 of this 'Goods Set' were made, 90% of which sold during the year and 65% of the sets were sold overseas. The loco was green and the set retailed for 19/6d.

RB1M This was almost certainly the same as RB1 but with a different lid indicating that it was in the Miniville range. 39,000 were made in 1970 and almost all sold abroad that year.

RB1MS 6,000 of this variation were also made in 1970 and all were sent overseas. How it differed from RB1M is not known.

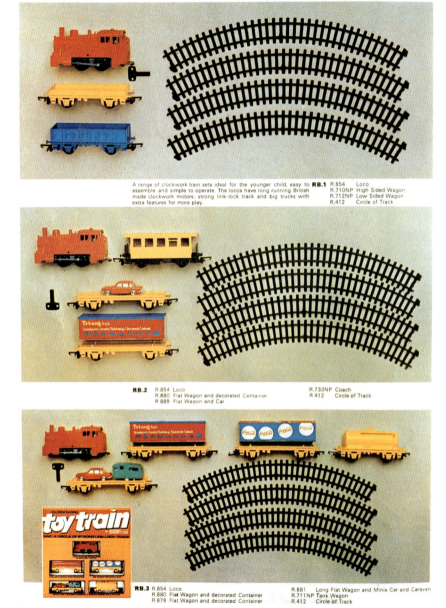

A range of clockwork train sets ideal for the younger child, easy to assemble and simple to operate. The locos have long running British made clockwork motors, strong link-lock track and big trucks with extra features for more play.

RB.1 R.854   Loco
R.710NP   High Sided Wagon
R.712NP   Low Sided Wagon
R.412   Circle of Track

RB.2 R.854 Loco
R.880 Flat Wagon and decorated Container
R.889 Flat Wagon and Car
R.733NP   Coach
R.412   Circle of Track

RB.3 R.854 Loco
R.880 Flat Wagon and decorated Container
R.879 Flat Wagon and decorated Container
R.881   Long Flat Wagon and Minix Car and Caravan
R.711NP   Tank Wagon
R.412   Circle of Track

Catalogue illustration of the RB1, RB2 and RB3 clockwork sets with plastic track, 1970.

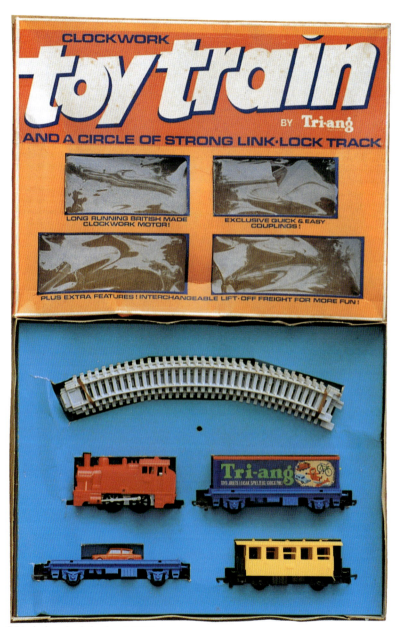

RB2 clockwork wholesale set, 1970.

RB2 Clockwork Continental 0-4-0 (R854), R733NP coach, R889 integral long flat wagon with Minix car, R880 integral long flat wagon with 'Tri-ang Toys' 30' container & circle of grey R412 track.

This 'Mixed Traffic Set' was also released in 1970, this time priced £1-4s-6d. 30,000 were made, all but 1,150 of which went abroad.

RB2S Presumably a variation of RB2. A little over 4,000 of these sets were made and sold overseas in 1970.

RB2M Probably an RB2 set in a Miniville box.

RB2MS Presumably the same as RB2S but in a Miniville box. Nearly 5,000 were made in 1970 of which 3,500 went abroad.

RB3 Clockwork Continental 0-4-0 (R854), R880 integral flat wagon with 'Tri-ang Toys' 30' container, R879 integral long flat wagon with 'Coca-Cola' 30' container, R881 integral long flat wagon with a Minix car and caravan & R711NP integral petrol tank wagon and circle of grey R412 track.

The Coca-Cola container had a red roof and blue sides with four Coca-Cola discs on each side and was carried on a yellow wagon. The Tri-ang Toys container had a blue roof and red sides and was also carried on a yellow wagon. A little under 24,000 were made in 1970 of which over half were sold overseas in the first year. All had been sold by about the middle of 1971. Called a 'Large Goods Set' it sold for £1-9s-6d.

RB3M A Miniville version of the RB3 set which it had replaced by the time the January 1971 trade price list was published.

RB3T This appeared in the trade price list in September 1971 but details of it are not known except that it retailed at £1.57.

RB4 Clockwork Continental 0-4-0 (R854A), two integral wagons (high-sided R710 & low-sided R712) & circle of grey R412 track.

The set contained the same items as the RB1 set had the year before but it was packed in the new type of box with a window lid. The loco was in SR green with red wheels and the number '7321' on its sides.

About 10,000 were sold in 1971; almost all abroad. It had been planned that one section of the track would be a special moulding in red incorporating a platform and level crossing. A signal and a pair of lifting booms were to have been moulded in yellow. These features were intended for both this and the RB5 set but the idea was abandoned in May 1970.

RB4M A Miniville version of the RB4 of which 16,000 were sold throughout 1971, mainly in the UK. 2,000 of these were sold for mail order. It retailed at £1.3s.0d.

RB5 Clockwork Continental 0-4-0 (R854A), R887 integral low-sided wagon with a Minix car, R875 integral long flat wagon with a 30' 'Coca-Cola' container, R398 curved platform with a signal, R397 level crossing & circle of grey R412 track.

The set was sold under the Toy Train label instead of Tri-ang Hornby. The loco was made to a new specification used in 1971 and had an SR green body with '7321' on the tanks. The Coca-Cola container wagon is particularly sought after by collectors and was the 30' version of the container carried in a pale yellow integral flat wagon of the same length. The other wagon was moulded in a bright blue known as Rainbow blue. The platform was a lemon yellow, known in the factory as 'BP yellow' while the base of the level crossing was in a colour known at Margate as 'Toy red' and the gates were plain white and not printed (*see* RB4 above). The set first appeared in July 1971 and 11,000 were sold that year, mostly in Britain. The recommended retail price was £1.50.

RB5M As RB5 but with a Miniville lid to the box. It retailed at £1.10s.0d in 1971.

RB6 Clockwork Wild West 0-4-0 (R873 & R874), coach R733, four plastic figures (S8569 standing Indian, S8570 standing cowboy, S8571 riding Indian and S8572 riding cowboy) and circle of grey R412 track.

A sample set was on display at the Harrogate Toy Fair in January 1971 and July was the deadline for the sets to be ready for export. Although it had been thought to have real export potential, very few sold abroad. It seems that about 25,000 sets were made that year of which 800 went for mail order and most actually sold in 1972. In 1972 the set was renumbered R694 but I can find no record of any being made under this number.

Despite the large quantity made, few have survived and the loco, which was a hybrid between the Top Tank 0-4-0 and Davy Crockett, is one of the most sought after by the more serious collectors. The loco is described and illustrated in the 'Locomotives' chapter of this book. The set, which was not listed in early trade price lists that year, was sold to dealers in packets of six to retail at £1.50 per set. It came in a yellow cardboard tray in a card box with a window in the upper surface.

The Cowboys and Indians were from a range made in h.i. polystyrene by the Medway Tool & Moulding Co. Ltd of Paddock Wood, Kent, an associate company. In December 1970, Mr Tickle of this company wrote offering prices for the figures both painted and unpainted:

Mounted Cowboys and Indians
    Unpainted – 8/4d per 100
    Painted – 23/8d per 100
Standing Cowboys and Indians
    Unpainted – 4/2d per 100
    Painted – 12/- per 100

Delivery of unpainted figures could start in seven days but the painted ones would take four to five weeks. Cowboys and Indians had to be ordered in equal quantities because they were made in the same mould. They also had a peg under the base and the tools had to be modified to remove these. The 1971/72 Development Programme indicated that each set was to have had six figures and that they were to be painted. The number, however, was later cut to four and it is not known whether the set ended up with the painted or unpainted versions.

RB7 This did not appear until 1972 and so is outside the scope of this book. It should, however, be recorded that it had the same contents as RB5 but had System 6 rather than plastic track.

C. MAINSTREAM SETS

This section covers those sets which catered for children who wanted something more realistic.

Although not necessarily sold in the same numbers they were much better promoted than the toy sets. It was important for the company to maintain a high level of sales on these as they had a greater profit margin and their sale was more likely to lead to the purchase of accessories such as stations, lineside equipment and additional track and rolling-stock.

The following sets are known to have been produced during the period 1965-1971. Some Tri-ang Hornby sets of this period and type are a continuation of earlier Tri-ang Railways sets.

RS7 0-6-0DS (R152), utility van, MkI 1st class coach & oval of Super 4 track.

The train was in BR blue livery (although the diesel might turn out to be green) and made for 1969 only, retailing at £4-19s-11d. A little under 5,000 were made and orders to cover these were received during the year. The set should not be confused with set RS7 described in **Volume 1**. It was shown as a 'new' set in the May 1969 Trade Price List and did not appear in the catalogue.

RS8 Deeley 0-6-0 (R251) and tender, two maroon scale coaches, uncoupler & oval of Super 4 track.

'The Midlander' set arrived in 1965 and was last listed in 1968. About 24,000 were made in all and about 10% of these went overseas. The set box had an attractive pictorial lid. The loco was in Midland maroon livery. Not to be confused with set RS8 in **Volume 1**.

RS9 Co-Co Type 3 (R751), three Pullman cars, uncoupler & oval of Super 4 track.

During 1966 and 1967 about 20,000 'Inter-City Express' sets were made. Not to be confused with set RS9 in **Volume 1**.

RS10 'Nellie' (R355B), five wagons, Super 4 track and RP3E controller.

This was an export set of which about 750 were made in 1965. Sales were so slow that after two years there were still 220 unsold. One set has turned up with a sticker on the box saying "Distributed in Singapore solely by the Orchard Store." *See also* **Volume 1**.

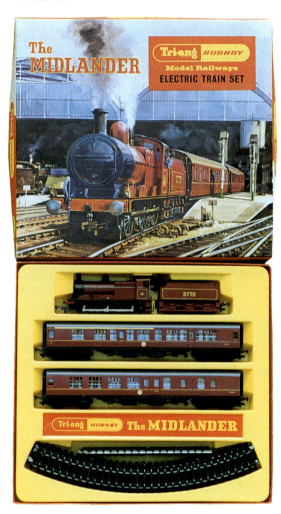

RS7 set which did not appear in the catalogue, 1969.

RS8 'The Midlander' set, 1966.

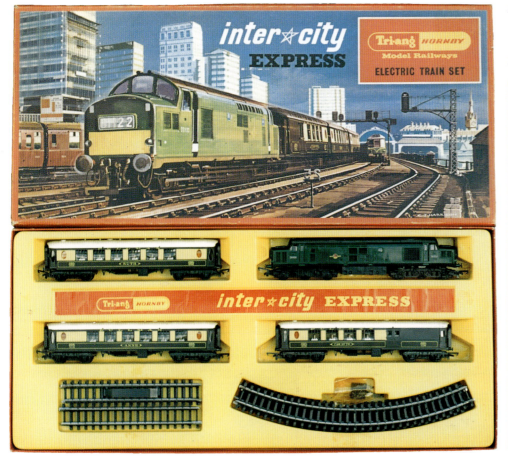

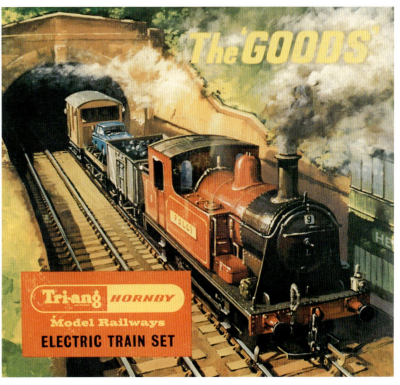

RS11 'Polly' (R355R), three wagons (mineral, flat wagon and Minix car, short brake), uncoupler, oval of Super 4 track and RP3 control unit.

This set is very similar to the RS24 'Pick Up Goods' set. Production of the set seems to have started late in 1965 with the main sales in 1966 and 1967. A little over 10,000 were made. Not to be confused with the other RS11 set described in **Volume 1**.

RS11A About 250 of these were made specially for Australia.

RS16 Battle Space Jinty (R558S), two wagons (R562 plane launcher, R568 assault tank transporter), uncoupler, oval of Super 4 track & 12 Commandos.

About 24,000 of this action set, called 'Strike Force 10', were made in 1966 and 1967 with, it seems, a further 1,500 sets being made in 1970. All but about

800 sold in the UK. (Also listed in 'Battle Space' section of 'Wagons' chapter)

RS17 Battle Space 0-4-0DS (R756), two wagons (R566 satellite launcher & R567 radar car), uncoupler, oval of Super 4 track & 12 Commandos.

This was a less popular set called 'Satellite Train' and, although it was available for the same five years, only 10,000 were made. These were produced in 1966 and 1967 and it was dropped from the catalogue in 1971. (Also listed in 'Battle Space' section of 'Wagons' chapter)

RS24 'Nellie' (R355B), three wagons (fish van, open wagon and short brake), R487 clip and oval of Super 4 track.

The 'Pick-Up Goods' set was a popular set from its introduction in 1962 until its replacement in 1970. In 1968 it was modified to become RS24A. About 30,000 of the original RS24 sets were released in Tri-ang Hornby packaging. Also covered in **Volume 1**.

RS24A This was the same as RS24 but 'Nellie' was replaced by 'Polly' and the wagons were a mineral, a ventilated van and an ER brake van. About 7,000 RS24A sets were made. It was discontinued in 1970.

RS24A *See also* Australian Sets.

RS29 B12 (R150) two scale maroon and cream coaches (R626 & R627), R487 clip, R488 uncoupling ramp and oval of Super 4 track.

Introduced in 1963, about 7,500 of the 'Holiday

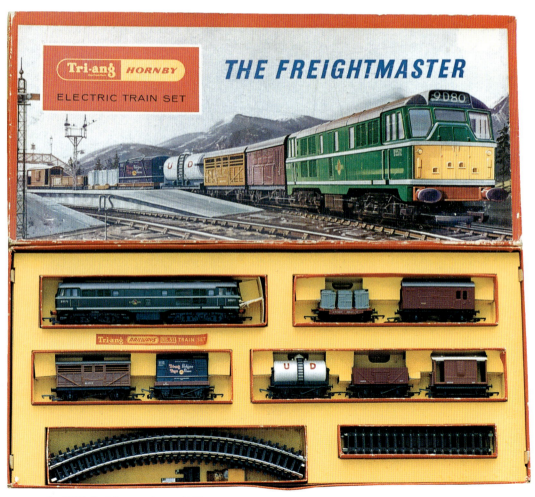

The ever-popular RS51 'Freightmaster' set, c.1966.

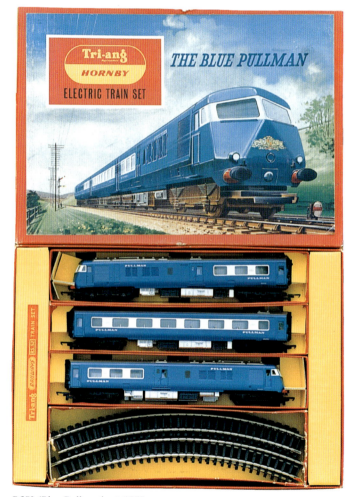

RS52 'Blue Pullman' set, 1965.

Express' sets were made in 1965 and 1,500 in 1966. How many of these were sold in Tri-ang Hornby boxes is not known. Also covered in **Volume 1**.

RS33 USA Special Set.

In 1970 about 650 of these were made. These were possibly the 'limited edition' sets assembled for AMRO (standing for American Rovex) and marketed by Mike Tagee as a corporate gift. The sets were specially packaged in America, the box being described as "a beautiful simulated wood carrying case". They contained the 'Flying Scotsman' in LNER livery, five Pullman cars mounted in the lid of the box, a station, 47 sections of track (making two linked ovals and two sidings), a power pack and the track plan book. They sold for $69.95. Not to be confused with the RS33 set described in **Volume 1**.

RS34 Double-Ended Diesel (R159), four TC blue
Series two passenger cars (R444, R445, R446, R447), R487 clip, R488 uncoupling ramp & oval of Super 4 track.

The production of this set for general distribution ceased in 1964 and is covered in **Volume 1**, but it remained available overseas.

RS34A A batch of a little over 1,000 'Blue Streak'
sets was made in 1965 for Australia. The coaches were probably marked 'TransAustralia'.

RS34F *See* 'French Sets'.

RS50 Green 0-6-0DS (R152), three military wagons
(R341 searchlight, R343 four rocket launcher, R249 exploding car), R487 clip, R488 uncoupling ramp & oval of Super 4 track.

Over 5,000 of 'The Defender' set were made in 1965 and packed in Tri-ang Railways boxes. Also covered by **Volume 1**.

See 'French Sets'.

RS51 A1A-A1A Diesel (R357), seven wagons (R15
milk tank, R113 drop-side wagon, R122 cattle wagon, R123 horse box, R340 3-container wagon, R561 Tri-ang Container wagon, R16 brake van), R487 clip, R488 uncoupling ramp & oval of Super 4 track.

The very popular 'Freightmaster' set continued in production throughout the Tri-ang Hornby period although it was renumbered RS651 in 1970 when equipped with System 6 track. About 12,000 were made in 1965, 16,000 in 1966 and 21,000 in 1967. In 1965, a special batch of about 800 was made for the Australian market but how these differed from the normal sets is not known. The 1968 set had a locomotive in experimental blue and that year more than 20,000 were made. In 1969 the loco changed again, this time to BR blue. 14,000 were made that year. For the earlier sets *see* **Volume 1**.

RS51A This set was assembled in Australia in 1968.
See Australian Sets.

RS52 Blue Pullman cars (R555 + R556), R487 clip,
R488 uncoupling ramp & oval of Super 4 track.

The 'Blue Pullman', also covered in **Volume 1**, was another very popular set which remained in production until 1970 although undergoing a number of changes. Over 18,000 were made in 1965 and a further 32,000 during the next two years. In 1968 the power cars were given yellow fronts and a further 17,000 sets were made that year. In 1969 the livery changed to grey and blue and 13,000 sets were made, the remnants of that year's production being sold in 1970.

RS52F *See* 'French Sets'.

RS52MP A special set made in 1967 for Myers of
Australia. Only 210 were made. It is not known whether this was a variation of the 'Blue Pullman' of whether the suffix 'P' stands for 'passenger' and the 'G' on the next set for 'goods'.

RS52MG Only 110 of this set were made for Myers
in 1967.

RS61 'Old Smoky' – *see* **Volume 1** for details.

The 'Old Smoky' set's main sales were in 1965 when about 5,000 passed through the stores. As it was dropped the next year it seems unlikely that it appeared in Tri-ang Hornby boxes. It was still available in Australia in 1966 where it is thought that some sets were assembled.

RS62 Jinty (R52S), two car transporters (tierwag)
each with six Minix cars, ER brake van, R487 clip, R488 uncoupling ramp & oval of Super 4 track.

The 'Car-a-Belle' set was a child of the transition year, 1965, and I do not know whether any came out of the factory in the old Tri-ang Railways boxes. The moulded plastic box in-fill may be found made of either white or yellow plastic but early sets had a traditional cardboard inner. Both forms may be seen in our picture.

The set remained in production until 1969 and available until 1970 when stocks dried up before the summer. Production amounted to almost 26,000 in 1965 and about 30,000 in total over the next four years. The set is also mentioned in **Volume 1**.

RS62A A batch of 505 sets were made for Australia
in 1966. These may have been minus a loco and track to be added on arrival.

RS89 Blue 0-6-0DS (R152), two wagons (bogie flat
with three Minix vans, bogie flat with two Freightliner containers) & oval of Super 4 track etc?

This set first appeared in 1968 when over 11,500 were made. A further 5,000 were made in 1969 and final stocks of the set were sold during the spring of 1970. Most of the sets were sold in Britain but about 1,500 went abroad.

RS62 'Car-a-Belle' sets; the early one with card decking and the later one with plastic decking, 1965/69.

RS89G A little under 1,000 sets were made to a special order for Australia in 1969 and were recorded under the name 'Ideal Freight Set'.

RS90 BR 'Albert Hall' (R759) three Pullman cars, level crossing, R487 clip, R488 ramp & oval of Super 4 track.

The 'Pullman Express' was made in batches in 1968 and 1969. Almost 8,700 were made the first year and about 6,000 the second. All but about 1,500

sold in the UK. Some of the surplus boxes at the end of 1969 were relabelled and used with a new inner tray for the 1970 RS604 'Night Mail Train Set'.

RS92 Battle Space Turbo Car but other contents not known.

This set was listed in 1968 as the 'Battle Space Turbo Car Race' but only 580 sets were ordered and so it appears not to have been made.

RS93 'Blue Train' Set.

This is thought to have been made for the South African market but I have no details of the contents. Only one batch of about 1,200 was made in 1968 and 1,057 were ordered from abroad. A little over 300 sets remained in store at the end of the year suggesting that about 900 were sent. The fate of the remaining sets is unknown.

RS94 South African Freight Set.

Made in 1968, this was obviously a partner for RS93. About 1,250 were made and orders for 1,015 were received. About 360 were left in store at the end of the year, again suggesting that around 900 were sent.

RS94M Myers Special Set.

In 1968 a special version of the RS94 set was sold to Myers of Australia and all 309 sets made for this customer were sent.

RS 100 A clockwork special set made for Waltons of Australia in 1971, 3,000 were sent but the contents are not known.

RS101 TC Diesel A unit in red (R0550), three TC 2nd series passenger cars in silver and red 'Transcontinental' livery (diner R4430, coach R4400, observation car R4410, power clip, uncoupling ramp and marker & large oval of System 6 track.

The coaches of the 'Transcontinental Streamliner', which were incorrectly numbered in the catalogue,

were in a sprayed silver livery with the red carried above the windows. The loco, which had a working headlight, was bright red and, like the coaches, inscribed with 'Transcontinental'. It also carried the number '1404'. The set was made for Australia in 1971 and that year 1,275 were exported and the remaining 600 sold in the UK. They were packed in a No3 size box.

RS101A Called 'The Overlander Set', this was also made for Australia, but not until 1972 (by which time it had been renumbered R590) and so is outside the scope of this book. It had the same contents as R101 except that the Buffet Car was replaced with an operating mail car. The set was also available in the UK that year.

RS102 Revamped TC 4-6-2 (R54NS), five wagons (Polysar tanker (R3490), container wagon with three ACT containers (R7344), yellow stock car (R1260), red pulp wood car (R2350), ER brake van (R16), uncoupling ramp and marker, power clip & oval of System 6 track.

The set was called the 'Express Freighter' and was made for export to Australia which possibly explains the strange mix of wagons. It was also available in the UK and was packed in a No3 box. Around 1,650 sets were made and sold in 1971, two thirds of them going overseas. The original intention had been for the pulp wood car and stock car to be in Australian liveries and the tank wagon to be finished in Shell livery, but the small numbers required meant that this was not economically viable at the time. The set was offered again, unchanged, in 1972 and was renumbered R592 the following year.

RS103 BR Jinty (R52S), four wagons (C&O refrigerator car (R1290), Wabash cement car (R1370), Johnny Walker whisky grain wagon (R648), ER brake van (R16), power clip, uncoupling ramp and marker & oval of System 6 track.

Another strange mixture designed for the

Australian market called the 'Steam Freighter'. The loco was the standard BR black version with smoke and the set was packed in a No2 box. Either the Johnnie Walker or the Vat 69 bulk grain wagon was used. In 1971 1,100 were dispatched from the works, 65% of them selling in the UK. For 1972 it had been intended to revamp the set with a Double-Ended Diesel fitted with new Australian bogies instead of a steam locomotive but it was listed as a steam set that year. In the event it was not made.

RS103A This was a special set made for Walton of Australia in 1971. 750 were dispatched that year. It had the same contents to RS103 but with a slight variation.

RS104 Red 0-4-0 Dock Shunter (R253), three wagons (SR brown open wagon (R10A), three container wagons (R340), ER brake van (R16), power clip, uncoupling ramp and marker & circle of System 6 track.

This was called the 'Husky Freighter' and was packed in the old RS614 box but with a new vacuum-formed tray. In 1971 a little under 1,000 sets were dispatched; almost 600 of them going abroad.

For 1972 the set was to have an R112A drop door wagon instead of the R10A open wagon and the loco would have been black, but none were made.

RS105 Double-Ended Diesel (blue and yellow) (R1590), five wagons (SR brown open wagon (R10A), ventilated van (R11), mineral wagon (R243), UD tank (R15), ER brake van (R16), power clip, uncoupling ramp and marker & oval of System 6 track.

Initially the loco had VR transfers to use up stocks and when these ran out, TR ones were used. This set was known as the 'Diesel Freighter' and sold equally well at home and abroad in a No2 box. Almost 1,300 left the factory in 1971, the only year it was made.

RS106 Myers Special Set.

This was another set made for Myers of Australia, this time in 1971. About 3,600 were sent.

RS600 Blue 0-6-0DS, two wagons (container wagon with Containerway and Pickford's containers, container wagon with three assorted 20' containers), power clip, uncoupling ramp and marker & oval of System 6 track.

The 'Junior Freightliner' was designed for those interested in container wagons. It reached the shops in February 1970 and of around 4,500 sets dispatched that year about 600 went overseas. None were made in 1971.

RS600A This was probably a special order for one customer but only 142 were supplied, all in 1970.

RS101 'Streamliner' set, 1971.

RS601 BR Jinty (RS52S), four wagons (red London Brick, bogie flat with three Minix vans, Johnny Walker whisky grain wagon, ER brake van), power clip, uncoupling ramp and marker, oval of System 6 track & level crossing.

The 'Steam Freight' set was also made in 1970 and around 8,500 were dispatched after it became available in February that year. It sold for £5-19s-6d. It had been intended to continue the set into 1971 with an LMS maroon Jinty but instead it was replaced by RS613 which had a small bolster wagon with car instead of the bogie bolster with three vans.

RS601A About 190 sets were made as a special order for one customer and dispatched during 1970.

RS602 Blue Hymek (R758), two wagons (container wagon with assorted 20' containers, container wagon with Fyffes and Manchester Liners containers), operating Freightliner depot crane, Tartan Arrow container, cti container, power clip, uncoupling ramp and marker & oval of System 6 track.

This was the 'Senior Freightliner' set which arrived in the shops in April 1970, 7,500 being made during the year. In January 1971 it appeared in the new style No3 lidded box and a further 7,000 were

produced of which 3,000 went for mail order. It had been planned to change the loco to green from January 1972 but the idea was abandoned in August 1971. However the set was renumbered R593, possibly only for export purposes, around 1972.

RS602A In 1970 approximately 150 sets were produced to a different specification for a particular customer, possibly with a different assortment of containers.

RS603 DMU in blue (R157), power clip, uncoupling ramp and marker & oval of System 6 track.

RS105 'Diesel Freighter' set, 1971.

RS602 'Senior Freightliner' set, 1970/71.

This was called the 'Local Diesel Set' and was another 1970 introduction, seeing the welcome return of the diesel multiple unit thanks to public demand. It was available from February and over 5,000 were sold. Production was to have continued into 1971 but in June 1970 the set was deleted from the 1971 programme. The set retailed at £4-19s-6d.

RS604 Blue Co-Co Type 3 (R751), three blue/grey coaches (TPO, lit MkII brake, sleeping car), power clip, uncoupling ramp and marker & oval of System 6 track.

This 'Night Mail Set' was scheduled for release in March 1970 but, due to tooling difficulties with adapting the lineside mechanism so that it could be used with System 6 track, the set did not appear until April. It retailed at £8-15s-0d and 6,350 sets sold in the first year and, in the new No3 box with a lift-off lid, a further 6,600 left the factory in 1971; over 3,000 of them for mail order. During the two years about 1,050 went overseas. It had been planned to fit lights to the loco for 1971 and renumber the set RS604A but this was dropped, presumably because it would have added further to the cost of the set at a time of rising inflation. As a Hornby Railways set it was renumbered R591 possibly for export.

RS604A In 1970 about 400 sets were sold under this code as a special order.

RS604B Another special order was met in 1970 consisting of about 380 sets.

RS605 LNER 'Flying Scotsman' (R855), two Thompson coaches (LNER composite, LNER brake), Pullman 1st, level crossing, signal, power clip, uncoupling ramp and marker & large oval of System 6 track.

The set arrived in April 1970 and approximately 8,500 sets were made in this form. When first dis-

RS603 'Local Diesel' set, 1970.

RS604 popular 'Night Mail' set, showing two versions of the box, 1970/71.

played at the Tri-ang House Toy Fair some traders expressed the view that a train of all LNER coaches would be more popular and so during the year production switched to RS605A.

RS605A The 'Flying Scotsman' set was available in the latter half of 1970 and differed from the RS605 set in having the Pullman car replaced by another LNER composite coach. As we have seen, this was a suggestion from retailers visiting the Tri-ang House Toy Fair that year. Almost 3,000 of these sets were sold. It was replaced in 1971 with the RS608 set which had only two coaches.

RS605B About 600 of these were made in 1970 of which 20 went overseas. Possibly a mail order batch.

RS605C 875 of these were sold in Britain in 1970, possibly to another mail order company.

RS605D This was a special set ordered for the USA in 1970. Only 48 sets were sent.

RS606 B12 in BR black (R150NS), four wagons (flat with Minix car, UD tank, mineral wagon [or R10 or R112], WR brake van [or R16]), power clip, uncoupling ramp and marker & oval of System 6 track.

The set was announced to retailers in August 1970. It was the first set to feature steam exhaust sound and was used in the Tri-ang Hornby national television advertising campaign which started in September that year. It was marketed as the 'Chuff-chuff Puff-puff Train Set' but for 1971 it was re-launched in January as the 'Express Goods Set'. In mid 1970 it was not known whether the set would become a normal set in the 1971 range and in case this was to be, a conventional box was used and the instruction was given to assemble the sets as orders came in rather than build up a stock of them.

The loco was already fitted with smoke and had been boxed for solo sales. To make up the RS606 set,

a supply of them was withdrawn from the finished goods store and had a modified tender chassis fitted, giving the loco exhaust sound. The removed tender chassis were dispatched to the R866S green B12 production line for re-use even though it meant removal of the tender wheels for painting green.

The wagons were also withdrawn from stores still packed in their window boxes and a new box lid had to be designed with a four-colour label. For 1971 the same No2 type box was used with the same label but a new definitive inner was developed.

It had been planned to use the R16 BR(ER) brake van but, as there were 5,000 of the R124 WR brake van in stock to dispose of prior to the introduction of the GWR version, this was chosen instead. The change came too late for the catalogue picture. Two other wagons that were being dropped were the R10

open wagon and the R112 drop door wagon and it was decided in June 1970 to use up the stocks of R10 first followed by the R112 in place of the R243 mineral wagon. The mineral wagon would then be used once these surplus stocks had been used up.

The track consisted of eight R605s and four R600s packed in a window box similar to the Canadian special train pack. The uncoupler and power clip were packed loose. A catalogue was also included in the box, together with a new leaflet prepared specially to include details of maintenance of the noise unit.

All 5,000 sets made in 1970 sold in the UK, presumably with the stock in their individual boxes as described above. The following year, 4,300 were sold in Britain, of which a number went to a mail order house, and a further 900 sold overseas. These 1971 sets would have been conventionally packed.

RS605 (1970) 'Flying Scotsman' sets.

RS607 M7 tank in SR livery (R868A), two scale BR coaches in SR livery (a composite and a brake), power clip & circle of System 6 track.

In June 1970 it was decided to produce two sets based on the R868 M7 tank. One was to have the tank in SR green with two Caledonian coaches in SR livery and the other would be similar but loco and coaches would be LMS maroon! Only the Southern set, called the 'Local Passenger Set', appeared and for this the BR Mk1 coaches in SR green were used. Scheduled for January, 4,200 were sold in the UK in 1971 and a further 720 overseas. They were packaged in the new No2 square, lidded, set box and the loco was the light green R868A version (ie. without firebox glow).

RS608 LNER 'Flying Scotsman' (R855N), two LNER Thompson coaches (composite and brake), power clip, uncoupling ramp and marker & oval of System 6 track.

There were plans to replace the larger 605A set of 1970 with an LMS one using the maroon Coronation but the idea was dropped in favour of a smaller 'Flying Scotsman' set. It was released in February in the new No2 lidded box. Nearly 10,000 of this smaller set sold in the UK in 1971 and a further 750 abroad. Of the British sales approximately 3,500 went for mail order. The recommended retail price was £8.95 instead of £8.47 for the larger set (RS605A) just one year before. Inflation had arrived!

RS609 LMS 'Princess Elizabeth' (R258NS), two LMS coaches (composite and brake Caledonians), power clip, uncoupling ramp and marker & oval of System 6 track.

This was a welcome return of the Princess locomotive which had not featured in a train set since the early 1960s and was now fitted with smoke and exhaust sound. The 'Express Passenger Set' was released in February 1971 in the new No2 lidded box and nearly 10,000 sold that year, mostly to a UK

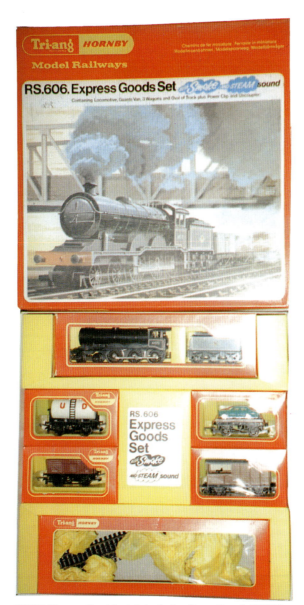

RS606 'Express Goods' set, introducing Steam Sound, 1970/71.

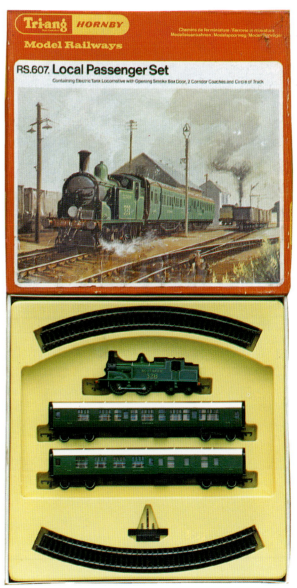

RS607 'Local Passenger' set, 1971.

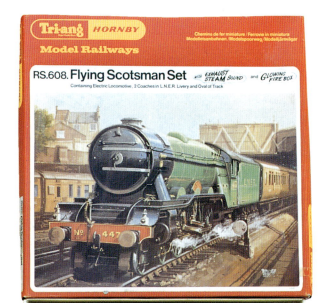

RS608 'Flying Scotsman' set, 1971.

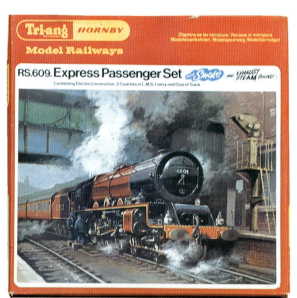

RS609 'Express Passenger' set, 1971.

market hungry for private company liveries. 1,700 went for mail order sales.

RS611 Green Industrial 0-4-0 (R355G), three wagons (drop-side, Express Parcels, ER brake van), circle of System 6 track & RP15 power controller.

It was planned in 1971 as an export version of RS612 'Local Freight' having an RP15 transformer/controller instead of an RP40 battery control unit. Although only the RS612 set was illustrated in the 1971 catalogue, reference is made to the set on the same page. It was not to be available in the UK or Australia.

The set was available from the second quarter of the year but only one set was sold and 875 were left in stores at the end of the year. It was to have had a black loco printed in white but this was later changed to the standard apple green one.

RS612   Same as RS611 but with an RP40 battery control unit instead of an RP15 power controller.

The 'Local Freight' was released in the second quarter of 1971 in the new type clear-view packaging. The box had to be strong enough for the power unit and required new artwork. Almost 2,000 were sold in Britain during the year and 500 overseas.

RS613   BR Jinty (R52S), four wagons (flat with Minix car, London Brick wagon, Whisky grain wagon, ER brake van), level crossing, power clip, uncoupling ramp and marker & oval of System 6 track.

Described as a 'Steam Freight', it arrived in the shops in January 1971 in the new standard No2 set lidded box. This was the same as RS601 of the year before but an R563 bogie bolster van transporter had been replaced by an R17C flat wagon with a Minix car. 5,700 were sold in the UK during the year and about 1,000 abroad. Approximately 1,000 of the sets sold in Britain were for mail order. The set became R594 in Hornby Railways days, possibly for export only.

RS614 Green Industrial 0-4-0 (R355G), three wagons (blue Insulfish, mineral, ER brake van), power clip & circle of System 6 track.

This was the 1971 version of the System 6 RS624 'Pick-Up Goods' set, slimmed down by the removal of two straight pieces of track and an uncoupling ramp. A new vacuum former was consequently required for the box inner and the window box inscription was altered. It was released in January and 7,700 sets were sold during 1971 retailing at £4.50 each. Almost half went overseas and the set continued the next year as a Hornby Railways set. In early sets the loco was not fitted with plated tyres; a feature not introduced until after January 1971.

It had been intended to change the loco to the black version (R359) from January 1972 but the idea was abandoned. However, the set was renumbered R595.

RS615 GN&SR Jinty (R377s), two coaches (maroon/white clerestory (R379), brown/yellow old time coach (R378), wayside station, power clip, uncoupling ramp and marker & oval of System 6 track.

The 'Railway Children Set' was a limited run. It was first advertised to retailers on 14 April 1971 with the announcement that it was due for delivery in the late spring. It was packed in the new No2 square box with a large lid label designed in the graphics department at Merton. This represented a scene from the film. In 1971 about 3,350 sold in Britain, 530 of them going for mail order. Only nine sold abroad as, presumably, the set did not mean much to children who had not yet seen the film on which it was based. Sets may also be found bearing the 'Hornby Railways' trademark.

RS616 The 'Take-a-Ticket Train Set' was planned for August 1971 but required a number of new components which were not ready in time. It arrived early enough to be initially packed in a Tri-ang

RS614 'Pick-up Goods' set, 1971.

Hornby box but will be covered in **Volume 3** as it did not arrive in the shops until 1972.

RS624 Green Industrial 0-4-0 (R355G), three wagons (blue Insulfish, mineral, ER brake van), power clip, uncoupling ramp and marker & oval of System 6 track.

The 1970 System 6 version of the 'Pick-Up Goods' set. Nearly 10,000 were sold during the year retailing at £4-7s-6d each. Production of this set was to have been extended into 1971 and its cancellation was taken after August 1970. It was replaced by RS614 in 1971.

RS651 Blue A1A-A1A diesel (R357), seven wagons (horse box, cattle wagon, UD tanker, Tri-ang container, three container wagon, drop-side, ER brake van), power clip, uncoupling ramp and marker & oval of System 6 track.

Here was the System 6 version of the ever popular 'Freightmaster Set' and 5,400 were sold in 1970. There was talk of replacing it in 1971 with a similar

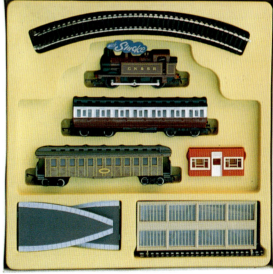

RS615 'Railway Children' set, 1971.

set but with only six wagons. RS651, however, remained and a further 7,400 were sold in 1971 including 4,000 for mail order. The 1971 sets did have one change. As production of the R561 Tri-ang container was ceasing, it was necessary to replace it and the R18 cable drum wagon was chosen for this purpose. In Hornby Railways days the set became R507 for home sales and records indicate that it was given the number R596 for export purposes.

RS651A   A special version of the set was made in 1970 with 280 sold in the UK and 50 going abroad.

RS652   3-car Pullman DMU, power clip, uncoupling ramp and marker & oval of System 6 track.

The train in the 'Diesel Pullman' set was in the grey and blue BR livery and the set was a System 6 version of the RS52 set. 4,500 were sold in 1970 but in June 1970 it was decided to drop it from the range the following year. Despite this, 100 sets survived in the stores and were sold in 1971.

RS801   Kay's Special Set.

605 of these were made in 1969 of which 400 went to Kay's. It is not known what happened to the 200 odd that remained at the end of the year.

D. PRESENTATION SETS

'Presentation sets' was a term given to sets which did not contain track. They were effectively 'add-on' sets for the person who already had a layout and wanted another train. More importantly from the manufacturers point of view, they were cheaper to produce and could be offered at a more attractive price than conventional sets.

The following is a list of presentation train packs made during the Tri-ang Hornby period:

R346   Stephenson's 'Rocket' (R651), one L & M coach in a window box.

The coach was usually shown as 'Experience' but it could have been one of any of the three made. From 1966 the smokeless version of the locomotive was used. The set had been in production since 1963 and is also covered in **Volume 1**.

Approximately 5,500 sets were made in 1965. It is not known how many were in Tri-ang Hornby boxes but about 450 of the sets with smoke survived into 1966. About 6,500 more were made before it was withdrawn from the home market at the end of 1967. It sold well overseas, including in France through Meccano France Ltd, some years exceeding home sales. The set was also assembled in Australia between 1963 and 1967 from parts sent out.

R346S   *See* 'French Sets'.

R346C   For 1968 it was decided to increase the size of the set by the addition of two more coaches. The three coaches each had a different name – 'Experience', 'Times' and 'Dispatch'. About 2,500 were made the first year but, as there were almost 3,000 of the old sets left over which had to be sold first, about 1,500 of the new sets remained for 1969. That year a further 1,500 were made, the residue of them being sold in 1970. The models were packed in a white pre-formed plastic tray which sat in another tray made of red cardboard which slid into a printed brown sleeve. The loco was without a smoke unit and the coaches had the simplified lettering.

R397   Red North British 0-4-0DS two wagons.

Planned for 1965 but appears not to have reached production.

R398   Battle Space Jinty two wagons.

Planned for 1965 but appears not to have reached production.

R399   Promotional Shunting Set.

Nothing more is known about this set, except that it was made for Christmas 1969. As this was about the time that Rovex made a promotional set which contained a flat wagon with a white Lyons Maid container on it, this could be the set in question.

R619   Littlewood's Special Set.

This was a goods set made in 1965, when 3,000 were manufactured, and again in 1968 when a further 700 were produced. Those in a later batch were numbered R619L.

R640   'Lord of the Isles' (R354), GWR clerestory brake 3rd & oil.

This set was made specifically for export. It first appeared in 1964 when almost 2,000 were made. About 800 were assembled in 1965 and a further 700 the following year. 150 more sets were assembled at the Moldex factory in Australia between 1965 and 1967.

See also 'French Sets'.

R641   'Davy Crockett' (R358S), old timer coach, oil & smoke.

Also made for export. The first batch in 1964 amounted to about 2,250 sets with a further batch of about 1,000 being made in 1965.

It is believed that some were sent to Meccano (France) Ltd to be sold with the Hornby Acho range. It featured in that catalogue from 1965 to 1970.

R641S   *See* 'French Sets'.

R644   E3000 (R753), three Mk 2 coaches (two composites & brake 2nd).

The E3000 locomotive in electric blue, was not selling well and so, in 1968, it was offered in the 'Inter City Express Train Pack'. Almost 5,000 were made and it was available in July that year. The loco

RS651 – the ever popular 'Freightmaster' set, 1970/71.

R644A and R645 presentation sets, 1969/70.

R346C 'Stephenson's Rocket' set, 1968/69.

STEPHENSONS ROCKET

R.346 'STEPHENSONS ROCKET' TRAIN. Fully detailed Locomotive has Crew and moving connecting rods. Tender and one Coach included in Train Pack. Track not included.

had yellow ends, double arrow insignia (BRe) and one pantograph. There is some question about the type of carriages it contained. The set may have contained Mk1 coaches in electric blue rather than the Mk2 coaches illustrated and these would have matched the colour of the loco. Certainly R644 sets have been found with the earlier type of coach. Stocks had virtually sold out by the New Year when a revised version of the set replaced it. This set was not popular abroad.

R644A In 1969, the loco was in BR blue and the coaches were fitted with lighting units. Some 4,000 were made that year which sold over the next two years. To help sales along a special battery-operated motorised-action stand was designed and offered free of charge to stockists (*see* 'Miscellaneous – Marketing'). This was listed as R718. The set was withdrawn early in 1971 having only been included in the catalogue that year to clear the remaining stock of sets.

R645 Hymek (R758), three Freightliner wagons.
'The Freightliner' set was advertised in the January 1968 editions of the model railway press and illustrated with the blue version of the loco. The first batch of 4,600 sets had made at the end 1967 but it is likely that they did not reach the shops until the New Year.
During 1968 a further 10,000 sets were produced but sales were slow and stocks lasted into 1970. It did not appear in the 1970 catalogue but was relegated to a black-and-white supplement printed during the year in an effort to shift old stock. In the meantime the set had been replaced by the RS602 Senior Freightliner Set which contained a freightliner depot. While early sets may have had the loco in electric blue, in later ones it was rail blue livery.

R695 BR 'Flying Scotsman' (R850), three coaches.
This was planned but not made.

R702 GUS Set. Contents unknown.
The set was made for GUS/Kays during the period 1968-70. Initially 5,000 were made in 1968. A further 400 were produced in 1970.

R703 Littlewood's Passenger Set. Contents unknown. 770 were made in 1968.

R704 International Import and Export Set. Contents unknown.
In 1969, 720 of this special set were made.

R792 Private Owner Set. Contents unknown.
This was a mail order set made in 1971. 851 were received by the customer.

R793 'Flying Scotsman' Set. Contents unknown.
Another mail order set made in 1971. 1,301 were received by the customer.

R795 Kay's Special Set. Contents unknown.
Another 1971 set of which 682 were made.

R796 Kay's Special Set. Contents unknown.
This was also made in 1971 and 700 were supplied.

R797 International Import and Export Set. Contents unknown.
In 1970 1,250 of these were supplied out of about 1,700 made. The last of the sets were used in 1971.

R798 International Import and Export Set. Contents unknown.
1,800 sets were supplied in 1970 out of a batch of more than 2,600. The remainder left the stores in 1971.

R799 International Import and Export Set. Contents unknown.

In 1970, 2,500 of these sets were made and supplied over a period of two years.

E. MOTORWAY SETS

The original combined Minic Motorway/Tri-ang Railways sets RMA, RMB and RMC were dropped in 1965 but in 1968, the year Minic Motorway production was transferred from the Canterbury factory to Margate, two further sets were produced.

RMD Dock Shunter, motorail car wagon, loading ramp, oval of motorway track and a siding, circle of Super 4 track and a siding, Minic Motorway car & controller.
Called the 'Moto-Rail Set', 3,580 were received by the Margate stores in 1968 and all but 120 going to UK dealers. By the end of the year only about 30 remained. In 1970 a batch of 180 was supplied to Debenhams.

RMZ Littlewoods Motorail Set.
A batch of 550 was ordered by Littlewoods in 1968 but it is not known how these differed from the RMD sets.

RM? Kelloggs Motorail Set. B12, Kellogg's container wagon, Minic car transporter, fish van, guards van, motorway parts including a figure of 8, chicane, roundabout, junction, car loader, station, oval of Super 4 track & a short siding.
It is thought that five sets were given away by Kelloggs as prizes in a competition held in 1967 or 1968. Competitors had to put into correct order the six factors most important to a good road/rail interface facility and say in 12 words why they thought that transporting cars by rail was a good idea.

RMD 'Motorail' set, 1968/70.

Tri-ang **HORNBY** **MINIC**

Road & Rail

...the excitement of hi-speed Motorways and Electric Railways in ONE set!

...ton Martin DB6, Oval of Motorway, Loading Ramp, Car Transporter, Lay-by Junction, Diesel Shunter & Circle of Track!

Motorail

DRIVE CAR UP RAMP & ON TO TRANSPORTER !!

MOTORAIL

RMD MOTORAIL SET. Contains Tri-ang Hornby Locomotive, Minic Car, Hand Controller, Car Loading Ramp and Transporter Wagon, Oval of Roadway with Junction and Circle of Railway Track with Point. Requires Power Controller with 12 volts D.C. output for Train and 12 volts un-controlled D.C. output for Car.

The set did not have a printed box. The parts came individually boxed in a large brown carton. There is no record in the factory census of these sets and so it is likely that they were not allocated a number but were just issued from stores as loose stock. The most interesting feature of the set is the Kellogg's container wagon, which, to the best of my knowledge, was not used anywhere else and is very rare.

F. CANADIAN SETS

Introduction

As we have seen, no models were made in Canada but the Canadian company assembled a wide range of sets from both boxed and unboxed models sent out to them from the Margate factory.

Despite the low level of production the range of sets being offered was constantly changed and there is reason to doubt whether all the sets that were illustrated in the Canadian catalogues actually reached production.

Standard sets which were imported into Canada from the UK, carried their normal 'RS' or 'R' prefixes. These have been dealt with already in this chapter under the appropriate number even where they were exclusive to Canada.

Sets assembled in Canada during the period covered by this volume were identified by a variety of prefixes:

1. Canadian Standard Sets with track and a transformer were given the 'CTS' code. This was abbreviated to 'TS' in 1967 and prefixes were eventually dropped in Canada in 1970.

2. Canadian Presentation Sets without track or a power unit. The Canadian company seems to have had difficulty in deciding how to code these. In 1965 the sets were given a 'CPS' prefix but in 1966 the code was dropped and just the number of the set, preceded with '#', was given. The coding changed yet again the following year when 'PS' was adopted and in 1969 Canada changed to the Margate custom where 'R', on its own, was used as the prefix for presentation sets. Finally, in 1970 all prefixes were dropped.

3. Canadian Mail Order and Special Sets which were produced as special orders for important customers.

The Canadians developed their own style of set packaging which was both variable and distinctive.

Packaging – Train Sets 1965-1967

Sets from the 1965-1967 period that have survived are in boxes measuring 18" x 30" x 2" made of heavy grade corrugated cardboard. The lid was rather crudely printed in red, grey, brown and black with a modified picture based on that used for the catalogue cover. It carried the inscription 'Canadian Tri-ang Railways Canadien' in the top right-hand corner and a sticker describing the set and its contents was stuck in the bottom right-hand corner on a drawing of a notice board.

The picture was later modified by the conversion of the passenger train to a freight train which was made up of the A Unit (R55CN), the auto transporter (R342CN), pulp wood car (R235CN), cattle car (R126CN) and caboose (R115CN). The engineer was depicted waving from the cab window to a father and son standing beside the track. The train crossing the bridge in the background had become a silver and red R55-hauled passenger train.

The inscription on the side of the box read 'Distributed in Canada by Meccano – Tri-ang Ltd, Toronto, Ont.' and on the inside box liner was a label, identical to that used on the lid, which described the contents.

Locomotives and rolling stock were individually boxed as they would be for sale separately. Track would have been boxed or bundled and the larger version Model 2200 transformer, which was made in Japan, was supplied with the set. Later sets had the smaller Canadian-made variant of this transformer.

Packaging – Train Sets 1968-1971
At sometime a decision was made to abandon these rather cumbersome sets. Instead, models were bought unboxed for assembly in the way they were at Margate. These new sets began to appear in 1968. The new box was 15" x 22" x 1.5" and made of corrugated cardboard and a thinner cardboard for the lid, similar to that used in the UK. The surface of the cardboard on the exterior of the box was white.

The sides of the box were decorated with illustrations from the 1969 Canadian catalogue. These were R138CN, R126CP, R440CP, R445CN, R128K, R735, R752, R129 (C&O), R235CN, R342CN, R137 (Wabash), R117 (Shell), R54S (Canadian Pacific) and R139 (Heinz). The picture on the lid top showed the

front view of two A Units side by side; one being the CN version and the other the pale blue/grey CP model. The CN loco was pulling the same freight train that we saw on the earlier lid illustration but the train of the CP unit was out of sight.

The lid carried the Canadian 'Tri-ang Railways' logo which was based on the 'Tri-ang Hornby' one used elsewhere, and also the inscription "by Meccano Tri-ang, Toronto, Canada" and "Electric Train Set HO Scale Chemin de Fer Electrique Echelle HO". Along the top of the picture on the right-hand side was a long sticker which changed according to the contents of the box. This carried a small illustration of the train within, together with the set number.

To hold the loco and rolling stock, a vacuum-formed styrene tray was inserted into the box and the track and transformer were then placed lengthways in the space that was left. Between 1968 and 1970 the trays were yellow but from 1971 to 1973 a pale green was used, the same as that used for the bases of the bubble packs in which solo rolling stock was sold at that time.

From 1968 to 1970 sets were issued with the smaller Canadian-made Model 2200 transformer, but from late 1970 to 1972 the similar Model 1000 black Japanese unit was used. The latter was illustrated on page 3 of the 1971 Canadian catalogue and it is interesting to note that the transformer may be found with the 'Hornby Railways' logo instead of the Canadian 'Tri-ang Railways' version, indicating that production of it continued after May 1972.

Packaging – Presentation Sets
These were packed in corrugated cardboard containers approximately 24" x 12" x 1.5" in size and printed appropriately according to the contents. Work Train sets of 1969-70 vintage were in olive drab containers, Dayliner sets in a pale shade of turquoise and the Container sets were in white packaging. The contents were in standard Margate

boxes as shipped out to Canada. Once assembled the container was wrapped in cellophane.

1. Standard Canadian Sets

1965-1966 Series Sets
The first six sets (CTS1-6), which were illustrated in the 1965 Canadian catalogue, were sold with the large early style Model 2200 transformer. A choice of power units (the PA102C and the PA200C) were available. The sets were listed again in the 1966 price list. It is not known whether all the sets were made. Much would have depended on the demand from retailers.

CTS1 'Hiawatha', three CN freight cars (box, cattle, caboose), oval of Super 4 track & uncoupling ramp.

The 'Steam Freighter Set' of 1965/66. It is not known whether any of these sets were made. The 1965 catalogue illustration showed the standard 'Hiawatha' locomotive and Canadian National freight cars of the period. The box car was brown, the cattle car yellow and the caboose red.

CTS2 CN A Unit, three CN passenger cars (day, diner, baggage/kitchen), oval of Super 4 track, ramp & 2200 transformer.

The 'Le Champlain' Diesel Passenger Set was illustrated in the 1965 Canadian catalogue as having the black, silver and red A Unit and standard Canadian National silver and black passenger cars. Examples of the set that have survived are in the early large-size set box with the models in individual boxes inside.

CTS3 CN Diesel Switcher, three CN freight cars (refrigerator, auto transporter, caboose), oval of Super 4 track, ramp & 2200 transformer.

The 'Diesel Freighter Set' of 1965/66 had a black

loco and standard CN freight cars. Examples of the set have been found.

CTS4 TC Yard Switcher, helicopter car, ambulance car, snow plough, oval of Super 4 track, ramp & 2200 transformer.

This was a Canadian version of the 'Snow Rescue Set' available from Margate, but the Canadians called theirs the 'Rocky Mountain Avalanche Train'. In the 1965 catalogue it was shown as having a green body on the helicopter car and a red helicopter. The diesel was red and the snow plough and ambulance car were also standard. No example of a locally-packaged set has come to my notice as yet. This may well have been sold in pre-1965 packaging.

CTS5 TC Yard Switcher, track cleaning car, caboose, oval of Super 4 track, ramp, cleaning fluid & 2200 transformer.

This was called the 'Work Train'. The illustration in the 1965 catalogue shows a standard red loco and black track-cleaning car and a red CN caboose. I am not aware of any examples of this set so cannot prove that it was made.

TS673 Canadian 'Pacific Steam Freight' set, 1967.

TS674 'The Canadian' CPR Diesel Passenger set, 1967.

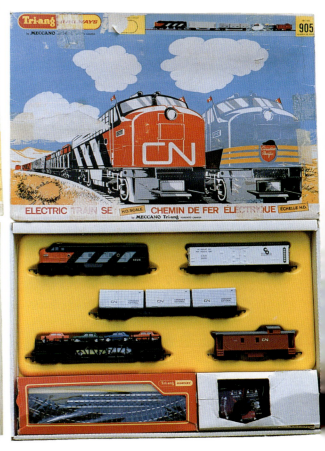

TS905 Canadian 'National Diesel Freight' set, 1969.

CTS6   TC silver/red A Unit, three silver and red passenger cars (diner, observation, baggage/kitchen), oval of Super 4 track, ramp & 2200 transformer.

Despite being called 'The Canadian Diesel Passenger Set', the livery shown in the 1965 Canadian catalogue was that of a train in standard Transcontinental livery. It is thought that no examples of the set have been found but a number of boxed R55s made in 1966 have come on the market in Canada as solo models.

1967-1968 Series Sets

The 1967 Canadian catalogue presented us with a completely new series of sets. These were numbered TS671–TS676. Each set contained enough Super 4 track to make an oval 452 mm x 322 mm and also contained a power connecting clip, an automatic uncoupling ramp and a Model 2200 transformer. In 1968 a further set (TS680) was added to this series.

TS671   TC Yard Switcher, CN gondola, crane car, CN caboose, oval of Super 4 track, power clip, ramp & 2200 transformer.

This was the 'The Mountaineer' which was shown with a red loco, brown gondola, green crane car with a brown crane and finally a red caboose. I have not heard of any surviving examples.

TS672   CN Switcher, four CN freight cars (pulp wood car, cement car, cattle car and caboose), oval of Super 4 track, power clip, ramp & 2200 transformer.

The 'Cross Canada Diesel Freight Set', which was also referred to as 'The Pioneer' contained a black diesel switcher, brown pulp wood and cattle cars, grey cement car and red caboose. This carried through into 1968 in the smaller box with the loco and rolling stock in a moulded styrene tray.

TS673   'Hiawatha', five freight cars (giraffe car, CN box car, CN auto transporter, CN oil tanker and CN caboose), oval of Super 4 track, power clip, ramp & 2200 transformer.

The set was called the 'Pacific Steam Freight Set' or 'The Iron Horse' and had all standard models. The box car was brown, the auto transporter black, the oil tanker blue or black and the caboose red. The track of eight Super 4 double curves and four straights was packed in a standard DP size window box inscribed on the end flaps 'R167 Track Pack'. This also survived into 1968 in the smaller set box.

TS674   CP A Unit, four CP passenger cars (day, diner, baggage/kitchen and observation car), oval of Super 4 track, power clip, ramp & 2200 transformer.

Strangely this was the only modern passenger set being offered in 1967. It was called 'The Canadian CPR Diesel Passenger Set' and contained models that were all in the early pale blue-grey and maroon Canadian Pacific livery. In the catalogue picture, the loco was shown with the beaver on the front.

TS675   'Davy Crockett', two old-time coaches, old-time caboose, oval of Super 4 track, power clip, ramp & 2200 transformer.

This was called the 'Centennial Steam Passenger Set' or 'Centennial Special' and contained all standard items. The loco was described as an "1867 type Pioneer Locomotive" and the coaches as "Clerestory Roof Passenger Cars". The locomotive was fitted with a smoke unit. This set appears to have been available in 1967 only and did not appear in the later packaging.

TS676   CN A Unit, Murgatroyd's tank, red ICI tank, CN caboose, oval of Super 4 track, power clip, ramp & 2200 transformer.

The 'Industrial Diesel Freight Set' or 'The Industrialist' was shown in the catalogue with the standard British bogie tank wagons (R247, R349) and the red CN caboose. It seems unlikely that the set was made, as it had been dropped from the price list the following year. Also, although 1,000 Murgatroyd's tanks went to Canada in 1968 there is no record of any being sent in, or before, 1967 when it is supposed to have been made. Furthermore the ICI tank wagon did not go out to Canada until 1969 and these are thought to have been blue ones.

TS680   Contents unknown.

As a late edition it was not included in the catalogue but it appears to have been a 'jumbo' set made up of a loco and eight freight cars. No examples have been found to prove that it was made.

1969 Series Sets

1969 saw another new series of train sets which replaced those that had gone before. These were numbered TS901 – TS906. Two of the sets (TS901 and TS906) had small ovals of Super 4 track measuring 40" x 32" and the others had the larger oval seen in earlier sets – 45" x 32" – achieved by the inclusion of two extra straights. Each set contained a power connecting clip and a power pack with speed control, and all but the TS906 had an uncoupling ramp.

TS901   CN A Unit, three CN freight cars (hopper, box car and caboose), oval of Super 4 track, power clip, ramp & 2200 transformer.

This was referred to as the 'Cross Canada Diesel Freight Set' and examples of the set have been found. The hopper car was illustrated as brown and the caboose as the later CN red. The box car was shown as bright red and yellow.

TS902   CP Switcher, Wabash cement car, three CP freight cars (box car, pulp wood car and caboose), oval of Super 4 track, power clip, ramp & 2200 transformer.

This set is also known to have been made and was called the 'Canadian Pacific Diesel Freight'. The loco was illustrated in the pale blue-grey Canadian Pacific livery and the box car and caboose in brown. The pulp wood car was black.

TS903 CP Pacific, Shell oil tanker, C&O refrigerator car, CN auto transporter, two CP freight cars (cattle car and caboose), oval of Super 4 track, power clip, ramp & 2200 transformer.

This was the first set to be illustrated with the Canadian Pacific version of the TC Pacific locomotive. It was called simply the 'Steam Freight Set' and had a black auto transporter, yellow Shell tanker, yellow cattle car, white refrigerator car and a brown caboose.

TS904 CN A Unit, four CN passenger cars (baggage/kitchen, day, diner and observation), oval of Super 4 track, power clip, ramp & 2200 transformer.

All of the models in this set were standard and it was referred to as the 'Canadian National Passenger Set'. Examples of the set have been found.

TS905 CN A Unit, C&O refrigerator car, three CN freight cars (3 x 20' container car, auto transporter and caboose), oval of Super 4 track, power clip, ramp & 2200 transformer.

An example of this set has been found. It was called the 'Canadian National Diesel Freight' and the freight cars and loco were illustrated as standard, the caboose being the later red CN version.

TS906 'Chugga' Continental 0-4-0T, three integral wagons (high-sided, low- and low with auto and house trailer), TC caboose, oval of Super 4 track, power clip & 2200 transformer.

This was called the 'Steam Switcher Freight Set'. The illustration showed the normal yellow loco with a TC maroon caboose. The low-sided wagons were dark green and the high-sided red. The 'auto and

house trailer' was a Minic car and caravan. Not surprisingly no boxed sets have yet been found. This is often the case with starter sets as the boxes quickly got destroyed. It does however seem very likely that it was made.

1970 Series Sets
The 1970 Canadian catalogue contained another series of train sets, most of them almost identical to ones offered the previous year. The new series was numbered 913-919 and for the first time contained some models in CP Rail livery. These also had ovals of Super 4 track, four with large ovals and two with small. Set 919 had sufficient track for a B-shaped oval 76" x 32".

913 CN A Unit, Rock Island Lines gondola, two CN freight cars (refrigerator car and caboose), small oval of Super 4 track, power clip & 2200 (or 1000) transformer.

This was clearly meant to be a replacement for the TS901 set of 1969. Even the name remained – 'Cross Canada Diesel Freight Set'. Examples of the set have survived.

914 CP Rail A Unit, four CP Rail freight cars (refrigerator car, paper car, box car and caboose), large oval of Super 4 or System 6 track, power clip & 2200 (or 1000) transformer.

Here was a completely new set effectively reflecting the long colourful trains of box cars to be seen in North America. It was called the 'CP Rail Diesel Freight Set' and survived into the 1971 Canadian catalogue. Sets made in 1971 would have had 12 pieces of System 6 track instead of Super 4. It seems probable that this set was made.

915 CP Pacific, Shell oil tanker, C&O refrigerator car, three CP freight cars (auto transporter, cattle car and caboose), large oval of Super 4 track, power clip, ramp & 2200 (or 1000) transformer.

Called just the 'Steam Freight Set', 915 continued on the path set by the TS673 Pacific Steam Freight Set of 1967/68 and the TS903 set of 1969, both of which were almost identical. Here was a large steam locomotive pulling five freight cars but the auto transporter was now in CP Rail livery. This set was made.

916 CN Switcher, Wabash cement car, three CN freight cars (pulp wood, refrigerator and caboose), large oval of Super 4 or System 6 track, power clip, ramp & 2200 (or 1000) transformer.

This was a replacement for the TS902 of 1969. The switcher, pulp wood car and caboose were now in CN livery and a CN refrigerator car had replaced the CP box car. The set was renamed the 'Canadian National Diesel Freight' and from 1971 was provided with System 6 track. An example of this set has survived.

917 CP Rail A Unit, four CP Rail passenger cars (day, diner, baggage/kitchen and observation), large oval of Super 4 track, power clip, ramp & 2200 (or 1000) transformer.

Again we have a replacement. The train is the same as that in the TS904 set of the previous year but in the new CP Rail livery and as one might expect it was called the 'CP Rail Passenger Set'. This set definitely was made.

918 'Chugga' 0-4-0T, three integral wagons (high-sided, tank and low-sided with auto and house trailer), TC caboose, small oval of Super 4 or System 6 track, power clip, ramp & 2200 (or 1000) transformer.

This replaced the TS906 set of 1969, the only change being a low-sided integral wagon instead of the integral tank. The other two integral wagons were blue (low) and red (high). The name remained 'Steam Switcher Freight Set' and the set survived into the 1971 catalogue where the illustration showed an unprinted yellow caboose and the high and low wagon were both blue. 10 pieces of System 6 track would have been used in sets made from 1971. Although an example has not been seen, as these starter sets were good sellers, it is reasonable to believe that this one was made.

919 CN Switcher, four container cars (2 x 30' Manchester Lines, 3 x 20' CN, 3 x 20' BP and 2 x 30' Canadian Pacific), CN caboose, very large oval of Super 4 or System 6 track, power clip, ramp & 2200 (or 1000) transformer.

While the catalogue illustration showed a switcher the description alongside referred to a CN A Unit. The set was called the 'International Container Express Set' and also survived into the 1971 Canadian catalogue. Sets made in 1971 would have had 24 pieces of System 6 track and according to the catalogue the BP container car had been replaced by one with three cti containers. Although not seen, it is thought to have been made.

1971 Series Sets

Four of these sets remained in the price list for 1971 but 913, 915 and 917 were slightly altered and renumbered 926, 927 and 928 respectively. That year also saw the return of the TS675 (now just '675') Davy Crockett set of 1967. The sets each had an oval (12 pieces) of System 6 track except the R926 which had 10 pieces, R919 which had 24 pieces and R918 which appears to have continued with its Super 4 oval. The sets now all had the Model 1000 power pack, and were marketed as 'Tri-ang Hornby'.

675 'Davy Crockett', two old-time coaches, old-time caboose, oval of System 6 track, power clip & 1000 transformer.

This was not illustrated in the 1971 catalogue but was listed in the price list that year. Few details were given and it is assumed that it received the 1000 transformer and the new System 6 track as did the other sets sold that year.

926 CN A Unit, three CN freight cars (hopper, refrigerator car and caboose), oval of System 6 track power clip & 1000 transformer.

This was a replacement for the 913 set of 1970 using a CN hopper car instead of the Rock Island Lines gondola. Examples have survived.

927 1542 Pacific, Shell tanker, CP Rail auto transporter, CP Rail box car, CN box car, CP caboose, large oval of System 6 track, power clip & 1000 transformer.

The 1971 version of the long-running steam freighter was now entitled 'Canadian Pacific Mixed Steam Freight Set'. The refrigerator and cattle cars had been replaced by two boxcars, one in yellow CP Rail livery and the other in red and yellow CN livery. The loco was shown as being the R054NS Pacific with the A3 tender but as this was not sent to Canada until 1972 it seems more likely that the CP version was used, if indeed the set was made.

928 CN A Unit, four CN passenger cars (day, diner, baggage/kitchen and observation), large oval of System 6 track, power clip & 1000 transformer.

This was the return of the TS 904 set of 1969 and the replacement of the CP Rail equivalent (917) of 1970. Examples have survived.

1973 Series Sets

The 1973 Canadian catalogue had yet another new series of sets this time numbered 7301–7306. 7301–7303 were quite new while 7304–7306 were variations of earlier sets (914, 928 and 927). These were the last sets to be designed specially for the Canadian market using models from the Margate factory. It is unlikely that any of them reached the shops.

7301 6042 Continental 0-4-0T, Rock Island Lines gondola, two CN freight cars (refrigerator and caboose), circle of System 6 track, power clip & 1000 power pack.

1971 Canadian set not listed in the catalogue.

This was the 'Side Tank Switcher Freight Set' which would have been the 'Chugga' set of two years earlier. The tank shown in the catalogue was the same black one, numbered '6042', that was used in starter sets in the UK at this time.

7302 British 0-6-0 Diesel Shunter, London Brick wagon, two CP Rail freight cars (refrigerator and caboose), circle of System 6 track, power clip & 1000 power pack.

The diesel in the Canadian catalogue illustration was rail blue with a yellow front and numbered '1520' but carried no logo. The edge of the running plate was picked out in yellow. It was described in the catalogue as a 'powerful Yard Switcher'. The brick wagon was the bright red one then in use in the UK and the CP Rail cars were also standard. The set was to be called the 'Diesel Switcher Freight Set'.

7303 CN A Unit, three CN freight cars (box car, caboose and snow plough), circle of System 6 track, power clip & 1000 power pack.

The box car illustrated was red and yellow and the snow plough green and grey. It was to be called the 'Canadian National Short Haul Freight Set'.

7304 CP Rail Switcher, four CP Rail freight cars (box, paper and refrigerator cars and caboose), large oval of System 6 track, level crossing, uncoupling ramp, power clip & 1000 power pack.

Here was an old favourite – the 'CP Rail Diesel Freight Set' which had been numbered 914 in 1970/1. The box car and caboose in the picture were yellow and the paper car was green. The refrigerator car was silver and the loco was bright red.

7305 CP Rail A Unit, four CP Rail passenger cars (day, diner, baggage/kitchen and observation), large oval of System 6 track, power clip & 1000 power pack.

This was now to be called the 'CP Rail Inter-City Express Set' and was a direct descendant of the TS674 of 1967, TS904 of 1969, #917 of 1970 and #928 of 1971. To the end, Canada offered only one modern passenger set at a time.

7306 1542 Pacific, International Logging Co pulp wood car, BOC tanker, Newsprint box car, CP Rail auto transporter and CP Rail caboose, large figure of 8 of System 6 track, power clip & 1000 power pack.

This was to be the 'Pacific Steam Freight Set' and a descendant of several large Pacific-hauled freight sets that had gone before it in the Canadian catalogue. Each had a slightly different assortment of freight cars. The loco shown was the R054NS with the A3 tender and the pulp wood car was bright red. The BOC tanker was the British 100T bogie tank and the newsprint car was the final version with the names of newspapers printed on its sides. The CP Rail cars illustrated were standard.

2. Canadian Presentation (Gift) Sets

1965-1966 Series Sets

The first four presentation sets (CPS101-401) were listed in the 1965 price list but not illustrated in the Canadian catalogue that year. Only two were eventually illustrated the following year but were now referred to as '#1' and '#2'. No examples of sets carrying the CPS numbers have surfaced as yet.

In 1966, besides the No1 and No2 sets described above there were two presentation sets made up only of wagons. These were No4 and No5.

CPS101 (#1) CN A Unit, three CN passenger cars (day, diner and baggage/kitchen).

This first appeared in 1965 as CPS101 but was shown as #1 or (Set No1) in 1966. It consisted of the four models in standard solo packaging but held together with a cardboard former. The contents all appear to be standard CN models of the period. It is not known how many sets were made but I have not heard of any examples being found.

CPS201 (#2) CN Switcher, three CN freight cars (box, refrigerator and caboose).

This was listed as #2 (or Set No2) in 1966 and shown in similar packaging to CPS201 above. Again, I am not aware of any examples having survived. Standard CN models were illustrated.

CPS301 CN A Unit, three CN freight cars (auto transporter, cattle and caboose).

No further details known and I have no evidence of the set having been made.

CPS401 CN A Unit, two CN freight cars (oil tanker and caboose).

No further details are known and there is no evidence of the set having been made.

#4 Track cleaning car, CN hopper car & CN oil tank car.

The illustration in the 1966 catalogue shows the standard R344 track cleaning wagon and standard CN hopper and oil tank wagon. It was also recorded as 'set No4'. I have no evidence yet of the set having been made.

#5 CN pulp wood car, CN cattle car, CN auto transporter & CN caboose.

The caption to the illustration (of 'Set No 5') describes only three models and mistakenly calls the pulp wood car a 'bulb-nosed car'! I have seen no evidence yet of the set having been made.

1967-1968 Series Sets

In 1967 a new series of three presentation sets replaced those previously offered. These had the prefix 'PS' and were numbered PS167–PS367. The '67' suffix presumably referred to the year of production.

PS167 Three CN passenger cars (day, diner and baggage/kitchen car).

These appear in the catalogue as standard models in the silver and black CN livery. The pack was referred to in the catalogue as the 'Passenger Presentation Set' but elsewhere was called 'Le Champlain'. Examples have been found.

PS267 Track cleaning car & three CN freight cars (hopper, snow plough and flat car).

This was called 'The Work Train Presentation Set' or 'The Track Gang'. The CN freight cars illustrated in the catalogue were all brown and the track cleaning car green. Examples of this set exist.

PS367 Four Battle Space cars (helicopter launcher, 4 rocket launcher, exploding car and searchlight car).

The cars in the 'All-Action Presentation Set' were all illustrated in khaki with Battle Space markings and stickers. It is unlikely that this set was made, at least not as illustrated. Military models in Battle Space livery are quite scarce in Canada and the green NATO predecessors to the series were in the shops in Canada well into 1970. There are no records of batches of either green or khaki versions of these four wagons being dispatched to Canada.

1969-1971 Series Sets

The 1969 catalogue in Canada showed the PS267 and PS367 presentation sets renumbered. The latter had become R367 while the former contained reliveried models and was numbered R469. The other two presentation sets listed at this time were numbered R467 and R468. Other previous sets were dropped.

In 1970 the range of presentation sets was reduced to just R467 and R469. These were sometimes referred to without the 'R' prefix. After 1971 even these disappeared from the price lists.

R367 Four Battle Space cars (helicopter launcher, 4

R468 Canadian 'Dockside Containerisation Switcher' gift set, 1969.

rocket launcher, exploding car and searchlight car).

This was the PS 367 set of 1967 with a new code. The notes about the PS 367 set apply here.

R467 CN Budd Power Car & two CN Budd Dummy Cars.

Planned for the Canadian market for 1969 and 1970, this was probably boxed in Canada from parts sent out, bulk packed, in 1969. Mysteriously 1,882 power cars were sent and 805 non-powered cars, according to stores records, and yet the sets required twice as many dummy cars as power cars. Could these possibly have been recorded the wrong way round? If so, it seems likely that 800 sets were made.

This offered the modeller a three-car Budd RDC unit and there is evidence of this being available in

the shops. It was called the 'Canadian National Dayliner Diesel Train'. The packaging was grey and the contents standard CN Budd models.

R468 Dock Shunter, three container cars (3 x 20' cti, 2 x 30' Canadian Pacific and 2 x 30' CP Sea Containers).

Assembled in Canada in 1969 and called the 'Dockside Containerisation Switching Set', there is some doubt about the exact assortment of container wagons used. The example I have seen had Containerway and Pickfords containers instead of CP Sea Containers. The red dock shunter (not yard switcher) provided the motive power and all were in their solo boxes, set out in a grey cardboard tray covered in cellophane.

R469 Crane car, three CN freight cars (hopper, track cleaning car and snow plough).

This loco-less set was also planned for release in Canada in 1969 and 1970. The 'Work Train Presentation Set' contained a brown CN hopper, green and grey CN snow plough, green and black track cleaning car and the usual green crane car with a brown crane, all in a khaki-coloured container.

3. Canadian Mail Order and Special Sets

Not much is known about this side of the business. Canada was such a large country for salesmen to cover that it was easier to deal with mail order companies. The high discount the companies demanded was offset by the savings in salesmen's overheads. **M-T Set** *See* B. Starter Sets (2. Miniville Sets) above.

CTS65?? Contents unknown.
This was the Eaton's exclusive set for 1965.

CTS662R CN A Unit, side tipping car set, caboose, oval of Super 4 track, transformer with leads and catalogue.

This was the set offered exclusively in the T. Eaton Co. catalogue in 1966. It sold for $39.95. The caboose was in CN livery and the transformer was probably the large 2200.

TS67?? CN Switcher, box car, gondola, bogie bolster with three Ford vans, caboose, oval of track and a siding in Super 4, connecting clip.

The set appeared in Eaton's catalogue for 1967 where it was claimed to be exclusive to the company. The price was $34.98. Offered with it was a 12-piece track extension set priced $5.98. As there is no record of the set having been made at Margate, it may have been a TS672 set assembled in Canada but with a different assortment of freight cars and some

extra track. The set seems to have survived until at least 1969 or 1970.

The box carried a blue sticker which confirmed that it was "exclusively for Eaton's". Eaton's had been selling Tri-ang sets since the 1950s.

TS681? Contents unknown.
The Eaton's set for 1968.

TS9?? Contents unknown.
The Eaton's set for 1969.

900 Contents unknown.
The Eaton's set for 1970.

G. NEW ZEALAND SETS

Introduction
All sets listed in the New Zealand illustrated price list were assembled at the Auckland factory using both home-manufactured models and models bought in bulk packed from Margate.

Many of the sets made during the Tri-ang Railways years remained in production after May 1965 and these included the following:

NZ50P 'Express Passenger'
NZ50G 'Express Goods'
NZ3DG 'Diesel Express Freight'
NZDBS 'Blue Streak Diesel Express Passenger'
NZTCF 'Transcontinental Freight Express'
NZST 'Suburban Train'
NZRC 'Rail Car'
NZ3RG 'Local Goods'

These carried an 'NZ' prefix which continued to be used on New Zealand-made sets until 1969 when it was reduced to 'N' and a second letter added to denote whether the set was electric or clockwork. Thus from 1969 the sets was prefixed 'NE' or 'NC'

followed by 1, 2, 3, etc.

The sets contained home-produced Series 3 track and a power clip, while the loco and rolling stock were a mixture of home made and imported parts.

Packaging
The sets were in packaging made in New Zealand to the style of the UK sets. In 1968, a display box was adopted to show off the contents better and correspondence that has survived from the time suggests that other forms of packaging were being considered under advice from Margate.

One of the changes was the adoption of slab-expanded polystyrene like that used in Australia in the mid 1960s. The instruction was that all rolling stock was to be 'level' with the length of the box and locomotives to be, where possible, in the top left-hand corner. Another consideration was vacuum-formed set trays which were cheaper to produce where small quantities were required.

New Zealand Sets 1965-68

The following is a list of the sets produced in the Auckland factory from 1966 to 1968. The number of sets sold each year was extremely low as the following table shows:

SET	1966 SALES	1967 SALES
NZ50P	155	128
NZ50G	184	246
NZ3DG	204	258
NZDBS	418	330
NZTCF	510	489
NZ3RG	823	630

NZ50P Black 'Princess', three red short TC coaches, 12 curved and six straight Series 3 & power clip.

The 'BR Express Passenger Set' had been introduced in 1962 and remained available until the end of 1968. The loco had smoke, Magnadhesion and Walschaerts valve gear. Only the loco chassis was imported, all the rest of the set being made in the Auckland Tri-ang factory. The set was priced £NZ 12.1s.0d in 1966 and this changed to $NZ 23.95 in 1967. Poor sales lead to the deletion of this set, and NZ50G, in 1969.

NZ50G Black 'Princess', open wagon, van, petrol wagon, cable drum wagon, short brake van, 12 curved and six straight Series 3 & power clip.

The goods version was added in 1964 and remained in the price list until the end of 1968. It was priced the same as NZ50P and the loco fitted the same specification.

NZST set.

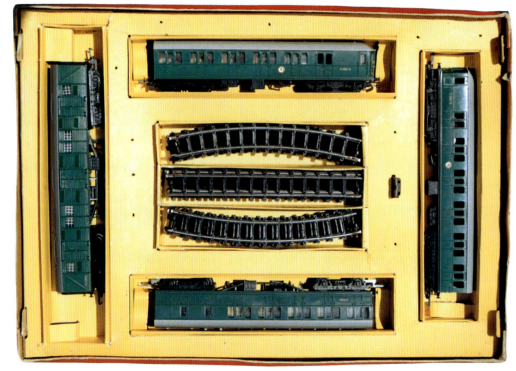

NZ3DG Maroon Double-Ended Diesel, gondola, small box car, oil tanker, caboose, 12 curved and six straight Series 3 & power clip.

This was called the 'Diesel Express Freight Set' and cost £NZ 10.9s.6d in 1966. It was an old set believed to date from 1958 and it remained in the illustrated price list until the end of 1968. For 1969, the set was reduced in size by the deletion of the R114 box car and given a new number – NE4.

NZST BR(SR) EMU powered and dummy ends, suburban brake, utility van, 12 curved and two straight Series 3 & power clip.

The 'Suburban Train Set' dated from 1964 and remained available until the end of 1967. The EMU

and the suburban brake bodies and roofs were made in New Zealand but the Utility Van was imported from the UK. Unlike the earlier British version, the suburban brake was given BR coach decals the same as the EMU cars. The set cost £NZ 9.8s.6d in 1966.

NZRC BR(SR) EMU powered and dummy ends, 12 curved and six straight Series 3 & power clip.

This was simply an EMU 2-car set and track. Like the NZST set, it first appeared in 1964 but it was dropped from the range a year earlier, at the end of 1967. The set cost £NZ 7.1s.6d.

NZDBS Blue Double-Ended Diesel, blue 2nd series Transcontinental passenger cars (R444, R445, R446 and R447), 12 curved and six straight Series 3 & power clip.

This was the New Zealand version of the famous 'Blue Streak' set which was popular in Britain in the early 1960s. It first appeared in the New Zealand price list in 1964 and remained there for five years. While the diesel body and track were made in the Auckland factory, the passenger cars were all imported. In order to cut the cost of the set for 1969, the observation car was dropped and the set renumbered NE5.

NZ3RG Black BR Jinty 0-6-0 tank, van, open wagon, short brake van, 12 Series 3 curves & power clip.

The 'Local Goods Set' was referred to as both the NZ3RG and NZR3G. It had been introduced in 1960 but this selection of wagons dated from 1963. It was dropped at the end of 1968. The price of the set in 1966 was £NZ 6.5s.6d. Only the chassis of the loco and wagons were imported.

NZTCF 'Hiawatha', stock car, refrigerator car, long box car, caboose, 12 curved and six straight Series 3 & power clip.

The 'Transcontinental Freight Express' was made

NZRC set of February 1966.

NZDBS New Zealand set. Although in a Tri-ang Railways box, the set was assembled in November 1966.

up of imported rolling stock and locomotive with the exception of the caboose. It had been introduced in 1964 and remained in the range until the end of 1968. In 1966 it was priced £NZ 12.1s.0d.

NZC3 C/W 0-4-0DS (R557), open wagon, petrol tanker, short brake van, 12 Series 3 curves, two keys.

This clockwork set was new in 1966 when 918 sets were sold and was made for three years. Sales in 1967 were just over 1,000 and total sales over the three years were probably about 3,000. Sets in the final year may have had the instruction manual and service dealer leaflet replaced with the New Zealand illustrated price list.

NZC4 C/W 0-4-0T Top Tank (R660), open wagon, milk tanker, short brake van, 12 Series 3 curves, two keys.

This clockwork set was new in 1966 and was made for three years. This set retailed at $NZ9.95 and sales were 1,200 in 1966 and 1,900 the following year. Plans for 1969 included a new box design and a simplification of the catalogue number to 'NC4'.

NZHYF BR blue Hymek, two bogie bolster wagons with load of three Minix vans, gondola, liner train with three Freightliner containers, short brake van, 12 curved and six straight Series 3 & power clip.

The cost of introducing new models was so high that the New Zealand company had to resort to making up sets with imported items. This was the first set to be thus assembled and it was sold under the name of 'Freightmaster Train Set'. It cost $NZ 25.45 when it first appeared in 1968. The price list illustration showed a TC bogie flat car with three containers which is thought to have been a mock-up for the photograph. The set was replaced the following year with the NE6 train set which was the

same but for the replacement of the liner train wagon and three containers with a bogie bolster wagon carrying two containers.

NZDSH BR 0-6-0DS, coal truck, open wagon, short brake van, power clip, point, uncoupler, buffer stop, 11 curved and two straight Series 3 track.

The 'Diesel Shunter Set' was introduced in 1968 and was renumbered the following year as NE3. It cost $NZ 14.95 and is believed to have survived until 1971.

New Zealand Sets 1969-71

The sets were not provided with a power unit and by 1969 the choice units available to customers was a P149 or P15A. The sets had Series 3 track until 1972 when they were advertised as having System 6 track. Between 1972 and 1978, a large range of System 6 sets were advertised assembled in New Zealand using imported Hornby Railways parts. These all carried 'NE' or 'NC' prefixes. In several cases a set's code number was reused for a new set with different contents in which case it was given an 'A' suffix. When used yet again for a different set the suffix changed to 'B'.

NE1 Continental 0-4-0T tank (R852), three integral wagons (low-sided with Minix car, tanker, high-sided wagon), 12 Series 3 curves & power clip.

It was advertised that the train could be run from the Tri-ang Hornby power unit or from a battery box controller which would be available in August or September that year (1969). The set was introduced as a low-priced electric set to try to interest first time buyers and was the first New Zealand set to use the new R852 electric tank engine. The set, which was also intended for sale to wholesalers, survived at least until 1973.

NE2 Black BR Jinty 0-6-0 tank, two integral wagons (tanker and low-sided with Minix car), short brake van, 12 Series 3 curves & power clip.

This was a replacement for the NZ3RG 'Local Goods Set' and was available until 1973. It was unusual to find integral wagons mixed with other types of wagon (in this case the R16 brake van) or with other than starter set locos.

NE3 BR 0-6-0DS, coal truck, open wagon, short brake van, power clip, point, uncoupler, buffer stop, 11 curved & two straight Series 3 track.

The 'Diesel Shunter With Siding Set' was a replacement for the NZDSH set dropped at the end of the previous year. The NE3 was made for two years and cost $NZ 15.45. This was also around in 1973.

NE4 Maroon Double-Ended Diesel, gondola, oil tanker (or short box car), caboose, 12 curved and six straight Series 3 & power clip.

The set replaced the NZ3DG of the previous year but had one less freight car. It retained the name 'Diesel Express Freight Set' and remained in the illustrated price list until the end in 1972 although in the final year the oil tanker gave way to the short box car. The set cost $NZ 20.95.

NE1 New Zealand set, 1969.

NE5 Blue Double-Ended Diesel, blue 2nd series Transcontinental passenger cars (R444, R446 and R447), 12 curved and six straight Series 3 & power clip.

This replaced the NZDBS Set but had lost one passenger car, the R445 Observation Car. It was listed as the 'Blue Streak Diesel Passenger Express'.

NE5 New Zealand set, 1972, with System 6 track.

NE5 Greeny-blue Double-Ended Diesel, red R27 Vista Dome and two R26 coaches, 12 curved and six straight Series 3 & power clip.

In its final year the imported TC coaches had run out and NE5 was listed with the R27 Vista Dome and two R26 coaches with BR bogies. The set was now called 'Diesel Passenger Express' and examples seen had a turquoise blue loco and red coaches. Late sets also had 'Aust and NZ' made System 6 track instead of Series 3.

NE6 BR Blue Hymek, two bogie bolster wagons with load of three Minix vans, gondola, bogie bolster with two Freightliner containers (or pulp wood car), short brake van (or gondola), 12 curved and six straight Series 3 & power clip.

This was a replacement for the NZHYF set of 1968 and retained the name 'Freightmaster Train Set'. In 1969 it was listed as having a container wagon and R16 brake van but by the following year these had changed to a pulp-wood car and a caboose. The set cost $NZ 26.45 and survived at least until 1973.

NE7 'Hiawatha', stock car, bogie chlorine tank wagon, bogie bolster with three Minix vans, caboose, 12 curved and six straight Series 3 & power clip.

The 'Transcontinental Freight Express' was a replacement for set NZTCF of the same name. The refrigerator car and long box car had been replaced with the British chlorine tank wagon and the bogie bolster with three vans. The set, at $NZ 25.45, remained on the price list at least until 1973. The loco was R54S but was wrongly listed as R545.

NC1 C/W Continental 0-4-0T (R854) (or C/W R660 Top Tank 0-4-0T), two integral wagons (low-sided and high-sided trucks), 12 Series 3 curves & two keys.

The set was available for at least five years between 1969 and 1973, although the track may

have changed to System 6 at some time. When first introduced it was priced $NZ 6.95 but this had risen to $NZ 9.80 four years later. Inflation had struck. The 1969 price list indicates that the R660 Top Tank 0-4-0T was used at first but was later replaced by the R854 Continental 0-4-0T. If the top tank 0-4-0T was used it must have been old stock as, by the time the set was being assembled, the Continental 0-4-0T was being sent out to Margate. Reference to the R660 may have been a composition error when the price list was being prepared.

NC3 C/W 0-4-0DS (R557), three integral wagons (R710 high-sided, R712 low-sided with Minix car, R711 tank wagon), 12 Series 3 curves & two keys.

Although introduced in 1969 we do not have a record of when production of it ceased. Initially it cost $NZ 7.50.

NC4 C/W Continental 0-4-0T (R854) (or R660 Top Tank 0-4-0T), three integral wagons (R712 low-sided with Minix car, R711 tank wagon, R710 high-sided wagon), 12 Series 3 curves & two keys.

Introduced in 1969 the set remained in the price list at least until the end in 1973. During this time its price rose from $NZ 7.95 to $NZ 11.50. The contents of this set is not certain. Again, the Top Tank 0-4-0 (R660) is listed whereas the Continental 0-4-0T (R854) is a more likely choice and indeed is illustrated. The illustration in the price list also shows the wagons described above but lists an R16 short brake van instead of the high-sided wagon.

H. AUSTRALIAN SETS

Introduction
Tri-ang Railways sets had been assembled at the Moldex factory in Melbourne, for the Australian market, for a number of years using a mixture of home produced and imported parts. Production of

some of these sets continued after the change to Tri-ang Hornby and the following is a list of these:

R4D	'Diesel Passenger'
R4F	'Diesel Freight'
R4Y	'NSWR Suburban'
RS20	'Freight Set'
RS21	'Passenger Set'
RS24	'Pick-up Goods'
RS37	'The Frontiersman'
RS61	'Old Smoky'
R346	'Stephenson's Rocket Train'
R640	'Lord of the Isles Train'
R641	'Davy Crockett Train'

Some of the sets were of Australian origin carried over from the 'Tri-ang Railways' days while others were an Australian version of a British set using boxes printed in Australia and carrying the Moldex name along with the country of origin. In price lists, these Australian 'copy' sets were identified by the addition of an 'A' suffix to the number, although this was not always shown. Thus the Australian assembled RS51 'Freightmaster Set' became RS51A. Its box and track were Australian made but the loco and all of the wagons had been imported from the UK. The RS24A Pick-Up Goods Set, on the other hand, contained an imported loco but the wagons were made by Moldex.

The Moldex factory in Melbourne continued to assemble its own Tri-ang Hornby sets long after it had ceased production of models in 1968. It imported the locos and rolling stock from Margate but made its own boxes and track. This was the principal reason that the company was supplied with the tools to make System 6 track but sadly, no records survive of the numbers made.

Packaging
Early Australian sets of this period continued to have Tri-ang boxes with the old lid illustrations.

Inside they had a rather crude white expanded-polystyrene tray instead of the earlier card interior. Stuck onto the surface of the tray was a Tri-ang Hornby sticker which had the set code number stamped onto it. There was also a Moldex sticker. Later Australian sets had a map lid and a yellow (early) or white (late) plastic vacuum-formed inner tray. The 'map lid' is so called because it carried an illustration of a map of Australia floating on a blue sea. The map, which was diagrammatic, showed the main railway routes across the continent and on it was superimposed a picture of a Princess loco pulling a train of silver and red short TC coaches at speed.

On the smaller set boxes the illustration occupied the whole of the upper surface of the lid but on larger sets there was room for the Tri-ang Hornby logo, the '00/H0' symbol, the word 'RAILWAYS' in the style of the former Tri-ang Railways logo and 'ELECTRIC MODEL RAILWAY SET'.

Track

Between 1965 and 1970, Australian sets would all have had home-made Super 4 Track. By 1970 and 1971 almost certainly the sets contained System 6 track made at the Melbourne factory. There was no mention of this in catalogue supplements and price lists at the time and so we can not be sure when the change from Super 4 took place. Where there is doubt in the following details about contents, reference to the type of track is left out.

Numbering

The numbering of sets during the Tri-ang Hornby period is very confusing as some of the earlier codes based on foundation letters continued alongside those with the British 'RS' and number system.

The letter-based codes included the following:

R4A	R4DA	RS4FA	R4YA
R4B	RS4DA	R4N	
RS4CA	R4F	R4W	
R4D	R4FA	R4Y	

Sets that followed the British RS system were:

RS3A	RS18	RS24A	RS51A
RS4A	RS19	RS24XA	RS70
RS5A	RS20A	RS43E	RS85
RS6A	RS21A	RS49	RS88A

The sets are listed below under three headings:
1. Australian Lettered Sets
2. Australian RS Numbered Sets
3. Australian Presentation Sets

1. Australian Lettered Sets

The following sets continued the traditional Australian system of set numbers based on a foundation letter:

R4A BR Jinty 0-6-0T (R52A), open wagon, closed van, short brake van, Super 4 (eight double curves & two straights) & power clip.

The 'Steam Freight Set' appeared in the 1966 and 1967 catalogue supplements, after which it was dropped. The loco was black and it is likely that all parts of the set with the exception of the loco chassis were made in the Moldex factory. It seems that later sets also carried an uncoupling ramp and an oil sachet.

R4B BR Maroon 'Princess Royal' (R258&R34), two R26 maroon and cream short TransAustralia TC coaches, Super 4 (eight double curves & two straights) & power clip.

The 'Countryman' Set, which was also illustrated in the 1965 Australian catalogue supplement as RS 21, was based on the British RS21 set but unlike the Australian RS21 set of 1962/63, it now had a maroon loco (which was without smoke or Walschaerts valve gear) and Australian short coaches. Only the loco chassis would have been imported. The set was priced £A 8-19s-6d.

In 1966 the set was renumbered R4B and this it remained until it was dropped in 1968.

RS4CA TC Pacific (R54S), five wagons, oval of System 6, power connecting clip, uncoupling ramp.

The set was available in 1970 only and was described as having the above contents. An example was found in the store at Margate containing a BR 'Flying Scotsman', stock car, cti container car, pulp-wood car and a WR brake van. The box had a 'map' lid and there is no record of how many were made.

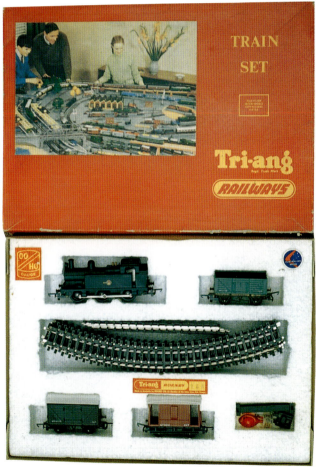

R4A Australian 'Steam Freight' set, 1966/67.

The locomotive looked uncomfortable in the box and the shape of its compartment suggests that the plastic inner tray was designed to take the R54S 'Hiawatha' or one of its variants. The compartments for the stock car and brake van are too large implying that they were made for other models. There are, however, only four compartments for rolling stock and so if this box was made for the RS4CA set described in the 1970 price list, one compartment must have been shared by two small wagons.

R4D Double-Ended Diesel (R159A), three R26 blue TransAustralia short TC coaches, Super 4 (eight double curves & two straights), power clip & uncoupling ramp.

The 'Trans-Australia Express' was a direct descendent of the RAXD set first made by Moldex in 1958. This version dated from 1962 and remained in production until 1967. In 1965, however, the number of coaches was increased from two to three and now carried the TransAustralia road name.

R4DA Double-Ended Diesel (R159A), three R444A blue TransAustralia TC coaches, oval of Super 4 or System 6 (eight double curves & two straights), power clip & uncoupling ramp.

In 1968 the 'Transcontinental Express Set' replaced the R4D of the previous year, and differed in having imported full-length coaches and an uncoupling ramp. It was replaced the following year by the RS4DA.

RS4DA Double-Ended diesel (R159), three second series TC passenger cars (coach, diner and observation car), circle of Super 4 or System 6 curves and two straights, power clip & uncoupling ramp.

In the summer of 1968 plans were being made for two special Australian sets for 1969. One of these was for RS4DA to replace R4DA with basically the same contents but with one of the R444A coaches changed for an R445A observation car. The livery was the normal blue with yellow lining and the loco

was probably the standard one for the period with TR shields on its sides. It is also thought that the coaches were Transcontinental rather than TransAustralia although in the case of Australia it is dangerous to be too sure of anything!

By 1970 Rovex had difficulty in supplying parts for a special train set Moldex were preparing. As stocks of the VR blue plastic dried up Rovex had changed to BR blue as the best alternative. Unfortunately locos and coaches in the two shades arrived as part of one order and Moldex had great difficulty in matching up sets. Eventually Rovex had to grind down some surplus mouldings in VR blue to produce material for extra bodies to complete a balanced order. In the meantime Moldex had resorted to spraying some of their surplus silver coaches in VR blue to make up shortages.

RS4DA R55 Single-Ended Diesel (A Unit), three second series TC passenger cars (coach, diner and observation car), circle of Super 4 or System 6 curves and two straights, power clip & uncoupling ramp.

The 1970 Australian price list describes the RS4DA set as having the R55 loco instead of the R159 Double-Ended Diesel and, indeed, this version of the set, also with silver and red coaches (sometimes red painted silver) to match the loco, was made; at least, some 1,750 R55 locos were sent out to Australia for this purpose. The set came in the usual long flat box with a 'map' lid. Inside, the contents were laid out in a white plastic vacuum-formed tray. The rolling stock was made in the UK but the couplings fitted in Melbourne where the track was also made. The loco and coaches had the Transcontinental road name. Later sets had System 6 track.

RS4A *See* 'Australian RS Numbered Sets'.

R4F Double Ended Diesel (R159A), open wagon, goods van, coal truck, UD tank wagon, short brake van, Super 4 (eight double curves & two straights)

& power clip.

This was the 'Diesel Freight Set' which made its debut in 1965 and remained in the price list for three years. It was replaced by the R4FA. The loco was in TransAustralia blue livery with a locally-made body and Australian wagons.

R4FA Double-Ended Diesel (R159A), open wagon, goods van, coal truck, UD tank wagon, short brake van, Super 4 (eight double curves & two straights), power clip & uncoupling ramp.

This was the same as R4F of the previous year but now the loco was in TC (shield) livery and an uncoupling ramp had been added. The short brake van had been replaced by an imported BR(ER) brake and the other wagons may well have also been imported. The set was illustrated in the 1968 catalogue supplement.

RS4FA Double-Ended Diesel (R159A), open wagon, goods van, coal truck, UD tank wagon, BR(ER) brake van, Super 4 or System 6 (eight double curves & two straights), power clip & uncoupling ramp.

The RS4FA replaced the R4FA in 1969. This should have had an R55A diesel in NSWR colours but without the name on the sides, but the 1969/70 Australian catalogue supplement shows the set with an R159 in up-dated TC livery; the same as used in the RS4DA set. The catalogue supplements further show a coal wagon instead of the mineral wagon originally proposed. All the wagons illustrated were imported as production of locos and rolling stock at the Moldex factory had now ceased. The set remained in the Australian price list until the end.

R4N 'Hiawatha' (R54), small box car, gondola, caboose, Super 4 (eight double curves & two straights), power clip & uncoupling ramp.

This was called the 'Interstate Freight Set' and was available from 1966 for two years. It is likely that the loco was bought in from Margate unfinished and the freight cars would almost certainly have been made by Moldex from the tools they received in the mid-1950s.

R4W A Unit (R55), two Short TC Southern Aurora coaches, one short TC Southern Aurora buffet car, Super 4 (eight double curves & two straights), uncoupling ramp & power clip.

The 'Southern Aurora Set' was based on the crack Sydney-Melbourne express but the loco was said to resemble the Commonwealth Railways GM-12 rather than the Class 44 diesel-electrics used by the NSWR on this route.

It was illustrated in the 1966 and 1967 catalogue supplements after which it was dropped. Both examples I have seen had a Tri-ang Hornby label inside the box but a Tri-ang Railways lid with the Gladioli picture of late 1963 onwards. The silver and red loco carried no markings on its sides but the TR shield and speed lines on the front. The items in the box were set into cut-outs in a block of expanded polystyrene, the top surface of which had been painted yellow.

R4Y NSWR Suburban Electric Motor Car, Trailer Car, Dummy Motor Car (R450, R451, R452).

This set did not contain track and should really be dealt with under 'Australian Presentation Sets'. It is included here, however, for the sake of continuity. The set, which was made in 1965 and 1966, was

RS4DA Australian 'Transcontinental Express' sets, 1969 – spot the difference!

R4W Australian 'Southern Aurora' sets, 1966/67.

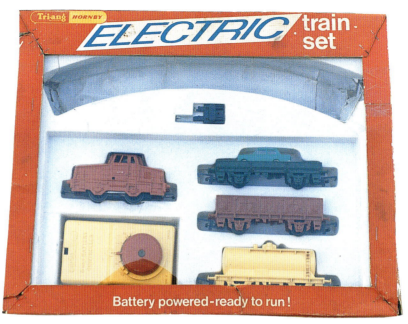

RS5A Australian 'Electric Powered Diesel Freight Train', 1969/70.

R4YA Australian 'Sydney Suburban' set, 1968.

tray. The models had the small lettering and the lighter colour plastic suggesting that it was a late set.

R4YA NSWR Suburban Electric Motor Car, Trailer Car, Dummy Motor Car (R450, R451, R452), oval of Super 4 track.

The set appeared in the 1968 price list and catalogue supplement but was dropped after that. It was a replacement for the R4Y set described above. The models had the small printing and lighter colour plastic typical of later production and were laid out in a yellow vacuum-formed tray in the box.

2. Australian 'RS' Numbered Sets

These sets were based on the RS numbering system and they were mostly copies of British sets of the time:

RS1EN Continental 0-4-0T (R852), two integral wagons, oval of track, power clip, battery controller.

This was the 'battery powered steam freight set' of 1970 and 1971 which used a UK loco and rolling stock and Australian track and packaging.

RS3A C/W Continental 0-4-0T (R854), four integral wagons (low-sided with Minix van, low-sided with Minix car, tank, high-sided), oval of Super 4 track, level crossing & two keys.

The track was not mentioned in the catalogue supplement description and the set was listed as coming from the Australian factory only for 1969.

RS4A Continental 0-4-0T (R852), two integral coaches, oval of Super 4 track, power clip, battery controller.

This 'Battery Powered Steam Passenger Set' was based on the British version but presumably sold in the Australian packaging and with locally-made track. It was listed in 1969, 1970 and in 1971 as RS4E.

RS5A Swedish 0-4-0DS (R853), three integral wagons (low-sided with Minix car, high-sided, tank), eight double curves, power clip, battery controller.

This was sold as the 'Electric Powered Diesel Freight Train' and was available in 1969 and 1970. This too would have been made up from imported models but used Australian track and packaging.

RS6A Swedish 0-4-0DS (R853), four integral wagons (low-sided with Minix van, low-sided with Minix car, tank, high-sided), oval of track, level crossing, power clip & battery controller.

This was distinguished from RS5A by calling it the 'Large Electric Powered Diesel Freight Train'! Its parentage was probably the same as that of the last three sets. It too was available for only two years.

RS11A Industrial 0-4-0T, Three wagons &?

Just 250 sets were made by Moldex for an unidentified customer. The date is not known.

RS18 C/W Top Tank 0-4-0T (R660), operating wagon & circle of Super 4 track.

This was listed in the 1967 Australian catalogue supplement suggesting that sets were being assembled by Moldex at Melbourne. The previous year this set had been obtainable direct from Margate. Its manufacture at Melbourne needs to be confirmed.

RS19 Clockwork Top Tank 0-4-0T (R660), bogie bolster with three Minix cars & circle Super 4 track.

This set had been available direct from the UK but in 1967 it was listed in the Australian catalogue supplement suggesting that sets were being assembled by Moldex that year. I have had no confirmation of this.

RS20A BR black Jinty 0-6-0T (R52), open wagon, closed van, short brake van, Super 4 (eight double curves & two straights) & power clip.

The 'Freight Set' illustrated in the Australian 1965 catalogue supplement was based on the equivalent British set but had all Australian parts with the exception of the locomotive chassis. The set was apparently available in 1964 and 1965 but was replaced by the R4A 'Steam Freight Set' the following year. It was priced £A 6-19s-6d.

RS21A *See* R4B

RS24A Australian 'Pick-up Goods' set, 1968/71.

RS24A Continental 0-4-0T (R852A), Insulfish, open wagon, short brake van, Super 4 (eight double curves & two straights), power clip & uncoupling ramp.

Initially the 'Pick Up Goods Set' was imported from the UK but by 1967 it was being assembled in the Moldex factory and in 1968 carried an 'A' suffix to the number in the price list. This was the same as the British version of the set but packed in an Australian box with Australian parts except for the loco. The parts were laid out in a white plastic vacuum-formed tray. The list price was £A 5-19s-6d.

The set has also been found in Australia with 'Nellie' (R355B) as the loco.

RS24XA Continental 0-4-0T (R852A), fish van, open wagon, BR(ER) brake van, Super 4 or System 6 (eight double curves & two straights), power clip & uncoupling ramp.

The set replaced the RS24A in 1969 and survived to the end. It differed from the earlier set in having all imported wagons including the BR(ER) brake van. The illustration in the 1969 catalogue supplement suggests that the loco and rolling stock carried no markings.

RS38 Dock 0-4-0DS, snow plough, helicopter car, ambulance car, large oval of Super 4 track & ramp.

This appears to have been assembled by Moldex in 1966 and 1967 using parts bought in from the UK.

RS43E C/W N British 0-4-0DS (R557), two wagons & circle of Super 4 track.

This set was possibly assembled in Australia in 1966 but I have no firm evidence of this.

RS49 C/W 0-4-0 Top Tank (R660), coach & circle of Super 4 track.

This had the previous year been available from Margate but it was listed in the 1967 Australian catalogue supplement suggesting that sets were being assembled in Australia that year. I have had no confirmation of this.

RS51A BR Blue A1A-A1A (R357B), drop side wagon, a three-container wagon, UD tanker, Insulfish van, cattle wagon, horse box, BR(ER) brake van, Super 4 (eight double curves & two straights), power connector & uncoupling ramp.

This set was originally imported ready-assembled but by 1968 Moldex were assembling their own. The

RS51A Australian 'Freightmaster' set, 1968/69.

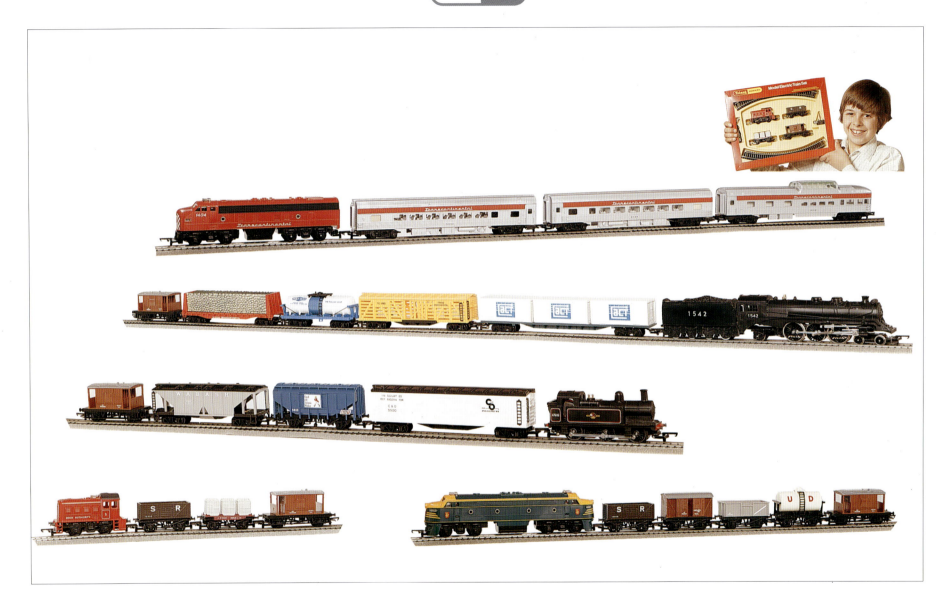

Australian sets for 1971.

RS102 'Express Freighter' set for Australia, 1971.

3. Australian Presentation Sets

R346 Stephenson's 'Rocket' (R651) and one L & M coach in a window box.

This set was assembled in Australia between 1963 and 1967 from parts sent out.

R640 'Lord of the Isles' & GWR clerestory coach.

Up until 1965 this was imported as a complete set but there is reason to believe that it may have been assembled by Moldex in 1966.

R641 'Davy Crockett' & old-time coach.

This was assembled from parts imported from Margate. Up to 1965 the set was imported ready-assembled but in 1966 it was listed as being available from Moldex. During 1965 and 1966, 500 were recorded by Moldex as having been assembled in the factory for solo sales.

R4Y *See* under Mainstream Sets.

4. Australian Special Sets

Many of the special sets for stores are difficult to find and all are poorly chronicled:

MR94A Jinty, two open wagons, closed van, short brake van, small oval of Super 4 track & power clip.

This is a set which did not appear in the price list and may have been a special set for a local customer, possibly Myers who bought a similarly numbered set (RS94M) from Margate. The loco in the only set I have seen, and of which a picture is provided here, is black, unlined or numbered but with a late BR logo. One of the open wagons may have had a coal insert. The box was the Tri-ang Hornby 'map' lid type. The yellow vacuum-formed tray suggests the 1965/66 period.

RS12 Special set for Myers of Australia – contents unknown.

loco was in BR blue livery and together with the wagons were imported but 'finished' in some way so that they received an 'A' suffix. This may have meant that the couplings were fitted on arrival. The items were set into a white plastic tray in a long box with a map-type lid. Like its British counterpart the set was called 'The Freightmaster'.

RS61 Deeley 0-6-0, utility van, coach & circle of Super 4 track.

This was assembled by Moldex in 1966 for one year only, using imported locos and rolling stock.

RS62A Jinty and 3 wagons.

This was the Australian-assembled version of the British Car-a-Belle set. It was made in 1966.

RS70 C/W 0-4-0 Top Tank (R660), open wagon, Insulfish van & circle of Super 4 track.

This was listed in the 1967 Australian catalogue supplement suggesting that sets were being assembled in Australia. The previous year retailers were having to obtain their supplies from the UK. I have had no confirmation of its manufacture in Australia.

RS85 C/W 0-4-0 Top Tank (R660), seven plank open wagon, short brake van, circle of Super 4 track, level crossing & two Minix cars?

This clockwork set was listed in the 1967 Australian catalogue supplement suggesting that sets were being assembled in Australia but this needs to be confirmed. The contents listed here are those that appeared in the British set and may be different.

RS88A C/W 0-4-0 Top Tank (R660), open wagon, bogie bolster with three Minix cars, Shell Lubricating Oil tank, drop side wagon, Super 4 (eight double curves & two straights) & level crossing (R495A).

The 'Clockwork Goods Set' was assembled in Melbourne in 1968 for one year only. It would have been the same as the British set but packed in a white plastic tray in an Australian 'map' box. The models would have been bought in, bulk packed, from the UK and 'finished' in the Moldex factory. The set was unusual in not having a brake van.

I. FRENCH SETS

These were included in the Hornby Acho catalogue where they looked incongruous amongst the French sets. Some of the subjects chosen made the Tri-ang Hornby range look a little gimmicky.

Details of what was sent over to the French company do not exist before 1967 and can only be guessed at. The sets themselves were standard British issue but with the Super 4 track having been replaced with Hornby Acho track and Acho couplings provided. In most cases a small French label was fixed to the box.

1. Train Sets

The three train sets chosen were boxed up at Margate but sent without track. Acho track was added to the set on arrival:

RS34F Double-Ended Diesel (R159), four TC blue Series 2 passenger cars (R444, R445, R446, R447), R487 clip & oval of Hornby-Acho track.

Batches of this popular 'Le Transcontinental' set were made for sale in France between 1965 and 1970. Three batches amounting to 900 sets are known to have been sent. These were featured in the Hornby-Acho catalogues and it is assumed that French labels for the boxes and Acho track were added on arrival at the French factory. The loco and rolling stock, according to the catalogue illustration, seem to have retained their Tri-ang tension-lock couplings.

RS50F Green 0-6-0DS (R152), three military wagons (R341 searchlight, R343 4-rocket launcher, R249 exploding car), R487 clip & oval of Hornby-Acho track.

This was another set offered through the Hornby-Acho catalogue, and called 'Le Train Militaire'. About 775 special sets were made in 1967 which were sent over to France in two batches in 1967 and in 1968. Again, they would have been provided with Acho track and a French label. The set remained in the Meccano Tri-ang French catalogue until 1970.

RS52F Blue Pullman cars (R555, R556), R487 clip & oval of Hornby-Acho track.

The 'Blue Pullman' was one of the first Tri-ang sets to be added to the Hornby-Acho catalogue when Lines Bros. acquired Meccano France in 1964. In 1965, 500 sets were sent to the Calais factory for distribution through Meccano France stockists as Le 'Pullman Bleu'. The boxes were sent out without track and Acho track was added at Calais. As far as we know, all RS52F sets had the standard Tri-ang tension-lock couplings. The set was still available in 1970.

2. Presentation Sets

These were trackless sets and so required nothing adding on arrival in France bar identification labels. The sets were:

R346 Stephenson's 'Rocket' (R651) & one L & M coach in a window box.

R346S Stephenson's 'Rocket' (R651S) & one L & M coach in a window box.

The presentation set, with loco and one coach, was listed between 1965 and 1970 in the Hornby Acho catalogue as 'Train "Rocket" de Stephenson'. France received 1,000 sets in 1967 and a further 500 in 1968. Both batches were listed as R346S, indicating that they had the smoking version of the locomotive, but the R346 smokeless version was also listed in the Hornby Acho catalogue from 1967 to 1970.

R640 'Lord of the Isles' (R354), GWR clerestory brake 3rd & oil.

The set was advertised in the Hornby Acho catalogue between 1965 and 1970. It was listed as 'Train XIX siecle "Lord of the Isles" '. The models in the set appear not to have been fitted with Acho couplings but the box probably received a French label, printed and affixed in France.

R641 'Davy Crockett' (R358), old-timer coach, oil & smoke.

R641S 'Davy Crockett' (R358S), old-timer coach, oil & smoke.

In 1967 and again in 1968 batches of 500 sets were made for France and advertised in the Hornby-Acho catalogue from 1967 as 'Train Western Davy Crockett'. These would have been sent to France ready boxed to receive a French label on arrival.

MYSTERY TRAIN SETS

Too little is known about the following sets to determine where they belong:

R? Lyons Maid Set.

It is known that a set was issued around 1968 containing a 'Lyons Maid' container wagon. As there is no separate record of such a set entering the stores it may have been that a very small number were required and stocks of a known set containing the small container wagon, such as the Freightliner Set RS51 or the RS87 starter set, were taken from the stores and the standard wagon exchanged for a specially made Lyons Maid version.

RS1051 Export Set.

Less than 100 of these were made in 1965 and it is thought that all went abroad.

INTRODUCTION

Hornby-Dublo Integration

Those who hoped that the Tri-ang Hornby range for 1965 would contain a good smattering of Hornby-Dublo locomotives were to be disappointed when they received their copy of the 1965 catalogue. Indeed, even the following year, when Hornby-Dublo locos did appear, their presentation was such as to suggest that they were a late addition and there was no indication of a planned integration. It has since become evident that the inclusion of the Hornby-Dublo 'Barnstaple' and Co-Bo in the 1966 Tri-ang Hornby catalogue was an attempt to use up the large stocks of these models found in the Meccano factory when Lines Bros. took over. They were, however, not the only Hornby-Dublo locomotives to be sold with Tri-ang Hornby labels stuck on the box as we will see later.

In time, only one model locomotive could claim to have been absorbed into the Tri-ang range and that had to undergo substantial modifications first. Rovex had planned to introduce a mainline electric Class AL3 locomotive as will be seen from the illustrations in the 1964 and 1965 catalogues. Furthermore, the 1965 illustration suggests that a pre-production model had been made. This, however, was abandoned in favour of the Class AL1 previously modelled by Meccano Ltd.

The Adopted Tri-ang Railways Range

After 15 years of development the Tri-ang railways range of locomotives was nicely balanced by 1965. The livery, with minor exceptions, was strictly early 1960s British Railways green or black.

In the BR steam depot there were three express

locomotives: 'Britannia', 'Winston Churchill' and 'Princess Elizabeth' – the last being in kit form only by then. Mixed traffic locomotives included the L1, B12 and the 2-6-2 standard tank and amongst the smaller engines were the Deeley 3F, the Jinty and the industrial tank engines: 'Nellie', 'Polly' and 'Connie'.

In the goods yard, modern traction was represented by the green 0-4-0 and 0-6-0 diesel shunters and the dock shunter. Only one mainline diesel, the A1A-A1A Brush Type 2 (now green again), and one electric ('Electra') were available but an English Electric Type 3 Co-Co was promised, as too was the West Coast mainline electric referred to above.

There were two diesel multiple units: the Metropolitan Cammell Class 101 and the Blue Pullman.

Even this late in the day there were four Transcontinental locomotives in the British catalogue. These were the record-selling Double Ended Diesel, the Continental Prairie, the Yard Switcher and, for many people the most attractive of the TC range, the Budd RDC.

Finally there were three classical locomotives: the Caledonian Single, 'Rocket' and 'Davy Crockett'.

In 1965 there were not many model railway systems in the world that offered a choice of 24 locomotives all available from their stores.

NEW MODELS

Plans

The 1966 catalogue spoke of interesting things to come and gave no hint of the growing recession in the model railway market. On the contrary, the cover carried an illustration of the class AL1 which was still awaited but would come out later that year.

Turning to the first page we were immediately introduced to two completely new locomotives, both representing the hitherto-neglected Western Region. For steam enthusiasts there would be a model of a Hall Class loco which also helped to plug the gap between prestigious express locomotives and the smaller mixed traffic engines. For the modern traction enthusiasts there was to be a model of the elegant Hymek diesel hydraulic which had started to replace the Halls on the real railways. "To be released in 1966" said the caption and in this Rovex were as good as their word. The other two awaited locomotives also arrived in 1966: the Type 3 Co-Co and the Class AL1.

The First Tri-ang Hornby Locomotive

There has long been disagreement as to which was truly the first Tri-ang Hornby locomotive. Certainly the first to be released after 1st May 1965 was the Co-Co but it had been planned, designed and tooled up before this date. As we have seen an AL3 had been planned for some time but the original proposal was dropped in favour of adapting the Hornby-Dublo AL1 model. Some, therefore, claim that this was the first and it seems appropriate as, with the exception of the Coronation tender, it is the only example of a Tri-ang Hornby model that borrowed from both the former systems.

The Hall and the Hymek were not announced until after the change of name and, although research for these two models may have been started prior to 1st May 1965, their development seems to have taken place after this date.

Other Locomotive Changes in 1966

Besides the announcement of the two new locomotives and the inclusion of the two overstocked

1966 catalogue cover specially painted by Terence Cuneo. The AL1 electric locomotive, more than any other model, signifies the marriage between Tri-ang Railways and Hornby-Dublo.

Hornby-Dublo models from the Binns Road stores, a number of new liveries appeared on locomotives used in sets. The most notable of these was that of the Deeley 0-6-0 3F which was shown in BR maroon. Although doubts about its release in this livery abound, it certainly did appear in MR maroon livery during that year.

As we have seen in the chapter on sets, with the change of the military models to Battle Space livery, two sets were produced each with a loco in Battle Space livery. The first was a khaki 0-6-0 Jinty tank and the other the 0-4-0 diesel shunter in bright red. Both carried Battle Space stickers.

1966 was also the year that the BR electric blue livery made its appearance on certain diesel and electric locomotives. While it was to be expected on the AL1 electric, it was also shown on both the kit and ready-to-run versions of 'Electra' and as an alternative livery for the A1A-A1A.

The year, however, had seen the loss of almost all in the Transcontinental range. Indeed, of the four models illustrated in the 1965 catalogue, only the Double-Ended Diesel remained.

Economies of Design
The new 1967 catalogue cover introduced us to another new locomotive – an M7 tank. Like the B12 and the Hall, its selection had almost certainly had something to do with economics. With the absence of valve gear, the M7 provided a larger tank that was relatively cheap to manufacture and was seen as a replacement for the 2-6-2 standard tank. Ironically, it was the standard tank that had displaced the large numbers of M7 tanks on the Southern Region of British Railways.

Page 2 of the catalogue was devoted to the model's exciting new features: a glowing firebox in the cab and a smokebox door that opened to reveal the boiler tubes. These showed that Rovex had lost none of their competitive edge despite the recession.

While 1966 gave an increased range of models and liveries, 1967 saw a consolidation of the range. The

two Hornby-Dublo locos had been dropped as had the last of the Transcontinental models: the Double-Ended Diesel. The 2-6-2 Standard tank and the Metropolitan Cammell DMU were also dropped that year along with the lined blue A1A-A1A which had arrived only the year before.

'Flying Scotsman'
The exciting news in 1968 was the expected arrival of a brand new express locomotive. The preservation in running order of 'Flying Scotsman' had made it a natural choice for modelling and both Rovex and Trix had decided to produce one. The race was on to see who could get their model in the shops first. For Rovex it was to be their most successful locomotive.

Most notable in 1968 was Tri-ang Hornby's move to electric blue livery with the new white double arrow insignia for its diesel and electric locomotives. Models receiving this livery included the A1A-A1A, Hymek, Co-Co, 'Electra' and the AL1. These were not additional liveries but replacements for those previously carried by the models. From this date the Pullman blue DMU was given yellow cabs.

The year also saw the addition to the kit range (now called assembly packs) of a blue Hymek and a BR black B12. The livery of the 'Electra' kit was also changed to blue.

Two of the Transcontinental locomotives were once again featured in the 1968 catalogue. These were the Double-Ended Diesel and, in Canadian National livery, the R55 Single-Ended Diesel (or A Unit). Models dropped in 1968 included the BR green 0-4-0 diesel shunter, 'Nellie', the Caledonian Single and the L1 in BR green.

Pre-Nationalisation and Rail Blue Liveries
Although there were no completely new loco models in 1969, the year was an important turning point. It saw the first of a series of steam engines in pre-nationalisation livery with the announcement of 'Flying Scotsman' in LNER apple green, complete with corridor tender.

Another important stage was reached that year when the diesel and electric locomotives appeared in BR rail blue instead of the electric or experimental blue previously used by Rovex. Benefiting from the new liveries were the AL1, 'Electra', Hymek, A1A-A1A, 0-6-0DS, Co-Co and in grey and rail blue the Pullman DMU. Once again the new liveries replaced the old ones.

Surprise of the year was the reappearance of the Continental Prairie remodelled to look more Germanic than French. It now sported two domes and red chassis and wheels. With this model came a return of the model that shared its chassis, the 2-6-2 standard tank. Another surprise return was 'The Princess Royal' in BR maroon as a ready-to-run model.

This was also the year that 'heavy die-cast metal locomotives built for Tri-ang Hornby by Wrenn' first appeared in the catalogue. A page was devoted to them and the following five models were offered: 'Cardiff Castle', 8F (No.48073), 'City of London', 2-6-4 tank (No.80033) and 'Barnstaple'.

Absentees were the B12 assembly pack and the Deeley 3F.

Further New Liveries
What was started in 1969 came to fruition in 1970 with the almost complete transformation of the locomotive stud to pre-nationalisation liveries. For the LNER there were 'Flying Scotsman' and the B12 in apple green; the Southern was represented by 'Winston Churchill' and the M7 tank both in malachite green; the GWR was represented solely by the Hall, while the LMS was celebrated with a maroon 'Princess Elizabeth' and a completely new model: a streamlined 'Coronation' in blue. The latter made use of the 'Flying Scotsman' chassis and the former Hornby-Dublo Duchess tender base.

BR steam continued to be represented by no fewer than six models: 'Britannia', 'Flying Scotsman', B12, 2-6-2 tank, M7 tank and 0-6-0 Jinty and the industrial scene by 'Polly' and a cousin in apple green with

Tri-ang Hornby locomotives. New models introduced after April 1965 in BR livery: Co-Co, Hall, AL1, Hymek, M7, Scotsman and 'Evening Star', and in pre-nationalisation livery: Hall, Scotsman, B12, Princess, Battle of Britain, Jinty, M7, blue and red Coronations and pannier tank.

An advertisement to introduce two new locomotives.

Alternative names and numbers, in the form of transfers and self-adhesive stickers were available with some models from 1970.

Tri-ang HORNBY
Model Railways

ALTERNATIVE LOCOMOTIVE NAMES AND NUMBERS.

To change the name and number of locomotives you should proceed as detailed below for your particular model.

R.258G & R.386G " PRINCESS " CLASS
Remove name and number from each side of the loco and cut out the new nameplates and numbers. Wet the back of transfer sufficiently (do not soak) to allow printed film to slide on gum. Remove and place in position pressing down firmly.

Princess Louise	6204
Lady Patricia	6210
Duchess of Kent	6212

R.259S " BRITANNIA " CLASS
Cut out the new cab-side number. Wet the back of transfer sufficiently (do not soak) to allow printed film to slide on gum. Remove and place in position over existing number, which is printed, and press down firmly.
The new smoke box numbers and nameplates are self-adhesive and may be affixed over the existing name and number.

Robert Burns	70006
Owen Glendower	70010
Oliver Cromwell	70013

R.759G " HALL " CLASS
Remove name and numbers from each side of loco and cut out the new number. Wet the back of transfer sufficiently (do not soak) to allow printed film to slide on gum. Remove and place in position pressing down firmly.
The new nameplates are self-adhesive labels.

Crumlin Hall	4916
Garth Hall	5955
Burton Hall	6922

R.864 " CORONATION " CLASS
Remove name and number from each side of the loco and cut out the new number. Wet the back of transfer sufficiently (do not soak) to allow printed film to slide on gum. Remove and place in position firmly. The new nameplates are self-adhesive labels.

Queen Elizabeth	6221
Queen Mary	6222
Princess Alexandra	6224

R.869S " BATTLE OF BRITAIN " CLASS
Remove name, crest and number from each side of the loco and cut out the new nameplates, crests and numbers. Wet the back of transfer sufficiently (do not soak) to allow printed film to slide on gum. Remove and place in position pressing down firmly.

NOTE : The new crest suits any of the three new names.

Biggin Hill	21C157
Fighter Command	21C164
Hurricane	21C165

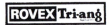

ROVEX Tri-ang
ROVEX TRI-ANG LIMITED
WESTWOOD, MARGATE, KENT

04/072/NN/61493/2   Printed in England

Instruction sheet issued with sets of alternative names.

Locomotive kits and accessories from the 1970 catalogue.

Tri-ang HORNBY
LOCOMOTIVE KITS AND ACCESSORIES

R.388U CLASS EM2 Co-Co LOCOMOTIVE "ELECTRA". Supplied complete with screwdriver, the only tool you need. No previous experience is needed to assemble this model and building it will help you to understand how the locomotive operates and will stimulate your interest in engineering.

R.386G 4-6-2 LOCOMOTIVE "THE PRINCESS ROYAL" with crew and tender. Decorated in authentic L.M.S. livery, a choice of three names is included with the locomotive. All motor parts are already wired and pre-soldered. Just assemble, with screwdriver included in the pack.

R.396 "HYMEK" B-B DIESEL HYDRAULIC LOCO-MOTIVE. The kit includes sufficient finished parts for you to assemble a replica of this modern B.R. diesel locomotive. Fully detailed instructions and screwdriver are included and no expert knowledge is required.

R.281 5 TRAIN FIGURES. Driver, Fireman, Motorman and 2 Guards.

R.413 LOCOMOTIVE CREW. Available for your older locomotives.

R.521 CAPSULE OF SMOKE OIL. Locomotives with Syncrosmoke are supplied complete with one capsule.

R.297U POWER CLEAN BRUSH. Plugs into Super 4 Power Connecting Clip or System 6 Track to make locomotive wheels revolve while brush simultaneously removes dirt.

Diesels and electric locomotives were turned out in BR Rail blue from 1969.

'27' on its tanks. A further important feature in 1970 was the introduction of alternative names for four of the locos. These were: the Hall, Princess, Battle of Britain and the Coronation.

Having gone through major livery changes over the last two years, the diesel and electric models stabilised in 1970. The only changes to the range of locomotive kits was the adoption of LMS maroon for the Princess.

The Wrenn Range Expands
1970 also saw an expansion of the Tri-ang Wrenn locomotive range with the addition of the N2 tank in BR black and the same model in LNER green, A4 'Mallard' in BR green, A4 'Sir Nigel Gresley' in LNER blue and 'City of Stoke-on-Trent' in post-war LMS black. Three of the previous year's models had been replaced: 'Barnstaple' replaced by 'Dorchester', 'Cardiff Castle' by 'Devizes Castle' (now in GWR livery) and the BR 8F by one in LMS livery.

Deletions in 1970 were few. Except for those models replaced by new liveries, the only exclusions from the catalogue were the Transcontinental models including the Prairie tank, that had survived for just one year in its revised form as it had failed to secure sufficient orders.

Finally, an addition that went almost unnoticed but was to provide the highest volume of sales of any model locomotive made by the company, was the new tank engine developed for starter sets. An 0-4-0 with a continental-style body, the R852 (electric) and R854 (clockwork) replaced the top tank 0-4-0, the tool for which was required to make the wild west loco for the RB6 set. The Continental 0-4-0 was to become the longest surviving of all the cheap starter set engines and was still to be found in sets in the late 1990s.

Another, but short-lived, starter set loco made its first appearance in 1970. This was the centre cab diesel believed to have been modelled on a Swedish prototype.

'Evening Star'
After the immense development of locomotives in 1970 it is tempting to say 'follow that'. The supposed challenge was answered with a locomotive that would be a milestone in the history of British model locomotive design. This was the 2-10-0 Class 9F 'Evening Star' which, when it finally arrived in 1972, had been many months in the making. This was a totally new model incorporating new concepts, most important of which was the tender drive based on an earlier Fleischmann design. Although it probably was not realised at the time, this model was to shape the future of Rovex and Hornby Hobbies locomotive design.

The pre-nationalisation range was expanded further in 1971 with the addition of an LMS maroon streamlined Coronation, with alternative names, and the 0-6-0 3F Jinty, for the first time, in LMS maroon. The problem of introducing pre-nationalisation liveries lay with the GWR as Rovex had no models of small GWR locomotives. What was needed was a model of the familiar and much-loved pannier tank. Work, consequently, was put in hand to produce one. In the meantime the 'Lord of the Isles' was reintroduced to plug the gap, and, as it shared the same chassis, it was decided to bring back the Caledonian Single, although this reintroduction was delayed.

As the pre-nationalisation range grew, the BR steam stable was weakened further with the loss of the B12 and M7. 'Polly' was replaced by 'Nellie' although the former remained available in a set. The maroon Coronation and both the LNER 'Flying Scotsman' and 'Britannia' would now have nickelled tyres on the wheels. In the modern traction range, 1971 saw the re-emergence of the Metropolitan Cammell DMU, now in BR rail blue livery but without the centre car. In the assembly packs the Princess was replaced by a kit of the B12 in LNER apple green.

The Tri-ang Wrenn models were again featured in the catalogue. The BR N2 tank had disappeared but both LMS and BR versions of the 8F were available. Not all the locomotives were illustrated.

Planned for 1972
Although the product name changed from Tri-ang Hornby to Hornby Railways on 1st May 1972, no immediate change was seen in the model range. Indeed, the 1972 catalogue and its contents had been designed in mid-1971 when any thoughts of name changes were far away. For this reason the locomotive range for 1972 is recorded here even though most of the new models would have been issued in the new packaging.

Completely new to the catalogue was the R51S GWR Pannier Tank, although it had reached the shops before the end of 1971, and it is covered in detail later in this chapter. Also described as 'new' was the L1 now in Southern dark green livery and the M7 tank changed to the darker Southern livery. The BR version of 'Flying Scotsman', however, had been dropped in 1972. In the diesel and electric depot the dock shunter had returned to black and the A1A-A1A to BR green. The electric locomotives and multiple units had all gone.

In sets there were two new Transcontinental variations in the form of the Continental Pacific with a black Scotsman tender and the A Unit in bright red.

In the Wrenn range two models had been dropped but the 2-6-4 and 0-6-2 tanks were to be released in LMS livery and the R1 0-6-0 tank was a new addition in BR black or Southern green.

Back in 1970 thought had been given to the possibility of producing a long-bodied, clockwork-powered locomotive which was allocated the number R884. In February that year the possibility of a reversing mechanism for it was investigated, but further consideration of this was deferred until 1972. In January 1971, there was reference to it as an 0-6-0 with a matt black body with a white number and red wheels and a proposed release date of April 1972. However, no further references to it can be found.

CHECK LIST OF LOCOMOTIVES

The codes used in this Check List are as follows:

Tenders. Where a model has a tender or dummy power unit with its own 'R' number, this is given in brackets after the name of the locomotive.

Colours. Where two colours are shown separated by a plus sign it indicates that the model is bi-coloured. Where they are separated by a comma it indicated that the model may be found in either colour.

Dates. The dates given are those when stocks of the model are known to have been made even though they may have been listed and illustrated in the catalogue in other years.

Decals. British Rail livery codes used refer to the insignia carried by the model as follows:

BRb   lion astride a wheel
BRc   lion holding a wheel (1957)
BRd   adaptation of BRc for AL1 loco
BRe   double arrow

Up until around 1958/59, the lion on BR totems faced the front of the engine but, following a complaint from the Royal College of Arms, BR ensured that the lion always looked to the left whichever side of the loco it was. Rovex acknowledged this change by altering their transfers to one type only.

For ease of reference, the locomotives have been separated under the following headings:

1. British Prototypes
2. For Starter Sets
3. Continental
4. Transcontinental – General Issue
5. Made for Canada
6. Made for America
7. Used in Australia
8. Made for France
9. Made for Italy
10. Used in New Zealand
11. Former Hornby-Dublo Stock Disposed of By Rovex

1. British Prototypes

The following locomotives were already in production when Tri-ang Railways became Tri-ang Hornby. The list includes subsequent models based on these originals but does not include locos made for starter sets which, for convenience, have been listed separately in Section 2 below:

No.	Livery	British Prototypes	Colour	Dates
		PRINCESS		
R386	BRc	46201 'Princess Elizabeth' CKD Kit	green	62-69
R386G	LMS	6200 'The Princess Royal' CKD Kit	maroon	70
R258	BRc	46200 'The Princess Royal'	maroon	69
R258G	LMS	6201 'Princess Elizabeth'	gloss maroon	70
	alt.name	6204 'Princess Louise'		
	alt.name	6210 'Lady Patricia'		
	alt.name	6212 'Duchess of Kent'		
R258NS	LMS	6201 'Princess Elizabeth', noise	gloss maroon	71-74
	alt.name	6204 'Princess Louise'		
	alt.name	6210 'Lady Patricia'		
	alt.name	6212 'Duchess of Kent'		
		JINTY		
R52S	BRc	47606 Jinty with smoke	black	65-72
R52SR	LMS	7606 Jinty	gloss maroon	70-73
R377S	GN&SR	Jinty (from Railway Children set)	gloss brown	70-72
R558S	BS	*see also* Battle Space	khaki	66-67
		BR 2-6-2 TANK		
R59	BRc	82004 Class 3 2-6-2 Tank	green	60-66
R59S	BRc	82004 Class 3 2-6-2 Tank	green	69-72
		0-6-0 DIESEL SHUNTER		
R152	BRb	D3035 0-6-0 Diesel Shunter	green	59-68
R152	BRe	D3035 0-6-0 Diesel Shunter	rail blue	69-73
		DOCK SHUNTER		
R253	PO	No.3 0-4-0 dock shunter	red	62-72
		DIESEL RAILCAR		
R157	BRc	DMU (& R158) M79079/M79632	green	58-67
R157C	BRe	DMU 2 car unit M79079/M79632	rail blue	70-71
		CLASS 3F TENDER LOCO		
R251S	BRc	43775 Deeley 0-6-0 3F (& R33)	black	58-66
R661S	BRc	43775 Old Smoky	black	65
R251	BRc	43620 Deeley 0-6-0 3F (& R33)	maroon	66
R251	MR	3775 Class 3F 0-6-0	maroon	66-67

No.	Livery	British Prototypes	Colour	Dates
		L1		
R350	BRc	31757 L1 4-4-0 (& R36)	gloss green	60-68
		BRITANNIA		
R259	BRc	70000 'Britannia' (& R35)	green	65
R259S	BRc	70000 'Britannia' (& R35) with smoke	green	61-70
R259SN	BRc	70000 'Britannia' (& R35N) smoke, noise	green	71-72
	alt.name	70006 Robert Burns'		
	alt.name	70010 'Owen Glendower'		
	alt.name	70013 'Oliver Cromwell'		
		INDUSTRIAL TANK		
R355B	PO	No.7 'Nellie'	blue	60-67
R355B	PO	No.7 'Nellie'	bright blue	71-72
R355R	PO	No.9 'Polly'	red	63-70
R355G	PO	No.57	light green	69-72
		DEAN SINGLE		
R354	GWR	3046 'Lord of the Isles' (& R37)	matt green	61-65
R354	GWR	3046 'Lord of the Isles' (& R37)	matt green	67
R354	GWR	3046 'Lord of the Isles'	gloss green	70-73
		EM2		
R351	BRc	27000 'Electra'	green	65
R388	BRc	27000 'Electra' CKD Kit	green	65
	alt.name	27002 'Aurora'		
	alt.name	27006 'Pandora'		
R351	BRc	27000 'Electra'	electric blue	66-67
R388	BRc	27000 'Electra' CKD Kit	electric blue	66-67
	alt.name	27002 'Aurora'		
	alt.name	27006 'Pandora'		
R351	BRe	27000 'Electra'	electric blue	68
R388	BRe	27000 'Electra' Assembly Pack	electric blue	68
	alt.name	27002 'Aurora'		
	alt.name	27006 'Pandora'		
R351	BRe	27000 'Electra'	rail blue	69-71
R388	BRe	27000 'Electra' Assembly Pack	rail blue	69-70
	alt.name	27002 'Aurora'		
	alt.name	27006 'Pandora'		
		BATTLE OF BRITAIN		
R356S	BRc	34051 'Winston Churchill'	green	61-69
R869S	SR	21C151 'Winston Churchill'	gloss malachite	69-70
	alt.name	21C157 'Biggin Hill'		
	alt.name	21C164 'Fighter Command'		
	alt.name	21C165 'Hurricane'		

No.	Livery	British Prototypes	Colour	Dates
		TYPE 2 BRUSH DIESEL		
R357B	BRc	D5578 A1A-A1A	exp. blue	65-66
R357G	BRc	D5572 A1A-A1A	green	63-67
R357	BRe	D5572 A1A-A1A	electric blue	68
R357	BRe	D5572 A1A-A1A	rail blue	69-71
		0-4-0 DIESEL SHUNTER		
R559	BRc	D2907 North British 0-4-0DS	green	63-67
R756	BS	see Battle Space	red	66-67
		see *also* Starter Set locos		
		B12		
R150S	BRc	61572 with smoke	black	64-69
R150NS	BRc	61572 with smoke, noise	black	71
R359S	BRc	61572 with smoke Assembly Pack	black	68-69
R866S	LNER	8509 with smoke	gloss lt. green	70
R866NS	LNER	8509 with smoke, noise	gloss lt. green	70-74
R359S	LNER	8509 with smoke Assembly Pack	gloss lt. green	70-71
		DMU PULLMAN		
R555	BR Pullman	DMU Pullman (& R556)	blue & white	63-67
R555	BR Pullman	DMU Pullman (& R556) yellow front	blue & white	68
R555C	BR Pullman	DMU Pullman 2 car unit	grey & blue	69-72
		ROCKET		
R651S	L&MR	'Rocket' (& R652)	yellow	63-66
R651	L&MR	'Rocket' (& R652)	yellow	68-69
		CALEDONIAN SINGLE		
R553	CR	No.123 Caledonian 4-2-2 (& R554)	matt blue	63-66
R553	CR	No.123 Caledonian 4-2-2	gloss blue	71-73

The following models were developed from the mid-1960s onwards and did not appear until after May 1965.

No.	Livery	New British Prototypes	Colour	Dates
		CLASS 37		
R751	BRc	D6830 Co-Co Type 3	green	66-67
R751	BRe	D6830 Co-Co Type 3	rail blue	68-84
		CLASS 81		
R753	BRd	'E3001' with 2 pantographs	electric blue	66
R753	BRd	'E3001' with 1 pantograph	electric blue	67-68
R753	BRe	'E3001' with 1 pantograph	rail blue	69-70

No.	Livery	New British Prototypes	Colour	Dates
		HALL		
R759	BRc	4983 'Albert Hall'	green	66-69
R759G	GWR	4983 'Albert Hall'	gloss green	70
	alt.name	4916 'Crumlin Hall'		
	alt.name	6922 'Burton Hall'		
	alt.name	5955 'Garth Hall'		
R759N	GWR	4983 'Albert Hall', noise	gloss green	71-77
	alt.name	4916 'Crumlin Hall'		
	alt.name	6922 'Burton Hall'		
	alt.name	5955 'Garth Hall'		
		HYMEK		
R758	BRc	D7063 Hymek B-B	green	67
R758	BRe	D7063 Hymek B-B	electric blue	68
R758	BRe	D7063 Hymek B-B	rail blue	70-76
R396	BRe	D7063 Hymek B-B assembly pack	electric blue	68
R396	BRe	D7063 Hymek B-B assembly pack	rail blue	69-70
		M7 TANK		
R754	BRc	30027 M7 4-4-0T, firebox glow	black	67-70
R868	SR	328 M7 4-4-0T, firebox glow	gloss malachite	69-70
R868	SR	328 M7 4-4-0T, no firebox glow	gloss malachite	71
		FLYING SCOTSMAN		
R850	BRc	60103 'Flying Scotsman' red plate	green	68
R850	BRc	60103 'Flying Scotsman' blk. plate	green	69-70
R850	BRc	60103 'Flying Scotsman' corridor tender	green	69-70?
R855	LNER	4472 'Flying Scotsman' A3	gloss lt. green	68-70
R855N	LNER	4472 'Flying Scotsman' A3, noise	gloss lt. green	71-77
		CORONATION		
R864	LMS	6220 'Coronation'	gloss blue	70-72
	alt.name	6221 'Queen Elizabeth'		
	alt.name	6222 'Queen Mary'		
	alt.name	6224 'Princess Alexandra'		
R871	LMS	6244 'King George VI'	gloss maroon	71-74
	alt.name	6221 'Queen Elizabeth'		
	alt.name	6228 'Duchess of Rutland'		
	alt.name	6241 'City of Edinburgh'		
		9F		
R861	BRc	92220 'Evening Star'	green	71-75
		PANNIER TANK		
R51S	GWR	8751 0-6-0PT	green	72

2. For Starter Sets

These are locomotives that were made almost exclusively for use in starter sets.

No.	Livery	For Starter Sets	Colour	Dates
		TOP TANK 0-4-0T		
R659	-	Top Tank 0-4-0T	black	63-67
R660	-	c/w Top Tank 0-4-0T	black	65
R660	-	c/w Top Tank 0-4-0T	bright blue	66?
R660	-	c/w Top Tank 0-4-0T	mid blue	66?
R660	-	c/w Top Tank 0-4-0T	yellow	67
R660	-	c/w Top Tank 0-4-0T	dark green	68
		WILD WEST 0-4-0		
R873 (R874)	-	1863 Wild West 0-4-0 loco	red	71
		NORTH BRITISH 0-4-0DS		
R756	BRc	c/w North British 0-4-0DS	red	66
R557?	BRc	c/w North British 0-4-0DS	apple green	68?
R557?		c/w North British 0-4-0DS	violet blue	68?
		BARCLAY 0-4-0DS		
R858		Barclay 0-4-0DS	blue	69-71?
		CONTINENTAL 0-4-0T		
R852	-	7744 Continental 0-4-0T	blue	68-74
R852	-	7744 Continental 0-4-0T	black	?68
R854	-	c/w Continental 0-4-0T	red	69?-79
R854	-	1863 c/w Continental 0-4-0T	red	71?
R854	-	c/w Continental 0-4-0T	maroon	69?-79
R854	-	c/w Continental 0-4-0T	green	71
		SWEDISH 0-4-0DS		
R853	-	5771 Swedish 0-4-0DS	yellow & blue	69-71
R853	-	4718 Swedish 0-4-0DS	red & black	69-71

3. Continental

Only one model was based on designs of Continental Europe.

No.	Livery	Continental	Colour	Dates
R653	-	Continental Prairie	black & red	69

4. Transcontinental – General Issue

No.	Livery	Transcontinental	Colour	Dates
		TC PACIFIC		
R54S	TR	2335 'Hiawatha' (& R32)	black	62-69
R54S	TR	2335 'Hiawatha'	black	68
R?	-	2335 'Hiawatha' with A3 tender	black	70?
R1542N	-	1542 Pacific with A3 tender	black	70-72
R0542NS	-	1542 Pacific with A3 tender	black	72
R054	-	1542 Pacific with A3 tender	black	73
		A UNIT		
R55	TC	4008 A Unit	silver	70
R0550	TC	4008 A Unit	red	71
		ROAD SWITCHER		
R1550	TR	7005 Diesel Switcher	yellow	made?
		DOUBLE ENDED DIESEL		
R159	TC	5007 Double Ended Diesel	VR blue	68,70
		YARD SWITCHER		
R353	TC	TR20071 Yard Switcher 0-4-0	red	62-68
		DAVY CROCKETT		
R358S	TR	1863 'Davy Crockett'	red & yellow	62-67

1stNo.	2ndNo.	3rdNo.	Livery	For Canada	Colour	Dates
				ROAD SWITCHER		
R155CN	R1551	R370	CN	3000 Bo-Bo Switcher	black	65-72
R155CP	R1552	-	CP	3000 Bo-Bo Switcher	sky & maroon	69
-	R1553	R371	CP Rail	1553 Bo-Bo Switcher	red	71-73
				DOCK SHUNTER		
R253C	-	-	-	0-4-0 Dock Shunter	red	67
				BUDD RDC		
R352CN & R232CN	R3521	-	CN	101 Budd RDC	silver	65-67,69,71
R352SF & R232SF	-	-	-	Santa Fe Budd RDC	silver	made?
				0-6-0 DS		
-	R1520	R369	-	1520 0-6-0DS	elec. blue	73
				CONTINENTAL 0-4-0T		
R852CN	R8520	-	-	'Chugga '	yellow	69

5. Made For Canada

Canadian models were renumbered for the 1970 catalogue. Where a model has an old number this is given in the first column and the new number in the second. Some models in Canadian livery (such as R55CN) were also included in the UK catalogue but for the sake of simplicity they are only listed here.

1stNo.	2ndNo.	3rdNo.	Livery	For Canada	Colour	Dates
				TC PACIFIC		
R54SCP	-	-	CP	2335 Pacific 4-6-2(& R32CP)	black	69
-	R0542S	R054?	CP	2335 Pacific 4-6-2	black	70-71
-	R1542N	R542?	CP	2335 Pacific 4-6-2	black	71-73
				A UNIT		
R55CN	R0551	R325	CN	4008 A Unit	black & silver	65-66, 69-73
R55CP	R0552	-	CP	4008 A Unit	sky & maroon	67-69
R553	R0553	R326	CP Rail	1404 A Unit	red	70-73

6. Made for America

No.	Livery	For America	Colour	Dates
		BUDD RDC		
R829/ R825	Northern Pacific	303 Budd RDC	silver	68
R830/ R826	Santa Fe	3403 Budd RDC (see also Canada)	silver	68
R831/ R827	Chesapeake & Ohio	9003 Budd RDC	silver	68
R832/ R828	Reading Lines	510/513 Budd RDC	silver	68
		ROCKET		
R800	L&MR	'Rocket'	yellow	67
		TOP TANK 0-4-0T		
R660T	-	c/w Top Tank 0-4-0T	yellow	68
		CONTINENTAL 0-4-0 T		
R852T	-	7744 Continental 0-4-0 T	black	68-69
R854T	-	c/w Continental 0-4-0T	?	sent?

7. Used in Australia

Dates of when models were made or finished in Australia are difficult to ascertain as we have only records of when solo locos were made. Many of those finished at Moldex were for set production. The dates used in the table below should therefore be used only as a guide.

M = made in Australia, F = finished in Australia, R = sent ready made from UK.

No.	Livery	Australia	Colour	Dates
		JINTY		
R52	BRc	Jinty 0-6-0t (M)	black or red	59-67
R52	BRc	47606 Jinty 0-6-0T (R)	black	70
		TC PACIFIC		
R54	TR	2335 'Hiawatha' (& R32) (F)	black	63-67
R54	TR	2335 'Hiawatha' (R)	black	70
		A UNIT		
R55	TC	4008 A Unit (F)	silver & red	61-68
R55	TA	4008 A Unit (F)	silver & red	65-66
R55	TA?	4008 A Unit (R)	silver & red	70
		ROAD SWITCHER		
R1550	TR/TA?	7005 Diesel Switcher (R)	yellow	made?
		DOUBLE ENDED DIESEL		
R159	TC	5007 Double Ended Diesel (M)	VR blue	58-67
R159	VR	5007 Double Ended Diesel (R)	VR blue	68,70
R159	VR	5007 Double Ended Diesel (R)	rail blue	68,70
R1590	VR	5007 Double Ended Diesel (R)		71
R159A	VR	5007 Double Ended Diesel (R)	electric blue	post 71
		PRINCESS		
R258	BRc	46200 'Princess Royal' (& R34) (M)	maroon	63-67
		BUDD RDC		
R352	TC	31018 Budd RDC (& R232) (F)	silver & red	65-67
R352A	TA	Budd RDC (F)	silver & red	65-67?
		YARD SWITCHER		
R353	TA?	TR20071 Yard Switcher 0-4-0 (F)	yellow	61-67
		DEAN SINGLE		
R354	GWR	3046 'Lord of the Isles' (F)	green	63-67?
		BRUSH TYPE 2		
R357A	BRe	D5572 A1A-A1A (R)	electric blue	68
R357A	BRe	D5572 A1A-A1A (R)	rail blue	69
		DAVY CROCKETT		
R358S	TR	1863 'Davy Crockett', tender (F)	red & yellow	63-67

No.	Livery	Australia	Colour	Dates
		SYDNEY SUBURBAN UNIT		
R450	NSWR	Sydney Suburban Electric motor car (M)	tuscan red	63-68
R451	NSWR	Sydney Suburban Electric trailer car (M)	tuscan red	63-68
R452	NSWR	Sydney Suburban Electric non-powered motor car (M)	tuscan red	63-68
		ROCKET		
R651A	L&MR	'Rocket' (F)	yellow	64-69
		CLASS 3F TENDER LOCO		
R661S	BRc	43775 Old Smoky (F)	black	65-66?
		M 7		
R754	BRc	30027 M7 Tank (R)	black	69
		NORTH BRITISH 0-4-0DS		
R654	?	North British 0-4-0DS	?	66-67?
		TOP TANK 0-4-0T		
R659A	-	Top Tank 0-4-0T (R)	black	65-67?
R660A	-	c/w Top Tank 0-4-0T (R)		68
		CONTINENTAL 0-4-0T		
R852A	-	Continental 0-4-0T (R)	black?	68, 70
R854A	-	c/w Continental 0-4-0T (R)	red?	69-70
		SWEDISH 0-4-0DS		
R853	-	Swedish 0-4-0DS (R)		69

8. Made for France

No.	Livery	For France	Colour	Dates
R54SF	TR	2335 'Hiawatha' with smoke (R54SF & R32F)	black	67-69
R150SF	BRc	61572 B12 with smoke (R150S & R39F)	black	67-70
R259SF	BRc	70000 'Britannia' with smoke (R259 & R35F)	green	67-70

9. Made for Italy

No.	Livery	For Italy	Colour	Dates
R54SL	TR	2335 'Hiawatha' (R54 & R32L)	black	68
R358SL	TR	1863 'Davy Crockett' (R358 & R233L)	yellow & red	68

10. Used in New Zealand

Dates quoted in the table below are for when the models were available and not necessarily when they were made. (F = finished only)

No.	Livery	New Zealand	Colour	Dates
		BALTIC TANK		
R56	TR	4-6-4 Baltic tank	black	57-66
R56S	TR	4-6-4 Baltic tank with smoke	black	63-66
		DOUBLE ENDED DIESEL		
R159	TR	5007 Diesel	blue	58-72
R159	TR	5007 Diesel (R250)		
		(see separate table in Directory)	red	58-72
		PRINCESS		
R50S	BRc	46205 'Princess' (& R30)	black	62-72
R258AS	BR	46205? 'Princess'	maroon	72
		JINTY		
R52	BRc	47606 Jinty 0-6-0T	black	59-72
		0-6-0 DIESEL SHUNTER		
R152		0-6-0 DS (see separate table in Directory)		
		EMU		
R156	BR	EMU (see separate table in Directory)	green	64-70?
		TC PACIFIC		
R54S	TR	2335 'Hiawatha' (& R32) (F)	black	68-71
R54SN	TR	2335 'Hiawatha' (& R856N) (F)	black	71
		HYMEK		
R758	BRc	D7063 Hymek (F)	blue	68-71
		TOP TANK 0-4-0T		
R660	-	Top Tank 0-4-0T (F)	?	68-70
		CONTINENTAL 0-4-0T		
R852	-	Continental 0-4-0T (F)	?	69-71
R854	-	c/w Continental 0-4-0T (F)	?	71

11. Former Hornby-Dublo Stock Disposed of by Rovex

No.	Livery	Hornby-Dublo Stock	Colour	Dates
R2207	BR	0-6-0 Tank	green	65
R2217	BR	0-6-2 Tank	black	65
R2231	BR	0-6-0 Diesel Shunter	green	65
R2232	BR	Co-Co	green	65-67
R2233	BR	Co-Bo	green	65-67
R2235	BR	'Barnstaple'	green	65-66
R2250 (R4150)	BR	EMU	green	65-67

The 1970 catalogue.

DIRECTORY OF LOCOMOTIVES

The directory contains three main sections and three small ones. The main ones deal with British locomotives, those made for starter sets and the Transcontinental series. The chapter is therefore divided up into the following sections:

A. British Outline.
B. Locomotives for Starter Sets.
C. Continental.
D. Transcontinental.
E. Australian.
F. Former Hornby-Dublo Stock.

Battle Space

Locos have been included together with coaches, wagons and Battle Space sets in a section of their own at the end of the 'Wagons' chapter.

Australian and New Zealand Models

Both Australia and New Zealand made their own locomotives from mouldings they produced themselves and other parts sent out from the UK.

Australia abandoned production in 1968 but New Zealand continued to be reasonably self sufficient until 1972 after which sets continued to be assembled using imported Hornby models and track.

Even in the early days, not all the locomotives listed by these companies were made by them. Some were bought in bulk-packed for finishing and using in set production. This often meant little more than fitting couplings or attaching bogies.

With the exception of the NSWR Suburban Electric multiple unit which was made exclusively in Australia and is therefore listed in a section of its own, all Australian- and New Zealand-made or finished locos are listed under their Margate-made equivalents in the main sections of the Directory. This is so that comparisons can more easily be made with their UK equivalents.

Canadian and American Locomotives

All locomotives supplied to Canada and the USA were made in the Margate factory. The Canadian ones were handled by Lines Bros. Canadian company but the American ones were made to order for an independent company, American Train & Track (ATT). The models are listed under their UK equivalents.

When ATT persuaded Rovex to supply them with models for the American market they ordered four sets of Budd RDCs as well as starter set locos. The locos and rolling stock were fitted with the standard American NMRA couplings and were supplied bulk packed for boxing in ATT's own pale grey packaging. Residue ATT stock was later bought and repackaged by Model Power in blue boxes.

A. BRITISH OUTLINE

The locomotive models have been arranged, more or less, in date order. The first ones described were initially released as Tri-ang Railways models and have already been extensively described in **Volume 1** of this series. The details provided here deal with the changes that occurred to them while they were being sold under the Tri-ang Hornby label. Some subsequently survived into the Hornby Railways range and will appear again in **Volume 3** with further variations.

Later models described below were issued after 1965, are not in **Volume 1** and are consequently described in greater detail. A number of these will also appear in **Volume 3**.

PRINCESS 4-6-2

'Princess Elizabeth' had been the subject that launched the Tri-ang Railways system in 1952 and yet, as that system slipped into the Tri-ang Hornby years, the only Princess locomotive remaining was in the form of a CKD kit. This was in BR green livery and is dealt with below. There was, therefore, a long gap during the 1960s when no ready-to-run model of a Princess was made.

BR Green 'Princess Elizabeth' Kits R386

In 1965 almost 4,000 CKD kits of the green 'Princess Elizabeth' were made with further batches each year up to 1969. Models made from these kits would

1969 example of R258 'The Princess Royal'.

R386 Assembly Pack for the 'Princess Elizabeth'.

LMS 'Princess Louise', Duchess of Kent' and 'Lady Patricia'.

R258G LMS 'Princess Elizabeth'.

have been similar to the last of the ready-to-run models of 1961 with see-through wheels, improved Walschaerts valve gear and fluted coupling and driving rods. They would have had a one-piece die-cast chassis but no smoke unit. Perhaps the best feature for identifying the kit built models of the BR 'Princess Elizabeth' is the pair of safety valves which would by then have been the restyled scale ones. Crew figures were added in 1966.

During this period a total of 16,500 kits left the factory. The greatest number date from 1967 when 5,300 were produced. This was also the year that it became an Assembly Pack and was given new packaging in the form of an almost square tray with a cellophane cover. It seems likely that of the 16,500 kits made during the Tri-ang Hornby period, 7,700 were in CKD boxes and the remaining 8,800 in Assembly Pack trays.

LMS 'The Princess Royal' Kits R386G

In 1994 an example of the LMS version of 'The Princess Royal' was found in Australia. This appears to have been one of just 1,217 kits made for 1970 and illustrated on page 11 of the 1970 catalogue. I have not seen this but understand that it is very similar to the LMS 'Princess Elizabeth'. Why so few were made is a mystery. Recently released correspondence indicates that the stocks of these kits were used up by the summer.

'The Princess Royal' R258

In 1969, 'The Princess Royal' reappeared in the catalogue as a ready-to-run model. While much had been done to improve its appearance since the 1950s, the fundamental problem of its under-scale length remained. It was surprising, therefore, that a further 5,000 ready-to-run models of 'The Princess Royal' in maroon livery were made in 1969, six years after it had disappeared from the range. It was priced £3-9s-6d, had Magnadhesion and came with a crew.

During the intervening years there had been a number of specification changes that can be used to identify these 1969 models. Scale safety valves had been introduced to Tri-ang steam locomotives in 1962 but as no R258 maroon Princesses were made that year, models with this feature will almost certainly date from 1969. This late batch will also have been on chassis with wire pickups and large hook couplings neither of which featured on the earlier model.

The model was replaced the following year by the maroon LMS 'Princess Elizabeth' and, as far as we know, after the 1969 batch no more of the BR 'The Princess Royal' were made.

'Princess Elizabeth' LMS Maroon R258G

In line with the trend at the time, it was decided to make the Princess available in pre-nationalisation livery. The LMS maroon version which arrived in the shops in 1970 came complete with a selection of alternative names and numbers offering the following alternatives:

> 6204 'Princess Louise'
> 6210 'Lady Patricia'
> 6212 'Duchess of Kent'

The model was priced at £3-17s-0d and had nickelled wheel tyres. It also had Magnadhesion and a crew. The loco normally had a lacquered semi-gloss finish but has been seen in matt and with non-nickel tyres.

'Princess Elizabeth' LMS Maroon & Noise R258NS

A further batch of 17,400 LMS maroon models of 'Princess Elizabeth' was made in 1971. These were

R258 Australian 'Princess Royal' and R52 Jinty tank believed to date from the Tri-ang Hornby period.

in the shops by March that year and had 'Exhaust Steam Sound' installed in the tender and Synchrosmoke but no crew. 10,000 of them were used in the RS609 train set while the rest were sold solo. In order that customers, who had bought models the previous year, could convert them, a further 1,500 steam sound-fitted tenders (R34N) were available separately boxed for sale on their own. Examples with plated wheels date from 1972.

Post-1971 Models

The LMS maroon version of the Princess remained in production until 1974 although alternative names were dropped after 1972. It was also used in the R506 train set of 1973. 1974 saw the return of BR maroon, green and black versions of the model, the green one for mail order purposes only. These three versions, all called 'Princess Victoria', can be identified by the gold decals on the tender. After 1974, the tools went out of use. When the Princess class loco returned in 1984 it had been completely retooled as a scale model.

Australian – Maroon R258

We have little record of when changes in the appearance of the Australian-made Princesses took place, but it would appear that production changed from black models (R50) to maroon models (R258) in 1963. During the Tri-ang Hornby era, it is believed that only maroon models were made. The Australian 1965 catalogue showed a maroon Australian-made 'Princess Royal' with two maroon and cream TC short coaches on the inside pages, but the maroon 'The Princess Royal' featured on the front cover is clearly one imported from Britain. The following two years the inside illustration was repeated. Some were made with maroon mouldings instead of black ones sprayed maroon.

Production records from Moldex Ltd show that 3,075 solo models were produced, of which 1,400 were made after 1964 (800 were made in 1965, 350

in 1966 and 250 in 1967). We do not know how many others were manufactured for sets. The solo loco sold for £A6-12s-6d which included the tender. Australian production of the model ceased at the end of 1967. In their book *Building and Operating Model Railways*, Australian writers Alan Johnson and Bill James, said that the 'Princess Royal' made in Australia had neither Walschaerts valve gear nor smoke. The latter is believed to be untrue. According to the very limited records we have available, in 1963 and 1964 it was available either with or without a smoke generator but from 1965 only the smokeless version was available.

New Zealand – Black R50S

The model, which first appeared in the New Zealand price list in 1962, remained on sale with full gearing, smoke and Magnadhesion until the collapse of Lines Bros. in the early 1970s. It had smoke fitted throughout its life and carried the simplified name 'Princess' picked out in either white or yellow paint and the number '46205'. It is estimated that

about 2,000 were sold in sets but it is not known how many were sold solo. The solo loco sold for £NZ 4-19s-6d and the tender for NZ 19/-.

New Zealand – Maroon R258AS

A batch of 100 of the LMS maroon 'Princess Elizabeth' was sent out to New Zealand bulk packed in 1972. If not used in sets these may have been boxed up locally for solo sales.

JINTY 0-6-0T

BR Black '47606' R52S

This model, which is extensively described in **Volume 1**, remained in production throughout the Tri-ang Hornby years during which time some 70,000 were dispatched as solo models in lined BR livery and in Tri-ang Hornby boxes. The peak year for production was 1969 when over 16,000 passed through the stores. It had been planned to modify it for 1971 removing the smoke unit and Magnadhesion but it was probably found to be more economical to let it continue to share the same chassis as the LMS version which had these features.

The model, which carried the number '47606', was also used in four train sets during the period which accounted for a further 72,300 locomotives. The sets were: RS62, RS103, RS601 and RS613. In addition to being fitted with smoke from 1965 the model received Magnadhesion in 1966. The retail price of the Jinty in 1965 was £1.19s.6d. The model survived until 1972 and sometime in its final years a batch was released from the factory with cream-coloured BR decals. This version appears to be quite rare.

LMS Red '7606' R52RS

Production of the maroon LMS model started in 1970 but did not get under way in any quantity until 1971 when about 11,000 were made. It was reviewed by *Model Railway Constructor* in February 1971 when it was pointed out that while the livery was not authentic, it would appeal to youngsters. It has, no doubt, ended up as the station pilot on more than one layout. The model carried the number '7606'.

The test given the model by the magazine revealed that on nickel-silver track it would pull

R52RS LMS Jinty.

Two versions of the BR R52S Jinty. The top one was the normal form but at some time late in the Tri-ang Hornby days it was released with white-backed transfers.

R377S Jinty from the Railway Children set.

New Zealand R52 Jinty dating from 1965/66.

The R59S BR Class 2 Prairie tank as it appeared when first reintroduced in 1969.

four Hornby-Dublo super-detail coaches and five on steel track.

The model had Synchrosmoke and Magnadhesion but is not known to have appeared in any Tri-ang Hornby sets. It was sprayed with varnish to give the model a slight gloss finish and those made from 1972 onwards had plated wheel tyres. It retailed at £3-5s-0d and remained available until 1974.

GN&SR [Railway Children] R377S

This model was only available in the 'Railway Children Set' (RS615) of which about 3,300 were dispatched in 1971 and 2,400 in 1972. Taking into account stock remaining in the stores at the end of 1972 it is estimated that a little over 6,000 GN&SR Jinty tanks were made. It was fitted with smoke and Magnadhesion and the body was an ochre-yellow. These did not have plated wheel tyres.

Khaki Battle Space Version R558S

The model was used in the RS16 'Strike Force 10 Battle Space' set. It was finished in khaki with Battle Space stickers on the tank sides and was available only in the sets. 24,000 were made during 1966 and 1967 and possibly a batch of 1,500 in 1970.

Khaki BR Model

The Battle Space model may be found with late BR logos on its tank sides instead of the stickers. This is thought to be very rare.

Australian R52

Moldex in Australia continued to manufacture their own Jinty throughout 1965, 1966 and 1967 by which time it had an imitation lamp on the top of the smoke box. Examples of this period are largely unlined, numberless and include both black and maroon versions. Some also had no logo on the tanks.

Quantities were low and it is likely that just under 3,000 were made during this period This is in addition to more that 10,000 made before 1965. It sold for £A3-5s-0d and it is understood that none were made after 1967. In 1970 a batch of 940 Jintys was sent bulk packed from Margate for set assembly. These are likely to have been the BR black version and some may have been boxed up as solo models in Moldex boxes.

New Zealand R52

New Zealand continued to list the R52 tank up to 1972 but it is not known how many were made

locally or when the last batch was produced. Like the Australian model, which was made from the same tools, the Tri-ang Hornby batches would normally have been fitted with an imitation lamp on top of the smoke box. The New Zealand models were black and lined with the later BR crest, similar to the one made at Margate.

BR 2-6-2 TANK

The Class 3MT standard tanks were built between April 1952 and August 1955. '82004' was one of the first of the class having been completed in May 1952. Its last shed was Bath Green Park and was withdrawn from service in October 1965. Sadly, it was not a very successful design. Much of the work they were designed for was taken over by DMUs and they lived out an uneventful life mainly on the Southern and Western Regions. None of the class has survived.

See **Volume 1** for a detailed description of this model.

1960 Green '82004' R59

The smokeless version of the large tank engine continued in production with 4,500 being made in 1965 and 4,000 in 1966. It was then dropped in favour of the M7 tank which was cheaper to manufacture. At this late stage it had Magnadhesion but no smoke generator. The body was green with single lining and a late BR logo. It was priced £2-15s-0d.

1969 Reissued Model R59S

The green standard tank was reintroduced in 1969 to boost the BR models available and retained its original running number '82004'. It was now fitted with a smoke unit but otherwise looked the same as it did in 1966. The price, however, had gone up to £3-7s-6d. In the next three years 16,000 were made. Plated wheel tyres were proposed for 1972 but only 800 were made that year and it seems unlikely that the alterations were made for such a small number.

DIESEL SHUNTER 0-6-0DS

Green 'D3035' R152

The diesel shunter remained in BR green livery until 1968. Between 1966 and 1968 about 20,000 were made. These had Magnadhesion and sold for £1-17s-6d each. They all carried the earlier BR logo as the later one could not be fitted easily onto any of

the panels on the side of the loco.

Blue 'D3035' R152

In 1969 the model was changed to BR rail blue livery which resulted in sales rising to 11,000 for the first year. It remained in production until 1971 and in all 25,000 blue diesel shunters were sold solo. In addition the blue model was used in three sets RS7, RS89 and RS600 and these accounted for a further 26,000 locomotives. This version of the model also had Magnadhesion and in 1969, was priced £2-4s-6d. Plated wheel tyres were not fitted during the Triang Hornby period.

Australia R152

Records indicate that Australia made only the clockwork version of the 0-6-0 diesel shunter (R256) but electric versions have been found. None was made after 1962.

New Zealand R152

The locally-made diesel shunters were shown in the New Zealand catalogue until 1972. Some were sold with the BR early logo on their sides while others, particularly later on, were sold unprinted. The model sold for £NZ 2-9s-6d in 1965 and may be found in black, green or maroon livery. Other colours may also exist. The table below shows the wide variety of diesel shunters found in New Zealand:

'R' No.	Motor	Colour	No.	Logo	Wheels	Date	Notes
R152	elec.	black	none	BRb	solid	59?	
R152	elec.	black	none	BRb, small, long	solid	61?	
R152	elec.	black	none	none	solid	61?	
R152	elec.	light green	13005	TR shield	solid	62?	
R152	elec.	green	none	BRb	spoked	65?	green roof
R152	elec.	green	none	none	spoked	65?	
R154	c/w	black	none	BRb	solid	60?	
R154	c/w	black	none	BRb, longer	solid	62?	

R152 BR 0-6-0DS in green livery.

R152 BR 0-6-0DS in rail blue livery.

British made R152 boxed in New Zealand.

Canadian R1520 0-6-0DS.

Canadian 0-6-0DS R1520 (R369)

This was planned for 1972 but arrived in 1973. It had a body moulded in electric blue with a yellow line along the running plate and '1520' in yellow on the cab sides. It was sold in a flat Hornby Railways box with a lift-off lid and a yellow pre-formed plastic tray.

The model is quite hard to find and it is thought that no more than 500 were made. While the box is marked 'R1520', the computer could only handle three digit 'R' numbers and so, for the computer records, the Canadian models were given a new 'R' number R369. We know that Louis Marx ordered 500 in 1973 but that this number remained in the Margate factory at the end of the year. It was, however, included on the Louis Marx price lists for 1973 and 1974 priced $15 and $17 respectively.

R253 dock shunter.

New Zealand R152 0-6-0DS.

SR SUBURBAN EMU

New Zealand R156/R225

The story of the New Zealand-made Southern EMUs is told in some detail in **Volume 1** and will not be repeated here. The model, however remained an important member of the New Zealand range between 1964 and 1967. The story goes that the tool was in some way damaged beyond repair in 1967 and no more could be made.

Sadly production records have not been found but there were several minor variations. The models were normally green bodied with pale grey ribbed roofs and white seating unit and carried a circular BR emblem in the middle of each side. The cab interior could be white or brown. There appears to be little pattern with the numbering. While most were numbered 'S1257S' or 'S1052S', some had no number at all. The numbers could appear on either the motor car or the dummy and they were heat printed in yellow or gold.

DOCK SHUNTER 0-4-0DS

The dock shunter had first made its appearance in 1957 as a cheap additional model that made use of the TC motor bogie.

Red '3' R253

Between 1966 and 1971, 38,000 red dock shunters were sold solo and a further 4,800 were sold in sets. The model had originally been introduced in a black livery and had changed to red in 1962 (*see* **Volume 1**). It had a working light on the front and in 1965 sold for £1-14s-6d. During the Tri-ang Hornby days it reached its peak of production in 1968 and 1969 when more than 8,000 solo models were passing through stores each year. In 1972 it was replaced with the black version also carrying the number '3'.

It was shown in the Australian 1966 catalogue as 'R253 Dock Shunter' but a TC yard switcher was illustrated. In contrast, the 1969 Canadian catalogue showed a red dock shunter labelled 'R253 Yard Switcher'. As the two models were basically the same, this confusion overseas was not surprising.

Canadian R253C

In 1967 the red dock shunter was bulk packed to Canada for use in the TS671 'Diesel Freight Set'. The catalogue illustration of the set, however, showed the red yard switcher. The batch contained 2,500 models, and the 'C' suffix to the 'R' number suggests that the models were not standard but were adapted in some way for the Canadian market. *See also* 'R353 Yard Switcher' later in this chapter.

Australia R253

The dock shunter was not made in Australia but the almost identical R353 yard switcher was assembled there. Details of this may also be found later in the chapter.

DIESEL RAILCAR

The Tri-ang model of the Metropolitan Cammell Class 101 was very popular and sold in very large numbers. It is described in detail in **Volume 1**. The first units were delivered to BR in 1955 and were amongst the first types of diesel unit to enter service in Britain. 780 individual cars were built. Later vehicles were formed into 2-, 3- and 4-car sets, a few of which included buffet cars. The major difference between the batches made was in the front end design. Later units had a two character head code panel instead of headlights.

The original model made during the Tri-ang Railways and Tri-ang Hornby periods was based on the 1955 prototype but with a 1958 trailer car also available which in theory could not have operated with this set in real practice. When Rovex reintro-duced the model in 1975 it had a code box but it was set above the cab windows. This will be found in **Volume 3**.

Green R157 and R158

In its green livery it was sold up to 1968 and during this time the power car and dummy were sold separately. The power car was priced £2-8s-0d and the dummy car 14/6d.

In 1966 the model was given yellow front panels. By now it had a ribbed roofs and carried the numbers 'M79079' and 'M79632' respectively in small characters. It also had the new brass buffers used on the Mk1 coaching stock and the buffer stocks were part of the body moulding. While the bodies of the power car and dummy power car looked very similar, in fact, they were not the same. By means of a changeable insert in the mould it was possible to give guard or passenger compartment sides.

It was a good model although approximately 20mm too short due to a slight reduction in window width. All three cars had a nicely moulded interior unit and the end cars had moulded cab interior detail. The model came with a selection of self adhesive destination blinds. The correct Air-Vac ventilators were shown on the roof but missing details included the exhaust pipes on the ends of the motor coaches. On the centre car the door to the first class saloon had been replaced by a window.

About 7,000 powered cars and 5,300 non-powered cars were sold between 1966 and 1968, the last batch being made in 1967 (and not 1966 as stated in **Volume 1**). It is interesting to note the imbalance in numbers.

Blue R157C

The model was reintroduced in 1970 as a result of public pressure. From its introduction in April that year, the rail blue DMU was sold as a 2-car unit. It remained available for just two years during which time about 9,000 were made; 5,000 of these being

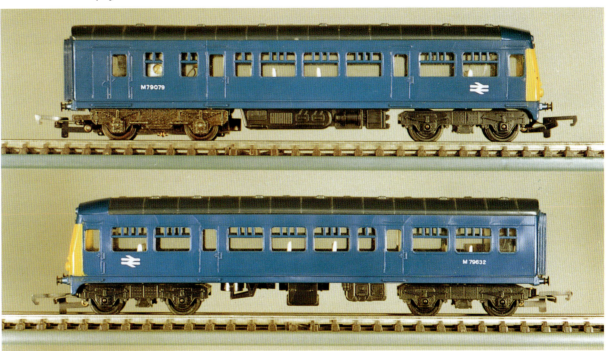

The R157/R158 DMU with a yellow panel and the destination displayed.

A blue R157C DMU.

sold in the RS603 set of 1970 and the rest solo. Sadly the centre car was not made in blue.

The model had a yellow front with 'Derby' on the destination boards and 'M79079' and 'M79632' respectively on the power car and dummy. The roof was dark grey and it had seat units fitted. It was priced at £3-19s-0d.

CLASS 3F 0-6-0 TENDER LOCOMOTIVE

BR Black '43775' R251S and R33

The black Johnson/Deeley model survived into 1966 but the last batch of 4,000 was made late in 1965. Thought was given to the possibility of reintroducing the model in 1977 but the idea was dropped.

BR Maroon '43620' R251

Doubt has long surrounded the existence of a BR maroon version of the model as illustrated in the 1966 and 1967 catalogues. Judging by the date of the drawings for the MR version (see below) it is unlikely that a batch of BR models was made. Models claimed to have been seen may have been pre-production models and a report in the *Railway Modeller* in January 1966 clearly states that the model was to be in Midland colours. This would have been based on information received from the factory late in 1965.

(NB The dates given for the BR maroon Johnson/Deeley 3F in **Volume 1** are incorrect and should all be advanced by one year.)

MR Maroon '3775' R251

The drawings for the tender printing die for this version of the 3F 0-6-0 were dated 11 November 1965. The model was made between 1966 and 1968 and a total of 35,000 were produced. Most of these were used in the RS8 train set but 8,000 were made as solo models. These sold for £2-8s-9d and had Magnadhesion and a crew but were not fitted with

Proposed BR maroon Deeley 3F as illustrated in the 1967 catalogue.

Two versions of R251 Deeley 3F in MR maroon livery. The darker one at the top has no black edge to the footplate and steps.

a smoke generator. They were packed in a DPL2 window box. The tender was R33.

Australian Maroon R251
There are indications that a maroon Australian-assembled model, with a smoke generator, was available up until 1966 and sold for £A4-19s-6d.

Old Smoky R661S
There are records of Old Smoky being assembled in Australia. This was the black version of the

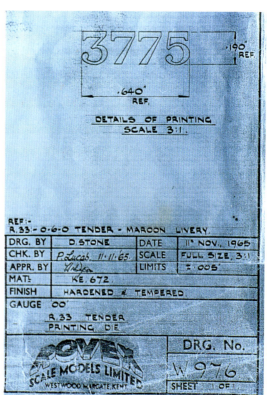

Drawing dated 11 November 1965 showing that the MR livery was planned before production of the maroon loco started. The catalogue picture, on the other hand, would have been prepared earlier in the autumn.

Johnson/Deeley 3F 0-6-0 but with a weathered finish. These records indicate that 150 were assembled in Melbourne, by Moldex, in 1965 and 125 more were made the following year. Margate has no record of these being supplied.

L1 4-4-0

BR Green '31757' R350

The model, described in some detail in **Volume 1**, entered the Tri-ang Hornby era in BR green livery (the most attractive form in which the model was made). From 1966 the loco and tender were sold together instead of separately. Batches were made in 1966 and 1967 and a total of about 7,000 were sold in Tri-ang Hornby packaging, most having been made in 1967. None were used in sets during this period, but the solo model remained in the catalogue until 1968.

Produced as a cheap-to-make mixed traffic locomotive, it was more popular with serious modellers than with children who preferred large express engines or small tank engines. It had Magnadhesion and came with crew. In 1966 the loco and tender retailed at £2-13s-8d.

In 1995 the pre-production model was discovered in the collection of an ex-employee of the company and was bought by a Tri-ang collector. The tender was missing and the cab had disintegrated.

Tools Offered to New Zealand

The New Zealand company were wanting other models to manufacture themselves and in December 1967 they were offered the L1 loco body tool and spraying mask. The tender tooling was not available as this was still required at Margate but in New Zealand it was proposed to pair the loco with a Princess tender for local sales.

The offer was accepted and a charge of £100 agreed. It was planned to ship the tools out in January 1968 and in the meantime some spare mouldings were run off. There was then some difficulty in the New Zealand company obtaining an import licence. In 1968, following a visit to New Zealand by Richard Lines, it was decided to update the New Zealand product range. The plans for shipping out the L1 tools appear to have been forgotten during this new upheaval. Needless to say they did not go.

SR '1757' R350N

The model was to be released in SR livery in August 1971. The livery was in fact the BR green plastic but lined out in white. The model would have plated wheels and an exhaust sound tender. Orders for 270 were received before the end of the year but none was made until 1972 by which time the idea of fitting exhaust sound to the tender had been dropped.

The model almost certainly appeared early in 1972 and as it dates to the Hornby Railways period, coverage appears in **Volume 3**. The model was eventually sacrificed when the tool was altered to make an LMS Class 2P.

'BRITANNIA' 4-6-2

Writing in the *Model Railway Constructor* in April 1968, Chris Leigh, in an article on improving the Tri-ang Hornby 'Britannia' said:

"As a proprietary model, the Tri-ang Hornby 'Britannia' is excellent, and the latest ones, at 89/6, are exceptionally good value. Over the years they (Rovex) have incorporated into the 'Britannia' all the improvements and popular features for which they are well known. These have included fretted-out scale size wheels, optional smoke-unit, Magnadhesion, crew and Synchrosmoke."

Despite its earlier prestigious standing in the Tri-ang fleet (*see* **Volume 1**), during the Tri-ang Hornby years 'Britannia' was not used in a set. Cheaper-to-manufacture locomotives were chosen instead in order to keep the cost of sets to the minimum. The original model without smoke generation was withdrawn at the end of 1965 leaving only the smoking version.

'Britannia' with Smoke R259S and R35

The early 'Britannia' had used the Princess chassis with the cylinder block clipped on underneath, solid die-cast wheels, double-lined tender, complete buffers (not set in tubes) and no Magnadhesion. For most of its life it ran with the wrong valve gear. That designed for 'Evening Star' would have been more appropriate and, indeed, was used on the model in later days.

By 1966, as we have seen, it had a solid chassis

R350 Southern class L1 in BR green livery.

Catalogue picture of R259S 'Britannia' and photograph of 'Robert Burns' (one of the alternative names available in 1971).

R355B 'Connie' made around 1971. This is unusual as 'Nellie' was the more common name at that time. The picture of 'Polly' is from the 1969 catalogue.

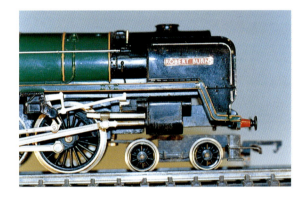

block, Magnadhesion, Synchrosmoke, see-through spoked wheels, crew, single lining on the tender, improved buffers (from 1963), was sold complete with the tender (from 1966) and was priced at £4-6s-2d.

About 36,000 were sold solo during the Tri-ang Hornby era before the sound tender was fitted at the end of 1970. Production peaked in 1968 when 8,000 models were made.

'Britannia' Smoke, Noise R259SN & R35N

Steam sound was fitted to the tender for 1971 and 1972 and about 6,700 of these were made. It first

appeared in the shops in May 1971. So that people could convert their old locos to steam sound an extra 1,350 steam sound tenders (R35N) were made in the first year.

Besides converting the tender to steam sound, considerable retooling of the inside was done to take a motor rendering it suitable for use with the new 'Evening Star' loco. The opportunity was also taken to close the axle boxes thereby improving its side view.

The new model also had nickelled tyres with matching bogie wheels but the tender wheels remained black. There was a problem with the bogie wheels being loose but this was quickly rectified. The crew had been dropped but there was a choice of alternative names and numbers. The names were printed in red on gold self-adhesive labels and the alternative numbers were transfers of the cabside to the outer limit of the lining. The alternative smoke-box door numbers came on black self-adhesive labels. The prototypes chosen were:

> 70006 'Robert Burns'
> 70010 'Owen Glendower'
> 70013 'Oliver Cromwell'

Consideration was given to making 'Britannia' a

tender-driven loco at this time but Rovex were keen that it should be fitted with steam sound and the two were not compatible. The model was sold in an open box with a white plastic vacuum-formed tray and an outer grey sleeve.

Australia R259

While price lists suggested that it was assembled by Moldex there is no record of any having been made.

France R259S and R35F

About 300 'Britannia' tenders were supplied with Hornby Acho couplings so that the model could be marketed in France by Meccano Tri-ang Lines Frères SA. It did not prove popular.

'NELLIE' AND FRIENDS 0-4-0T

1960 'Nellie' R355B

The blue 'Nellie', which first appeared in the shops in 1960, is described in some detail in **Volume 1**. It remained in production until 1967 with about 49,000 models being sold within the period covered by this volume. Most of these were used in the RS24 set of 1966/67. The price of 'Nellie' in 1965 was £1-9s-6d.

1971 'Nellie' R355B

The model reappeared in blue part-way through 1970. In 1971 a total of 6,500 were made. These late models tend to be of a slightly brighter blue. The model was not fitted with plated tyres until 1972 by which time it was not available as a solo model, but only in the RS616 'Take-a-Ticket Train Set'.

'Polly' R355R

'Polly' was always the red engine and was made up to 1970 having first appeared in 1963. It was to have been made in 1971 but in August 1970 it was decid-

ed to return to the blue 'Nellie'. 35,000 solo models of 'Polly' were made after 1965 and a further 75,000 were used in sets during this period. Production reached its peak in 1968 when almost 10,000 came off the production line.

'57' R355G

Between 1969 and 1972 the Industrial 0-4-0T appeared in apple green with the number '27' in black on its tanks. 13,000 were sold as solo models and 21,000 were sold in sets. The solo model was priced £2-3s-0d. From July 1971, the model was to be fitted with plated tyres but tooling for these was not available in January of that year and they were delayed until 1972.

Black BR R359

A black version with a BR crest was to have been issued with plated coarse scale wheels in 1972 but the idea was dropped in August 1971 and 'Nellie' (R355B) and the green version (R335G) remained in production but with plated-wheel tyres.

Later Models

Bright red versions of the industrial tank, with a telephone number on the side tanks, and with or

without a large silver dome, date from 1973-75. A blue tank with '7178' on lined-out tanks dates from 1976-78. These versions were Hornby Railways models and will be covered in **Volume 3**.

Australian Models R355

None were assembled at Moldex after 1965 and so any appearing in Tri-ang Hornby boxes would have been from imported stock.

DEAN SINGLE 4-2-2

1961 Issue R354 + R37

The original issue of 'Lord of the Isles' survived until 1965 and is well covered in **Volume 1**. It was also available in an export presentation set between 1964 and 1966. For this 800 locos were made in 1965 and 700 in 1966.

1967 Special Run R354

Stocks of the Tri-ang Railways model probably ran out early in 1966 but there were still some orders coming in during 1967. Under pressure from certain retailers, it was decided to make a further batch of 1,000 that year to fill these. The models were given a gloss finish and boxed in DPL2 window

The R355G Industrial Tank is shown in the two different shades of green used during its life.

R354 'Lord of the Isles' in the gloss green of the early 1970s.

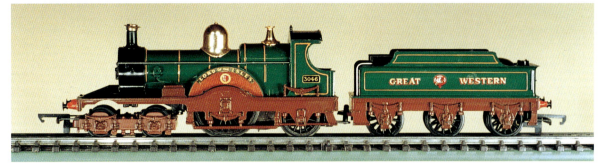

boxes which carried the name 'Rovex Industries Ltd' and a printed label on one end only inscribed 'R354 GWR 4-2-2 Locomotive 'Lord of the Isles' and the code '(03/244)'. My own example has a IIIc coupling on the loco and a IIId on the tender.

They were advertised by retailers such as Lou Nadin of Warrington, Eames of Reading and Hobbytime of West Wickham as a 'Limited Run' and priced £2-14s-0d and £3-16s-9d with its tender (R37). By November that year, Lou Nadin were offering the loco and tender for £2-17s-6d in an attempt to clear their stocks.

Australian 'Lord of the Isles' R354

Very small numbers of the model were assembled by Moldex of Australia between 1965 and 1967 for use in sets and to be sold as a solo model. During this time only 150 were made for solo sales. It is not known how many were made for the sets and there is no record of unfinished models being sent out to Australia during the period covered by this volume.

1970 Reissue R354

With the change to pre-nationalisation liveries at the start of the 1970s the 'Lord of the Isles' was reissued to strengthen the GWR range. It had been planned to reintroduce it in 1971 but in May 1970 the decision was taken to release it in September that year and so it appeared in the September Trade Price List. This drew attention to its 'new authentic high

gloss finish' as the reissued loco was lacquered. Early models from the factory with this lacquered finish were poor. They suffered from a 'treacle' effect which was later overcome.

These later models may also be distinguished from the earlier ones by their IIId couplings. They were released in window boxes and without smoke generators. Of those made at this time, about 8,600 would have been sold in Tri-ang Hornby boxes. The model was priced in 1970 at £3-17s-0d.

The model may be found in plain white boxes with a Tri-ang Hornby sticker at each end. It remained in production in 1972 and 1973 but with nickelled plated wheel tyres, except for the tender wheels.

Tampo Printed Model

The model sometimes turns up with a paint finish and tampo printed splashers. This is from a special presentation set issued in the mid 1980s and will be covered in **Volume 3**.

EM2 CLASS ELECTRIC CO-CO

The class was built between 1953 and 1955 and was intended to pull passenger traffic on the Manchester to Sheffield line that passed through the Woodhead tunnel. After their withdrawal from service in 1968 the whole class was sold to the Dutch Railways.

Three have been preserved including 'Electra' which is owned by the EM2 Locomotive Society.

The EM2 model, which is described in detail in **Volume 1**, entered the Tri-ang Hornby era available in both ready-to-run and kit-form. All of the models had working pantographs and Magnadhesion.

Kits R388

The ready-to-run model was 27000 'Electra' but in the R388 CKD Kit it had the alternative names 27002 'Aurora' and 27006 'Pandora', the names being on red stickers and the numbers on blue ones to blend with the body colour of the loco. The kits came com-

The cover of the first issue of Transport Age, *a British Transport Commission publication for industry.*

R351 'Electra' in electric blue showing the 1966 and 1968 versions and R388 'Pandora' and 'Aurora', in rail blue, built from assembly packs. Note that 'Pandora' has green number labels.

plete with two masts and a length of catenary wire and were priced £2-17s-6d in 1966. In 1967 the CKD kits became Assembly Packs.

All Tri-ang Hornby kit models were in either lined or plain electric blue or rail blue livery but today they frequently turn up with green number stickers instead of blue ones. This suggests that the correct colour stickers were not available when the blue kits were released or, at some time stocks of them ran out and old green stock was used up. I have examples of both shades of blue plastic body with the green stickers. Whatever was the reason for this practice it did tend to spoil the appearance of the model.

Lined Electric Blue R351, R388

The green version had been available up until 1965 and the first blue version of the 'Electra' appeared in 1966. This was in electric blue with a 'ferret and dart board' BR logo, white lining and yellow cab fronts. About 4,000 ready to run models (R351) and 3,700 kits (R388) were made with this finish. The model was priced £3-9s-6d and was made for just two years.

Electric Blue R351, R388

In 1968, for one year only the lining was dropped and the BR arrows logo (BRe) used on the electric blue plastic body. This was probably non-authentic as the change to arrow logos on the prototypes was accompanied by a change to rail blue. It is estimated that about 5,000 were made of which 2,000 were sold in Assembly Packs. Ready-to-run locos (R351) were priced £3-9s-11d and assembly packs (R388) were 10/- less.

Rail Blue R351, R388

By 1969, BR rail blue had been adopted for the model moulding (if not by the prototype) but otherwise the model resembled that of the year before. It had been proposed to drop it from the catalogue in

1971 but sufficient stocks remained for it to be featured. This, however, was the last year it was seen and, although instructions were given that all spare components and raw materials should be packed up, labelled and preserved as it might be reintroduced in a few years time, it was not to be seen on the production line again. In the last two catalogues, the model is shown with an 'E' prefix to the number (E27000). As far as I can recall, all examples I have seen had no 'E'.

Only about 2,000 ready to run models (R351) were made in this shade of blue but there were also 3,000 Assembly Packs (R388) made. The kits were the last to sell out, 650 of them surviving in the stores at Margate until 1972. In 1970 the assembly pack was renumbered R388U to reflect the track fitting masts inside.

BATTLE OF BRITAIN 4-6-2

110 of these light-weight locomotives were made and they were 9 tons lighter than their Merchant Navy cousins which gave them a much wider route availability.

While the model's body was a good likeness to the prototype, the following features let it down:
1. The lack a crosshead – corrected in 1995.
2. The wrongly shaped tender sides – the curve on

them was at the top instead of at the bottom.
3. The head-on view of the smoke deflectors which were fixed to the body, to protect them from damage, instead of being separate.
4. The sandbox filler holes in the side of the body were too high – corrected in 1995.
5. The excessive number of rivets – corrected in 1995.
6. The Bulleid-Firth-Brown wheels were too close together although they were the correct size and nicely detailed.
7. The front buffer beam was too shallow.

BR Green 'Winston Churchill' R356S

'Winston Churchill' in BR green livery is well covered in **Volume 1** but continues here as a Tri-ang Hornby model.

From 1966 onwards the locomotive and tender (R38) were sold as one unit in a single box priced £4-1s-0d. In this form, batches were made each year between 1966 and 1969. Over this period about 18,500 were manufactured, all as solo models.

The model came with Magnadhesion on the rear driving wheels and on steel track could haul 11 coaches. It also came with crew and the chassis which, while still based on that of the Princess, was as redesigned in June 1964 when Synchrosmoke was added. It was still a scale 2'6" too short. The cou-

pling and connecting rods were correctly fluted but it did not receive plated wheels. It remained in this form throughout its Tri-ang Hornby years. The BR lining was correct for an engine with an un-rebuilt tender but the background to the coat of arms panel should have been light blue and not green.

SR Green 'Winston Churchill' R869S

The BR Churchill was not made after 1969 but that year the first 1,600 models in so called 'SR malachite green' arrived priced £4-17s-6d. The shade was probably a little too dark to be accurate as the plastic used was that of the BR(SR) Mk1 coaches. It was finished with a lacquer spray to give it a glossy look and furnished with red nameplates. It came in an open box with a yellow plastic tray and a grey outer sleeve.

The model, which was featured in the September 1969 Tri-ang Hornby advertisement, sported the number 21C151 which, with the name 'SOUTHERN' on the tender (R870) sides was correctly in the 'sunshine' style of the period but the gold lettering had red shading instead of black. While this was normal for the model, rare examples of this early model can be found with the correct black shading. The yellow lining, which was applied to raised beading by girls with mapping pens, was very fine on this model. Sadly it did not feature the 'SOUTHERN' ring in the smokebox door or the locomotive number above the front buffer beam.

While the Battle of Britain model in Southern livery came carrying the name 'Winston Churchill', it was the first Tri-ang model to be released with alternative names and numbers. To help owners to position the number on the cab sides two pips were added to the surface by an alteration of the moulding tool. The transfers provided the following choice:

21C157 'Biggin Hill'
21C164 'Fighter Command'
21C165 'Hurricane'

A sectioned 1966 model found in the factory.

Right: R356S 'Winston Churchill' in BR livery with Tri-ang Golden Arrow stickers.
Below right:The R869S Battle of Britain class loco in Southern livery as 'Winston Churchill' and altenatively named 'Fighter Command'.
Below: Battle of Britain advertisement.

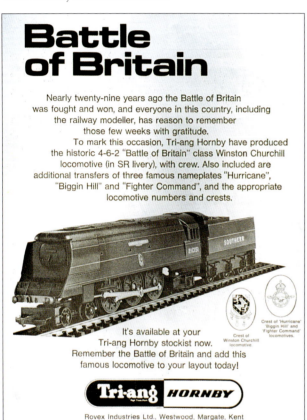

Under *MRC* test, the maximum speed was a scale 120 mph running light and the minimum speed was as low as 3 mph with a transistor controller. On steel track it would pull 7-8 of the heavy Hornby-Dublo coaches. The loco was noisy due to the body shape but could be quietened by putting a sheet of foam rubber inside the body. It came with cab crew.

In this form the model was made until 1972 and it is estimated that during the four years it was in production approximately 10,000 were made; half as Tri-ang Hornby models and half as Hornby Railways.

Steam Sound R869NS

It had been intended to give the model steam sound in 1971 but it had proved impractical to convert the tender which was to have been R870N if sold on its own. It therefore remained R869S for that year but was supplied without a crew for the last two years of production.

The A1A-A1A Brush diesel in green (R357G), early blue (R357B) livery, electric blue (R357) and rail blue (R357).

Reissued Model

The SR model in a better malachite green and with a correctly decorated front end, was to return in 1981 with a different selection of alternative names. Unlike the 1969 model these did not have a gloss finish. The Battle of Britain also appeared as '41 Squadron' in BR green livery in 1985 and a Golden Arrow SR model was released in 1987. These will be covered in **Volume 3** along with versions of the improved model that dates from 1995.

TYPE 2 BRUSH DIESEL A1A-A1A

By 1965 the Type 2 Brush diesel, later to be known as a Class 31, was still the only mainline diesel in the Tri-ang fleet. It had sold in large numbers and paid off the development cost within a very short time. It had also taught the company how to make a successful motor bogie and the lessons learnt were used to develop other models.

All the Tri-ang Hornby Brush A1A-A1As were fitted with Magnadhesion. The headcodes varied between '9D80' and '4C01'. They were sold in window boxes with a yellow card insert until probably some time in 1968 when vacuum-formed plastic trays replaced the card. For more information about the earlier models see **Volume 1**.

The two headcodes used on the A1A-A1A diesel.

Experimental Blue 'D5578' R357B

An electric or experimental blue model with a white roof and two white lines along its sides had arrived in the shops in 1965 and remained in production for a period of two years, during which time 10,000 models were made. The model was based on that used on the locomotive that pulled British Rail's XP64 train. It was priced £2-9s-6d in 1965. This model should not be confused with the 1962 model that had duck-egg blue window surrounds and no stripes.

Green 'D5572' R357G

This is the most common colour one finds on second hand stalls and it will quickly be noticed that the finish varies from model to model. Early green examples had a matt finish and were made between 1963 and 1967. About 50,000 were produced of which about 20,000 would have been packaged in Tri-ang Hornby boxes.

These figures are for solo locos only. In addition, during the Tri-ang Hornby period, a further 50,000 were used in sets. One of the company's most successful sets was the Freightmaster which used the A1A-A1A as its motive power.

The 'R' number was given a 'G' (for 'green') suffix in 1966 and 1967 to distinguish it from the blue version. The 1966 price was £2-10s-2d.

The model was to return to green in January 1972 and be fitted with an additional power pickup on the trailing bogie. This model will feature in **Volume 3**.

Luminous Code Labels

It was decided to bring back the green livery at the end of 1971 and this is illustrated in the 1972 catalogue. It is interesting to note the code carried by the model photographed for the illustration; a picture that would have been taken around September 1971. The code box of the model reveals a sticker (which is peeling off) with the code '6L32'. I have not seen a model with this particular number. At the time it

was proposed that, when these self-adhesive code labels were reordered, the new ones would be luminous. Could this have been a sample label used for the picture?

Electric Blue 'D5572' R357

Plain electric blue with blue roof and white arrow logos (BRe) was the livery carried by the model in 1968. A total of 28,600 models were made that year of which 20,000 were sold in 'Freightmaster' sets. By the following year the shade of the plastic had changed to rail blue. The electric blue model continued to carry the number 'D5572' and the codes '9D80' and '4C01' used on the green model but, unlike the green version, it had an all-yellow cab front. When first introduced, the price was £2-9s-11d.

Rail Blue 'D5572' R357

For all intents and purposes the rail blue model was the same as the electric blue one of the year before except for the shade of plastic used and the fact that it had oval buffers instead of round ones. It was made between 1969 and 1971 during which time over 42,000 were made, all but 15,000 of them selling in sets.

Australian R357A

For Australia the model was sent out bulk packed for use in sets. 1,500 went in 1968 and a further 1,100

the following year. These appear to have been the electric and rail blue versions. Later catalogue inserts show the model in NSWR maroon and yellow livery but these all fell outside the period covered by this volume and so will be found in **Volume 3**.

'NORTH BRITISH' 0-4-0 DIESEL SHUNTER

BR 'D2907' Green R559

The model is described in **Volume 1**. It remained in production until 1967 and in the last two years 4,500 were made. The 1965 price was £1-9s-6d. The body tool was altered late in 1968 to produce the body for the electric R858 (Barclay) 0-4-0DS.

Battle Space Version (R756)

This was bright red and carried Battle Space stickers on its sides. Although 10,000 were made in 1966 and 1967, few have survived in good order. It was only available in the RS17 set.

Starter Set Locos

See Starter Set Locomotives towards the end of the chapter.

Australian R654

It is known that 200 of the 0-4-0 diesel shunters were sent out to Australia in 1966 with a further 100 following in 1967 but it is not known whether these were clockwork or electric nor what colour they were. It is thought that they were for either the RS71 or RS43E sets made up by Moldex for the local market.

B12 4-6-0

BR '61572' with Smoke R150S

The model arrived in 1963 and has already been described in detail in **Volume 1**. From 1966 the loco and tender (R39) were sold together, priced £3-0s-4d, and fitted with Synchrosmoke. It had Magnadhesion and came with a crew. *See also* 'Assembly Packs' below. Consideration was given to bringing it back without smoke or Magnadhesion at some later date. As a smoking loco, however, it remained in production until 1969 and over this period about 20,000 were made, the peak of production being reached in 1966.

At the end of 1969 about 5,200 remained in the store. About 1,800 of these were sold in 1970 and the rest appear to have been converted to R150NS models (*see below*).

BR '61572' with Smoke and Noise R150NS

In 1970 about 3,400 of R150S black BR B12s were taken out of the Finished Goods Store, had the tender chassis removed and a modified one fitted. The new chassis had an exhaust sound mechanism. These locos were used in the RS606 'Chuff-Chuff Puff-Puff' train set which was promoted on television that Christmas. As 5,000 sets were made up that year a number of additional locos must have been made to make up the numbers.

A further 5,200 were made in 1971 and during that year just 60 black BR B12s were released as solo models with exhaust sound fitted to the tender (R39N).

LNER '8509' Green with Smoke R866S

Caught up in the fashion to change models to pre-nationalisation liveries, the B12 was released in 1970 in LNER apple green with '8509' on the cab sides. It was in the shops by February having been slightly delayed as the transfers did not arrive on time. Just 5,000 were made that year, and loco and tender (R867) sold for £4-2s-6d. Some of the tender chassis (maximum 3,400) were recycled from converted R150S black BR models as described above. These had to have their wheels removed for painting green.

The model had Magnadhesion, smoke and a crew and the following year exhaust sound was fitted. The model without exhaust sound was released again in 1978 and the Hornby Railways model may be distinguished from the earlier one by its nickel wheels. *See also* 'Assembly Packs'.

LNER '8509' green with smoke and noise R866NS

In January 1971, the LNER apple green B12 was released with exhaust sound in the tender and no crew. Only 1,500 were made that year but this ver-

The R389S B12 Assembly Pack.

The B12 as sold in France (R150SF).

An R866S LNER apple green B12.

sion of the model survived into the Hornby Railways era, remaining in production until 1974. From 1972 it was fitted with nickel-plated wheel tyres.

The revised model was reviewed in the *Model Railway Constructor* in March 1971 and the appearance of the model in its LNER apple green was described as 'very good indeed'. The model was found to be very close to scale and a good performer. It pulled five heavy coaches and seven Tri-ang Hornby coaches on nickel-silver track at a scale speed of 60 mph.

The tender with noise (R867N) was available separately in limited quantities. The NE black model with exhaust sound followed in 1976.

BR '61572' Black Assembly Packs R389S

The black BR version of the B12 was available in kit form in 1967 and 1968 and in total about 7,200 were made. Price £2-13s-10d.

LNER '8509' Green Assembly Packs R389S

The LNER green version was available from May 1970 as an assembly pack and remained available into 1971. It was priced £3-15s-0d but only 1,500 were made.

French Version of B12 '61572' R150SF

Some B12s were fitted with Hornby Acho couplings on the tender for sale in France through the French member of the Lines Bros. Group, Meccano Tri-ang. These were sent to France in a standard Tri-ang Hornby DPL2 size window box where a blue sticker, 65 x 7 mm in size, with the text 'LOCO. VAPEUR 4.6.0 REF. 150S', was fixed on one end over the label carrying the 'R' number. It is strange that '4.6.0' was used as, under the French classification it should have been '2.3.0'. It is not known how many went to France as, for some unexplained reason, these were not separately listed in the factory census. Their existence is confirmed by examples seen.

DMU PULLMAN

Blue and White R555 & R556

The DMU Pullman had been a great success since its introduction to the Tri-ang Railways range in 1963 (*see* **Volume 1**). The model that entered the Tri-ang Hornby period, priced £2-15s-0d, was as originally made, with a Pullman coat of arms on the cab front, and was perhaps the most attractive of all those made. An article in the January 1967 edition of *Model Railway Constructor* described how the models could be improved and a full train made up using both Tri-ang and Kitmaster parts.

The power car was numbered 'W60095' and the trailer car 'W60097'. The centre cars were available numbered either 'W60745' or 'W60747'.

During 1966 and 1967 about 42,000 power cars were made and a similar number of the dummy ends. All but 10,000 of these were sold in the RS52 train set. A similar number of parlour cars were also made over this period. The price these centre cars command today might suggest that they are something of a rarity but this is certainly not the case. The price has been falsely inflated by the demand for them from collectors and model railway operators seeking to make up a scale-length train.

Blue and White with Yellow Front R555 & R556

In 1968, in line with British Rail practice, the beauty of the cab front was spoilt by painting it yellow, which to some critics was not even that accurate. It remained in this condition for one year only before giving way to the grey version. On BR the yellow front survived only a matter of months before the change to corporate grey and blue. The cars continued to carry the same numbers as above.

About 22,000 models were made with this finish of which 17,000 were sold in the RS52 set. According to the catalogue illustrations the yellow-fronted blue set was also made for one year only. Almost 10,000 more parlour cars were also made this year. These of course did not differ in appearance from those previously made.

The only variation to this livery I have seen had the roof sprayed grey instead of silver.

Grey and Blue R555C

In 1969 Rovex followed BR practice and turned out the model in pale grey livery with a blue window strip and dark grey roof. The chassis was all-over black and from 1969 the power car and dummy were sold together as a single unit. The model remained in this colour scheme until 1972, the last

The R555C diesel Pullman in grey livery and a photograph of the interior of the prototype when first built.

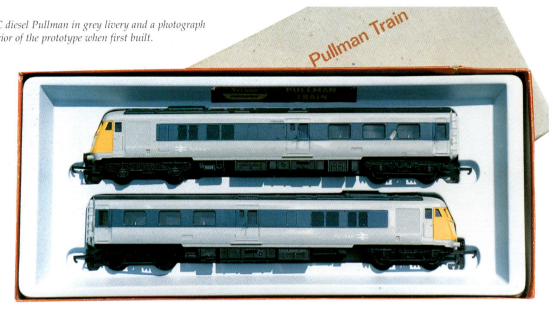

batch being made in 1971 and being sold as a train pack. The power and dummy cars continued to carry the same numbers as the previous blue and white versions.

About 7,800 solo sets were sold and the RS52 set of 1969 and the RS652 set of 1970 accounted for a further 17,600. The unit was priced £4-9s-6d when it first appeared in this livery.

Late Version
The Pullman DMU was to appear in a fourth and final livery in the 1970s when it was released in blue and pale-grey with crests heat-printed in white on the sides of the units. This fell within the Hornby Railways era and will be covered in **Volume 3**.

'ROCKET' 0-2-2

The 00 scale model of 'Rocket' had not proved to be a great success in Britain but it was popular abroad. No solo locos were advertised in the catalogue but it is thought likely that more than 12,000 locomotives were sold in Tri-ang Hornby sets between 1966 and 1969. The tender carried the part number R652 and the model had Magnadhesion and crew. It is described in greater detail in **Volume 1**.

Australia R651A
Between 1964 and 1969, the Australian factory assembled 535 R346 presentation sets from parts sent out bulk packed. Of these, 325 were made between 1965 and 1968.

USA R800
Tager of America ordered the 'Rocket' with the R652T tender which was fitted with NMRA couplings for the North American market. 4,920 were supplied in 1967 and these were sent over to the States bulk packed to be sold under the American Train and Track trade mark and in ATT packaging.

The sets had a silver-coloured box.

Later Models

The model was reissued in the mid 1980s in a special presentation set. This will be covered in **Volume 3**.

CALEDONIAN SINGLE 4-2-2

The Caledonian Single was the only Scottish-owned 4-2-2. It was designed by Drummond of the Caledonian Railway and built in 1886 by Neilson as a one-off.

All versions of the model were fitted with the XT60 motor.

1963 Model R553

The model is based on the locomotive after it was modified in 1924 with its later boiler. It first appeared in 1963 and is covered in **Volume 1**. The first version was last made in 1965 but stocks were not cleared from the factory until 1967. The model had Magnadhesion and a crew and sold for £2-5s-0d for the loco and 7/11d for the tender.

1971 Model R553

As we have seen above, Rovex decided in 1971 to re-release the model in August. It made sense anyway

An R553 Caledonian Single in the 1963-66 and the brighter 1971-73 liveries.

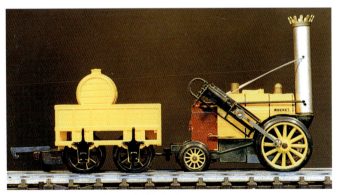

R651 Stephenson's 'Rocket'

A sectioned Caledonian Single found in the factory.

as the chassis was already in production for the 'Lord of the Isles'. About 5,100 were made in this late batch and although it was in the shops late in 1971 it was not illustrated in the catalogue until 1973. It was dropped from the catalogue in 1974 and stocks were finally cleared from the shops in 1976. Unlike the earlier model the loco and tender (R554) were sold as one and the price was £5.90; more than twice what it was six years earlier.

The reissued model could be easily distinguished from the original production model by its brighter gloss blue finish and nickel wheels. The Indian Red chassis was a dark brown instead of the correct red brown. The chimney had gone bulbous at the top.

LMS Version R553M (R553R)
Under pressure from retailers for the return of the Caledonian Single, there were plans to introduce the model in 1971 in Midland or LMS livery (R553M) but by May 1970 the idea had been dropped in favour of 'Lord of the Isles' because another GWR loco was needed. The idea was reawakened for the 1972 range and was to have had coarse-scale plated wheels (R553R). This was again dropped in favour of reintroducing the Caledonian version.

A maroon version of the model was eventually made as an LMS Class 1P in 1983 and was sold in a presentation box. This will be covered in **Volume 3**.

CLASS 37 CO-CO

The English Electric Type 3 Co-Co was the first locomotive to be introduced after the change to Tri-ang Hornby, arriving early in 1966. In reality, it was a Tri-ang Railways model as was evidenced by the inscription on the body moulding. The drawing of the body had been completed by Rovex Scale Models Ltd at the end of June 1964.

Diesels were easy to make. Once Rovex had dis-covered how to make a good power bogie using a lay shaft with worms mounted on it, it did not matter whether the power bogie had four or six wheels. The manufacturing principles had been established and they could manufacture a wide range of diesel and electric classes. The Class Type 3 Co-Co made use of the existing bogies of the A1A-A1A model and was therefore relatively cheap to develop.

It would appear that an illuminated head code was planned as two little holes to take leads were provided in the end of the chassis. This, however, did not come about. The model was fitted with oleo buffers.

Prototype
The Class 37, as it was later to be called, was first introduced in 1960. By the end of production in 1965, 309 had been built by English Electric and they formed the largest class of BR standard medium power mixed traffic diesel locomotives at that time. The loco had many features of the earlier English Electric Class 40 including its body shape.

It was first seen in the Eastern Region, replacing steam in East Anglia but in more recent years they were associated with Scotland where it repeatedly proved its worth and versatility having replaced less powerful locomotives. The one modelled was from a batch made for the Western Region. By this time, however, the doors in the cab fronts of the prototype had been done away with and a single four character code box fitted.

Green 'D6830' R751
The green diesel, as it originally appeared, was one of the most attractive versions Rovex made. It was illustrated in the January 1966 edition of *Railway Modeller* and manufactured in this livery between 1966 and 1967 during which time about 20,000 solo models were made and a further 20,000 were sold in the RS9 train set. The original price of the solo model was 55/- and was first featured in Tri-ang Hornby advertisements in April 1966.

The model carried the running number 'D6830' and the head code '8H22' and was one of a batch allocated to Landor and Cardiff Canton depots. It had yellow panels on the cab fronts which were applied as self-adhesive labels. It was fitted with Magnadhesion and was quiet running.

A test track review was published in the May 1966 edition of *Model Railway Constructor*. This described the performance as 'quite good' reaching a scale speed of 150 mph and controllable down to a scale speed of 13 mph. On steel track it romped away with 14 coaches. The model was said to have an exceptional amount of detail in the moulding almost all of which was faultless. Criticism was made of the protruding tail lights and the thickness of the wind-screen pillars. The correct shape buffers and the lou-ver and grill detailing was praised. As we have seen, the bogies were the same as those used on the A1A-A1A and did not look right. It had the 3-pole motor with worm and gear on the outer motor bogie axles. In summary it was considered a very worthy addi-tion to the Tri-ang Hornby range and at £2.15s.0d was considered 'superb value'.

The *Model Railway News* track test showed that it could take a 1 in 13 incline with ease with one coach but struggled beyond that. On the flat it could man-age the scale equivalent of nine coaches and the addition of 0.75 lb of lead to the roof improved the performance fantastically pulling three times the original weight.

The proving model of the body was sold by Hornby Hobbies in 1994 and is now in a private collection.

Green with Yellow Ends
Sadly, the Co-Co was not produced in green with yellow ends, but alterations to the drawing of the model show that spray mask lines were added to both ends late in May 1967. Some green bodies were used for testing the masks eventually used for

Right: One of the yellow fronted Co-Co diesels found at the factory.

Below right: The most common version of the R751 Co-Co diesel showing two types of packaging in the late 1970s.

Below: Pre-production sample (top) and production model of the Co-Co diesel.

putting the yellow ends on the blue version of the model and two of these sprayed bodies were still in the factory in the mid 1970s. One is now in a private collection.

Blue 'D6830' R751

From 1968 until 1984 the model was made in rail blue livery. It does not appear to have been made in electric blue as it was still in green livery in 1967. It is not known how many were made in all but it must have been a very large quantity. We do know that 25,000 would have been sold as solo models in Tri-ang Hornby boxes and a further 13,800 in the RS604 set of 1970/71. Its price when first released in 1968 was £2-17s-6d.

By mid-May 1972, when the name on the packaging should have changed to 'Hornby Railways', the model was being packed in a plain white box with a Tri-ang Hornby sticker on one end and a Tri-ang Hornby instruction leaflet inside.

The chassis was originally 'plugged' into the body using the code box at the No.2 end as a location point. This meant that one code box at that end was a hole and the code board formed part of the chassis. This was altered in the late 1960s to provide a fixed code box at both ends and, instead, the chassis location point was a slot in the lower panel at the No.2 end.

This caused confusion on the production line and so the body was altered in May 1971 so that it could be fitted to the chassis either way round. This meant the addition of a chassis locating slot at the No.1 end and a second boss for the fixing screw was added to the inside of the roof at the same time. Later that same year the slot at the No.2 end was modified to match the one at No.1 end. To add to the difficulties for collectors trying to date their models, two years later in April 1973, the slot at the No.1 end and one of the bosses were deleted.

Like the green version before it, the model carried the number 'D6830', and the head code '8H22', both of which it kept until 1976. The latter may be found with large or small characters the smaller ones being later. From 1977 the model carried the running number '37 130'. In 1971 luminous code box labels were planned and 1972 promised fine-scale wheels. There had also been plans to fit lights to the model in 1971 but this idea was abandoned presumably for fear of increasing production costs at a time of rapid inflation.

Later Versions

During the 1980s, variations in livery blossomed to reflect the changes being seen on the real railways. This included the named locomotive 'William Cookworthy' and both green and blue models in non-authentic livery for starter sets. Code boxes were added to the fronts of the cabs in 1986.

CLASS 81 (E3000)

Prototype

Under the 1955 modernisation plan, it was proposed that the LMR main line between Crewe and Euston would be electrified with 25kV ac overhead current collection. 100 locomotives were ordered from five different manufacturers providing a mixture of passenger and freight units. So tight was the specification that the five resulting classes of locomotive looked almost identical to one another. Initially they were classified as AL1 – AL5 but under the TOPS Scheme these became Classes 81 – 85.

The Model

The E3000 was first illustrated in the 1964 Tri-ang Railways catalogue. The picture showed a locomo-

tive numbered 'E3000' carrying the route code '1T91'. The artists' impression was in fact of the AL2 (Class 82) one of a class of only 10 locomotives built by Beyer Peacock of Manchester.

It seems likely that Rovex had a testing model made – the one used for the illustrations in the 1964 and 1965 catalogues. Unfortunately this model has not been found although a cab, modelled in clear plastic, was discovered in 1995 in a collection of pre-production models kept by one of the company's model makers. The reason it was abandoned in favour of a new model, based on the AL1 (Class 81) modelled by Hornby-Dublo, is not known. The pre-production model that has survived, and been sold to a private collector in 1994, is an adaptation of a Hornby-Dublo 'E3002' body with a much altered roof and a new chassis.

'E3001' with 2 Pantographs R753

The 1964 catalogue described the locomotive wrongly as a Co-Co Class E3000 electric with working pantographs and Magnadhesion and working from the high-low dual control system, either from the track or, at the flick of a switch, from the overhead power supply system.

No models were made in 1964 or indeed in 1965. In 1966 E3077, an AL5 (Class 85), adorned the cover of the catalogue in a painting by Terence Cuneo and on pages 10 and 11 we at last had a picture of a model based on the AL1 but, for some reason, carrying the late red lion (BRc) logo on its sides instead of the simplified and enlarged version (BRd) actually used on the production model.

Despite its late arrival, I have heard of at least one example of the model turning up in a Tri-ang Railways box, correctly labelled. This may have been an early model ready for dispatch before the boxes for them arrived or the company may have had to resort to old box stocks when supplies of the correct ones ran out.

It featured in the Tri-ang Hornby advertisement

R.753 Co-Co Class E.3000 Electric Locomotive. **WITH WORKING PANTOGRAPHS** and **MAGNADHESION.** Blue livery. Operates from the High-Low Dual Control System, either from the track, or, at the flick of a switch, from the Overhead Power Supply System. 8¾" (22·2 cms.) long. *Available later*

Catalogue picture of proposed AL2.

The Hornby-Dublo E3000 was used as the basis for the Tri-ang Hornby model and here we see an example of the Hornby-Dublo model, the altered sample and the finished Tri-ang Hornby production model. There is also a picture of the reverse side of the converted mock-up.

The earliest version of the Tri-ang Hornby R753 model with two pantographs.

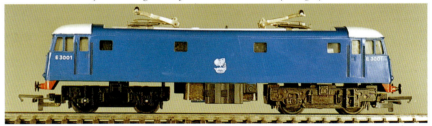

from April 1966 described as a 'Bo-Bo Class E.3001' and priced 67/6d. It should be noted that the model used here, and the one used by the *Model Railway Constructor* for their track test and that in the *Railway Modeller* review, had the correct large (over-scale) silver lion with wheel logo (BRd).

The running number chosen, 'E3001', was the first of the class, and the lower half of the cab front now correctly had a yellow panel. It is thought that it was made with two pantographs for one year only (1966) and, if this is true, less than 5,500 will have passed through the stores. The pantographs were live providing an alternative to current collection from the rail and making possible the running of two locos on one track.

The changeover plug from overhead catenary to track collection had been replaced by a neat unobtrusive three-position switch and the raised 'A' and 'B' on the Hornby-Dublo model had been erased from the pantograph mounting plates. In the 'TK' position the loco picked up power from the track and set on 'OH' it took the current from the overhead catenary returning it through the track. The third position was marked with an 'N' and with the switch in this position the locomotive remained dead.

The power bogie side frames were embossed with an 'I' on one side and 'R' on the other. When placing the model on the track it was important to place it with the 'I' on the side with the catenary posts.

The track test review in the June 1966 edition of *Model Railway Constructor* pointed out that, like the Hornby-Dublo model, it had a head-code indicator which was too small and that it should have had white window surrounds. The buffers and buffer beams had, however, been altered and were now of scale appearance. The Hornby-Dublo Ringfield motor had been replaced by a conventional Tri-ang motor bogie with serrated wheels. The wheel centres had been extended and were now to a scale of 9ft 6ins and the bogie sides had been remodelled and were correct for the prototype. Unfortunately the open axle boxes on the non-powered bogie rather spoilt the effect.

The model, which cost £3-8s-5d, also had Magnadhesion. It was disappointing to find that while the Hornby-Dublo model had hauled a 16-coach train at a scale speed of 90 mph, the Tri-ang Hornby version slipped with five coaches and would haul only four coaches without slip at a scale speed of 120 mph. It is assumed that the tests were carried out on nickel-silver track denying the model the benefit of the Magnadhesion.

The *Railway Modeller* declared that:
"The new Tri-ang Hornby 00 gauge AL1 25kV 'Blue Electric' loco No. 'E3001' is a very satisfactory model of the earliest of the locomotives for the LMR electrification. The proportions are correct and there are some very pleasing small touches – for example the characteristic large buffers of the prototype are fitted. The performance and hauling power are good, but the loco, being fitted with knurled wheels, is rather noisy... at £3-8s-5d it represents excellent value."

The reviewer went on to criticise the absence of white window surrounds, what he thought was an obtrusive changeover switch, the open end axles on the trailing bogie and the fact that Rovex had not gone as far as raising the surface of the silver crest on the locomotive's sides.

Electric Blue 'E3001' with One Pantograph R753

When running, the prototype locomotives used only one pantograph, usually the rear one, and the other remained collapsed. British Rail soon found that the collapsed pantograph needed as much maintenance as the one in use and so, to reduce the maintenance bill, one was removed from each locomotive. No doubt happy to reduce the shelf price of the model, Rovex fitted only one pantograph to their models from 1967 onwards but despite this, the price had now risen to £3-9s-3d.

The model in the 1967 catalogue was again shown as 'E3001' with the standard red lion (ferret and dart board) (BRc) logo but sported only one pantograph. In fact in 1967, the Tri-ang Hornby models continued to carry the original large lion and crown logo (BRd) in silver on electric blue body sides, changing to the 'arrow' type (BRc) in 1969. About 8,000 of these were sold solo and a further 5,000 were made up into presentation sets. The remainder of the old stock was being offered to retailers at 37/2d each early in 1970 under the reference number R753B.

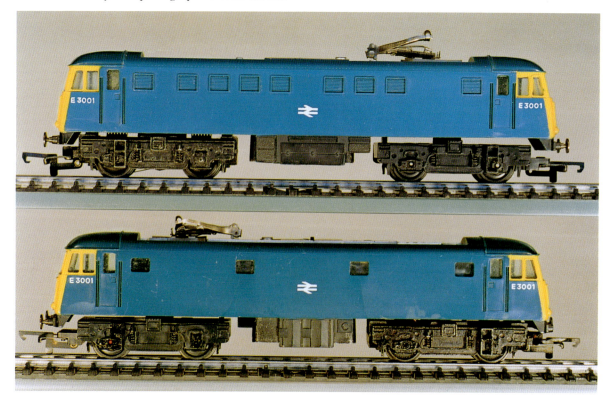

A single pantograph was fitted to the R753 E3000 from January 1967 and this shows the model in both electric blue and rail blue.

Rail Blue 'E3001' R753

The numbers and logo on the sides of the real locos were raised metal castings and these were changed to raised metal arrow logos in the late 60s and the body colour changed to rail blue, with blue cab roofs. Rovex followed suit with the model which first appeared in the new rail blue livery in 1969 and remained in production until 1971. About 2,500 of this version were sold as solo locomotives and a further 4,000 were sold in sets.

Considering the importance of the E3000 to the development of electric traction in Britain the model had a comparatively short life. It fell victim to the rationalisation that took place after the takeover of Rovex by Dunbee Combex Marx. From an economic point of view it was a natural choice for the axe as it had never sold in large quantities, in all 25,000 were made, and its going meant that the company could abandon production of the overhead power equipment.

It was to have been axed from the 1971 catalogue but there were sufficient stocks of the model to warrant including it for one more year. The remaining components would then have been packed up and put into storage against the day when the model would be reintroduced.

The Class 86

In 1965 the first of a new class of 100 electric locomotives started to arrive on British Railways. The AL6 (Class 86), a second generation of AC locos, took the best features of the five earlier classes. In the late 1970s Hornby Hobbies felt that the time had come for the reintroduction of a mainline electric but the Class 86 and not the Class 81 was appropriate to the day. They therefore had one of the Class 81 models modified as '86 225 Lion'. The result was not very pleasing and it was dropped in favour of a completely new tool. The proving model of 'Lion' survived and is now in a private collection. The full story will be found in **Volume 3**.

HALL 4-6-0

With the notable exception of the 'Lord of the Isles', the Tri-ang range of model locomotives had so far virtually ignored the Western Region and the GWR. By 1966 the time had come for the balance to be redressed. Hornby-Dublo had already modelled the Castle Class and Graham Farish had modelled the Kings. Rovex went for one of the classes of mixed traffic locos which included the Halls, Granges and Manors.

Sub-express passenger locos of reasonable size were not well catered for at that time and a loco was required that was not expensive to manufacture and could make use of the B12 chassis. Thus it was the Hall that was chosen.

The Hall had few frills being a simple and rugged two-cylinder 4-6-0 and as a model this may well have counted against it in the popularity stakes. To a child it did not have quite the charisma of 'Britannia' or 'Flying Scotsman', nor was it a very attractive model. Rovex were missing the wisdom of John Hefford.

The Prototype

The real 'Albert Hall' was one of an early batch of locos built in 1931 and withdrawn from service in December 1963. The Halls were descendants of the Saint Class, a point not missed by Hornby Hobbies in the 1980s when, with a bit of judicious retooling, they turned the Hall into a Saint to regenerate interest in the model which hitherto had been flagging. We have referred above to the Hall being a passenger engine but in fact it was an important early application of the concept of a mixed traffic engine.

BR Model R759

This was first seen in the model railway press in January 1966 when the pre-production model, which is now in my own collection, was displayed.

Pantograph maintenance.

First adverts for it appeared in December 1966 with the question:

"Where does the 'Albert Hall' hit the track? At your Tri-ang Hornby shop. Look! The locomotive that modellers have been eagerly awaiting has arrived from Tri-ang Hornby."

Railway Modeller in their January 1967 edition described it as:

"One of the most attractive locomotives to emerge from Margate. It is fully detailed, with many small, delightful touches, even the smokebox steam lance valve, often omitted on handmade models, is included, also the characteristic narrowing of the valances immediately in front of the cab."

An R759 BR 'Albert Hall' as originally made and press advertisement announcing its arrival.

The model faithfully reproduced the Swindon features. The safety valve and the chimney were accurately reproduced. The latter when first made had a brass chimney but this was replaced very soon after with a copper-plated steel one of a different shape. Unfortunately the capuchon which was present in the plastic moulding was not clearly visible as it was lost below the copper ring. This fault has never been rectified on this model even when it was converted to a Saint Class in the mid 1980s.

The smokebox front was well reproduced and included lamp bracket and step detail. The livery was late BR. After Nationalisation the prototype would have been finished in black with red and grey lining but the model was in lined Brunswick green and carried the later BR decal with the red lion holding the wheel (BRc). The lining was simplified to a single orange line instead of the orange and black on the prototype.

The review in the *Model Railway News* suggested that attention should be given to the slide bars, vacuum pipes and brake gear. Their track test showed that at three quarters full speed the loco ran at a scale speed of 84 mph and on a gradient of 1 in 13 it could push 26.25 oz from a standing start.

The *Model Railway Constructor* review in the January 1967 edition of the magazine, found the footplate moulding disappointing. As usual Rovex, to simplify moulding, had filled in the space between the boiler and the footplate. The cab was clear of protruding motor thus allowing cab interior detail to be provided. This included a driver and fireman. The hand rails, with the exception of the vertical ones on the cab, were part of the body moulding, a defect which I always felt spoiled the appearance of the model but a necessity to keep the cost of production down. The later Saints had this eyesore removed. The body was secured to the chassis by a single screw at the rear of the cab

The chassis, as we have seen, was that developed for the B12 and this, the *Railway Modeller* reasoned, was why the loco scaled 1' 3" too long. To overcome this the smokebox was slightly extended. The front drop of the footplate was also slightly under-scale. The magazine felt that the most serious error, if one was looking for things to criticise, lay in the cylinders and motion. The fluted side-rods were wrong, the cylinders too wide, the crosshead and slide-bars pure Tri-ang rather than Swindon.

It was driven by an XO4 motor onto the leading coupled axle through a brass worm and gear and weighed 6.75 oz. The model had Magnadhesion on the rear driving wheels but did not have a smoke generator, possibly due to lack of space in the very narrow boiler barrel.

The R760 tender was modelled on a Collett 4,000 gallon prototype and was a good, well-detailed model. It had spoked wheels with metal axles and closed axle boxes and contained a metal weight held in place by a block of foam rubber.

The *Model Railway Constructor* track test revealed that the loco performed badly on nickel-silver track, managing only four coaches, but 'amazingly well' on steel track with 14 coaches before slipping occurred! Running light, the top speed was a scale 200 mph.

Railway Modeller found their sample performed superbly:

"5 rather stiff coaches were handled on nickel-silver track and without the benefit of Magnadhesion, rushed at

a scale speed that made 'City of Truro' appear a sluggard and negotiated a 1 in 25 gradient."

The 'Albert Hall' in BR livery arrived in the shops late in 1966 priced £3-1s-0d and remained in the price list until 1969. It was not long before the knife men got to work and the February 1967 edition of *Model Railway Constructor* carried an article by Chris Leigh showing how the footplate could be reduced to the correct length and the distance between the loco and tender could be shortened. The following month, an article in the same magazine showed how the Tri-ang Hornby Hall could be converted into an early Saint.

About 25,000 'Albert Halls' in BR livery were made for sale as solo models with production reaching a peak in 1967. That year over 10,000 passed through the stores at Margate. In 1968 and 1969 the model was also used in the RS90 set and over these two years another 14,700 were made.

GWR R759G

In 1970, the model in GWR livery, priced £3-19s-0d, replaced the earlier BR one. Production was slightly delayed due to the late arrival of the transfers but the model was in production by late January. About 3,200 were made that first year. The purchaser was given the opportunity of renaming their model with self-adhesive nameplates and transfers for the number plates. Whether this actually sold more models is questionable as, today, it is very rare to find renamed examples on swapmeet stalls. The alternative names were :

> 4916 'Crumlin Hall'
> 6922 'Burton Hall'
> 5955 'Garth Hall'

By 1971 the model had a gloss finish but still no nickel tyres which did not arrive until 1972. Models with a gloss finish and no 'steam sound' were probably made late in 1970 or early in 1971.

GWR with Noise R759N

In keeping with a number of other models at the time, the Hall tender was fitted with a noise generator in March 1971; a feature it kept until the model

In GWR livery, R759G 'Albert Hall' with alternative named 'Burton Hall' and 'Crumlin Hall'.

One of the extra R760N Hall tenders with steam sound fitted which were initially available for customers to upgrade their existing model.

was dropped at the end of 1977. A little under 3,800 models were made with steam sound in 1971 together with a further 1,500 steam sound tenders (R760N) for people who wanted to convert a model bought the previous year.

During this period the model underwent minor changes including the fitting of nickel wheels in 1972, the dropping of crew in 1971 and the alternative names in 1973. In 1978 it was replaced by 'Kneller Hall' with its unattractive waxy blue- green body moulding.

CLASS 35 B-B (HYMEK)

It was interesting that Rovex should introduce a Hymek diesel hydraulic so soon after they launched the Hall Class loco, as the Hymek, when planned by the Western Region, was intended to replace the ageing Halls on the real railways.

The Western Region Hymeks, in their original green livery, were one of the most attractive of all the modern British locomotives thanks to their designer E.G.E. Wilkes. The locos which were built by Beyer Peacock between 1961 and 1964, made up a class of 101. Despite their excellent performance, their service with British Rail lasted only 14 years as BR had decided to standardise on diesel electrics. Four Hymeks have been preserved. They were 1,700 bhp Type 3 locomotives and worked every type of train from expresses to branch line freight.

Green 'D7063' R758

The model had been planned for 1966 but did not arrive until mid 1967. It was first seen by the public at the 1967 Toy Fair although it was not available for distribution at that time. It was the feature of the Tri-ang Hornby advertisements in the model railway press in June that year under the title "Great Galloping Diesels! Tri-ang Hornby have done it again".

It was initially made in green for one year only but did revert to this livery in 1977, although a return in

1972 had been planned but later abandoned. Most green Hymeks around today belong to these later Hornby Railways batches which will be properly covered in **Volume 3**. The 1967 models may be recognised by the large exhaust vent on the roof being flush with it instead of being raised.

'D7063' was the running number used on all green and blue models up until 1979 and the head code in the box on the cab ends was '1A20' which was for a London-bound express passenger train. The loco carried the later BR logo (BRc) and had raised guidelines on the moulding to aid the girl whose job it was to attach these in the factory.

The white window surrounds and the lime green bottom band along the sides emphasised the good lines of this locomotive which were lost when it was made in all-over blue with yellow ends. The windows were a separate moulding in white which was clipped in place and the yellow indicator panel on the front was applied as a self-adhesive label.

The black plastic chassis moulding consisted of the floor of the loco, the bogie housing and fuel tank. Red buffer beams were cemented onto the chassis at both ends. One end of the chassis plugged into the head code box and carried the head code for that end mounted on it. The cab steps formed a fixing for the body at the other end. A single screw provided additional stability.

Details of the Type IIIB motor bogie will be found in the 'Components' chapter where it is explained that the wheelbase was too short. The same applied to the trailing bogie but it was a very neat moulding. The latter had pin-point axles and carried a weight to aid stability.

11,680 were made in 1967 and sold for £2-19s-3d.

The proving model was seen in a photo in the June 1966 edition of *Railway Modeller* but its whereabouts today is not known.

The Hymek was the subject of a *Model Railway Constructor* track test in November 1967. This found the livery 'commendably' accurate and the ventilator louvers 'particularly neat'. Criticism included

the too deeply recessed windows and the front lights printed onto the warning panel instead of being moulded. Performance was not very good on nickel-silver track where it managed 180 mph travelling light but slow running was difficult due to the high gear ratio. With five Tri-ang coaches the speed dropped to 120 mph as slipping started and the model could not pull three Hornby-Dublo coaches without difficulty. Transferred to steel track, where the Magnadhesion would come into play, the loco managed to pull 11 Hornby-Dublo coaches.

The *Railway Modeller* track test found that the model could pull five Tri-ang coaches up a 1 in 50 gradient and would just take seven on the flat. The *Model Railway News* track test included incline performance and the model was said to handle two coaches on a 1 in 18 incline from a standing start without slipping.

The model was priced £20-19s-3d when first released and weighed 7.5 oz. It was the subject of an 'improving' article in the December 1967 edition of *Model Railway Constructor*.

Electric Blue 'D7063' R758

This was the wrong shade of blue for mainline diesels and no doubt was the cause of criticism at the time. It was, however, easier to use a blue plastic already in use in the factory than to buy in a new shade. The model had yellow clip-in window mouldings and sprayed-on yellow indicator panels on the cab ends. The large exhaust vent on the roof now had a raised rim. As indicated above, it carried the number 'D7063' and the head code '1A20'.

The model was made in this shade of blue in 1968 and a total of about 11,300 were sold as solo models and a further 14,600 in R645 sets of which about 4,600 had been made late in 1967. The set was not in the 1967 catalogue and, therefore, one can only guess at the colour of the loco used in 1967 sets. It is assumed that the model was blue and not green.

In addition, a further 5,000 were made for assembly packs and a batch of models was made for export to New Zealand of which details are given below. By 1968 the price had dropped to £2-17s-6d for the ready-to-run model and £2-9s-11d for the kit.

Rail Blue 'D7063' R758

Rovex decided to change to a more accurate rail blue for their locomotives with effect from 1969, but a batch of models was not made in 1969. As with previous versions of the model, the rail blue Hymek was numbered 'D7063' but at some point the 'D6830' tool for the Co-Co was picked up in error and used on a few blue Hymeks which passed through quality control unnoticed. During the Triang Hornby period 27,300 were made in rail blue of which 20,000 were used in sets. Assembly pack numbers are given below. Luminous head code labels were planned in 1971.

The model remained in production in rail blue until 1976 despite plans to change it back to two-tone green for 1972. Instead, the model was given an additional power pick-up on the trailing bogie and plated wheels. The pre-production sample has survived and is in a private collection.

Assembly Pack R396

The model was available in kit form between 1968 and 1971. As indicated above, 5,000 of these assembly packs were made in 1968 when the body colour was electric blue. A further 1,500 were made in 1969 and about 1,000 in 1970. These last two batches would have had the rail blue body.

Production for New Zealand

In 1968 the New Zealand company launched their NZHYF set for local consumption. This included the blue Hymek as its motive power unit. These models were made in Margate and sent out to Auckland bulk packed. It is not known whether the models were complete when received or had to have couplings fitted.

In 1968 a batch of 400 was sent followed by 750 in 1969. A further 50 went in 1970 and again in 1971

R758 Hymek in green and rail blue livery, and model railway press publicity.

The much publicised R754 M7 tank.

although a greater number had been ordered. It is assumed that the last three batches were fitted with rail blue bodies but I have no evidence of this.

Later Models

Between 1978 and 1980 the model, now green again, went through a number of changes and improvements both mechanical and aesthetic. An orange and black version was made for the Irish market and belongs to 1976-78. The model was also re-released in 1994 in blue livery with white window surrounds. These later versions of the model will be covered in **Volume 3**.

CLASS M7 0-4-4T

Prototype

In 1896, Dugald Drummond of the London South Western Railway was asked to seek quotations for 25 new 0-4-4 tanks. The quotations received were unacceptable and so the engines were built in the company's own workshops at Nine Elms at a cost of £1,445 each! They entered traffic the following year. So successful were they that by 1911, 105 of them

had been built and in 1948 all but one of them survived into British Railways ownership.

Prior to the spread of electrification on the Southern Railway the M7s handled most of the suburban commuter traffic. After that they were to be found on the branch lines often undertaking push-and-pull operations. In their final years in the early 1960s they were to be found on the branch lines in the Bournemouth area or pulling heavy trains of empty stock in and out of Waterloo station. The M7 was one of the great tank engines of Britain's railways. Sadly, only two have been preserved one of which is in the National Collection and the other, after a trip to the USA was appropriately repatriated on the Swanage Railway.

The Model

The cover of the 1967 Tri-ang Hornby catalogue featured a painting by Terence Cuneo of an M7 tank in a steam depot during the days of the Southern Railway. This same picture was the subject of a jigsaw made by Arrow Games, another member of the Lines Bros. Group of companies. By March 1967 the new model was the star of Tri-ang Railways adver-

tisements in the model railway press.

The M7 was the only new model locomotive that year but its introduction was clearly based on two policy decisions. On the one hand the company had decided to soldier on with new models despite the severe loss of business due to changes in market demand and on the other hand they needed to reduce production costs. The M7 was seen at the time as a cheap to manufacture, 'larger' tank locomotive that provided an alternative to the more expensive to make 2-6-2 standard tank which had an excellent body moulding but a terrible looking chassis.

This latter policy was cleverly disguised by the

inclusion of gimmicks like the opening firebox door and the glowing hearth which instead of implying regression, suggested that Tri-ang Hornby models were getting ever better. These attractions were played on in the advertisements for the model.

On the whole it was a good model although a bit on the light side and the wheels and splashers were criticised for not being quite the right size. The spectacle plate, however, was incorrect, an error that was never put right. The spectacles on the prototype were circular while those on the model were arched. Other than these discrepancies, the detail was good. It had a moulded hand rail in accordance with Rovex practice and this was never replaced.

BR Black '30027' R754

Its first appearance was made at the 1967 Toy Fair, where the pre-production model performed on a circuit for the public. This was detailed on one side only and a photograph of its 'good' side appeared in the *Model Railway News* in March that year.

A pre-production model with chassis, finished for the catalogue photograph, was sold by Hornby Hobbies to a private collector in 1994 and the following year the original pre-tooling model of the body turned up having been kept by the modeller himself. This too is now in a private collection.

The model remained available until 1970. It carried the number '30027', which belonged to a locomotive built in 1904 and withdrawn in 1959. It was one of a batch of 20 built with a longer frame. The model was lined out for passenger service and carried the later 'ferret and dart board' (BRc) logo. The lining emphasised the forward splasher-mounted sand box which was such a feature of the prototype. The M7 in the advertisements at the time was numbered 30021, and had lining missing from the leading sand box and splasher, confirming that it was the pre-production model.

The smokebox door that opened to reveal imitation flues within and the firebox that glowed, when

the model moved so that the inside of the cab lit up, were certainly popular features with children at the time. The smokebox door was a separate moulding integral with the dummy smokebox interior which was force fitted into the locomotives front. The door hinges were two fine slivers of plastic and looked fragile. Despite this, I have yet to find a model on which the door is missing. It was a nice touch.

Like other models of that period it had Magnadhesion and a crew. The cab had interior detail and the boiler top had a fine turned brass whistle. The bunker was 'filled' with coal but there were only three rails around the top of the bunker, instead of the five there would have been in BR days.

The chassis was new and not the L1 chassis reversed. The normal Rovex compromises to keep the cost down resulted in driving wheels that were 2 mm over-scale and bogie wheels that were a little too small. It had an X04 motor mounted low in the

Pre-production sample (top) and production model.

frame driving off the front axle. The bogie was very light and was forced down onto the track with a piece of sprung phosphor bronze.

The firebox glow fitment was clipped to the chassis. This was a plastic moulding incorporating the cab floor and an imitation firebox lit by a small bulb covered by a piece of yellow plastic. The glow had a tendency to vary according to the amount of power being fed to the motor.

The *Model Railway Constructor's* track test showed that the model could pull five Tri-ang Hornby coaches with ease on nickel-silver track but on steel it romped away with 12 coaches. 15 coaches proved too many. The August edition of the magazine contained an article which showed how the model could be improved.

The *Model Railway News* found the model just 1 mm longer than the 'Roche' drawing but the driving wheels were 6' instead of 5' 7". All other dimensions they thought were good but the leading splashers could be slightly reduced. In their track test the model behaved perfectly, handling reasonably-sized trains from a standing start and at slow speeds.

The model cost £2-10s-9d and despite over 20,000 having been sold, all as solo models, it is one of the less common models today. It was sold in an open box with a yellow plastic tray and an outer sleeve. A further 1,000 were bulk packed to Australia (*see below*).

SR Light Green '328' with Glow R868
Besides the black BR version, stocks of which had not yet sold out, in 1969 buyers were also offered an M7 tank in Southern malachite green livery with pre-1940 lettering and white lining. The loco carried the number '328' which was one of the last batch made and therefore also with the lengthened frame. The blackened steel-tyred wheels had the spokes painted a matching green. This model also had Magnadhesion, crew, an opening smokebox door and a firebox glow.

It was made for two years only with about 700 being produced in 1969 and 5,200 in 1970. The 1969 price was £3-5s-0d.

SR Light Green '328' without Glow R868A
This 1971 model was the same as that of 1969/70

but, in order to reduce the cost of the model, the firebox glow was not installed. Despite this, the price rose that year to £3-8s-0d. During 1971, 3,150 solo models were made and a further 5,000 were used in the RS607 set. The solo loco was packaged in an open box with a yellow plastic tray and a grey outer sleeve.

SR Dark Green '245' R868
Early in 1971 plans were made to change the livery of the M7 tank from the light green plastic of the coaches which was meant to echo the malachite green of the Southern Railway to the darker green of the BR locos, but lined in white, which was meant to represent the earlier olive green used on the Southern. It had already been decided to bring out the L1 in dark green and the M7 would compliment it. As far as we know, the new livery was not used until the following year and so it will be detailed in **Volume 3**. This volume will also carry details of the much later '249' olive green version.

Australia R754
In 1969 a batch of 1,000 of the M7 tank was bulk packed and sent to Australia. As they were recorded under the number R754 it is assumed that they were the BR black version. There is no explanation why these should have been sent without boxes. At the time Rovex resisted sending models bulk packed unless they were for making up sets. In this case there is no evidence of the Moldex factory assembling a set with an M7 as its motive power.

CLASS A3 4-6-2

Prototype
As one of the best-known steam locomotives on British Railways and the subject of an interesting preservation agreement, 'Flying Scotsman' was an obvious subject to model. In 1934, while on the London-Leeds run, it became the first British

The R868 M7 in Southern Railway pseudo-Malachite green livery.

BR (R850) 'Flying Scotsman' models showing tender and nameplate variations and the LNER (R855) version.

Press advertisement for the two 'Flying Scotsman' models.

locomotive to be officially recorded as travelling at 100 mph.

The Class A3s of the LNER, which took their names from racehorses, were a development of the A1s which were designed for the Great Northern Railway by Nigel Gresley in 1922. At the 1924 British Empire exhibition the 'Flying Scotsman' was displayed next door to the much smaller 'Caerphilly Castle' of the GWR. Much to the annoyance of Gresley the Castle was claimed to be the most powerful locomotive in Britain. Unhappy with this, the

LNER agreed to a locomotive exchange which sadly for them proved the GWR claim to be correct. Work immediately started on improving the design of the A1 and the A3 was the result. All existing A1s, with the exception of one, were converted to A3s by the end of 1948 and the final 27 of the class were built as A3s. In this form the loco had the boiler pressure raised from 180 lb/sq in to 220 lb. One of the most visible changes was the new banjo steam dome fitted to the A3s and which was nicely reproduced on the original Tri-ang Hornby model.

'Flying Scotsman' was the third of the class to be built. It was completed at Doncaster in February 1923 and was rebuilt as an A3 in January 1947. It was withdrawn from BR service in January 1963 and was bought by Alan Pegler with an agreement that it could be used on BR tracks for a set number of special runs. When handed over by British Rail, 'Flying Scotsman' was given a corridor tender of the type used by the A1/A3s working the non-stop trains between London and Edinburgh.

Trix v Tri-ang Hornby

This great racehorse of the LNER was to take part in a more bizarre race in 1967 and 1968. News of the proposal to manufacture a scale model of 'Flying Scotsman' came at the 1967 Toy Fair but the news came from the British Trix stand. A year went by without any more information but at the 1968 Toy Fair, dealers were informed that the model was nearing production. Late in 1967 Rovex had announced that they also intended to introduce the model but in both LNER and BR livery. The former was to have a corridor tender and the latter a non-corridor one.

Only the BR version was illustrated in the new Tri-ang Hornby catalogue and so the inclusion of an LNER version must have been a late decision. The catalogue showed the mock-up of the BR version in a layout display on page 5 but, at the London Toy Fair early in 1968, the pre-production models of both versions were shown and the message was that both would be ready by September that year. By August both Tri-ang Hornby and Trix had their models in the shops.

Price differences and an ability to deliver sufficient models to meet demand ensured that the Tri-ang Hornby model out-sold the Trix version many times over. The Trix advertisement in August showed their LNER version and said that 4,287 models (at the last count) were available to buy. By the following month the number had dropped to 1,759.

The BR and LNER versions of 'Flying Scotsman' were the subject of the Tri-ang Hornby advertisement in October, making it clear that both were now available. The same advert was shown in November and December but in December three versions of the Trix model were being advertised – LNER, BR and a two-tender version of the LNER model. Hornby were not to produce their two-tender model for another twenty five years.

Over the years 'Flying Scotsman' became one of Hornby's best-selling locomotives and at the time of writing (1997) it has been in almost continuous production for over quarter of a century.

BR Brunswick Green Livery R850

Throughout the 1950s and 1960s any current steam engine Rovex modelled was issued in British Railways livery. Pre-nationalisation liveries were kept for the classical locomotives such as 'Lord of the Isles' and the Caledonian Single. It was therefore natural for the people at Rovex to think of a BR livery for the A3 class locomotive they were planning to introduce and this was their original plan. As we have seen, the LNER version was an afterthought.

When the pre-production models were seen at the London Toy Fair early in 1968, the press had commented on the fact that the BR version was shown with a corridor tender. It is not known whether this was the plasticard model or one of the first mouldings off the tools painted up for the occasion. The plasticard model of the loco body was sold to a collector in 1995 having remained in the possession of the model maker until then.

The model, in semi-matt Brunswick green with the 'demi-lion rampant' emblem on the tender sides (BRc), arrived in August 1968 and, to many people's relief, had a non-corridor tender (R851) as promised.

It was a model of the prototype after the change to left-hand drive and before the fitting of the double chimney and was accurate for the period July 1957 to December 1958. The recommended retail price was £5-5s-0d.

A review by *Model Railway News* said:

"The moulded plastic superstructure is well done, and all surface detail is sharp and clear. Although the handrails are integrally moulded, they are picked out in silver and are so fine that they really do look like separate, free-standing fittings. There seems to be no details missing, and the snifting valve behind the chimney and the 'blisters' on the smokebox which accommodated the ends of the superheater header are particularly effective pieces of moulding."

In their review and track test written up in the August edition of their magazine the *Model Railway Constructor* took a tape measure to the model. This found the model to be a good reproduction of the prototype. The overall length was just 3 mm too long due to a slight lengthening of the front platform to clear the bogie on sharp curves. The thickness of the plastic moulding resulted in the spectacle windows being under size.

While these faults were necessary in the production of a model that could work and survive use by children, other errors could not be so easily justified in the opinion of the reviewer. The rear down curve of the running plate was slightly too far back resulting in the base of the cab and firebox being fractionally too short. The nameplates were printed with a red background instead of black, an error later corrected by Rovex. The lining and number on the cab side were applied as one-piece transfers and the numbering was too heavy. This was also improved on later models.

On the plus side, the rivet and piping detail were praised and the single orange line was neatly applied. The cab interior had firebox glow when the model was running and the separate safety valves and buffers were good.

The model used a completely new chassis, details of which will be found in the 'Components' chapter .

The tender was the first eight-wheel (non-bogie) type to be produced by Tri-ang Hornby and was an excellent model of the LNER 5,000 gallon prototype. It was exactly the right length and had the right wheel spacing, however, its disc wheels, which had sleeved axles and were all flanged, were 8" underscale. The tender underframe was well detailed and inside it carried a weight and two foam rubber packing blocks.

In the track test it achieved a maximum scale speed of 128 mph and a minimum controllable scale speed of 20 mph. On nickel-silver track the maximum load it could manage without slipping was three Hornby-Dublo coaches. On steel track, with

A page from the 1970 catalogue showing the 'Flying Scotsman' and 'Coronation'.

the advantage of Magnadhesion, this was increased to 10 Hornby-Dublo coaches. This compared with 'Albert Hall' which pulled 14.

The Trix LNER 'Flying Scotsman' was track tested in December and proved to be a much better performer pulling 13 Hornby-Dublo coaches with ease on nickel-silver track. It was however fitted with traction tyres.

Thought today by many to be a scarce model, the early BR version of the Tri-ang Hornby 'Flying Scotsman' is in fact quite easy to find. 12,000 were made in 1968 and all but 270 of them sold during the year. This was not the end of this version, however, and a further batch of 10,000 was made in 1969. The decision was then taken early in 1970 to not make any more batches when the existing stocks ran out and it was included in the 1971 catalogue merely to clear the remaining stocks. A BR version of a much-altered and improved model of 'Flying Scotsman' was marketed from 1993 onwards with a variety of names and these will be found in **Volume 3**.

BR Corridor Tenders

For the LNER version, described below, a different tool insert for the tender back end was used to produce the corridor tender. Sometime in 1969, after a batch of tenders for the LNER version had been made, some BR tenders were required and these were made without changing the insert. Consequently an unknown number of BR Scotsman's were issued with corridor tenders.

As we have seen, early models were also given the same nameplate stickers as the LNER version i.e., red instead of black. These are variations which may be of interest to collectors.

LNER apple Green Livery R855

By the late 1960s there was a growing interest in the preserved prototype of 'Flying Scotsman', which was in LNER apple green livery. It was one of the Lines directors who suggested making the model in LNER livery and so this was announced late in 1967. It appears that it arrived in the shops, priced £5-7s-6d, at the same time as the BR version, i.e. August 1968 and it has remained in production virtually ever since. Its success lead to pre-nationalisation liveries being applied to other locomotives in the range followed by many of the coaches and wagons.

The LNER Scotsman had a semi-matt finish with single white lining. The first batch had apple green wheels (except on the tender) with black tyres and Magnadhesion. After that they were fitted with nickel tyres. They also had firebox glow and a crew. The loco came with an R856 corridor tender. The smokebox door carried a numberplate blank that would not have been on the prototype at that time but it correctly had red nameplates over the centre splashers, as did the preserved loco.

In the second half of 1968, over 12,000 solo models were sold and during the Tri-ang Hornby period over 32,500 were sold without steam sound tenders, a third of them in sets.

LNER with Noise R855N

For 1971 exhaust sound was fitted to the tender (R856N), nickelled-tyred wheels for the loco and white-rimmed ones for the tender. The bogie wheels were also plated and fitted with plastic inserts but early examples of these tended to fall off. There was no crew and the body was also now in a glossier finish. It had been planned to have the new version of the model ready by 1 September 1970, and to use up the old stocks of 'Flying Scotsman's' in sets being made for Christmas, but the new version was not available until the January 1971. Only the first few of this revised model had firebox glow as this was left out of models from October that year.

At least 16,000 of the revised model were sold before the change to Hornby Railways, 10,000 of them in sets. The solo models included about 1,100 made for one of the mail order houses in 1971. It is not known how these differed from the standard model. There were also an additional 1,700 steam sound tenders made in 1971 for those who wanted to convert earlier models. The tenders did not have nickelled tyres.

Later Models

This version of the model was made up until 1977. The locomotive went on to become 'Gordon' in the World of Thomas The Tank Engine & Friends series in the mid 1980s. In 1993 it appeared in four different versions in the catalogue but this was a completely new model.

CORONATION 4-6-2

The LMS built two classes of express Pacifics both designed by Stanier. The first of these were known as 'Princess Royals' after the first member of the class 'The Princess Royal' and had been modelled by Rovex since 1950. The second class of Stanier Pacifics were the 'Princess Coronations'. The first five of the class arrived in 1937, had streamline casing and were painted blue and silver. They were to haul the 'Coronation Scot' train from Euston to Glasgow. The first of these was 6220 'Coronation' which gained the British speed record of 114 miles per hour down Madeley Bank near Crewe.

While 'Princess Coronation' is the correct name for the class, around the Rovex factory the name 'Princess' was reserved for the 'Princess Royals' and 'Coronation' for the streamlined 'Princess Coronations'. The non-streamlined members of this class, which were named after duchesses of the royal household and British cities, were not mod-

elled by the company until the 1980s and these were referred to as 'Duchesses'. For ease of reference these three names have been used in these books.

Initially, it seems surprising that the model had not appeared before, bearing in mind the high status of the prototype. Tri-ang had been successful with the streamlined Bulleid Pacific and the Hornby-Dublo A4 had been one of the most popular models of the 1950s. Trix had modelled the Coronation before the War but none of the big three post-war British model railway manufacturers had thought it

a worthy subject, possibly because it had not survived into BR livery and BR livery was what people were thought to want during the 1950s and 1960s.

Due to the difficult curves employed in the casing which were not easy to represent in drawing form only, Rovex had a large-scale wooden pattern made for the body from which the toolmaker could take measurements. This survived in a store room at the factory for many years and was eventually sold to a private collector in 1994. The model had no front coupling.

Alternative-named 'Duchess of Rutland', 'City of Edinburgh' and 'Queen Elizabeth'.

R864 'Coronation' and alternative-named 'Princess Alexandra' and 'Queen Elizabeth'.

'Coronation' R864

The model, finished in LMS blue livery, first appeared in the shops in September 1970, later than originally planned. The delay was caused by difficulty in decorating the models which was done partly by hot-foil printing and partly by hand lining. It had proved awkward to get the two methods to blend together in a satisfactory manner but by trying many different paints they eventually overcame the problem. It had Magnadhesion and crew and came complete with its tender (R865). The model was supplied with a choice of three alternative names:

6221 'Queen Elizabeth'
6222 'Queen Mary'
6224 'Princess Alexandra'

It was designed to fit the 'Flying Scotsman' chassis and, as a result was approximately 1 cm too short. The bogie wheelbase was 6 mm short and the bogie set 4 mm too far back from the buffers. The rest of the dimensions were on the whole good and the general shape of the casing was pleasing.

The valve gear did not look right to the railway enthusiast but it is easy to forget that Rovex were toy makers and not model makers. Their success had largely resulted from keeping the cost of their products below that of their competitors. The skill was in knowing how far one could go in improvising. One saving was the use of the former Hornby-Dublo tender chassis which can clearly be seen on the proving model of the tender which is now in a private collection. These were supplied by G & R Wrenn Ltd and the reviewer in the *Railway Modeller* claimed that this made it "the first true Tri-ang Hornby model"! If it was, it was certainly the last. The reviewer also pointed out that it facilitated the fitting of a Peco coupling to the tender.

The model scored an important first. It was the first Rovex model to be fitted with separate handrails. The nameplates were silver and blue embossed self-adhesive labels and looked quite effective. These early versions of the streamlined

Studio picture of R871 'King George VI'.

Princess Coronation were finished in self-coloured plastic and lacquered to make them shine. Some early models left the factory before this process had been perfected and so were in less than adequate finished condition. When the model reappeared in 1983 it had a paint finish and looked altogether better; furthermore, matching blue and silver coaches were then available. Had matching coaches been available in the early 1970s it seems likely that the loco sales would have been greatly increased. Unfortunately, at that time, a suitable coach had not yet been developed.

In a test carried out by *Model Railway Constructor*, running light, the model reached a scale speed of 128 mph on nickel-silver track. When four heavy old Hornby-Dublo coaches were added it slipped badly. With five Tri-ang Hornby coaches it slipped slightly on starting but managed to pull the load and maintain a scale speed of 65 mph. On Tri-ang Hornby steel track, the Magnadhesion allowed the model to take two more coaches. The loco was controllable at all speeds and the slow running was quite good.

The original blue 'Coronation' sold for £4-19s-0d and was made for two years. It arrived around July/August 1970 and 2,300 models were made that year and a further 3,700 in 1972 despite plans to drop it from the range. A batch of 800 models was supplied to at least one mail order house in 1971.

It was to reappear in the catalogue in 1973 but without alternative names and still with blackened steel rimmed driving wheels. This was possibly because it was being used to finish off stocks of the

old 'Flying Scotsman' chassis left over when that was converted to nickelled-plated tyres. There is no record of any more models being made.

'King George VI' R871

After building five Coronations in blue and silver livery, the LMS built a further five in maroon and gold; followed in 1939/40 by another ten. The Tri-ang Hornby lacquered maroon version, 'King George VI', was available from September 1970 and was priced £5-15s-0d which showed a considerable increase in cost in just one year.

It had been intended to release the model in 1971, and indeed it was not illustrated in the 1970 catalogue, but the decision was taken the previous May to bring forward its release date. One wonders whether this was as a result of comment in the model railway press when the blue version was reviewed that the Coronation would be more popular in maroon and gold.

The model was fitted with nickel tyres on the locomotive wheels but not on the tender (R872). An early fault with this model was the loose fitting of the plated front-bogie wheels. This was quickly rectified.

This model also had a set of three alternative names:

6221 'Queen Elizabeth'
6228 'Duchess of Rutland'
6241 'City of Edinburgh'

The illustration in the 1971 catalogue was of the pre-production model which is now privately owned. It had a different shaped front to the production model, the vertical curve being of a smaller radius. The model had been made in clear plastic and sprayed maroon and had the gold lining and numbers stuck on. This suggests that the original plan was to release the model as the maroon 'King George VI' but for unknown reasons it was decided to change to 'Coronation' in the blue and silver livery after the pre-production model had been made; possibly because 'Coronation' was first of the class or because the men at Rovex knew that blue locos sold well. A maroon and gold loco would have been lost amongst the other LMS maroon locos being made at that time. To prove this theory wrong, it will be necessary to find a pre-production model in blue with the wrongly-shaped front end and, preferably, labelled 'Coronation'.

The model was available for four years. A little over 7,300 were made in 1971 of which 500 went to a mail order house and 1,000 went abroad. Over the next three years a further 4,100 were released. The Tri-ang Hornby version was sold in an open box with a yellow plastic tray and a grey outer sleeve.

Variations may be found. There are, for example, two distinct shades of maroon, one darker than the other. There are also number transfer variations. Some had numbers like those provided on the sheets for renumbering with the alternative names and some will be found with slightly larger numbers. The latter may date from 1973-74 when the alternative names and numbers were no longer provided.

Coaches

The reviewers of these models in both *Railway Modeller* and *Model Railway Constructor* drew attention to the lack of suitable coaches. The *Railway Modeller* even suggested reintroducing the old Tri-ang 9″ mainline coaches in the correct livery. At this

time, Rovex did not respond but when the model was reintroduced in the 1980s the Stanier coaches in production at the time were painted up in both blue and silver livery and maroon and gold. Only those in blue and silver livery went into production.

CLASS 9F 2-10-0

Prototype

It is well known that 92220 'Evening Star' was the last steam locomotive to be delivered to British Rail and the only one of its class to be finished in lined-out green and given a name. Several serious railway enthusiasts and modellers believe that too much attention is paid to this one locomotive at the expense of the rest of the class.

The 9Fs were the most powerful of a range of standard locomotives designed and built especially for British Railways between 1954 and 1960. While originally intended for heavy freight traffic, the class took on express work as well. They were a very successful design and 251 were built. Sadly they came too late and withdrawals started in 1964. Happily nine of the class have been preserved.

Origins of the Model

The Tri-ang Hornby model of 'Evening Star' came about largely as a result of a chance meeting Richard Lines had with WE Ward Platt, formerly a designer with Meccano Ltd. In the late 60s the latter was running the shop, Lucas's (Hobbies) Ltd in Liverpool, and bumped into Richard Lines at a Tri-ang Provincial Toy Show on 23 May 1968. Ward Platt was an enthusiast of Rivarossi's Big Boy and took Richard home to see his large layout where Big Boys roamed freely. He remarked on how well this complex and expensive model sold and suggested that Tri-ang Hornby ought to do something similar. They discussed the possibility of a 9F.

It seems that the decision to pursue this course of action was taken very quickly for the day after the

visit Richard Lines wrote a letter to Ward Platt in which he said,

"You will appreciate that it will take us considerable time to gather together all the basic information we need but we certainly intend to go ahead with the project and I am certain I shall be coming back to you for further assistance during the course of the designing."

On the 31 May he wrote to Henden at Margate saying:

"I want to start a preliminary study with a view to bringing out, in probably 1970, a very deluxe version of the 'Evening Star' locomotive which may retail at up to £9 . . . I have chosen the 'Evening Star' as in addition to having a nostalgic appeal to all British train buyers it will also have a considerable export potential.

The particular features that will be required in order to make our model a class above anything else produced in the UK would be that the moulded body would be in two or three separate parts, probably based on the colour split-up, thus permitting the correct detail beneath the boiler. It would have separate handrails, parts of the valve gear would be die-cast rather than pressed and it would have all other fittings which we normally abbreviate on our 'regular' models.

I do not think it would be necessary to have a different motor or method of drive. I think in view of the ten driving wheels we would have to specify that it would not be suitable for operating on the 14⅝″ radius curves. I believe the prototype, in fact, does not have flanges on each wheel and we should try and reproduce the flange arrangement of the original accurately.

In order to study the problem of wheels you should obtain a sample of the Marklin model No. 3046. You should also have an elaborate Rivarossi sample to study and I would be glad if you would make arrangements to obtain these samples so that you can start studying the overall concept. I would envisage that whilst tooling will be expensive, we would anticipate recovering something like 30/- per model in order to get our tooling back within, say, two years."

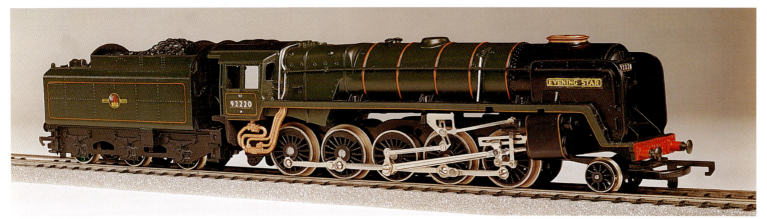

R861 'Evening Star' – the first in a new era of model locomotive production.

Expensive Model

It was known that it would be an expensive project but the company decided to throw caution to the wind and make the model irrespective of the fact that the finished article would have to sell at above the normal Tri-ang price for a loco of its size. As it happened the shop price was initially £8.00 compared with £6.25 for 'Britannia'.

The pre-production loco body, which was made in a clear plastic, and the tender-mounted Ringfield motor were kept by the model maker when he left the company and are now in a private collection.

It was in January 1971 that the prototype model was displayed at the Toy Fair after which Richard Lines wrote to Ward Platt to thank him for his part in the venture. He went on to say:

"It has certainly taken us rather a long time to get to this stage but it has been a somewhat involved project… We do not, as yet, know how many further problems we shall encounter when we get to the pre-production stage, but the project is quite well advanced already and we are hoping very much that we will be able to start production so as to be able to execute all our orders in good time before Christmas 1972 [he presumably meant 1971]."

In fact it was released late in 1971 with only just over 1,000 models, out of 13,000 ordered, being despatched that year.

The reviewer for the *Railway Modeller* said that "the detailed improvements are very promising although the general features are a little toy-like" and finished the report by referring to the model as a 2-10-0T!

A Revolutionary Design

Since 1950, Rovex locomotive models had been driven by a motor mounted in the main body and since 1961 had had the benefit of Magnadhesion to increase their power. Richard Lines' predictions, in his memo to Henden in May 1968, about the design features of 'Evening Star' proved to be incorrect on a number of points. The wire handrails were not to come until later, it was a single-piece body moulding, a non-conventional motor was fitted, the valve gears were not die-cast and the model was not unsuitable for 14⅝" curves.

All of this was evidence of the considerable changes that take place in a model as it goes through the development phase and problems are weeded out. The metamorphosis of the pre-production 'Evening Star' was to have far reaching effects on the whole of the locomotive range in the years to come. This was the first of a new breed of Rovex models and set a path which was to lead to the superior models of the 1990s.

The Locomotive

The body had good lining out although still only single lines were used except on the cab side. The finished body was sprayed with lacquer to give the model a gloss finish. The rivet detail was particularly pleasing and the nameplates were well printed self-adhesive labels and amongst the most realistic done by the company.

The earliest models had realistic handrails on the edge of the cab sides but girls handling them to detail and assemble the models were breaking the handrail off and so the gap was filled in. It seems that this was done early on in the life of the model and can be used to help with dating them.

Press Reaction

'Evening Star' was reviewed in both *Railway Modeller* and *Model Railway Constructor* in December 1971. The *Modeller* found it difficult to criticise confirming that it was:

"one of the best scale models Tri-ang have yet produced".

The *Constructor* gave the loco a more thorough

testing and came to the conclusion that it was:

"a model of which we can be proud and it will stand comparison with the best".

They said that it was:

"a near perfect 4mm scale representation of the original",

and they felt that it was:

"one of the finest proprietary models manufactured in the British Isles".

Criticisms were naturally few but included:

1. The size of the gap between the top of the wheels and the bottom of the running plate which was 2 mm too much.

2. The gap between the loco and tender was 3.5mm too great. It was recognised that this was normal to allow for negotiation of tight curves.

3. Moulded hand rails.

4. Over conspicuous white and copper pipework. This needs toning down for realism.

5. Lack of glazing in the cab windows.

6. A little noisy, a problem partly solved by placing foam in the tender, under the coal moulding.

Performance

The track test carried out by *Model Railway Constructor*

provided spectacular results coming up to the standard of Continental models. Running light, the model produced a scale speed of 100mph but could be reduced to a realistic crawl.

The haulage capacity was found to be greatly superior to any other Tri-ang Hornby model. They had it pulling a train of 14 heavy coaches and even managed a 1:40 gradient with this load. It was claimed that it would have taken a further four or five vehicles. With a heavy load it was also very realistic at slow speeds, something that serious modellers like to achieve.

The model was also found to negotiate 17" curves although with some impairment of its performance.

Production and Packaging

'Evening Star' was featured on the cover of the 1971 catalogue, in a picture painted by Terence Cuneo in 1963.

Over 12,000 orders were received that first year but, despite predictions, less than 500 came from abroad. As it was late in the year that the model finally arrived, only 1,120 were released in 1971. Over 12,000 were despatched the following year. In its original form the model was made up until

1975 but it was to undergo a number of improvements over the years which will be covered in **Volume 3**.

Due to the uncertainty about the future of the company, from the start, it appears that 'Evening Star' was packaged in white boxes with a blue vacuum-formed tray; some were in plain red boxes.

Black 9F R550

The black 9F was not illustrated in a catalogue until 1973 but the models had been in the shops the previous summer. The model will be covered in **Volume 3**.

CLASS 5700 0-6-0PT

GWR Pannier '8751' R51S

This was in the planning stage as early as January 1971 and was scheduled for release twelve months later but in fact arrived in the shops late in 1971. It was a very sensible choice being based on a popular prototype many of which had been preserved and were operating on preserved lines around the country.

The class chosen, the Collett 5700, was the largest of the pannier classes and was built over a period between 1929 and 1950 during which time 863 were made. Few classes of locomotive in the world can claim such a long period of production. '8751', the prototype modelled, was built at Swindon in 1933. The class has been claimed to be the best six-coupled tank engine ever designed. They were both easy to handle on shunting duties and free running on passenger trains. Some were later sold to London Transport and remained in steam until 1971.

Samples of the model arrived in time for Christmas in 1971 and it was reviewed in the January 1972 edition of *Model Railway Constructor* and February edition of *Railway Modeller*. Despite this no stock for sale was made in 1971.

It was planned from the start to have fine scale plated driving wheels and a smoke generator. The

R51S GWR pannier tank which was released in the closing days of Tri-ang Hornby.

company hoped to use existing wheel tools but make a new tyre tool. The *Model Railway Constructor* review found the wheelbase only slightly too large, but the driving wheels to be almost spot-on. It did, however, find the over-all length 4 mm short and the height 2 mm too much. *Railway Modeller* remarked that the wheel flanges were a trifle too deep.

The finish was gloss Brunswick green with 'GWR' on the tank sides. The model had a copper-coloured chimney top and a well-shaped dome. The hand rail was moulded and this and the rivet detail were effectively done. The general view was that it gave a good representation of the class and would be welcomed by GWR fans with the exception of real scale constructors.

On test, the model pulled five Tri-ang Hornby coaches on nickel-silver track but managed seven on steel. It was quiet and managed slow running quite well.

BR 'KESTREL' DIESEL

Also in January 1971 plans were in hand for another large mainline diesel and the Development Programme had it listed as R863 BR 'Kestrel' Diesel. 'Kestrel' was an experimental loco built by the Brush/Hawker-Siddeley group. It was a very powerful locomotive which BR operated from 1968 to 1971 and it was subsequently exported to the USSR. Notes at the time indicated that the new diesel model was to have the 'Evening Star' mechanism and reversing head and tail lights. What Rovex eventually produced was a Brush/Sulzer Co-Co Type 4 later to be known as the Class 47 and this will be covered in **Volume 3**.

B. LOCOMOTIVES FROM STARTER SETS

Rovex had long known that the model railway world was a cut-throat business and the secret to

Various clockwork starter set locomotives: 0-4-0 diesel shunters, R660 top tanks and R854 Continental 0-4-0 tanks. The bottom row contains two damaged models which have been included because of their scarcity. The red Continental tank in the third row is the most common. See page 166 of Volume 2 for other variations which were also used during this period.

success was to catch children while they were young in the hope that they would then stay loyal to that particular model railway system. Starter sets were often sold as loss-leaders to capture that first-time buyer.

For cheap starter sets one needed cheap locomotives and this was an area in which Rovex were masters. Over the years Rovex developed a good range of models to fit the bill. They included both clockwork and electric versions but of these the clockwork ones were by far the most prolific. The cheap clockwork mechanism designed in the early 1960s was to power many millions of locomotives over the years and was still selling in the mid 1990s having changed very little over the 30 years it had been manufactured.

This, and the various electric chassis developed over the years, carried many different bodies but, judging by the number made, it seems that the steam-outline models were much more popular than the diesels.

'TOP TANK' 0-4-0 TANK

One of the most prolific of the starter set locos was the box tank or top tank loco which first appeared in this role in December 1962. This was the clockwork R657 and was followed the next year with the R659 electric version. Such was the need to compete on the price that the production cost of the clockwork model was further reduced by the removal of its 'connecting rods' in 1965, the revised model being R660. This was the version that saw the Top Tank loco through into the Tri-ang Hornby era.

The loco got its name from the large box 'tank' which was placed on top of its boiler in order to accommodate the clockwork spring. It ended life as the American loco used in the RB6 Wild West set of 1971.

Electric R659

The electric version of the Top Tank loco was used in sets between 1966 and 1968. The model was usually black but other colours may exist. About 100,000 were made for sets over this period. The chassis was the same one used in 'Nellie' and the electric 0-4-0 diesel shunters. It had see-through wheels and wire pickups.

Electric for Australia R659A

This model was supplied to Australia in an incomplete form between 1965 and 1967. The Australian company completed the models for use in sets (possibly the RS20). According to Australian records they received 375 locos over the three years but the Margate factory census did not record any of these.

Clockwork Special R660

Although called the 'Special' when introduced as a cheaper form of R657, it was the standard clockwork starter set loco between 1965 and 1968. During this period over 830,000 were made and less than 1% of them were sold as solo models.

They can be found in a number of colours including black, yellow, bright blue, mid blue, and dark green. The models mostly had red wheels but some were issued with black ones.

USA Clockwork R660T

Tager of America ordered 60,000 Top Tank locos with NMRA couplings to be used to make up starter sets for sale under the American Train and Track (ATT) label in the USA. In fact 54,650 were sent in 1968. It is thought that these were yellow.

Australia R660A

In 1968, Australia received 7,000 clockwork Top Tank locos bulk packed for use in sets. These may have been sent without couplings and been fitted with Moldex-made couplings on arrival.

It seems that there was a problem with a lot of the motors at this time. In 1969, 1,000 mechanisms were sent out to Melbourne of which 825 were used to replace faulty ones. It is interesting to note that these were supplied by Rovex at 1/- (5p) each.

New Zealand R660

New Zealand received three batches of the clockwork Top Tank loco for assembling the 'NC' sets. These were bulk packed and may have required completion on arrival. 2,600 were sent in 1968, 6,200 in 1969 and 3,150 in 1970. Twice as many keys as locos went out with each batch; evidence that each set carried two keys.

WILD WEST 0-4-0 LOCOMOTIVE

1863 R873

This is perhaps the rarest and most sought after of the starter set models. It appeared in only one set

R873 Wild West locomotive.

and as far as we know, none was sold as a solo model. The set in which it appeared was the RB6 of 1971 of which 25,000 were produced. Despite the large number made, few remain as, like other starter locos, they were normally thrown away when grown tired of, and not passed on to the next generation.

The model was designed in March 1971 from the Top Tank 0-4-0T loco which had been replaced by the Continental 0-4-0T in 1969. The body had a red Crockett cow-catcher added at the front and a spark arrester chimney and lamp in black. It was coupled to a tender made up from a red integral low-sided wagon with the Crockett tender top mounted on it.

The loco was red with a yellow number '1863' on its sides, the number having been taken from that already displayed on the side of the Crockett tender and enlarged. The rear end of the cab had been retooled to extend the roof backwards and to alter the appearance of the bunker. The impression of double doors in the rear of the bunker was added to indicate how the crew could get to the tender for logs.

Electric starter set locomotives: two R853 Swedish 0-4-0DSs, R852 Continental 0-4-0T and R858 Barclay 0-4-0DS.

'NORTH BRITISH' 0-4-0 DIESEL SHUNTER

Clockwork R557

The model appeared in a number of liveries for starter sets. These included blue, apple green and red. Some had BR decals on their sides and a few were electric, using the 'Nellie' chassis. The vast majority were clockwork. These sets are very poorly recorded and so it is very difficult to apply dates to particular colour schemes or indeed to determine whether the different colour schemes were awarded different 'R' numbers.

It is also hard to say whether a particular model belonged to the Tri-ang Railways or Tri-ang Hornby period. Amongst those that almost certainly belong to the latter are the red, apple green and a deep, almost violet, blue – a colour used for starter set models in the late 60s. It is highly unlikely that any of these diesels were used in sets after 1968.

Electric R654

The Australian records indicate that Moldex turned out finished models in 1966 and 1967 but there is no record in the Margate census of these having been supplied. 200 solo models are supposed to have left the Moldex factory in the first year and 100 the following year.

'BARCLAY' 0-4-0 DIESEL SHUNTER

Blue R858

The North British 0-4-0 diesel shunter was withdrawn after 1967 and in November 1968 the body was redesigned to take the new electric chassis with its imported can motor. The old body was too small for the new chassis and so a large hump was created in front of the cab which gave the model the look of a Barclay diesel shunter. All examples seen have

had a bright blue body and a red chassis with black wheels. Other alterations to the body had included removal of the buffers leaving blind stocks, a widening of the bonnet, the removal of the chassis location holes in the back of the body and the provision of new chassis location holes in the lower sides.

The model had been planned for 1969 but almost certainly it came into use in 1970 in one of the Miniville sets. At least 27,000 of that set were made, but for the present little more about it is known.

'CONTINENTAL' 0-4-0 TANK

Referred to in the company's drawing office as the "inexpensive clockwork steam loco", the Continental 0-4-0 tank was designed as a replacement for the Top Tank loco possibly because the latter had insufficient international appeal. It used the same chassis and was to become the longest surviving of all the starter set locos being made well into the 1990s. Over the years there have been many versions of the model and it has appeared in many colours. Those known to have been available during the Tri-ang Hornby period are recorded here.

Electric Model R852

The model was first shown in the 1969 catalogue as well as being demonstrated at the Trade Show at Tri-ang House that year. In the catalogue it was illustrated as plain blue with a red chassis. This may have just been an artist's impression but the 1970 catalogue showed a bright blue loco with '7744' in white on its sides. All examples seen by me have had a black body and in some later models the chassis was black.

With the exception of the wheels and coupling rods which were from the 'Nellie', the red (or black) chassis was completely new. It consisted of a one-piece plastic moulding surrounding a large can motor. Copper strips provided electrical contacts on all four wheels. Half way along each side of the plastic chassis unit was a peg which clipped into a hole in the bottom edge of each side of the body moulding. The electric models had the key hole blanked off. As the same tool would have been used for electric and clockwork bodies, a pin must have been inserted into the tool when a key hole was required and replaced with a blank when not.

Almost 50,000 of this loco were made at Margate between 1968 and 1971 and all of these were used in the assembly of sets.

USA Electric with NMRA Couplings R852T

In 1969 Tager of America were supplied with 11,300 electric Continental 0-4-0 tanks fitted with an NMRA coupling, only at the back, to be used in the assembly of their ATT sets. A further 28,600 were supplied the following year. The model seen by me was black with '7744' in white on its sides and the

new black chassis fitted with a can motor as described above. The only inscription beneath the loco body was "Built in Britain". With the exception of the coupling it is thought likely that the models were identical to those being used in set assembly at Margate at the time.

Australian Electric R852A

6,000 electric Continental 0-4-0 tanks were bulk packed to Moldex in Australia in 1968. It is quite possible that these had to have couplings fitted on arrival. Another batch of 1,900 was sent out in 1970. These had the new chassis and appear to have had black bodies but without any markings.

New Zealand Electric R852

In 1969 and 1971 batches of the model, totalling 2,100 were sent bulk packed to New Zealand. It is possible that these had to be completed on arrival. They were supplied for use in assembling 'NE' sets.

Canadian 'Chugga' Electric R852CN or R8520

Canada was supplied with their own version of the electric 0-4-0 Continental tank. It was yellow and marked 'Chugga' and with a maple-leaf emblem in red. Unlike the other electric Continental tanks these used the 'Nellie' chassis instead of the new one with the can motor.

The first batch of 1,600 went out to Canada bulk packed in 1969 followed in 1970 by 600 more also unboxed and 500 boxed. A final batch of 1,580 boxed models went in 1971.

Clockwork Model R854

The clockwork version of the Continental 0-4-0 tank was predominantly red. Over the years the shade of plastic changed and examples can be found that vary from a light waxy red to pale maroon. The ones planned for use with the RB plastic track sets of 1971 were to be SR green with red wheels.

The chassis was the standard cheap clockwork

mechanism that had been used for all previous 0-4-0 clockwork locos and which would remain in production for at least another 25 years. The red-bodied locos had black wheels and a pseudo coupling/connecting rod.

There were some difficulties with the integral coupling bar at the rear of the loco not coupling correctly to rolling stock. The bar was moulded in a semi-flexible material which if packed badly could be left bent. This led to the metal coupling hook used on all

R852T USA version of the 0-4-0 electric tank with a single NMRA coupling.

R852CN Canadian 'Chugga' electric Continental 0-4-0 tank.

R653 Continental Prairie tank.

stock being modified thereby creating the type IIId coupling.

With one known exception, the clockwork Continental 0-4-0T carried no markings. The exception was a red-bodied loco that had been printed with the number '1863' in yellow. The numbering tool used was the same as that used on the Wild West clockwork 0-4-0 loco (R873).

In just four years between 1968 and 1971 about 720,000 continental clockwork tanks were made for starter sets and a further 27,000 to be sold as solo models. Over the years it must have become Britain's most manufactured model railway locomotive. It is not known how many million have been made but in the mid 1990s it was still in production!

Australian Clockwork R854A

This replaced the R660A clockwork Top Tank loco in Australian starter sets. 4,500 were sent out bulk packed in 1969 and a further batch of 4,100 went the following year. Some of the 1969 batch contained the faulty mechanisms referred to above. It is possible that these were sent incomplete, perhaps requiring couplings to be fitted on arrival.

New Zealand Clockwork R854

A batch of 4,250 was sent bulk packed out to New Zealand in 1971 but it is not known whether these were complete or whether any turned up as solo models in local packaging.

USA Clockwork with NMRA Couplings R854T

In 1969, 15,800 clockwork tanks were made for ATT of America but were not sent and all but a handful remained in the stores at the end of the year. As there was no record of these models being in the stores at the end of 1970 it seems that they were disposed of during the year. There is no indication that they were sold and so they may have had their couplings replaced and been absorbed into the normal R854 stock.

'SWEDISH' 0-4-0 DIESEL SHUNTER

UK Models R853

The model, which was demonstrated at the 1969 Trade Show at Tri-ang House, first appeared in starter sets in that year and had a life of only three years. It is not definite what the model was based on, but possibly it was the Class Z65 built by Kalmar between 1961 and 1967, as it had a distinctly Swedish look.

It is thought that 35,000 were made and came only in an electric form with coupling rods and a can motor. Two colour variations are known to exist: yellow with a blue roof numbered '5771' in black, and red with a black roof numbered '4718' in white. In both cases the chassis was black and fitted with 'Nellie' type wheels. This is not an easy model to find.

Due to the complicated curves required for the body, a large scale wooden pattern of the body was made. This is now in my own collection being one of a number of items I was able to purchase from Hornby Hobbies around 1990.

Australia R853

In 1969, 6,000 were bulk packed and sent out to Australia for use in starter sets assembled in the Moldex factory.

C. CONTINENTAL

CONTINENTAL PRAIRIE TANK LOCO

1969 Version R653

This was a re-release of the 1963 model originally designed for manufacture in France. It had not sold well originally and it was presumably only reissued now in order to help justify the reintroduction of the British standard Prairie which used the same chassis. The appearance of the original model had been much criticised and so the tools were modified to give the loco two domes and to remove 'SNCF' from the cab sides. It was also issued in a new livery making it look Germanic. The chassis was also modified. It had the brake blocks removed and a different collector plate screw spacing.

The model was not a great performer and in a track test by *Model Railway Constructor* it managed four Tri-ang coaches on nickel-silver track. It was quite fast when running light but the slow running was not particularly good. The front and rear pony trucks tended to oscillate, possibly due to the lack of a spring and the fact that the pivot point was below the level of the centre line of the axles.

It was illustrated in both the *Model Railway Constructor* and *Model Railway News* in March 1970, after it had been dropped from the price list. The *MRC* review in March 1970 indicated that it was already in short supply which explains why it is one of the harder models to find. We now know that just over 2,000 were made and all but a handful left the stores at Margate in 1969. In Britain it was priced £3-7s-6d and it was fitted with Magnadhesion and a Seuthe smoke generator.

D. TRANSCONTINENTAL

TRANSCONTINENTAL PACIFIC

'Hiawatha' R54S and R32

The Transcontinental Pacific had been given the name 'Hiawatha' in 1962 in order to give it a handle by which to refer to it in the factory. As 'Hiawatha' it became a Tri-ang Hornby locomotive in 1965 although by then it had been dropped from the British catalogue. Despite this it remained in production for overseas sales. The last batch of 2,000 for general export was made in 1966. After that, the only batches made were for specific customers. The Canadian company received about 6,000 boxed locos in small batches between 1965 and 1968 and in 1968 it also received 100 'Hiawatha' models, bulk

packed, for its TS673 set.

The basic model had a German smoke generator which was the only type that would fit this model. It also had Magnadhesion, see-through wheels and a light at the front. From 1968 the loco and tender were sold as one unit. It appears not to have been made in this form after 1968. 'Hiawatha' normally carried the number '2335' in white on both the tender and cab sides but the model may be found with the number on the tender in yellow instead of white. The latter may have been made at about the time production was changing over to the Canadian Pacific version (*see p.193*) which was detailed in yellow.

Australian 'Hiawatha' R54(A) & R32(A)

The model was said to have the grandeur of the C38 class of the NSWGR although only the bulky boiler, short chimney and twin domes resembled this prototype.

It was never made in Australia but was 'finished' there or assembled from parts sent out from the Margate factory bulk packed. In 1963/64 a total of 350 'Hiawatha's were finished in the Moldex factory for sale as solo models. Batches of 800 and 850 respectively were also assembled there in 1966 and 1967. How many additional ones were sold in sets in these early years is not known. In 1970 a batch of 640 'Hiawatha's was again sent out bulk packed and some of these may have been boxed up as solo models in Moldex boxes. It is strange to note that only 340 of the R32 tenders were sent with this last batch.

New Zealand R54S(NZ) & R32(NZ)

The model was not made in New Zealand either but was supplied bulk packed for use in sets. Between 1968 and 1971, about 1,500 TC Pacifics were sent out to New Zealand in small batches. While the 650 sent out in 1968/9 were almost certainly 'Hiawatha', the 1970 and 1971 batches, amounting to about 760 models, may have been the later R0542S with the

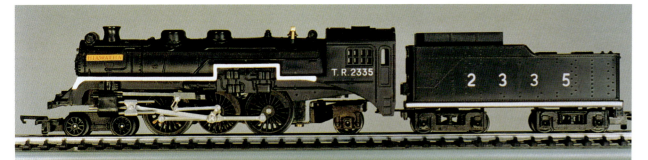

R54S 'Hiawatha'.

R54S(F) 'Hiawatha' sold through Hornby-Acho.

Scotsman tender (*see p.194*). However, they were listed as the original 'Hiawatha' and illustrated as such in the 1972 North Island price list. We know that an extra 450 tenders with steam sound were sent to New Zealand in 1971 for those who wanted to convert their models. The loco was never offered for sale as a solo model in New Zealand but was bought in to make up the NE7 Transcontinental Freight Express set. The yellow numbered version has turned up in New Zealand.

France R54S(F) & R32(F)

With the takeover of Meccano Ltd in 1964, Meccano France Ltd were obliged to incorporate models from the Tri-ang Hornby range in their catalogue. One of the models chosen was the 'Hiawatha'. 1,500 models were sold with Hornby Acho couplings for the tender so that they were compatible with the French rolling stock. 1,000 of these locos went out to France in 1967 and 500 the following year. All were boxed and a label was stuck on the end of the box on arrival at the Calais factory describing the contents in French.

Italy R54S & R32L

Possibly as part of the deal which lead to Wrenn marketing Lima N gauge model railways in Britain, Lima agreed to take two batches of Tri-ang Hornby locos fitted with Lima couplings. One of these was 1,800 'Hiawatha's which went out to Italy in 1968.

Canadian Pacific R54SC & R32C (or R0542S)

Between 1969 and 1971 the TC Pacifics sent out to Canada carried the Canadian Pacific road name. The locos were black and the same as 'Hiawatha' but for having no name. The 'CANADIAN PACIFIC' on the tender and '2335' on the cab sides were in yellow. The model was illustrated in the 1969 Canadian catalogue with '2389' on the cab sides but this is thought to be a picture of the pre-production mock-up.

R1542N 'Australian' TC Pacifics.

R54A Australian-assembled TC Pacific. Note the chromed eyelets used to fix the tender bogies.

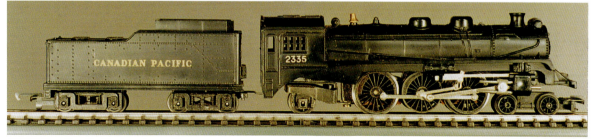

R54SC Canadian Pacific steam locomotive.

In all, about 5,380 of this version of the model were made. These were made up as follows:

R54SCP & R32CP (loco and tender sold separately) boxed – 1,120

R0542S (loco and tender sold as one unit) boxed – 2,000

R54SCP & R32CP (loco and tender sold separately) unboxed for sets – 1,410

R0542S (loco and tender sold as one unit) unboxed for sets – 850

There may have been confusion over the 'R' numbers. The model appears to have still been in stock in Canada as late as 1973 or even 1974 under the number R542. Some or all of these, however, may have been the so-called 'Australian' version with the Gresley tender, as stocks of this appear to have been sent out to Canada in 1972 and 1973. Indeed, pictures of the Australian version appeared in both the 1971 and 1973 Canadian catalogues, first as R0542S (1971) and later as R054 (1973). The latter was probably the correct picture for the model available at the time.

There was obviously a demand for the tender on its own to convert models of 'Hiawatha' already bought by Canadian modellers or to convert models still held by the Canadian company. Accordingly, 1,800 extra R32C tenders were sent out to Canada, 760 of them boxed and the remainder bulk packed.

'Australian' TC Pacific R1542N

This 1971 model was a strange hybrid consisting of an unnamed TC Pacific loco attached to a non-corridor 'Flying Scotsman' tender. It was initially made for Australia where the use of this tender was to make it look more like one of their steam locomotives. It also meant that the loco would have Steam Sound which was thought to be a good selling feature even though smoke could not also be provided as Rovex no longer stocked the German smoke unit which was the only one that would fit this model.

Both the loco and tender were in black with the number '1542' in white on the tender and cab sides.

Australian 'Hiawatha' Hybrid

The Australian version has also been found with the loco carrying 'Hiawatha' paper stickers and with the 'Flying Scotsman' tender numbered '2335' to match the number on the cab side. This is thought to predate the '1542' version.

Australian TC Pacific with Noise and Smoke R54NS

This was the model featured in the Australian catalogue supplement for 1971/72 as the motive power in the R102 'Express Freighter' set. A little over 600 ready-made models were sent bulk packed to Australia in 1970 for these sets. A British version of the set was to appear in the British catalogue in 1972 and the loco was available boxed up as a solo model that year. The Synchrosmoke unit had now been fitted but there was no front lamp although the metal eyelet was there.

The loco was also illustrated in the Canadian catalogue in 1971 but the note alongside said "not exactly as illustrated". It is believed that this refers to the fact that, in 1971, Canada received a further batch of R0542S CP Pacifics and not the new model, at least, possibly not until the following year when it was listed as R054NS. The factory census sheets definitely list the Canadian Pacific version of the model for 1971 and this is confirmed in the Programme of Deletions/Alterations Issue 3 which was published in May 1970. The latter publication listed the companies intentions for 1971 and besides referring to the model for Canada as R0542S it indicated that there was an intention to delete the smoke unit.

A memo sent from Stafford to Richard Lines in December 1970 makes the following reference to this matter:

"Despite the fact that the new 'Hiawatha' 1542NS is shown in the catalogue, Canada are not placing an order

at this moment. I consider, however, that it would be advisable to issue a parts list and prepare the necessary printing dies."

It is assumed that the use of the number 1542NS was an error but it is difficult to be sure.

In August 1971, it was planned to send a variation of the 'Australian' version, numbered R0540NS, to Canada. This was to have a smoke generator and no front light (or eyelet) on the loco. The tender was to have exhaust sound. At some stage it was decided to change the number to R54S and yet records show that almost 3,500 of the model with the R054NS number were made for Canada in 1972. A further 1,000 were made in 1973 but now they were referred to as R054! It's all very confusing!

4-6-4 TC BALTIC TANK

New Zealand R56 & R56S

This model is well covered in **Volume 1** and only in New Zealand did it remain in use into the Tri-ang Hornby years. Even New Zealand had dropped it by 1967 but the year before both the smoke and non-smoke versions were available.

TC 'A' UNIT

Supposedly based on a General Motors F7 diesel, the 'Single-Ended Diesel', as it was known to the men at Margate, was used in large quantities in Canada and was chosen by Rovex as a subject to model to help generate sales in Canada in the mid 1950s. It was available to retailers all over the world but its largest market was the UK. This was not because it was a popular model in the UK but because British sales far outweighed the rest of the world put together.

Transcontinental R55

The model in the 1950s had been silver and red and carried the Tri-ang Railways road name. In 1962, while the colour scheme and even the 4008 running number were retained, the model was given the 'Transcontinental' name printed on its sides. It was in this form that it passed into the Tri-ang Hornby era. By now the model was not being offered in the UK but it may have still been available for those overseas retailers who wanted small supplies. It was also still listed in the 1966 Canadian catalogue.

Australian-Assembled Model R55
In Australia it was felt that the model had some resemblance to the Commonwealth Railways Class GM-12 and so was passable in the absence of more accurate models of Australian prototypes.

At some point the standard Transcontinental model described above was sent out to Australia bulk packed for finishing and used in the Australian set RS4DA of 1968. There is no record of the R55 being bulk packed to Australia that year. However 800 R55s of some type or other were sent the year before and these may have been the models used. The models can be recognised by the chromium-plated eyelets used to attach the couplings. In 1970 a

batch of R55s was bulk packed to Moldex for use in sets but it is not known what these were like. It seems likely however that they were silver.

TransAustralia R55
For a short time in the mid 1960s Moldex were receiving unfinished silver and red A Units onto which they fitted their own couplings and it is likely that they also printed the road name which in this case was 'TransAustralia'. These did not look like the illustration in the 1965 catalogue Australian supplement but had the name printed in script and without the word 'Railways'. No doubt the same tool was used for the R159 Double-Ended Diesel and the Budd RDC which both carried this road name for a while.

This tool (63mm) was also used on the full length TC coaches while a shorter tool (54mm) was used on the short coaches. This smaller tool was also at some time used for the R55 diesel as both lengths of 'TransAustralia' may be found. It is not known how many of the TransAustralia version were made but it is thought to have been about 1,000 during 1965

and 1966. The model sold for £A6-9s-6d.

Canadian National R0551 (R55CN)
The A Unit was eventually made in an authentic livery in 1965 when it appeared in the Canadian catalogue in Canadian National black, silver and red and priced $15.95. Early examples of this model had the diagonals, logo and number printed in a light greenish grey instead of silver, and the red front end was scarlet rather than the cherry red used on later models. Despite the fact that the dummy version of the A Unit was not made in any of the Canadian liveries, the body still carried both 'R55' and 'R57' on the inside of the moulding and on the rear end. The model carried the same running number as the original TC model – '4008'.

Between 1965 and 1972, Canada received about 26,500 Canadian National A Units of which at least 8,400 were bulk packed for use in sets. For 1972 it was proposed that the wheels be nickel tyred on Australian bogies and the four eyelets (portholes) on each side of the loco be removed. Later this instruction was deleted.

New Zealand 4-6-4 tank dating from December 1966.

TransAustralia 'A' unit.

A further small batch of A Units went to Canada in 1973 but it is not known what livery these carried.

The model was also included in the UK range appearing in both the 1968 and 1969 catalogues and price lists at £2-17s-6d. For this purpose a further 4,000 were made in 1968 and 2,000 in 1969. With in excess of 30,000 manufactured, it is today the easiest to find of the locomotives in Canadian livery.

Canadian Pacific R0552 (R55CP)

The very pale bluish-grey and maroon Canadian Pacific model first appeared in the Canadian catalogue in 1967 and was not withdrawn until after 1970. The model can be found with or without a beaver above the Canadian Pacific shield on the front but it is not known which was the earlier version. The maroon was applied to the band at the bottom of each side and to the roof. The band carried the name 'Canadian Pacific' in script and the number '4008'. Some catalogue pictures showed it with '4028' but this may be just an artist's impression.

The model was priced $15.95 in 1967 and between 1967 and 1971 about 9,000 models were sent to Canada of which at least 2,600 were bulk packed for set assembly.

CP Rail R0553

The CP Rail version of the A Unit first appeared in May 1970 as R0553 and was probably made until 1972. It had a bright red body with the long line of louvers on each side painted black. At the rear end of each side was a large CP Rail logo and in white on each side was heatprinted 'CP Rail' and '1404'. The nose of the loco was heatprinted with white diagonal stripes.

Between 1970 and 1972 about 14,000 of the CP Rail version were sent out to Canada. Of these it is known that at least 5,500 were sent bulk packed for making up sets. It was used in the #914 and #917 sets of 1970/71 and the #7305 Inter-City Express set of 1973.

Red Transcontinental Version R0550

In 1971 some 1,900 of the bright red bodies made for the CP Rail version (described above) were printed with 'Transcontinental' on their sides for non-Canadian sales. These were used in the RS101 set made for Australia. The following year it appeared in the British catalogue as the Overlander Set of which 5,000 were made. It would therefore seem that about 7,000 of the red Transcontinental A Units were made.

The model was to have been sent out to New Zealand, bulk packed, in 1972 for set production but the order was cancelled although some silver and red coaches did go.

Transcontinental R0554

It had been planned to bring this model back in Transcontinental livery for 1972 with an extra pick-up on the rear bogie for the Australian and South African markets. It would have had the same paint finish as in 1971 and it was proposed that the wheels be nickel tyred, on Australian bogies, and the four eyelets (portholes) on each side of the loco be removed. Later these instructions were deleted.

TC BO-BO ROAD SWITCHER

The Transcontinental Switcher was not available in the UK range during the Tri-ang Hornby era but it did go on to provide some interesting Canadian and Australian liveries during the later half of the 60s and the early 70s.

Transcontinental R115, R1550

In 1970 just 24 of the original Transcontinental locos were sent unboxed to Australia. It is assumed that these were old stock. They were, however, referred to as R1550.

Canadian National R155CN or R1551 (R370)

The black-liveried Canadian National version is one of the more common of the later versions. It was priced $15.95 and carried the number '3000' and the 'CN' wet spaghetti logo in silver on its sides. The line along the bottom of the running plate was also in silver. The noses at the cab end were painted red and the model had a working light at each end. The nose colour tended to vary ranging from vermilion red through rose red to purple depending on who mixed the paint.

The first batch was made in 1965 when 3,000 were sent to Canada. In the following years up to 1971, a further 12,530 boxed models and 4,000 unboxed locos were sent there. Of the boxed models about 4,200 carried the later number R1551 on the box, this number having been first used in 1969. It seems that a further 1,200 boxed locos were also sent to Canada in 1972 and 800 in 1973.

In Hornby Railways days the model was printed using a different technique. This was most notable in the style of the number '3000' on each side of the hood. This version is illustrated in the 1973 Canadian catalogue. For 1972 the light and eyelet at the opposite end to the cab were removed.

Canadian Pacific R155CP or R1552

The body of this model was moulded in a very pale bluish-grey with a maroon cab roof and line along the side of the running plate. The ends were yellow and in yellow on each side were printed 'CANADIAN PACIFIC' and '3000', the latter printed with the same tool as that used on the CN version.

Only one batch of these models was made. They were manufactured in 1969 and 2,200 boxed models went out to Canada that year leaving about 200 in the stores at Margate. Boxed locomotives must have been used for the TS902 set, if it was made.

Two bodies from CN 'A' units showing colour differences.

CN (R0551), CP (R0552)
and CPRail (R0553) 'A' units.

R0550 Red TransContinental 'A' unit.

CP Rail R1553 (R371)

This was bright red with the CP Rail logo at the front end of each side. Also printed on the sides were 'CP Rail' and '1553' and the ends were printed with diagonal stripes. All of the printing was in white except for the logo which was black and white. As we have seen above, for 1972 the light and eyelet at the cab-less end of the loco were removed.

In 1971 the first batch of 1,710 was made in the second quarter of the year and all but 70 were sent out to Canada in boxes. What was left over went to

Canada in 1972 along with a further 1,500 manufactured that year. In 1973 the 'R' number was changed to R371 and a further 3,000 were made and sent to Canada in Hornby Railways boxes.

TransAustralia Model R155

This had a certain similarity with the Class 40 heavy-duty diesels used by the New South Wales Government Railways. The model had the normal yellow moulded body with black diagonal stripes and '7005' in red on the cab sides but in place of

the usual Transcontinental inscription it had 'TransAustralia' in red. No record has survived of the number of these that were made. In 1970 a batch of just 24 locos was bulk packed and sent to the Moldex factory but it is not known whether they had TC or TA inscriptions.

The model was also released in NSWR maroon livery and VR blue but these will be found in **Volume 3**.

DOUBLE-ENDED DIESEL

VR Blue R159 (R1590)

The model that transferred from Tri-ang Railways to Tri-ang Hornby had the TR shields on its sides. It had first been introduced in 1958 principally with the Australian market in mind but the largest quantity of them was sold in the UK. By 1965, however, the heyday of the double-ended diesel had past. After 1965 there was a break of two years. 5,500 had gone into the stores at Margate in 1965, half of which were sold that year and the remainder in 1966.

Its price at this time was £2-19s-6d.

In Australia, Moldex had their own tool (which they shared with New Zealand) and manufactured bodies to fit to imported mechanisms. (*See* **Volume 1**). In 1968 Australian production ceased and Moldex were forced to import ready-made models from the UK. These were sent over bulk packed for the assembly of sets at the Moldex factory.

In 1968 the double-ended diesel reappeared in the UK catalogue and to match the orders received from retailers, 2,500 were made. That year a further 5,100

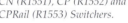
CN (R1551), CP (R1552) and CPRail (R1553) Switchers.

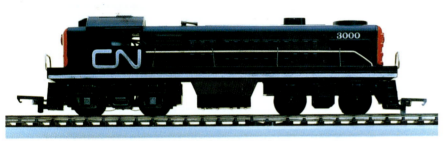

Below: Rail Blue R159 double-ended diesel from the late 1960s.

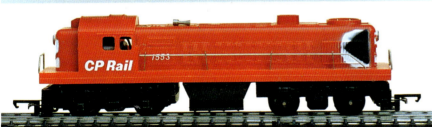

were bulk packed to Australia for sets or to be boxed up and sold as solo locomotives. It is likely that these had couplings fitted after they arrived in Australia.

In 1969 another 2,000 were made mainly for home consumption and in 1970 a further 3,660 were sent to Australia for use in the RS4FA set.

In 1971 the loco was used in the RS105 set made specially for Australia. The loco carried the number R1590 and initial models off the production line had VR transfers but when these ran out TR ones were used.

Rail Blue R159

In 1970 Rovex were having difficulty supplying Australia with these models as they had run out of VR blue plastic and could not order less than 2 tons of it under the supplier's contract. Australian orders were too small to warrant such a large purchase. For this reason Rovex changed the colour of the loco to rail blue. This soon brought complaints from Melbourne where Moldex staff were having difficulty in matching the colour of coaches and locos in their sets.

Electric Blue R159A

The 1968 Australian supplement suggests that the model was at that time in electric blue but this seems unlikely from the evidence available. It was, however, made in this colour plastic in the early 1970s and this version will be covered by **Volume 3**.

Australian-Made Model R159

As we have seen, Moldex of Australia had their own tool for moulding R159 bodies. The double-ended diesel is the most common of the Australian-made locomotives. It was no doubt popular because it resembled the Victorian Railways Class B-60. Between 1958 and 1967 over 30,000 were made for sale as solo models and an unknown number were made for sale in a variety of sets based on the model.

In 1965 3,500 solo models were made but after that annual production figures dropped as more British-made ones were introduced and Australian sales of Tri-ang Hornby dropped. The production figures for 1966 and 1967 were 1,900 and 2,500 respectively. Again there is no record of how many more were made for sets.

The Australian model was usually blue and yellow and it seems to have spent its last three years carrying 'TransAustralia' on its sides; both sizes of tool were used at sometime. After production had ceased at Moldex in 1968 they received their stocks from Margate and bulk packs of R159s went out in 1968 and 1970 as mentioned above. The first batch consisted of 5,100 locos and the second of 3,660. Some of the models were in rail blue as the original VR blue plastic had run out and was too expensive to buy in the small amounts required.

Transcontinental R1594

It had been planned to bring this model back for 1972 with an extra pick-up on the rear bogie for the Australian and South African markets. It is possible that it was not made.

New Zealand Red R159 & R250

As we have already mentioned, New Zealand shared a body tool with Australia and manufactured their own R159s. Indeed production in New Zealand appears to have continued well after that in Australia. New Zealand double-ended diesels looked quite different from those made in Australia or the UK as they had no yellow paint work. A red/maroon version was also made because locomotives on the real railways in New Zealand ran in carnation red.

The maroon or red version appears to be more common than the blue one. The model was advertised in the New Zealand catalogue as late as 1972. The R250 non-powered car was also available in red and but was dropped sometime between 1969 and 1972. A number of different colours were used for the roof and pilots including yellow, white, pale grey and stone.

Variations of the New Zealand red and blue double-ended diesels appear below:

New Zealand Blue R159

The blue version was available in powered form only. It was also different in appearance to the British and Australian made models as it did not have the yellow markings and the roof and pilots

'R' No.	Body	Roof & pilots	Road name	Colour	Front	Date
R159+R250	red	grey	transfers	yellow	TR+stripes	58?
R159+R250	red	white	transfers	yellow	TR+stripes	59?
R159+R250	red	grey	transfers	yellow	TR+stripes	59?
R159+R250	red	grey	print	yellow	TR+stripes	61?
R159+R250	maroon	yellow	print	yellow	TR+stripes	62?
R159+R250	maroon	stone	print	yellow	TR+stripes	63?
R159+R250	maroon	yellow	print	white	TR+stripes	68?
R159+R250	maroon	white	print	white	TR+stripes	68?
R159+R250	dark maroon	grey	print	yellow	plain	66?
R159	blue	yellow	print	yellow	TR+stripes	60?
R159	blue	yellow	print	yellow	TR shield	61?
R159	turquoise	light grey	print	yellow	TR+stripes	71?

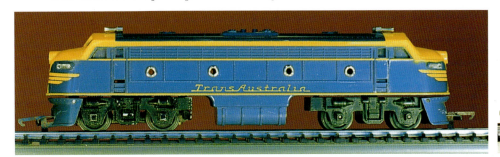

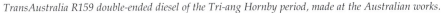

TransAustralia R159 double-ended diesel of the Tri-ang Hornby period, made at the Australian works.

Rail blue R159 double-ended diesel from New Zealand.

were a distinctive colour. These was generally yellow but grey or white can be found. At the very end of production there must have been some difficulty in obtaining correct colour plastics. This resulted in late double-ended diesels being turquoise instead of blue.

YARD SWITCHER

Yard Switcher R353

The model was last available in the British catalogue in 1965 when it cost £1-15s. It was featured in the Canadian catalogue, however, until 1968. This was a red model with the small TR shield and the number T.R.20071 on its sides.

In 1966 and 1967 a total of 3,680 red yard switchers were made and sent out principally to Canada, the last 400 being sent there bulk packed for sets in 1968.

Dock Shunter R253C

The red dock shunter was sent out to Canada.

Australian R353

Australia had the yard switcher between 1961 and 1967 assembling it from parts sent out to the Melbourne factory from Margate. During the Tri-ang Hornby years there was no record of unfinished models sent out but 1,600 were assembled by Moldex and catalogue pictures suggest that all Australian models were yellow and inscribed TransAustralia.

BUDD DIESEL RAILCAR

The Transcontinental Budd RDC was made between 1961 and 1965 and is described in **Volume 1**. There were both powered (R352) and non-powered (R232) models.

It was also assembled in Australia, both power and dummy units being available. This seems to be even rarer than the TransAustralia version

described below.

Canadian National R352CN & R232CN

The Canadian National model was the same as the Transcontinental version but was in silver and black livery with 'CN' wet spaghetti logos and red end panels. Unlike the TC model it was sold in a set consisting of one power car and two dummy cars.

The first batch of Canadian National Budd RDCs was made in 1965 and virtually all 1,000 of them went out to Canada that year. Further batches were sent in 1967, 1969 and 1971 making a total of about 2,200 power cars and 2,700 dummy cars dispatched to Canada during the Tri-ang Hornby period. Not all were boxed, many of them going bulk packed for use in sets.

R353 Transcontinental yard switcher.

R353A Australian-assembled yard switcher with the 'TR' logo without a shield. 1,600 were made over the three years 1965 – 1967.

Northern Pacific R829 & R825

These were made especially for Tager of America. They were the usual silver colour typical of all versions of the Tri-ang Hornby model and were inscribed 'NORTHERN PACIFIC' in black on the long panel over the windows on each side. The ends of the unit were marked with two round black stickers. The running number, which was printed in black on silver self-adhesive labels, was '303'. Instead of the usual tension-lock couplings, the units were fitted with NMRA couplings that were standard in America.

The batch of 4,000 power cars and 4,000 non-powered units was bulk packed to the USA in 1968 where they were boxed up in grey ATT boxes; ATT standing for American Train & Track Corp. It is known that 84 power cars and 175 dummy units did not go but had been disposed of overseas by the end of 1970. It is also known that some of these went out to Australia in plain Tri-ang Hornby BG size boxes with the 'R' number rubber stamped on one end.

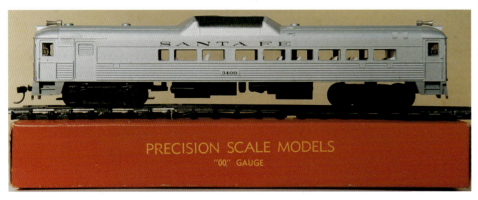

ÉCHELLE H.O.

Tri-ang HORNBY

LE RÉSEAU FERROVIAIRE CANADIEN EN MODÈLES RÉDUITS

PRECISION SCALE MODELS
"00" GAUGE

Top left: R352CN/R232CN Canadian National Budd rail diesel cars.
Middle left: R232CN CN non-powered Budd rail diesel car in Canadian bubble pack.
Left: R830 Santa Fe Budd rail diesel car dispatched to New Zealand to sell.
Above: R829/R825 Northern Pacific and R830/R826 Santa Fe Budd rail diesel cars made for ATT of America.

Santa Fe R830 & R826

In every respect except livery the Santa Fe models were like the Northern Pacific. On the panel over the window was 'SANTA FE' in black and the ends carried a pair of large square stickers with the Santa Fe logo in yellow on red. One running number used was '3403' and was heat printed in black onto the number panels provided for it on the car's sides.

In 1968, 3,000 power cars and 5,900 non-powered units were sent bulk packed to Tager in the USA. 420 power cars remained in the stores at Margate and their fate is not known. It seems likely that they would have been sold off as a job lot to a retailer or wholesaler overseas.

Canadian Santa Fe Model R352SF & R232SF

The 1969 Tri-ang Railways Canadian price list recorded a model as 'R352SF Santa Fe Budd Rail Car'. There was also the corresponding non-powered version listed under 'Coaches'. While it would be logical to think of some of the surplus models made for ATT being sent out to Canada for sale, it seems strange that they were not referred to as R830

R352A TransAustralia Budd rail diesel car.

Below: R831/R827 Chesapeak & Ohio and R832/R828 Reading Lines Budd rail diesel cars made for ATT of America.

and R826. R352 and R232 were the numbers of the original TC version and the Canadian National models were R352CN and R232CN.

The logical explanation which fits the thinking at Margate is that the R352SF and R232SF were to be fitted with standard tension-lock Tri-ang couplings and not the NRMA couplings used on the stock sent to ATT. Although listed in the Canadian price list, sadly no record of the manufacture of this version of the model has been found.

Chesapeake & Ohio R831 & R827

Here was another model similar to the Northern

Pacific but with 'Chesapeake and Ohio' in blue on the panel over the windows and the 'C and O for Progress' logo on the window panel. There were no end markings and one running number was '9003'. This was heat printed in blue along with the above inscriptions.

In 1968, Tager received 2,300 power cars and 2,940 non-powered units bulk packed. 770 power and 35 non-powered cars remained in the Margate stores at the end of 1968 but had gone by the end of 1970.

Reading Lines R832 & R828

The last of the special Budd RDCs made for Tager of

the USA was in Reading Lines livery. It had 'READING COMPANY' in black over the windows and no markings on the ends. The running numbers in black on silver were '510' and '513'.

Tager received 3,920 power cars and 4,070 non-powered units in 1968 and a further 80 units were disposed of from the stores in 1970. Despite the numbers made the Reading cars appear to be harder to find than the others. No explanation has been found for this.

TransAustralia R352A

A rare version of the model is that finished as a TransAustralia unit. Budd RDCs were sold to Australia and ran on the New South Wales lines and the Commonwealth Railways. Daylight expresses were made up of three or four power cars and a trailer.

The model looked exactly like the Transcontinental version but carried the TransAustralia name instead. It is not known whether this was applied in Britain or Australia. The running number was '31018' and the raised centre of the roof was not picked out in black. No record has been found of a non-powered version and there are no records of the number made.

'DAVY CROCKETT'

UK Model R358S

1965 was the last year that 'Davy Crockett' appeared in the UK catalogue but the locomotive remained in production for export both as a solo model and in the R641 presentation sets. For these, 4,250 models were made between 1965 and 1968 of which 1,000 had smoke units. Another 500 went over to France in 1967 in the French version of the R641 set made for Meccano Tri-ang. Canada received 1,200 solo models boxed in 1967.

R385S 'Davy Crockett'.

Lima Crocketts R358SL

In 1968, 1,500 solo locos were made specially for Lima of Italy. These had a Lima coupling fitted to the rear of the tender. Strictly speaking the loco remained R358S and the tender was R32L. It is presumed that this order was one of the side agreements made at the time that Lines negotiated with Lima to market their N gauge system under the Wrenn label.

Australian R358S

While the non-smoking version of the model was not assembled in Australia after 1963, the smoking version remained in production until 1967. In actual fact production figures were very low because of the limited market. 200 were made in 1965, 300 in 1966 and just 25 in 1967! Australian Crocketts are therefore very difficult to find. In all, 2,525 models were assembled by Moldex including both those with and those without a smoke unit.

Crocketts were also assembled by Moldex for their own version of the R641 presentation set but it is not known how many.

GM CO-CO Diesel

There was a proposal to design a new GM diesel for the Canadian and US markets. R843, as it would have been, was to be designed to fit the R751 Co-Co chassis and to have been released in 1971 (later postponed to January 1972). It was also to have fine scale wheels, lights at both ends and power pick-up from both bogies.

It was deferred pending the introduction of the HP type motor on 'Evening Star' but there was a suggestion that it should operate in conjunction with a Pifco siren to be activated by trackside contacts.

By August 1971 it was planned to introduce three versions of the new diesel. These were to be:

R8431 CN (later deleted)
R8433 CP (Rail?)
R8434 Transcontinental

The model did not materialise although a GM diesel was made for Australia in the mid-1970s.

E. AUSTRALIAN

NSWR SUBURBAN ELECTRIC

NSWR Set R450 and R451 & R452

The 'Red Rattlers', as they were affectionately called, served Sydney for more than 60 years and the last was withdrawn on 11th January 1992. At the time of writing 23 cars have been preserved, including at least one in full working order.

The model, which was designed at Margate but made at the Moldex factory in Melbourne, has been described in some detail in **Volume 1**. It remained in

NSWR Suburban Electric.

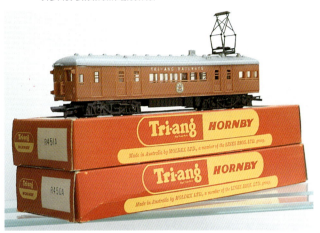

production until 1966, the last two batches of solo models being 350 in 1965 and 150 in 1966. Sadly we do not have the production figures for those used in sets which may have been made after this date.

These later models would have had the 'Tri-ang Railways' in small lettering. The solo models were individually packed in window boxes. It is understood that the correct placing of the cars, which ran in three-car units, should be with the power cars at each end with their pantographs at the inner ends of each car and therefore adjacent to the centre car.

VR Set R550 and R551 & R552
In **Volume 1**, reference was made to a blue version of the NSWR Suburban Electric set and a picture of it in the 1962-63 Australian Catalogue was also shown. Since the publication of **Volume 1** more research on this subject has been attempted with little increase in knowledge. One version of the story seems to fit the normal practice for new models and so seems more feasible than others. The story goes that Moldex took a standard two car NSWR set and painted it blue with yellow lining for an Australian toy fair. Visitors reacted badly and insufficient

orders were received for it to put it into production. The picture in the catalogue is of the pre-production model.

That is not the end of the story as, late in 1993, a box turned up in Australia containing the Australian R4D set (R159 Double-Ended Diesel and three short blue coaches all in TransAustralia livery) which had an R4X sticker on the lid. Over the 'X' was stuck another label with a 'D' on it. R4X was the number of the proposed blue Suburban Electric set. It is reasonable to assume that, as special labels were printed both to number the box originally and again to change the 'X' into a 'D', this can not have been an isolated example. Could it be that the boxes for the VR blue suburban set had been ordered before the fateful toy fair and that when the model was cancelled the boxes were re-labelled instead of scrapping them?

No doubt this story will run and run.

F. FORMER HORNBY-DUBLO STOCK

When Rovex management visited the Meccano factory in Liverpool, following the takeover of Meccano Ltd by Lines Bros., they found large stocks of Hornby-Dublo in the stores. All saleable stock was subsequently given 'R' numbers which were all four figured and based on the original Hornby-Dublo catalogue numbers. The following locos were listed:

Green 0-6-0 Tank R2207
Listed in 1965 price lists only.

N2 Tank R2217
Listed in 1965 price lists only.

0-6-0 Diesel Shunter R2231
Listed in 1965 price lists only and boxed examples have been found with Tri-ang Hornby label in the box.

'Barnstaple' R2235
These received Tri-ang Hornby stickers on the box and were offered for sale through the 1965 price list and the 1966 Tri-ang Hornby catalogue. In 1966 there were just under 1,200 models in the stores and orders for them from retailers exceeded 1,450, 126 of them coming from abroad. All were disposed of before the end of the year priced £5.16s.7d.

CO-BO R2233
There was also stock of the Met-Vic Co-Bo diesel which was listed in the 1965 price list and also advertised in the 1966 catalogue. Only 725 models had been taken into stock but this proved to be far less popular and 120 were left in the stores at the end of 1966. They were priced £4.7s.2d.

CO-CO R2232
The stock taken over was too small to be worth illustrating in the 1996 catalogue but it was included in the 1995 and 1996 Tri-ang Hornby price lists. Margate were trying to sell their own Co-Co diesel and would not have wanted to offer the Hornby-Dublo one in competition. About 225 entered the stores in 1966 and there were still 120 there at the start of the following year.

EMU R2250/4150
Again, the small quantity of stock absorbed made this not worth advertising. Only 61 EMU sets were to be found in the stores in 1967, the first and only year the model appeared in the Tri-ang Hornby census. This possibly means that they had not been transferred from the Binns Road store until then.

INTRODUCTION

Adoption

As with so many aspects of the Tri-ang Railways system the company entered the Tri-ang Hornby era with a series of coaches which were well moulded, realistic in appearance, flexible in construction with the potential to expand the range at comparatively little cost and reasonably priced in the shops. This was the recipe for market domination.

There were three rakes of scale length Mark 1 coaches each in a different regional livery and all sharing the same chassis, bogies, carriage ends and set of roof mouldings. In maroon livery there was a composite, brake end, buffet, sleeping car and all-passenger brake. The composite, brake end and buffet car were also available in green livery as well as maroon and cream. The only other rake of British coaches was the Pullman train consisting of a first class car dating from 1958, with a choice of four different names, and a brake second car of 1960 vintage.

The five maroon Mk1 coaches were also available in pairs in CKD kits.

Odd coaches included an engineer's clerestory coach in black, a Continental sleeper in blue (based on the Pullman parlour car moulding) and an ambulance car with its origins in the second series Transcontinental baggage/kitchen car. There were also centre cars for both the green DMU and the blue Pullman DMU.

In addition, classical coaches included the Caledonian composite and brake composite (both extensions of the modular coach system described above), the rocket coach with a choice of three names and the old time coach for the Davy Crockett loco. In Britain, the Transcontinental range was reduced to a rake of four blue second series passenger cars.

The Old Smokey train set contained two coaches unique to it. These were the red and yellow 9" composite and the green and red utility van. The short Royal Mail TPO was also still available in maroon in the R402 Operating Mail Coach Set.

A Taste of Blue

This then was the range of coaches available as the packaging changed from Tri-ang Railways to that of Tri-ang Hornby. The rake of maroon and cream coaches did not survive beyond the end of 1965, however, and was dropped in 1966 along with the Continental sleeping car and the old time coach. The CKD kits were reduced to just the composite and brake second but were still sold in pairs. The latter was a prelude to introducing a new series of construction kits which were in preparation.

Important additions to the catalogue in 1966 were the first coaches in the new blue livery adopted by British Rail. These had in fact been made the previous year as a late decision. The models were the Mk1 composite and brake second and the blue was the light electric blue and the roofs were grey.

From 1966 onwards, the clerestory engineer's coach was dark green instead of black.

New Kits

With the new blue and grey livery spreading like a rash on the real railways, the public were demanding more models to match. The Mk1 buffet car was added to this range in 1967 and the Mk1 composite, brake second and buffet car in blue and grey livery were available in new assembly packs, which had replace the CKD kits; each kit containing enough parts for one model.

The catalogue also saw the reappearance of an old favourite, the utility van in BR coach green livery, to supplement the green Mk1 coaches. Missing from the range of coaches this year was the rake of blue Transcontinental cars.

There was one more new coach in 1967 but this was lost on the pages of the Battle Space section. Here we could find the former TC mail car in the guise of a command car moulded in khaki-coloured plastic.

Blue Revolution

1968 saw a blue revolution sweep through the locomotives and coaching stock at Margate. To the three Mk1 coaches that had already succumbed to the blue and grey livery, the full parcels brake and sleeping car were added. The utility van was also changed to blue livery and the first new coaches for some time, the Mk2 Inter-City second and brake second, were promised although their arrival was to be delayed.

The maroon and green Mk1 coaches were reduced to just a composite and brake second in each case. Also dropped from the range were the Caledonian coaches, the engineer's coach, DMU centre car and the ambulance car.

Added to the catalogue were two TC second series passenger cars in Canadian National silver and black livery. These were the observation car and the coach.

Rail Blue Livery and Mk2 Coaches

In line with locomotive development the coaches changed to BR rail blue and grey in 1969 and the Pullman DMU centre car to the same colours in reverse. The carriage roofs were now dark grey and it is assumed that the parts in the coach assembly packs all changed to rail blue at the same time.

The Mk2 Inter City coaches arrived at last, but not until after the change in shade to rail blue plastic and so cannot be found in the lighter blue. As a special surprise the new coaches were fitted with lighting units which picked up power from the track.

1969 also saw the short Royal Mail TPO coach change to the new blue and grey livery. For the first time in about ten years there was also a new coach for starter sets. It was a short yellow four-wheeled vehicle on a brown chassis with balconies on each end giving it a dated appearance.

Catalogue advert for coach CKD kits. The picture shows the boxes carrying the Tri-ang Hornby logo but all examples seen by the writer were Tri-ang Railways.

Pre-nationalisation Liveries

The full range of blue and grey coaches remained in 1970 but all earlier BR liveries disappeared. In their place came a series of pre-nationalisation liveries mostly based on the BR Mk1 10" composite and brake end. These included Southern and GWR composite and brake thirds and LMS full third and brake first.

For the newly-introduced LNER Flying Scotsman there were two new coaches which remained basically the 10" Mk1 modular coaches with Thompson sides with their squared windows. A prominent feature of these was the oval windows at the ends of each side. While shown in the catalogue in a light orange brown, when they arrived the coaches were moulded in a simulated wood finish plastic to represent the teak finish of the prototypes. To many at the time, the catalogue illustrations looked better than the actual model and it might have been better if Rovex had concentrated on colour rather than wood grain effect. After all, models rarely benefit from close inspection. Their miniaturisation should represent the prototype seen at a distance and the teak LNER coaches seen at a distance on the real railways would have simply appeared a light orange brown.

Also returned to production in 1970 were the two GWR clerestory coaches although it is not clear why these were re-released a year before the 'Lord of the Isles' returned.

The 9" Pullman cars continued in the range but the Canadian National cars ceased to be offered in the British catalogue. The coach assembly packs also disappeared after a comparatively brief life. They had not sold as well as the ready-to-run models.

New Liveries for Caledonian Coaches

The blue and grey coaches remained unchanged in 1971 except for the dropping of the utility van and the return of the composite and brake second assembly packs.

Coach Assembly Packs

Coach assembly packs.

R.730 COMPOSITE COACH

R.731 2nd CLASS BRAKE COACH

R.732 BUFFET COACH

Build your own coaches in New British Rail Livery with only the aid of a screwdriver. Each pack contains sufficient parts to make a finely detailed coach complete with interior. Step-by-step assembly instructions included.

The pre-nationalisation range continued to expand with the addition of the two Caledonian coaches in both Southern malachite green and LMS maroon livery. In the case of the latter, the window panels were lined out in yellow.

A strange reappearance was the ambulance car in the Battle Space series. A new and eye-catching model was the Thompson brake coach in bright red for the breakdown train. Obvious by their absence were any BR Mk1 coaches in liveries of the 1950s and 60s. We would have to wait 25 years to see the return of these.

CHECK LIST OF COACHES

The following coaches were made during the Tri-ang Hornby era, i.e. between May 1965 and December 1971. The dates given are those when stock was manufactured and are not the dates the model was illustrated in the catalogue. Models sometimes remained in the catalogue a year or two after production had ceased in order to clear outstanding stocks. It is also possible that a coach remained in production for use in a train set after it ceased to be made for solo sales. The dates shown also extend back into the Tri-ang Railways period and forwards into the Hornby Railways days in order to give the reader a clear picture of the span of production. The couplings, bogies and wheels quoted, on the other hand, are those used only during the period covered by this volume.

For ease of reference, the coaches have been listed under the following headings:

1. Models of British Coaches
2. For Starter Sets
3. Transcontinental – General Issue
4. Made for Canada
5. Made for America
6. Used in Australia
7. Made in New Zealand

1. Models of British Coaches

The first list is of coaches that were either already in production when the name changed from Tri-ang Railways to Tri-ang Hornby or introduced during the Tri-ang Hornby era. They are described in detail in the 'A' section of the Directory of Coaches that follows.

No.	British	Colour	Dates	Cpls	Bogies	Wheels
	MAIL COACHES					
R402	BR TPO	maroon	55-69	IIb-IIId	b2b	MS4
R402M	BR TPO	blue & grey	70-72	IIId	b2b	MS4
	MAINLINE 9" COACHES					
R720	Composite	red & yellow	63-67	IIIc	c2b	DP1-DP2
R724	Composite	red & yellow	64-65	IIIc	c2b	DP1-DP2
R?	Composite	red	66?	IIIc	c2b	DP1
R?	Composite	red	66?	IIIc	c2b	DP1
	(no windows or buffers)					
	9" UTILITY VAN					
R726	Utility Van	green & red	64-65	IIIc	c2b	DP1-DP2

No.	British	Colour	Dates	Cpls	Bogies	Wheels
R226	BR Utility Van	green	67	IIIc	c2b	DP1
R226	BR Utility Van	electric blue	68	IIIc-IIId	c2b	DP1-DP3
R226	BR Utility Van	rail blue	69-70	IIIc-IIId	c2b	DP1-DP3
	9" SUBURBAN COACHES					
R121	BR Composite	maroon	67	IIIc	c2b	DP1
R223	BR Composite	green	67	IIIc	c2b	DP1
	9" PULLMAN CARS					
R228	1st 'Ruth'	choc & cream	58-73	IIIc-IIId	c2b	DP1-DP5
R228	1st 'Jane'	choc & cream	58-73	IIIc-IIId	c2b	DP1-DP5
R228	1st 'Mary'	choc & cream	58-73	IIIc-IIId	c2b	DP1-DP5
R228	1st 'Anne'	choc & cream	58-73	IIIc-IIId	c2b	DP1-DP5
R328	Brake 3rd	choc & cream	60-73	IIIc-IIId	c2b	DP1-DP5
	DMU CENTRE CARS					
R334	BR DMU	green	61-67	IIIc	d2b	DP1-DP2
	CLERESTORY COACHES					
R332	GWR Composite	choc & cream	69-72	IIIc-IIId	c2b	DP3-DP5
R333	GWR Brake	choc & cream	69-72	IIIc-IIId	c2b	DP3-DP5
R620	BR Engineer's Coach	black	63-65	IIIc	c2b	DP1
R620	BR Engineer's Coach	green	66-67	IIIc	c2b	DP1-DP2
R379	Railway Children	maroon & white	71-72	IIId	c2b	DP5
	ROCKET COACHES					
R621	'Times'	yellow	63-66	IIIc	-	SP1
R621	'Experience'	yellow	63-66	IIIc	-	SP1
R621	'Dispatch'	yellow	63-66	IIIc	-	SP1
	10" MK1 COACHES					
R339	BR Sleeping Car	maroon	61-67	IIIc-IIId	c2b	DP1-DP2
R422	BR Composite	maroon	62-69	IIIc-IIId	c2b	DP1-DP3
R423	BR Brake 3rd	maroon	62-69	IIIc-IIId	c2b	DP1-DP3
R424	BR Buffet	maroon	62-67	IIIc	c2b	DP1-DP2
R425	BR Full Parcels	maroon	62-67	IIIc	c2b	DP1-DP2
R626	BR Composite	maroon & cream	63-65	IIIc	c2b	DP1-DP2
R627	BR Brake 3rd	maroon & cream	63-65	IIIc	c2b	DP1-DP2
R628	BR Buffet	maroon & cream	63-65	IIIc	c2b	DP1-DP2
R622	BR Composite	green	63-69	IIIc-IIId	c2b	DP1-DP3
R623	BR Brake 3rd	green	63-69	IIIc-IIId	c2b	DP1-DP3
R624	BR Buffet	green	63-67	IIIc	c2b	DP1-DP2
R727	BR Composite	electric blue & grey	65-68	IIIc-IIId	c2b	DP1-DP3
R728	BR Brake 2nd	electric blue & grey	65-68	IIIc-IIId	c2b	DP1-DP3
R729	BR Buffet	electric blue & grey	67-68	IIIc-IIId	c2b	DP1-DP3
R730	BR Composite-kit	electric blue & grey	67-68	IIIc-IIId	c2b	DP1-DP3
R731	BR Brake 2nd-kit	electric blue & grey	67-68	IIIc-IIId	c2b	DP1-DP3
R732	BR Buffet - kit	electric blue & grey	67-68	IIIc-IIId	c2b	DP1-DP3

No.	British	Colour	Dates	Cpls	Bogies	Wheels
R727	BR Composite	rail blue & grey	69-71	IIId	c2b	DP3-DP4
R728	BR Brake 2nd	rail blue & grey	69-71	IIId	c2b	DP3-DP4
R729	BR Buffet	rail blue & grey	69-71	IIId	c2b	DP3-DP4
R730	BR Composite-kit	rail blue & grey	69	IIId	c2b	DP1-DP3
R731	BR Brake 2nd-kit	rail blue & grey	69	IIId	c2b	DP1-DP3
R732	BR Buffet - kit	rail blue & grey	69	IIId	c2b	DP1-DP3
R339	BR Sleeping Car	electric blue & grey	68	IIId	c2b	DP1-DP3
R425	BR Full Parcels	electric blue & grey	68	IIId	c2b	DP1-DP3
R339	BR Sleeping Car	rail blue & grey	69-70	IIId	c2b	DP3-DP4
R425	BR Full Parcels	rail blue & grey	69-71	IIId	c2b	DP3-DP4
R422A	LMS Composite	maroon	69-77	IIId	c2b	DP3-DP5
R423A	LMS Brake 3rd	maroon	69-77	IIId	c2b	DP3-DP5
R622A	SR Composite	yellow-green	69-71	IIId	c2b	DP3-DP5
R623A	SR Brake 3rd	yellow-green	69-71	IIId	c2b	DP3-DP5
R743A	GWR Composite	choc & cream	69-77	IIId	c2b	DP3-DP5
R744A	GWR Brake 3rd	choc & cream	69-77	IIId	c2b	DP3-DP5
	CALEDONIAN COACHES					
R427	CR Composite	maroon & white	62-64*	IIIc	c2b	DP1-DP2
R428	CR Brake 3rd	maroon & white	62-64*	IIIc	c2b	DP1-DP2
R427	CR Composite	maroon & white	71-73	IIId	c2b	DP4-DP5
R428	CR Brake 3rd	maroon & white	71-73	IIId	c2b	DP4-DP5
R747	LMS Composite	maroon	71-73	IIId	c2b	DP4-DP5
R748	LMS Brake	maroon	71-73	IIId	c2b	DP4-DP5
R749	SR Composite	olive green	71-73	IIId	c2b	DP4-DP5
R750	SR Brake	olive green	71-73	IIId	c2b	DP4-DP5
	PULLMAN DMU CAR					
R426	centre car	blue & white	63-68	IIIc	c2b	DP1-DP3
R426	centre car	grey & blue	69-70	IIIc-IIId	c2b	DP3-DP4
	Mk2 COACHES					
R722	BR 2nd Class	rail blue & grey	69-83	IIId	e1	DP4
R723	BR Brake 1st	rail blue & grey	69-83	IIId	e1	DP4
	THOMPSON COACHES					
R745	LNER Full 3rd	teak	70-77	IIId	c2b	DP3-DP5
R746	LNER Brake 3rd	teak	70-77	IIId	c2b	DP3-DP5
R740	BR Breakdown Crew	bright red	71-73	IIId	c2b	DP5

* Although last made in 1964, they remained in the catalogue until the end of 1966.

Made for Battle Space
See separate Battle Space checklist.

2. For Starter Sets
Further information on coaches for starter sets will be found in section 'B' of the Directory that follows.

No.	For Starter Sets	Colour	Dates	Cpls.	Bogies	Wheels
R733	4 Wheel Old Time Coach	yellow	69-72	IIId	-	DP3-DP4

3. Transcontinental - General Issue
These were coaches sold worldwide during the Tri-ang Hornby years. More detail about them will be found in section 'C' of the Directory of Coaches that follows.

No.	Transcontinental	Colour	Dates	Cpls.	Bogies	Wheels
	TC 2ND SERIES CARS					
R440	Passenger	bright silver & red	70-73	IIId	t3b	MP1-MP2
R441	Observation	bright silver & red	70-73	IIId	t3b	MP1-MP2
R443	Diner	bright silver & red	70-73	IIId	t3b	MP1-MP2
R444	Coach	blue	62-72	IIIc-IIId	t3a-t3b	MS4-MP2
R445	Observation	blue	62-69	IIIc-IIId	t3a-t3b	MS4-MP2
R446	Baggage/Kitchen	blue	62-72	IIIc-IIId	t3a-t3b	MS4-MP2
R447	Diner	blue	62-72	IIIc-IIId	t3a-t3b	MS4-MP2
	OLD TIME COACH					
R448	Coach	yellow	62-70	IIIc-IIId	B2d-B3b	MS4-MP2
R378	Railway Children Coach	teak	71-72	IIId	B3b	MP2

4. Made for Canada
From 1965 Canada was supplied with Transcontinental coaches in authentic Canadian Liveries. While a few ended up in other countries, almost all went to Canada either ready boxed or bulk packed for set assembly. For further detail of these turn to section 'C' of the Directory of Coaches that follows.

No.	For Canada	Colour	Dates	Cpls.	Bogies	Wheels
	TC 2ND SERIES CARS					
R444CN	Passenger Car-CN	silver & black	65-71	IIIc-IIId	t3B	MP1-MP2
R445CN	Observation Car-CN	silver & black	65-71	IIIc-IIId	t3B	MP1-MP2
R446CN	Bag/Kitchen car-CN	silver & black	65-71	IIIc-IIId	t3B	MP1-MP2
R447CN	Diner - CN	silver & black	65-71	IIIc-IIId	t3B	MP1-MP2
R440CP	Passenger Car-CP	sky & maroon	67-68	IIIc-IIId	t3B	MP1
R441CP	Observation Car-CP	sky & maroon	67-68	IIIc-IIId	t3B	MP1
R442CP	Baggage Car-CP	sky & maroon	67-68	IIIc-IIId	t3B	MP1

No.	For Canada	Colour	Dates	Cpls.	Bogies	Wheels
R443CP	Dining Car-CP	sky & maroon	67-68	IIIc-IIId	t3b	MP1
R4403	Passenger Car-CP	rail silver & red	70-71	IIId	t3b	MP1-MP2
R4413	Observation Car-CP	rail silver & red	70-71	IIId	t3b	MP1-MP2
R4423	Bag/Kitchen-CP	rail silver & red	70-71	IIId	t3b	MP1-MP2
R4433	Dining Car-CP	rail silver & red	70-71	IIId	t3b	MP1-MP2

5. Made for America

These were also made in authentic colours but this time based on US railroads. With the exception of the Rocket coaches which will be found in section 'A', the American passenger cars are detailed in section 'C' of the Directory of Coaches which follows these check lists.

No.	For America	Colour	Dates	Cpls.	Bogies	Wheels
	ROCKET COACH					
R621	'Times'	yellow	67	NMRA		SP1
R621	'Experience'	yellow	67	NMRA		SP1
R621	'Dispatch'	yellow	67	NMRA		SP1
	OLD TIME COACH					
R801	Pullman Car	green & black	67	NMRA	B2d	DP1
R802	Central Pacific Car	yellow & brown	67	NMRA	B2d	DP1
	TC 2ND SERIES CARS					
R803	Coach - Pennsylvania	silver & maroon	67	NMRA	t3b	MP1
R807	Diner - Pennsylvania	silver & maroon	67	NMRA	t3b	MP1
R811	Bag/Kitchen - Pennsylvania	silver & maroon	67	NMRA	t3b	MP1
R815	Observation - Pennsylvania	silver & maroon	67	NMRA	t3b	MP1
R804	Coach - Burlington	silver	67	NMRA	t3b	MP1
R808	Diner - Burlington	silver	67	NMRA	t3b	MP1
R812	Bag/Kitchen - Burlington	silver	67	NMRA	t3b	MP1
R816	Observation - Burlington	silver	67	NMRA	t3b	MP1
R805	Coach - Santa Fe	silver	67	NMRA	t3b	MP1
R809	Diner - Santa Fe	silver	67	NMRA	t3b	MP1
R813	Bag/Kitchen - Santa Fe	silver	67	NMRA	t3b	MP1
R817	Observation - Santa Fe	silver	67	NMRA	t3b	MP1
R806	Coach - Baltimore Ohio	silver & blue	67	NMRA	t3b	MP1
R810	Diner - Baltimore Ohio	silver & blue	67	NMRA	t3b	MP1
R814	Bag/Kitchen - Baltimore Ohio	silver & blue	67	NMRA	t3b	MP1
R818	Observation - Baltimore Ohio	silver & blue	67	NMRA	t3b	MP1

Some of the residue USA coaches were sent to Australia already boxed and fitted with MkIII couplings.

6. Used in Australia

These were models made (M) at the Moldex factory in Melbourne, models sent to Australia bulk packed and finished (F) at the Moldex factory or models sent bulk packed to Australia, complete and ready (R) for use in train set assembly. The British outline models will be found in section 'A' of the Directory that follows and the Transcontinental types in section 'C'.

No.	Australia	Colour	Dates	Cpls.	Bogies		Wheels
	SHORT COACHES						
R26	Coach - TransAustralia	VR blue	65-68	IIIcA	Aust	M	DS2
R27	Vistadome - TransAustralia	VR blue	65-68	IIIcA	Aust	M	DS2
R26	Coach - TransAustralia	red & crm	65-68	IIIcA	Aust	M	DS2
R26A	Coach - Southern Aurora	silver	65-68	IIIcA	Aust	M	DS2
R421A	Buffet - Southern Aurora	silver	65-68	IIIcA	Aust	M	DS2
	TC 2ND SERIES						
R440	Coach - TransAustralia Cars	silv & red	62-63	IIIcA	t3a	F	MS4
R441	Observation - TransAustralia Cars	silv & red	62-63	IIIcA	t3a	F	MS4
R442	Bag/Kitchen - TransAustralia Cars	silv & red	62-63	IIIcA	t3a	F	MS4
R443	Diner - TransAustralia Cars	silv & red	62-63	IIIcA	t3a	F	MS4
R444	Coach - TransAustralia Cars	VR blue	62-63	IIIcA	t3a	F	MP1
R444A	Coach - TransAustralia Cars	VR blue	68-71	IIId	t3b	R	MP1
R444A	Coach - TransAustralia Cars	BR blue	69-70	IIIc-IIId	t3b	R	MP1
R445	Observation - TransAustralia Cars	VR blue	62-63	IIIcA	t3a	F	MS4
R445A	Observation - TransAustralia Cars	VR blue	69-71	IIId	t3b	R	MP1
R445A	Observation - TransAustralia Cars	BR blue	69-70	IIIc-IIId	t3b	R	MP1
R446	Bag/Kitchen - TransAustralia Cars	VR blue	62-63	IIIcA	t3a	F	MS4
R447	Diner - TransAustralia Cars	VR blue	62-63	IIIcA	t3a	F	MS4

7. Made in New Zealand

All of these coaches were made in the Tri-ang factory in New Zealand for selling locally. For further detail of them you will need to look in section 'A' of the Directory of Coaches for the SR coaches and section 'C' for the short coaches.

No.	New Zealand	Colour	Dates	Cpls.	Bogies	Wheels
	SHORT COACHES					
R26	Coach - Tri-ang Railways	red	57-72	IIIc-IIId	t1, Aust	MS4
R27	Vistadome - Tri-ang Railways	red	57-72	IIIc-IIId	t1, Aust	MS4
	SR COACHES					
R222	Suburban Brake Coach	green	65-72	IIIc-IIId	c1b, c2b	DS1-DP3
R226	Suburban Utility Van	green	65-72	IIIc-IIId	c1b, c2b	DS1-DP3
R220	Brake Coach	green	67-72	IIIc-IIId	c1b, c2b	DS1-DP3
R229	Restaurant Coach	green	67-72	IIIc-IIId	c1b, c2b	DS1-DP3

DIRECTORY OF COACHES

The directory is presented in three sections to acknowledge the difference between the British coaches and those of the Transcontinental series. Only one coach was modelled exclusively for starter sets (but the 9" mainline composite did also serve this purpose) and these are to be found in a section of their own in the middle of the directory, which is therefore divided up as follows:

A. British Outline Coaches
B. Starter Set Coaches
C. Transcontinental Passenger Cars

Battle Space
Battle Space coaches have been included together with locos, wagons and Battle Space sets in a section of their own at the end of the chapter on 'Wagons'.

Australian and New Zealand Coaches
Most Australian production was during the Tri-ang Railways period but it did continue until 1968. The short Transcontinental coaches have been described in **Volume 1** but at the end of the Transcontinental section of this chapter, details are given of those Australian and New Zealand short coaches thought to have been available during the Tri-ang Hornby years.

The longer Transcontinental coaches with Australian markings are listed under their standard equivalents in the Transcontinental section.

New Zealand had the tools to make some of the 9" British coaches and these have been listed under their British equivalents.

Canadian and American Passenger Cars
Nearly all Transcontinental passenger cars made during the Tri-ang Hornby years were based on the four second series cars. These were produced in a rich array of Canadian, American and Australian liveries and all are listed under their standard equivalents in the Transcontinental section.

When American Train & Track Corp. (ATT) persuaded Rovex to supply them with models for the American market they ordered four sets of these coaches and the old time coach in two different liveries. All of them were fitted with the standard American NMRA couplings and were supplied bulk packed for boxing in ATT's own pale grey packaging.

Residue ATT stock in America was later bought and re-boxed by Model Power in blue boxes but back home in Britain 'frustrated' stock was offered to retailers. This included 12 of the range of passenger cars which were sold for about 4/- each. In the last 25 years their value has increased by 20,000%! This frustrated stock was sold in standard coach boxes with the 'R' number of the contents hand written or rubber stamped on the end flaps.

A. BRITISH OUTLINE COACHES

Many of the coaches manufactured under the Tri-ang Hornby label were already in the Tri-ang Railways range and therefore underwent only a change of packaging. These will have been dealt with in detail in **Volume 1** and so are only briefly referred to here.

TRAVELLING POST OFFICE

Maroon TPO R402
The maroon TPO had first appeared in 1953 and remained in production until 1968. During this time it went through several changes which are described in **Volume 1**. At the time that it became a Tri-ang Hornby model it had largely completed its metamorphosis. It was a deep maroon colour with the detail heat printed in yellow; the stock number 'M30224' being in small characters. The bogies were the X310 type or b2b using the collectors' classification.

About 28,000 maroon TPOs were made during this period with sales remaining around 7,000 per year.

Blue & Grey TPO R402M
In 1969 the colour scheme for the TPO was changed to match the other coaching stock and, until the model was finally withdrawn at the end of 1972, it was rail blue and pale grey.

A fall-off in sales coincided with the colour change. 6,500 were sold in 1969, 3,750 in 1970, 4,600 in 1971 and 3,700 in 1972.

In 1973, there was no TPO in the catalogue for the first time in twenty years. The next year the Transcontinental version was available in BR livery as part of a train set. The best model of a travelling post office was not to appear until 1978. Both these later versions will be covered in **Volume 3**.

9" SUBURBAN COACHES

Under pressure from retailers, Rovex did a special run of both maroon and green suburban composites in 1967. The brake end could not be made as the tools had gone out to New Zealand for manufacture there.

These 1967 models can easily be recognised as they have X531 bogies with closed axle boxes and pin-point axles. Earlier suburban coaches made in the Margate factory had sleeved axles and open boxes. The late suburbans are quite difficult to find.

They were available from a limited number of dealers priced 14/3d each but by November they were being advertised at 12/- each in order to clear stocks.

Maroon Suburban Composite R121

This was a mid-maroon with the normal grey roof and matching grey interior. It carried a small '41007' as its stock number. Orders for a little over 800 were received in 1967 and 1,207 were made.

Green Suburban Composite R223

The 1967 batch of green suburbans totalled 1,021 against orders for 684 from retailers. These were made with the yellow-green plastic of the 10" coach-

es but had the normal grey roof and matching interior. The number ('S3153S') heat-printed in small characters on its sides was white rather than yellow.

New Zealand Suburban Brake R222(NZ)

As we saw in **Volume 1**, the tools for the Suburban Brake Second body and roof were sent out to New Zealand in 1964 where a green version of it was produced, initially for the NZST train set, but it was also available as a solo model. As such it remained

in the illustrated price list until 1972. Those used in the train set had a round BR coach decal on their sides.

9" MAINLINE COACHES

Composite for Starter Sets R720, R724 etc.

Until the introduction of the 'integral' 4-wheeled coaches in 1969, starter sets contained a red and yellow or plain red coach from the 9" mainline com-

Late travelling post office in blue and grey livery. Below: inside the real thing.

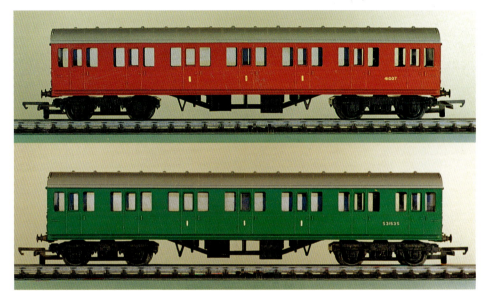

1967 late release of suburban coaches.

Red starter set coach.

Above: R226 green, electric blue and rail blue utility van.
Top left: New Zealand-made R222 suburban brake dating from around 1969/71.
Left: R226 utility van boxed in New Zealand.

posite coach tools. These coaches varied according to the price of the set. Besides the dropping of the yellow colouring, the model at times was cheapened further by leaving out the interior, the buffers or even the windows.

The earliest red and yellow coaches date from 1963 and were the same moulding as the earlier mainline coaches but, at some time, the tool was altered to raise the window panel and thus simplify the painting process. As the all red coaches are thought to have been made after this change, it is likely that they all have the raised window panel.

The different versions would have been allocated different 'R' numbers for reference purposes within the factory but, as these do not appear in published literature we do not know what all of them are. We do know that both the R720 and R724 were red and cream.

New Zealand Mainline Brake R220(NZ)

This is described adequately in **Volume 1** and there is nothing further to add. It remained available in New Zealand throughout the Tri-ang Hornby years.

New Zealand Restaurant Car R229(NZ)

The tools for the body and roof were sent to New Zealand in 1967 and the model remained in the illustrated price list until 1972. Initially bogies were bought in from Margate but later Auckland received the discarded c1b open axle box bogie tool. This means that the coaches with closed axle box (imported) may be older than those with the open axle box! Australian metal bogies were also used.

The coaches varied in detail over the years. The stock number was S34245 and heat-printed in white but for much of the time coaches were issued without a number. The 'kitchen' windows were sometimes white and sometimes clear and the coach ends were sometimes painted black and sometimes left unpainted. The most recent example (with IIId couplings) I have seen, had no number, or clear windows and non-painted ends, suggesting that these details were abandoned in later years.

9" UTILITY VAN

Green with Red Doors R726

This was made for the 'Old Smokey' set and was briefly covered in **Volume 1**.

BR Green R226

The green utility van had been withdrawn in 1964 but reappeared for one year only in 1967 when 8,700 were made. These would have had the c2b bogies and IIIc or IIId couplings and were numbered 'S2357S' in yellow. The yellow green plastic of the 10" Southern Region Mk1 coaches was used and the roofs were in the normal platform grey.

There was a proposal to reintroduce the utility van in 1977 but this did not come to anything and it was not until the mid 1990s that the green utility van was to be revived in two different Southern liveries. Details of these will be found in **Volume 3**.

New Zealand Suburban Utility R226(NZ)

The New Zealand 'Suburban Utility' was not made in the Auckland factory but nearly complete models were bulk packed over to Lines Bros. (NZ) Ltd where couplings were fitted. After the initial supply (no record of this) for the NZST train set which was

first listed in 1964, no records can be found of the R226 Utility Van having been sent to New Zealand until 1970 when just 200 went out there.

Electric Blue R226

The blue utility van was available from 1968 and the 10,000 made that year were in electric blue. These had the platform grey roof and 'S2355S' or 'S2357S' heat printed in white on their sides. They may be found with either IIIc or IIId couplings and DP1 pinpoint axles.

Rail Blue R226

The following year rail blue plastic was used and, assuming that the shade change occurred after December 1968 and before any of the 1969 production commenced, about 6,000 solo models were made. The model, which had IIId couplings, a very

dark grey roof and 'S2355S' in white on its sides, was available again in 1970 but there was sufficient stock left over from the previous year for further production to be unnecessary. The model was dropped in 1971 presumably because of lack of demand.

Besides the green utility van with red doors which was produced solely for the 'Old Smokey' set, the rail blue version was the only one to be used in a British train set. This was the RS7 of 1969 which was not shown in the catalogue. 5,000 of this set were made which brings the total production of rail blue utility vans up to 11,000.

9" PULLMAN CARS

Standard Models R228 and R328

The Pullman First, with its choice of four names

R228/R328 Pullman coaches with white-rimmed wheels and much reduced table lamps.

('Anne', 'Ruth', 'Mary' and 'Jane'), had been introduced in 1958 and had been followed by the brake 3rd in 1960. They seem to have been based on the cars used by the SR on their 6-PUL and 6-CITY electric units. The models had been an instant success and sold in their thousands.

This popularity survived throughout the Tri-ang Hornby years when the models would have had c2b bogies with closed axle boxes and IIIc couplings later to be replaced by the IIId type. DP3 wheels were fitted until the end of 1970. For a brief period in August that year, as an experiment, the rims were painted white to see how the public reacted. These coaches with white-rimmed coarse scale wheels (DP3a) are naturally scarce. From January 1971, fine-scale white-rimmed wheels (DP5) were fitted. Both cars were very much under scale and had freelance interior units.

The adoption of white-rimmed wheels in 1971 lead to the addition of an 'A' suffix to the 'R' number; thus R228 and R328 became R228A and R328A. Both cars were withdrawn at the end of 1974 and replaced with the scale length version, still in production in the 1990s.

There is always dispute as to when Rovex stopped painting the lampshades in the Pullman cars.

Catalogue illustrations continued to show them pink until 1968, but as we have observed on numerous previous occasions, catalogue illustrations are not always accurate. The lamps were to change completely in the 1970s as instead of a lamp with a shade, a simple brass spike with no shade was fitted.

During the Tri-ang Hornby years over 56,000 of the parlour car were made and 34,500 of the brake 3rd. In addition both were used in train sets. Total sales during the seven years were as follows:

	DATES	PARLOUR	BRAKE
Solo Sales	55-71	56,000	34,500
RS9	66-67	40,000	20,000
RS90	68-69	29,400	14,700
RS605	70	8,500	-
TOTAL		133,900	69,200

9" DMU CENTRE CAR

BR Brunswick Green R334

As far as we know the DMU Centre Car remained unchanged and was withdrawn in 1968. It was Brunswick green and carried the stock number 'M59120' heat printed in yellow on its sides. It had open axle box bogies throughout and any found

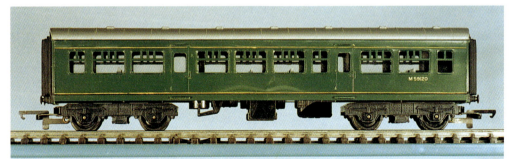

R334 DMU centre car.

Model Railway Constructor extracts dated December 1966 and May 1967.

news desk

We have heard from Rovex Scale Models Limited, that although the GWR 2nd class clerestory coach (R332) is no longer shown in the current Tri-ang Hornby catalogue, they are planning to produce a limited number of these models during the early part of next year. The coaches will not be available until approximately February, and only obtained from the Service Department at Margate at a price yet to be announced. As a limited run is being made, it would be helpful if readers would kindly write to the Editor *mrc* stating how many they would be prepared to purchase, so that we can assess the demand and pass the information to Tri-ang in order that they can determine the number to be produced. Will readers kindly write to the Editor as soon as possible, and *not* to Tri-ang or Rovex. When the models are available the information will appear in these columns.

Readers who wrote in stating that they would be willing to purchase the Tri-ang clerestory cloaches, following our announcement in the *Newsdesk* feature of the December issue are advised that these are now ready. The cost of the clerestory composite, and the clerestory brake composite coach is 13s 0d each, plus 1s 6d for postage and packing on one coach. Orders for more than one coach are sent post free. Readers who intimated their wish to secure these coaches should now write to Rovex Scale Models, Service Department, Westwood, Margate, Kent, enclosing the necessary remittance. Other readers may be able to obtain some of these coaches from a few selected Tri-ang dealers who have also been sent small supplies, and who are advertising in our pages.

with closed axle boxes are Hornby Railways models belonging to the period 1976-1979. Between 1965 and 1967 (the last year of manufacture) a little over 8,700 DMU Centre Cars were made.

CLERESTORY COACHES

Original Issue R332 and R333
The GWR clerestory 3rd class and brake coaches had been introduced in 1961 and are described in **Volume 1**. Including solo models and those used in sets, it is estimated that about 60,000 were made before the models were withdrawn after 1965. Over 35,000 of these would have been the 3rd class coach. The initial batches of solo models were made in 1961 and 1962 but they were slow to sell and so stocks lasted until the end of 1964. From 1962 the coaches had closed axle boxes. Under pressure from retailers a further 3,700 3rds and 1,760 brakes were made in 1965. These also sold slowly.

1967 Limited Issue R332 & R333
Despite the poor sales, pressure from the public through the *Model Railway Constructor* continued and in December 1966 it was announced that a limited edition run of the 3rd class clerestory (R332) was planned in February 1967. These were to be available only from the Service Department of Rovex Scale Models Ltd. An entry was made on the factory census sheets but crossed off with no production numbers set against them. This suggests that none were made and there is no evidence that there was old stock in the stores that could have been sent out instead. Despite this, in the May issue, EAMES of Reading were advertising both coaches (R332 and R333) as being 'in stock' and priced 12/9d each and the *Model Railway Constructor* was inviting readers who wanted them to write to the Service Department at Rovex Scale Models!

We can only speculate as to what happened. Rovex may have had difficulty in finding a tool, or

1969 and 1971 versions of the R332 clerestory coach and R379 coach from the Railway Children Set.

perhaps a tool needed work done on it which made the project non-viable. Advertising had to be with the publisher well in advance of publication and E.A.M.E.S may have jumped the gun. This, however, does not explain the *MRC* announcement, which

was repeated in the May edition, that the coaches were ready, priced 13/0d plus postage, or the fact that Lou Nadin of Llandudno were still advertising them in June by which time they were priced 11/-.

It seems more likely that there was an error on the

census sheets, and the coaches were made without record of them. If so, it is reasonable to assume that they were made in a similar quantity to the suburban coaches reissued at the same time – i.e. 1,000 of each. It seems likely that these would have looked like those made up to 1965; only the packaging would give them away.

1969 Reintroduction R332 & R333
In 1970 the coaches reappeared in the catalogue but the decision to reintroduce them had been taken part way through 1969. Indeed, by the end of 1969 about 8,500 of the 3rd class car and 8,100 of the brake had passed through the stores. About 7,200 and 6,600 respectively were made in 1970. These would have differed from the 1967 coaches in having large hook (IIId) couplings.

1971 Version R332A & R333A
By 1971, the locomotive range had been going through a period of transformation from BR models to pre-nationalisation liveries and to boost the GWR choice the 'Lord of the Isles' locomotive had been re-launched in a gloss livery. To go with the locomotive and boost the pre-nationalisation coaching stock, from January 1971 the GWR clerestory coaches were smartened up with fine-scale white-rimmed wheels. The example in my own collection has GWR crests on its sides but these were not to be seen on

the examples in the catalogues.

In 1971 some 7,600 of the 3rd class car and 6,500 of the brake were made in this new improved form. 2,400 and 2,100 respectively were made in 1972 but after that it was dropped from the catalogue.

The coaches were to reappear again, this time with metal wheels, in a 1980s Hornby Railways presentation set which will be covered in **Volume 3**.

Railway Children Coach R379
This was moulded in maroon plastic with the upper half painted white. The roof was black. It was available exclusively in the 'Railway Children Set' R615 which was based on the film of the same name. About 6,000 were made.

LMS Clerestory Coaches R24 & R25
These were initially planned for January 1972 but were later replaced in the programme by LNER teak coaches marketed under the same 'R' numbers. This, however, did not occur until 1973. The tools were eventually used for LMS clerestory coaches, but not until 1986.

ENGINEER'S COACH

Black Version R620
The black engineer's coach was made until 1965 and is covered in **Volume 1**.

Green Version R620
The engineer's coach made use of the GWR clerestory brake coach tools and had been introduced in 1963. In the 1966 catalogue the colour of plastic for the body had changed from black to Brunswick green but it seems likely that the colour change had occurred sometime during 1965.

In 1965, about 4,000 were made but over 3,000 of these were left at the end of the year. As a result, no more were made in 1966. By the end of 1966 the stocks had gone and so a further batch of a little over 3,000 was made in 1967; the last year it was included in the catalogue. If both 1965 and 1967 production were in green plastic, then there were 7,000 made and this seems likely.

'ROCKET' COACH

General Issue
These were mainly used for export during the Tri-ang Hornby period. In July 1969, *Railway Modeller* carried an article by Mike Sharman on converting and detailing these coaches.

USA
A supply of these went to Tager of America for assembly of their 'Rocket' set which contained three coaches. These appear not to have been any differ-

R620 green engineer's coach and a prototype engineer's coach.

ent to the general models as NMRA couplings were not fitted.

SLEEPING CAR

Maroon R339

Released in 1961, the Sleeper was the first of the new range of scale length coaches which were to compete with the Hornby-Dublo Super Detail stock and the Kitmaster kit coaches which were proving very popular with the public.

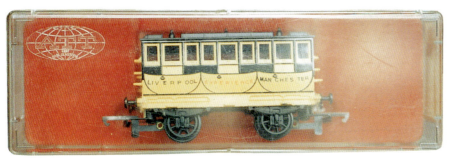

Maroon Mk.1 coaches.

R621 Rocket coach.

R621 Rocket coach sold in America by ATT.

Green Mk.1 coaches.

R731 2nd brake Assembly Pack.

Electric blue Mk.1 coaches.

The coach carried the stock number '2510', had a BR coach decal on each side and remained in the Triang Hornby range until the end of 1967. Coaches numbered '2511' are from CKD kits and '80658' has also been found.

Only 1,600 solo models were made in 1965 and 900 of these remained unsold at the year end. Another 1,750 were made in 1966 and 3,300 in 1967. None were used in sets during this period. The last CKD kits of this model were made in 1964.

To help sell the sleeping cars the 1967 catalogue contained a diagram which showed how a train of sleepers should be made up.

Electric Blue & Grey R339

With the maroon sleeping car being dropped at the end of 1967, a blue and grey version replaced it in 1968. The first version of this was moulded in electric blue plastic with a platform grey roof. It carried the words 'SLEEPING CAR' in white in the middle of each side and a small '2510' in white at the right-hand end. The cars were fitted with c2b bogies and IIId couplings.

In this form 6,666 were made all of which were sold as solo models. After the introduction of the blue sleeping cars the tool for the body sides was modified to provide a raised line high up as a guide for applying the grey window band.

Rail Blue & Grey R339

In rail blue and grey livery with a very dark grey roof and the same heat printed detail as its predecessor, the model first appeared in 1969 and survived for four years although the last year (1972) it was available only in a set.

About 5,000 solo models were made in 1969 and a further 3,800 the following year. Between 1970 and 1972 the model was included in the RS604 Night Mail Set and this accounted for about 20,000 more sales. Models made from 1971 probably had finescale wheels.

The model reappeared in 1974 with chrome window surrounds and, later on, aluminium ones. The inscription on the side was also shortly to change to 'Inter-City Sleeper' but the number it carried remained '2510'. These later versions, which had e1 type bogies, are outside the scope of this book and will be dealt with in **Volume 3**.

FULL PARCELS BRAKE

Maroon R425

The model had been introduced to the range in the spring of 1962 and, by 1965, was numbered '80657', with '80658' as the additional CKD model. (The latter has also been found with the number '1823' from

Mk.1 coaches in the later rail blue and grey with dark grey roofs.

the Buffet Car). In 1965, about 2,600 were sold and another 1,500 in 1966. The following year, the last that it was made, a batch of 3,900 was produced.

The pre-production model which has survived in a private collection was made in clear plastic.

Electric Blue & Grey R425

In 1968 a batch of 10,000 full parcels brakes was made in electric blue and pale grey livery. These had a mixture of IIIc and IIId couplings, DP1 wheels and c2b bogies. The roof was platform grey and the stock number carried in white on its sides was '80657'. About 3,000 of the models were unsold at the end of the year and would have been sent out to shops the following year.

As with the sleeping car, after the introduction of the blue livery the tool for the body sides was modified to provide a raised guide line for the mask used in painting the grey window band.

Rail Blue & Grey R425

This replaced the previous version in 1969. Like the other coaches in this livery it had a very dark grey roof. The model was dropped in 1972 but between 1969 and 1971 about 7,000 were made. It was to have been dropped from the 1971 catalogue but was given a reprieve in order to use up remaining stocks of boxes. Models made in 1971 may have had fine-scale wheels.

10" MAINLINE COACHES

Maroon R422 & R423

The 10" BR maroon composite and brake end Mk1 coaches remained in the catalogue until the end of 1969 after which they were replaced by an 'LMS' version. The models are well described in **Volume 1** and other than the use of type IIId couplings from 1967 onwards it is thought likely that little change occurred during the Tri-ang Hornby period when the models would have been numbered '15865' and

R422A/R423A BR Mk.1 coaches in LMS livery.

'15863' for the composite and '35024' and '35025' for the brake 2nd.

Batches for solo sales of the composite were made every year from 1965 to 1969 and every year except 1969 for the brake end. This meant that about 37,500 maroon composites and 27,500 brake ends were made during this period. A further 24,000 of each were used in the RS8 'Midlander Set' made between 1965 and 1968.

In addition to the ready-to-run models about 6,000 of each will have been produced from CKD kits (R382 & R383) made and sold in 1966, possibly in Tri-ang Railway boxes as there appears to have been an excess stock of these.

Green R622 & R623

Again, these have been well described in **Volume 1**. They remained in production until the end of 1969 and between 1965 and 1969 a total of 27,000 composites and 25,000 brake ends were made. Production of the 'SR' version of the Mk1 coaches also started in 1969 and continued the following

year as replacements for the BR stock.

Maroon and Cream R626 & R627

These coaches were withdrawn after 1965 and are covered in **Volume 1**.

Electric Blue & Grey R727 & R728

These had the same colour scheme as the early blue and grey sleeping cars described above. The composite carried the number '15865' in white on its sides and the brake end was numbered '35024'.

Although these were not illustrated in the catalogue until 1966, the first batches were in fact made the year before when nearly 10,000 of each passed through the stores. Indeed, none were made in 1966 but about 11,000 of each were made in 1967 and 13,000 the year after. Following this the plastic changed to rail blue.

The models were also available in Assembly Packs each of which contained parts for one coach. The composite kit was R730 and the brake R731. In this livery approximately 8,000 composites and

R622A/R623A BR Mk.1 coaches in SR livery.

R743/R744 BR Mk.1 coaches in GWR livery.

7,000 brake ends were made.

Sometime after the introduction of the blue and grey livery, the moulds for the coach sides were modified to provide a raised line near the top as a painting mask guide.

Rail Blue & Grey R727 & R728

This replaced the electric blue set in 1969 and took the same stock numbers. They were to be made for just three years during which time 20,000 of the composite and 15,700 of the brake end were made for solo sales. In addition, about 4,000 of each were available in assembly pack form. These carried the same number as the ready-to-run coaches, were catalogued until the end of 1969 and featured in a 1970 stock clearance catalogue supplement sent out to retailers.

LMS Maroon R422A & R423A

These direct replacements for the BR maroon coaches arrived in 1969 although they were not illustrated in the catalogue until 1970. Only the composite and

brake end were so treated. The composite became a full third coach and the brake end a brake first. These carried the stock numbers '2257' and '5051' respectively and had platform grey roofs.

Those issued in 1969 and 1970 would have had DP3 wheels. Initially there were thoughts of dropping the coaches from the 1971 catalogue but instead it was decided in April 1970 to fit the DP5 type wheels with white rims for 1971. That year the 'A' suffix was dropped from their 'R' numbers.

Both coaches may be found with silver seal wheels but these belong to the Hornby Railways era. The coaches survived until the end of 1976 although at one stage it had been intended to replace them with LMS versions of the Clerestory coaches for 1972. Production figures for the Tri-ang Hornby period were 13,000 for the full 3rd and 12,000 for the 1st brake.

SR Green R622A & R623A

These arrived in 1969 and during the Tri-ang Hornby period they were moulded in BR coach

green. Until 1971 the roofs were grey but that year they were changed to white. Any found with dark green sides were made from 1974 onwards and will be found in **Volume 3**. Those moulded in BR green were short lived, surviving from 1969 to 1971. Initially there were thoughts of also dropping these coaches from the 1971 catalogue but, like the LMS ones, it was decided in April 1970 to fit the DP5 type wheels with white rims for 1971. The 'A' suffix was also dropped in 1971.

The coaches were presented as a first/third and a brake composite. The wheels followed the same pattern as the LMS coaches described above. The name 'SOUTHERN' was heat stamped in yellow in the middle of each coach side. The first/third was numbered '5015' and the brake composite '4351'.

Only 9,700 of the first/third were sold as solo models and 8,200 of the brake composite. In addition, however, the coaches were used in the RS607 set of 1971 which accounted for a further 5,000 of each coach.

GWR chocolate & cream R743 & R744

Once again the model went into production in 1969 but none left the factory until 1970. In this case the 'GWR' versions did not replace BR(WR) stock because sadly the 10″ Mk1 coaches were not made in BR(WR) livery (not until 1996). This is perhaps surprising as they could have been used with the Hall locomotive. Perhaps it was felt that the sales of the green coaches and the maroon and cream set had not been sufficiently high to justify introducing yet another livery.

Both coaches carried the name 'GREAT WEST-ERN' in the middle of each side. The GWR composite carried the number '5015' and the brake third had '5104'. The roofs were black and the wheels were originally DP3. In 1971 they received DP5 white-rim wheels and the 'R' number received an 'A' suffix to become R743A and R744A. There was a proposal to issue them with white roofs in 1972 and drop the 'A' suffix at the same time but with the Caledonian coaches being released in GWR livery that year it was decided to drop R743A and R744A from the range.

About 20,000 of the GWR composite and 11,000 of the brake third were made, most of them in 1970. Of these about 3,500 of each had white-rimmed wheels (DP4). None of the coaches were made for catalogued sets.

Red & Yellow R626 & R627

These were not released until 1972 and so are Hornby Railways products which will appear in **Volume 3**.

BUFFET CAR

Maroon R424

The maroon Buffet Car, which had first appeared in 1962, remained in the catalogue until 1967, the last year that a batch was made. The model was well described in **Volume 1** and other than the use of

R427/R428 Caledonian coaches with grey roofs.

type IIId couplings in 1967 it is thought likely that little change occurred during the Tri-ang Hornby period. It would have had a red classification line and carried the stock number '1823' or '1825'.

Sales in 1965 and 1966 amounted to about 2,200 a year. In the final year 3,300 were made.

Green R624

Again, this has been well described in **Volume 1** and was also not made after 1967 when it was dropped from the catalogue. It too would have had the red class line and it carried the number S1851.

About 2,000 were sold in 1965 and 3,000 the following year. In its final year a further batch of 3,720 were made.

Maroon and Cream R628

The last year this was available was 1965 when 2,400 were sold, half of which had been carried forward from the previous year's production.

Red & Yellow R628

These did not arrive until 1972 and will be included in **Volume 3**.

Electric Blue & Grey R729

The blue Buffet Car first appeared in the 1967 catalogue and a little over 7,000 were made that year. The model was in electric blue and pale grey with a platform grey roof and a red class line the length of the roof line. The word 'BUFFET' was in capitals off centre on the coach sides and it carried the stock number '1825'. The coach had DP1 wheels and c2b bogies. The couplings were types IIIc or IIId.

In this shade of blue the model lasted for two years and during this time about 17,000 were made. It was also sold in an Assembly Pack (R732) of which about 5,000 were made in this shade of blue. Models from these kits may be recognised by the screw fixing of the bogies.

Rail Blue & Grey R729

In 1969, the electric blue version was replaced by

R747/R748 Caledonian coaches in LMS livery.

R749/R750 Caledonian coaches in SR livery.

one in rail blue and pale grey livery with a very dark grey roof. This had the same markings on the side as its predecessor but had type IIId couplings and in 1971 it had DP4 fine-scale wheels.

About 10,000 were made between 1969 and 1970 with a further 1,000 in 1971. There were also 2,000 sold in Assembly Packs (R732). These carried the same stock number.

The model was dropped from the catalogue at the end of 1971 but shown as a new model in 1973. Any found with X680 (e1) bogies belong to the Hornby Railways period and will be covered in **Volume 3**. This includes all those with chromed or 'aluminium' window frames.

CALEDONIAN COACHES

Caledonian Livery R427 & R428

The Caledonian coaches were introduced in 1963 to support the Caledonian Single locomotive which was added to the fleet that year. The last batch of coaches made in 1964 was slow to sell and so, despite no more being manufactured at that time,

the coaches were retained in the catalogues for 1965 and 1966. Indeed, no more were produced until 1971 when a batch of 2,850 composites and 1,200 brakes were made. These had grey roofs and fine-scale wheels and may have been the only ones to have been sold in Tri-ang Hornby packaging.

Throughout their various liveries these coaches were not fitted with seat units. Rovex felt that the extra cost of tooling up units that fitted the window patterns could not be recouped as they were unlikely to sell in large numbers. At the time, no doubt, they were not planning to use the body tool to produce LMS, GWR and SR versions of the coaches otherwise a different decision may have been made.

LMS Livery R747 & R748

Both coaches were released in 1971 in a pseudo-LMS livery. The body was moulded in maroon plastic with the panel surrounds picked out in yellow, the latter being rather too heavy and too bright (this should have been gold or pale yellow). According to the *Model Railway Constructor* review, some of the lining worked out to be 3 scale inches wide instead

of three eighths of an inch. The *Railway Modeller* recommended drawing a black vertical line down the centre of the broad yellow bands to achieve a rewarding result. Both reviewers thought that the LMS crests were good.

The roof was moulded in black and, like the Caledonian coaches the LMS versions were fitted with BR bogies. These were the c2b and were fitted with type IIIc or IIId couplings and DP5 fine-scale white-rimmed wheels. R747 was a first/third and numbered '2643' in yellow and R748 was a brake composite numbered '2640'. The letters 'LMS' and the LMS coat of arms was applied to the centre of each coach side.

About 11,600 of the first/third were made as solo models in 1971 and all but 500 sold during the year. Nearly 8,900 brake composites were also made. In addition to these, 10,000 of each were sold in the RS609 train set that year. In all, the models were available for three years and were dropped from the catalogue in 1974.

SR Livery R749 & R750

The coaches were reviewed in the May 1971 edition of *Model Railway Constructor* and were thought to resemble ex-LSWR and SECR coaches absorbed into Southern stock at the Grouping. The scale length over headstocks was found to be 63' 6".

Over the windows, on the cant rail, was written 'SOUTHERN RAILWAY' in yellow and the classes were written on the doors as words rather than numbers. R449 was a first/third numbered 'S.1750' and R750 a brake composite numbered 'S.1774'. The magazine review pointed out that the 'S' was not added until BR days.

The coaches were in dark green with white roofs. The 1971 catalogue suggests that the coaches were to be released in the brighter BR green but this is unlikely. Certainly those sent to magazines for review were dark green. The bogies, wheels and couplings were the same as those described above for the LMS coaches. There were no interior units and, from release, the wheels were fine scale with white rims and were found to be free running.

In 1971, some 6,240 first/thirds were made and 7,680 brake composites. Nearly all had sold by the end of the year. None were used in a train set that year. The models were available for three years and dropped from the 1974 catalogue.

GWR Livery R26 & R27

These were not released until 1972 although early ones were sold in Tri-ang Hornby boxes but with a Hornby Railways label on the ends. They will be found in **Volume 3**.

BLUE PULLMAN PARLOUR CAR

Blue & White R426

This coach dates from 1963 and forms part of the very popular Blue Pullman DMU train. It carried the stock number 'W60745' or 'W60747' and remained in the catalogue until replaced by the grey version in 1969. The original trains ran in Nanking blue and the electric blue plastic used for the model did not seem too far out. The roof was silver and the win-dow panel, white.

Between 1965 and 1968 about 24,000 solo models were made. To these should be added 67,000 which were sold in the RS52 set. Despite the fact that so many were made compared with each of the blue coaches listed above, the Pullman centre cars sell at a much higher price than the others. They are much over valued.

Grey & Blue R426

In 1969 the pale grey Pullman DMU with a rail blue window panel replaced the blue and white version and with it came a matching parlour car numbered 'W60747'. It was made for three years as the Pullman DMU was dropped in 1972.

The model that returned in 1974 was in yet another different livery, closer to the original but with crests on the coach sides and will be covered in **Volume 3**.

In its grey livery, 11,000 parlour cars were made for solo sales and a further 13,000 were used in the revised RS52 train set.

Above: R722/R723 Mk2 coaches showing normal (with chromed window frames) and mail order versions.
Left: R426 DMU Pullman centre cars.

MK2 COACHES

These coaches acknowledged the changes taking place on British Railways at the time. The Mk2 coaches with their familiar rounded ends were becoming increasingly common on the main lines. The models were based on a Tourist Second Open (TSO) and a Brake First Corridor (BFK). The prototypes were built at British Rail's coach works at Derby between 1965 and 1967. The coil sprung B4 bogies fitted to these coaches, and for that matter some of the later Mk1s, were so clearly different in appearance to the ones that Rovex had been using that new tools had to be made to produce them. On the model, even the 'SKF' was visible on the axlebox cover.

Blue & Grey R722 & R723

They were first seen in the catalogue in 1968 but while the coaches may have been used in the RS644 set, none were made as solo models that year. By the time solo models were available in 1969, they had been modified to take internal lighting units. The lights operated from power picked up from the track using both bogies and all four sets of wheels to maintain a steady power supply. One side of the bogie had light-weight pressed metal wheels and the power was transferred into the carriage by way of copper wipers on all four axles.

The lights depended upon adequate voltage and so did not light when stationary or travelling slowly. The exception to this was when the loco was fitted with working pantographs switched to overhead power and a constant independent power supply provided to the track to light the coaches. Each coach had a consumption of approximately 120 milliamps and spare bulbs were available as part no. S5099.

The coach bodies were moulded in rail blue coloured plastic and the window panels spray painted grey. The *Model Railway News* pointed out

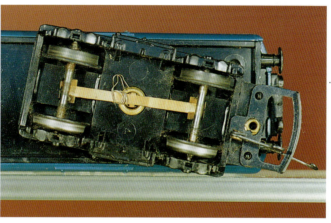

Coach lighting unit - extract from the 1969 catalogue, an advert and the bogie power pick-ups.

Instructions issued with lit coaches.

that the black moulded underframe should have been brown. Very early examples of the BFK had a yellow class line along the cant rail over the windows of the first class section. There was not much room for this and it could hardly be seen. Interior detail was fitted in the form of a separate white moulding.

Some models used for mail order sets did not have chromed or aluminium window surrounds. The TSO was numbered 'M5120' and the BFK 'M14052'. These were heat printed in white. The early models also did not carry the words 'Inter-City' on their sides. This was to have been added in 1972 but examples do not appear in the catalogue until 1973. The sides, ends and roof were moulded in one piece and the floor was held in place with flexible spigots in each corridor end. In 1971, the wheels were changed from DP3 to the fine scale DP4.

The 1969 review of the coaches in *Model Railway Constructor* claimed that they were very faithful reproductions of the prototypes.

For the R644 presentation set of 1968 about 10,000 of the TSO and 5,000 of the BFK were made. These did not have lighting but the rest of the Mk2 coaches made during this period were fitted with it. These included 15,000 of each coach made in 1969 for solo sales (solo sales had reached 23,000 for the TSO and 21,000 for the BFK by the end of 1971). In addition to these there were 8,000 TSOs and 4000 BFKs used in the R644A and a further 13,000 BFKs made for the RS604 set. These bring total Tri-ang Hornby production of Mk2 coaches to 41,000 TSOs and 43,000 BFKs.

Very large quantities of these coaches were made in years to come and several detail variations may be found. They were also to appear in at least four other liveries during the Hornby Railways period: orange and black of Ireland, Network South East, Regional Railways and the brake end only for the Royal Train. These will be covered in **Volume 3**.

THOMPSON COACHES

While Rovex felt that they could get away with using the BR Mk1 coaches as LMS and Southern (and later, GWR) passenger stock, when it came to coaches for their new Flying Scotsman locomotive to pull, only something reasonably authentic would do. In fact, they got away with the minimum of expenditure replacing only the coach sides as they had done with the Caledonian coaches. The result, if not authentic, was certainly quite pleasing with their Thompson-style oval windows at each end and they sold in their thousands even though they were 2 scale feet too long. The only regret is that Rovex did not at some stage produce them in BR maroon livery, although many people painted theirs to achieve the same effect.

R745A/R746A Thompson coaches.

R740 Thompson riding coach.

Teak R745A & R746A

The teak effect was created by adding a 'fizzer' to the plastic powder before it went into the mould. What was used for this purpose has been kept a closely guarded trade secret.

The first batch of models was made in 1970 and although scheduled for the summer were actually in the shops by February. They were numbered R745 and R746. The 'A' suffix did not come until 1971 when the coaches were given DP5 white-rimmed fine-scale wheels. R746 was a brake third and R745 a full third and so there were no first class compartments on the Tri-ang Hornby Flying Scotsman expresses. This was probably because the Thompson composite coach was built on a short chassis and would therefore have been too expensive to produce.

The catalogue shows the models in orange and while the early ones were not quite such a deep colour, they certainly were darker than the later ones. The heat printing was in yellow giving 'LNER' in small capitals at the left hand end and a stock number on the right. The full third was numbered '1010' and the brake third '1870'. The moulding on the guards and luggage doors was exemplary.

Unfortunately the window spacing meant that the standard interior units would not look right and so they were sold without interiors, as it was felt that the sales would not cover the cost of the tooling. The glazing sheet had the oval windows blanked out and on one side of the carriage had a white line, representing a corridor hand rail, printed on.

Despite the compromises that Rovex had to make, the reviewer for *Model Railway Constructor* felt that, at 16/- each they were excellent value and a good representation of LNER stock.

The production run in 1970 created 35,000 full thirds and 26,000 brake thirds. Half the coaches were used in the RS605 and RS605A sets. In 1971, about 16,800 of each coach were made of which about 6,000 of each were used in the RS608 set. The coaches remained available until 1977.

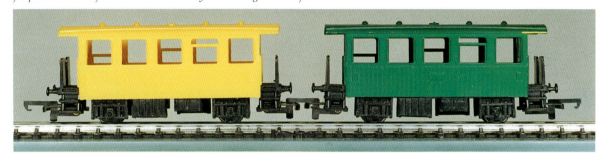

R733 coach designed for starter sets and a green version found in the factory, possibly prepared when considering an alternative colour for production. In fact the model remained yellow throughout its life.

Breakdown Riding Coach R740

This model only just slips into the Tri-ang Hornby story as 6,500 of the bright red coach were made in 1971. It lasted only three years in the range being dropped after 1973. All had plain fine-scale wheels.

It was introduced to serve as crew accommodation and eating facilities with the large Cowans Sheldon breakdown crane which had appeared in 1970. Everything about it was standard Tri-ang Hornby coach except for the red sides onto which was heat printed in white 'BREAKDOWN TRAIN UNIT (RIDING)' and the glazing sheets which had more windows blanked out than on the original Thompson brake third. The roof was white, the underframe and bogies black and the model had fine-scale wheels from the start.

LNER SLEEPER

During 1971 thought was given to producing an LNER Sleeping Car. This would presumably have been based on the Thompson composite tools but the idea was dropped. We had to wait until 1978 and the Gresley version before one was added to the range. This will be covered in **Volume 3**.

B. FOR STARTER SETS

FOUR-WHEELED COACH

Old Time Coach R733

In their perpetual struggle to master the beginners' market, in the late 1960s, Rovex designed a small number of models which would be very cheap to manufacture as they required the minimum amount of assembly. All of the rolling stock produced were freight types with the exception of a four-wheeled coach which is said to resemble early Swedish coaches.

The unglazed coach always had a yellow plastic body stuck onto either a black or dark brown simple chassis which provided it with verandas at each end and seats inside.

The coach was not available as a solo model during this period, but between 1969 and 1971 it was used in many starter sets a lot of which we have no record and so we can only guess at the total number of R733 coaches that were made. For just the few sets that we know about, a quarter of a million coaches were made in these three years.

In 1969 and 1970, Moldex in Australia received 5,000 coaches bulk packed for use in assembling their own sets and a supply of 3,000 was sent bulk packed to New Zealand in 1971 for the same purpose.

Studio picture for overseas literature.

C. TRANSCONTINENTAL PASSENGER CARS

2ND SERIES TC CARS

The baggage/kitchen car tools were modified in 1966 to produce vertical lines on the window strip to help locate the red cross stickers when the tool was used for ambulance cars.

Transcontinental Blue R444, R445, R446, R447

The blue second series Transcontinental passenger cars outlived the other colours and today are the most common. A small batch of them was made in 1966 and were illustrated in the UK catalogue but to all intents and purposes these were Tri-ang Railways models and are dealt with in **Volume 1**.

Transcontinental Green R335, R336, R337, R338

Green 2nd series TC cars were in stock in 1966 although not shown in the catalogue. None would have been made during the Tri-ang Hornby period. *See* **Volume 1**.

Transcontinental Silver R440, R441, R443

Some of the silver cars survived in the Margate stores until 1971 and possibly beyond but production had ceased by 1966. *See* **Volume 1**.

Reissued Transcontinental Silver R4400, R4410, R4430

In 1970, three of the silver and red Transcontinental passenger cars were reissued in bright silver, with a broad red band above the windows onto which 'Transcontinental' was heat printed in white. The cars were made exclusively for the RS101 Transcontinental Streamliner Set produced principally for the Australian market. These were the R4400 Coach, R4410 Observation Car and R4430 Diner. Nearly 2,000 of each car were made. The coach carried the number '70831' and the Observation car '91119'.

For 1972 it was planned to make the coach and observation car using new 'R' numbers (R4404 and R4414 respectively) but the factory census shows none were made. The specification for the models was the same as that for 1971.

Canadian National R444CN, R445CN, R446CN, R447CN

For a number of years Canada had been receiving Transcontinental rolling stock but in the mid 1960s it was decided to offer them trains in their own liveries in an attempt to increase sales.

The four TC second series passenger cars were therefore finished in the silver and black livery of Canadian National Railways, in 1965. These origi-

nally had a dark silver body moulding and a black window stripe which carried the letters 'CN' heat printed in silver. The stock number was heat printed in black on the panel provided for the purpose on the side of the vehicle. The coach and diner were either '300' or '303' and the observation car was '304'. The baggage/kitchen car was 304 or had no number.

In late 1969 or early 1970, Rovex stopped using the '300' series numbers on Canadian National passenger stock and reverted to the 1962-66 Transcontinental period identification. Sometimes the coach and observation car numbers would be mixed up, the coach carrying the '91119' and the observation car the '70381'. These are probably the

1971 TC R440/R441/R443 silver passenger cars for Australia.

Canadian National R444CN/R445CN/R446CN/R447CN passenger cars (unsprayed).   *Canadian Pacific R440CP/R441CP/R442CP/R443CP passenger cars showing unlined and lined versions.*

least common of the CN variations. By this time the cars were being turned out in a much brighter silver similar to that used for the American stock. The 'R' numbers also changed in 1970, the new numbers being indicated in the table below.

The numbers of each model made in the first year seems a little strange. There were, for example 3,000 of the coach, 8,000 of the diner but no observation

cars made. According to records the latter was not to be made until 1968 and yet it can be found in the early dark silver plastic although this form is rare in Canada. In 1968 and 1969 the Observation Car (R445CN) and the Coach (R444CN) were illustrated in the UK catalogue.

Total production of the CN passenger stock was approximately as follows:

NO.	1970 NO.	1971 NO.		QTY
R444CN	R4441	R4401	Coach	13,200
R445CN	R4451	R4411	Observation Car	8,900
R446CN	R4461	R4421	Baggage/Kitchen	12,100
R447CN	R4471	R4431	Diner	15,800

It was proposed to make more of the coach and observation car in 1972 but for some reason the number was changed to R4401 and R4411 respectively. In actual fact, none were made. Some of the coaches sent out to Canada for set production eventually found their way into bubble packs and some of these found their way back to the UK when warehouses were cleared.

Canadian Pacific R440CP, R441CP, R442CP, R443CP

The cars were a soft grey rather than the silver of the prototypes and had a broad maroon band of colour above the window line. Onto this was heat printed in yellow 'CANADIAN PACIFIC'. The roof was platform grey. Each car carried a number in yellow on a maroon sticker mounted on the number panel. The coach had '7752', the Observation Car '7420', the Baggage Car '7914' and the Diner '7525'. The final batch of Baggage Cars left Margate without stickers as presumably stocks had run out.

The Observation Car also had a special silver sticker on the rear end door reproducing the Canadian Pacific crest.

Earliest examples of Canadian Pacific passenger stock had yellow lines down their sides, above and below the panel under the windows. These had been abandoned by 1968.

The cars first appeared in Canadian Pacific livery in 1967 and were available only in Canada. A little over 1,000 of each were sent out boxed to Canada that year, followed by 1,000 of each bulk packed and 1,500 - 2,000 boxed the following year. Total production was as follows:

NO.	1970 NO.	TYPE	QTY
R440CP	R4402	Coach	3,800
R441CP	R4412	Observation Car	4,200
R442CP	R4422	Baggage/Kitchen	4,100
R443CP	R4432	Diner	4,000

C P Rail R4403, R4413, R4423, R4433

After a gap of a year in which no Canadian Pacific passenger cars were made, a batch of CP Rail cars appeared. These were a bright silver with a bright red band above the windows which carried 'CP Rail' and the logo. They may be found with either MP1 or MP2 wheels.

The cars did not carry a number but in its place was a destination name. Rovex had chosen the town of Banff but somehow this had been misinterpreted and the first batch of cars carried the name 'Danff'. This error can even be seen in the 1971 Canadian catalogue illustrations. The error was later corrected but there appear to be approximately equal numbers of models with each spelling.

The CP Rail passenger cars arrived in May 1970 and were made for only 2 years – 1970 and 1971. In the first year 2,000 of each were sent out bulk packed for set production at the Canadian factory and about 800 of each were sent out boxed. In 1971 only boxed cars were sent of which there were about 2,000 of

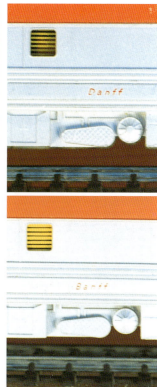

Banff and Danff on CPRail R4423 baggage cars.

CPRail R4403/R4413/R4423/R4433 Banff passenger cars.

each of the four models. Total numbers made were approximately as follows:

NO.	TYPE	QTY
R4403	Coach	2,760
R4413	Observation Car	2,780
R4423	Baggage/Kitchen	3,170
R4433	Diner	2,550

It was proposed to make more of the coach and observation car in 1972 but none were made.

The baggage/kitchen car seems to have sold badly as many, packaged in bubble packs for solo sales, were returned to the UK.

Pennsylvania R803, R807, R811, R815

The four rakes of passenger cars in American liveries with NMRA couplings were made for American Train & Track Corp. (ATT) in 1967. These were bulk packed and boxed up in America in ATT boxes. Some were later sold on to Model Power and re-boxed.

The Pennsylvania set were silver with a broad maroon band above the windows. On to this was heat printed 'PENNSYLVANIA' in yellow. On the sides of the cars the stock number was carried in black printing on silver stickers which were fixed to the number panels. The numbers used were as follows:

R803 Coach '7752'
R807 Diner '7525'
R811 Baggage/Kitchen Car '3101'
R815 Observation Car '3102'.

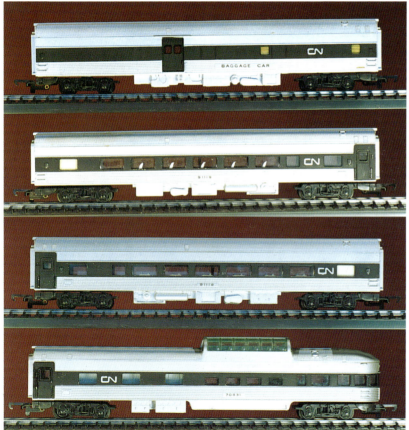

Sprayed versions of the CN cars.

Pennsylvannia R803/R807/R811/R815 passenger cars.

Production figures were as follows:

NO.	ATT NO.	TYPE	QTY
R803	200	Coach	3,750
R807	230	Diner	3,750
R811	220	Baggage/Kitchen	3,413
R815	210	Observation Car	3,750

About 100 of R811 and 150 of R815 remained in store in 1970 together with a very small number of R807. These were disposed of during the year, some, if not all, going to Australia. These were packed in Tri-ang Hornby coach boxes (windowless) with the 'R' number rubber stamped or hand written on the box end. Some of these have been returned to the UK by collectors.

Burlington R804, R808, R812, R816
This set was all silver with the name 'BURLING-TON' heat printed in black above the windows. The stock number was printed in black on silver stickers and the numbers carried were as follows:
R804 Coach '3100', R812 Baggage/Kitchen Car '3101'; R816 Observation Car '3102'; the R808 Diner carried just the word 'DINER' on the number panel.

All but a few went to ATT in America and some also ended up in Model Power boxes. In 1967, the only year they were made, the following were the production figures:

NO.	ATT NO.	TYPE	QTY
R804	201	Coach	3,756
R808	211	Diner	3,756
R812	221	Baggage/Kitchen	3,806
R816	231	Observation Car	3,750

Burlington R804/R808/ R812/R816 passenger cars.

Santa Fe R805/R809/ R813/R817 passenger cars.

Santa Fe R805, R809, R813, R817

These had a similar livery to the Burlington set but instead of the name Burlington over the windows they had 'Santa Fe'. Not only was the colour scheme the same but the cars appear to have carried the same set of numbers and again black on silver.

The following quantities were made in 1967:

NO.	ATT NO.	TYPE	QTY
R805	202	Coach	3,756
R809	212	Diner	3,756
R813	222	Baggage/Kitchen	3,693
R817	232	Observation Car	3,750

In 1970, 90 of R805 were still in the stores, 180 of R809, 50 of R813 and 10 of R817. All were disposed of during the year; most likely abroad.

Baltimore & Ohio R806, R810, R814, R818

This set had silver bodies with a broad blue band above the windows and looked very attractive. The name 'BALTIMORE & OHIO' was heat printed in yellow on the blue band. The stock numbers were in yellow on a blue sticker and seem to have been the same numbers as the above two sets.

The following quantities were made in 1967:

NO.	ATT NO.	TYPE	QTY
R806	203	Coach	3,850
R810	213	Diner	3,750
R814	223	Baggage/Kitchen	3,671
R818	233	Observation Car	3,714

By 1970 only 26 of R810, 48 of R818 and a handful of the other two cars remained. These were offered to retailers in February as 'frustrated' stock.

TransAustralia Silver & Red R440A, R441A, R442A, R443A

These were identical to the second series Transcon-

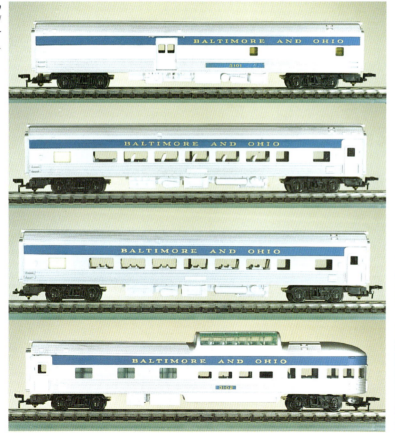

Baltimore & Ohio R806/R810/R814/R818 passenger cars.

TransAustralia silver coach (R440A) and observation coach (R441A).

TransAustralia coach (R444A) and observation car (R445A) in rail blue.

tinental version but had 'TransAustralia' as a road name, applied with the 63mm tool, instead of 'Transcontinental'. Silver TransAustralia coaches may be found numbered 31027, 70831 or 31018 while the Observation Car carried the number 9119. The rake may be found with either silver or black underframes.

Some may date from as early as 1962 but the whole range was not available all of the time. Confusion about dates of availability arises from the knowledge that Moldex were, at times, finishing and selling the ordinary 'Transcontinental' version as well; and even using them in some sets. A possible pattern, based on catalogue pictures and price list descriptions, follows:

All four cars were available, finished at the Melbourne factory between 1962 and 1964. We know that the 1964 models were the Australian version and can only guess about the earlier ones. The catalogue suggested that they too carried the TransAustralia name.

Only the R440A coach survived into 1965, 1966 and 1967; in the latter year it may have been the Transcontinental form. The only other record we have is of the R443 Diner being listed in 1970. This may have been old stock found in the factory, as otherwise it is difficult to understand why it should have been offered on its own.

Finally, the 1971 catalogue supplement showed the RS101 set with the bright silver Transcontinental R440, R441 and R443. It is assumed that these were not replaced with TransAustralia coaches for the Australian market.

The only year for which we have a record of silver TransAustralia passenger stock being sent bulk packed to Australia was 1970. That year 1,400 of R440A, 1,121 of R441A and 1,543 of R443A were sent.

TransAustralia VR Blue R444A, R445A, R446A, R447A

All four blue second series Transcontinental passen-

ger cars were assembled in Australia between 1962 and 1964 and, if the Australian catalogue picture is to be believed, these were finished as TransAustralia models. These were identical to the blue Transcontinental ones but with 'TransAustralia' in yellow above the windows instead of 'Transcontinental'. The interior units found in these passenger cars have been beige, white, blue or pink. The number carried on the observation car was 91119 and that on the coach was either 70831, 31027 or 61116 the latter being the observation car tool inverted. The set may be found with either silver or black underframes.

Once again we have little firm evidence about when they were assembled in Australia and when they were being imported from the UK ready to sell. In 1964 we were told that Australian production of these coaches would shortly cease and while all four remained in the price list, they were imported from Margate and were probably marked 'Transcontinental'. In 1967 only the R446 Baggage/Kitchen Car was available. Between 1968 and 1970, the R444 blue TransAustralia coach was listed and the R445 Observation Car in the same livery was available from Moldex in 1969 and 1970.

In the last year of production, as stocks of these

R444A/R445A/R446A/R447A blue passenger coaches. The blue plastic tended to vary in shade.

cars dried up, Rovex changed to BR blue as a best alternative. This caused problems for Moldex who could not have the loco in a different blue to the coaches and so Rovex had to grind down some surplus mouldings in VR blue to produce material for extra bodies to complete a balanced order. In the meantime Moldex, desperate for VR blue passenger cars, had resorted to spraying some of their surplus

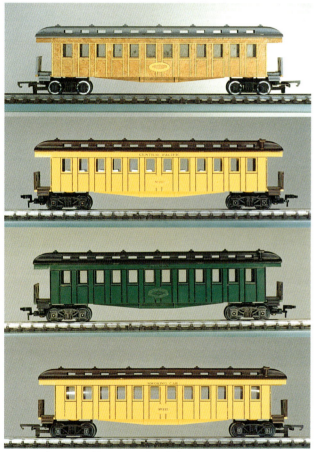

The R448 Old Time coach and three variations including two sold in America by ATT (R802 Central Pacific and R801 Pullman) and the teak finish one made for the Railway Children set.

silver coaches in VR blue to make up shortages. These were heat printed again to restore the detail.

There are records of R444A being bulk packed to Australia in 1968, 1969 and 1970, totalling about 6,800 models and R445A was sent out in this form in 1969 and 1970 accounting for a further 2,350 models.

OLD TIME COACH

The coach had been introduced in 1962 to partner the 'Davy Crockett' locomotive.

Yellow R448

Originally it was only available in the set but by 1964 it was also available as a solo model. Early examples may be found without panel lining. It is possible that it was originally made in this form but looked too plain and so the lining was added for later batches. The coach remained in production well into Tri-ang Hornby days but was not readily available in Britain, the last remaining 400 models for British sales almost certainly being sold in 1966.

Overseas it was to remain available throughout the remainder of the 1960s. Canada received a batch of 2,360 boxed ones in 1967 and in 1969 Moldex in Australia received 3,000 bulk packed and a further 2,100 the following year.

Railway Children Coach R378

This was released late in 1971 and during 1972 as part of the 'Railway Children Set' which was based on the film of that name. The coach was in mock teak with yellow lining. About 6,000 were made.

USA Pullman R801

In 1967 ATT of America ordered 5,000 Old Time Coaches in dark green and black livery. The walls and base of the coach were Brunswick green with the panelling lined in black. The name 'Pullman' was heat printed in black, in the middle of each side, over the windows and a black 'No.250' was printed on the number panel below. The coach had a black

roof, smoke stack and verandas and metal B2d bogies fitted with NMRA couplings. The seating unit was pale blue.

As far as we know they were all delivered. A few have returned to the UK for private collectors but it was not sold here by Rovex. ATT coded the coach 'No. 351' in their price list in 1967 and '251' in the 1969 list.

USA Central Pacific R802

ATT also took the yellow and brown version of the Old Time Coach but asked that it should carry the 'Central Pacific' road name. The name was printed in black in the centre over the windows and 'No.257' in black on the number plate. The coach side panelling was lined in black and the roof, smoke stack and verandas were chocolate brown. The coach also had a pale blue interior unit and B2d metal bogies with NMRA couplings.

As far as we know 5,000 of these were supplied unboxed in 1967. It was sold by ATT in 1967 as 'No.350' but by 1969 it had been re-coded '250'.

TRANSCONTINENTAL MAIL COACH

This model remained out of production during the Tri-ang Hornby years.

SHORT COACHES OF NEW ZEALAND AND AUSTRALIA

Tri-ang Railways R26, R27

New Zealand continued to make the short TC coach inscribed 'Tri-ang Railways' into the late 1960s and possibly beyond although it is not known when production at the Auckland factory finally halted. As we have seen in **Volume 1**, the coaches were normally red or maroon but blue ones can be found from early sets. It is believed unlikely that any blue coaches would have been made in New Zealand during the Tri-ang Hornby years.

R26 was the ordinary coach while R27 had the

Late New Zealand R26 short coaches with yellow printing.

The grey used for the roof was much lighter than that used at Margate and the coach bodies seem to have started red but changed to dark red or light maroon sometime before the couplings were changed from type IIb to IIIa. The roof seats in the vista dome, when painted, were red and the gangway green. Transfers were used on early models but heat printing quickly replaced these. The stock numbers sometimes appear on the wrong vehicle but this can be rectified simply by switching over the roofs.

Once the Auckland factory was importing BR bogies for the green (SR) coaches it was logical to use these on the short TC stock and so late models had this improvement. A further addition was the inclusion of an interior unit. The last pair listed in the above table seem to have been a late attempt at reducing the cost of the coaches.

Originally models were marked 'Made in New Zealand' by the insertion of a tablet in the moulding tool each time it arrived from Australia for a batch of coaches to be made. In the mid 1960s, to avoid the constant changing of the tablet at each end of the journey, one was permanently fixed in place which was inscribed 'Made in Aust & NZ'.

TransAustralia VR Blue R26, R27

The blue TransAustralia coach (R26) and vista dome (R27) had been available from 1962. By 1965 the R4D set had three of the coaches and the vistadome was available only as a solo model priced £1-6s-6d. Both carried the number '20711' at that time. They had grey roofs, a yellow stripe above and below the windows and all heat printing was in yellow. There is no evidence of any having been made after 1967.

TransAustralia Russet and Yellow R26

The R4B set of 1966/67 consisted of a maroon 'Princess Royal' and two russet red and yellow coaches. It is thought likely that this colour scheme was chosen as it could pass for the livery of the NSWGR or that of the Commonwealth Railways of Australia. The coach was numbered 20711 and this

and the road name were in yellow; the roof being grey.

Southern Aurora R26A

The R4W Southern Aurora set of 1966/67 contained two silver coaches and a silver buffet car. The coaches had seat interiors and the Buffet car had its own moulding inside. They were not numbered and it is assumed that they were not normally available as solo models. Although a vista dome version may be found, it is thought likely that these have been made up by owners switching roofs with other models.

Australian short coaches: TransAustralia coaches in blue and Indian red & yellow and Southern Aurora coach and buffet car.

vista dome roof fitted. Auckland did not make the Buffet Car.

The following are variations of the red New Zealand coach which have been found. While many of these predate the period covered by this book, the information was not available at the time that was published and so is included here for the reader's convenience:

'R' NO.	PRINT	NUMBER	BOGIES	INTERIOR
R26	black	10724	metal	none
R27	black	20425	metal	none
R26	none	none	metal	none
R27	none	none	metal	none
R26	black	none	metal	none
R27	black	none	metal	none
R26	yellow	10724	metal	none
R27	yellow	20425	metal	none
R26	yellow	none	metal	none
R27	black	none	metal	none
R26	yellow	none	BR plastic	white TC
R27	yellow	none	BR plastic	white TC
R26	none	none	BR plastic	–
R27	none	none	BR plastic	–

A young model plays with a train of wagons on a studio layout.

INTRODUCTION

The Wagon Range So Far

Wagons, as much as any part of the system, retained their links with the past. Looking at the pages of wagons in the 1965 catalogue which were about to cross the artificial line and become Tri-ang Hornby models, the vast majority of them dated from the 1950s and included the descendants of the original Trackmaster wagons of 1949 and the first Tri-ang Railways wagons of 1952-55. A few, such as the flat wagons and tankers, had been provided with improved bodies but others, like the cattle wagon, were to live on into the 1990s with little change to their body moulding.

Colour Changes

In 1966, the fish van changed from white to pale blue and the bolster wagon was given a Minix car as a load. The colour of the open wagon and closed van had stabilised as mid green and the Engineer's Department wagon and track cleaning car both changed from black to dark Brunswick green used for locomotives. The open wagon was also available as a converter wagon with a Tri-ang tension-lock coupling at one end and a Hornby-Dublo coupling at the other.

There was one completely new freight vehicle in 1966. This was the cement wagon which initially appeared in cement grey with the Blue Circle stickers but later, in Hornby Railways days, the body colour changed to a garish yellow.

In the 1966 UK catalogue, we said good bye to the Shell BP fuel oil tank wagon and all of the Transcontinental freight cars with the exception of the track-cleaning car, giraffe car and side-tipping car. The Trestrol wagon was also missing from the catalogue.

Battle Space Arrives

The Battle Space range took off in 1966 when all the existing military models were changed to khaki Battle Space livery and two new wagons, the catapult plane launching car and the assault tank transporter, were added for the 'Strike Force 10' train set.

A further two new wagons, the spy satellite launching car and the radar tracking command car, appeared in 'The Satellite Set', this time in blue and red livery.

A Herald of Things to Come

1967 saw the introduction of three more wagons to the BR range. The most exciting development and an indication of things to come was the liner train with three 20' Freightliner containers. Another new vehicle was the bogie bolster wagon with three Minix Ford vans. Two other wagons using the bogie bolster as a base came into being at the same time but neither of these featured in the British catalogue. These were the wagon with three cars and the same with two Freightliner containers. The final new wagon was a long-overdue scale-length standard BR brake van.

With the appearance of the three models based on the bogie bolster as described above, the grey bogie bolster was withdrawn. Another sad loss was the red ICI bogie caustic tank wagon which had always provided a cheerful presence to any layout.

A second converter wagon was added in 1967. This time the horse box was chosen for use on pas-

Converter wagons showing the Tri-ang coupling at one end and the Hornby-Dublo coupling at the other. Why the open wagon was sold in such a large box is not known. An instruction leaflet came with each wagon.

An illustration from the 1967 catalogue.

senger trains. The Battle Space series was extended further by the addition of four more new wagons in khaki livery. These were the sniper car, POW car, tank recovery wagon and the now rare G10 'Q' car.

With the Help of a Little Whisky

An exciting arrival in 1968 was the 'whisky' bulk grain wagon which was being offered with a choice of four different names: Johnny Walker, Vat 69, Dewers and Haig. Although no mention of the fact was made in the catalogue, the wagons were to be supplied by British Trix but fitted with Tri-ang couplings. The full story of the Trix 'whisky' wagons is told in detail later in this chapter and it is fair to say here that few arrived in the shops.

The closed van had been in production for nearly 20 years and was looking rather dated. In 1968 it gave way to a completely new ventilated van with sliding doors. The 1968 catalogue also saw the return of two Transcontinental freight vehicles: the pulp wood car and the oil tanker, the latter in Canadian National livery.

This year we had to say farewell to the well wagon, brick wagon and the short brake van of 1953

while the cattle wagon and horse box (with the exception of the converter wagon) were now available only in sets.

Containerisation

In the real world, a revolution was taking place in the transport industry with containerisation. This affected the shipping companies, the ports, railways and the roads, as well as the lives of hundreds of thousands of workers around the world.

For the model railway manufacturers it was a blessing, allowing them to produce a large range of colourful models for a comparatively low investment in new tools. There were only three units to make, the wagon, a 20' container and a 30' container. Then, by varying the liveries used on the containers and the mixture placed on the wagon the company was able to offer the customer considerable choice. Rovex went further than most by not stopping with just the two containers. They also made a liquid and powder tank and an open container neither of which became very familiar on the railways.

In the 1969 Tri-ang Hornby catalogue, plans did not stop at the wagons for, on page 3, a model of a container handling crane was shown. It resembled the type becoming familiar in the new container depots springing up around the country and came with a 30' container in the livery of Tartan Arrow. There was even an AEC articulated tractor and trailer planned, which sadly never saw the light of day.

During the year, five more combinations of container on the standard black liner train chassis were released. These were: 30' Pickfords and 30' Containerway; 30' Ford and 30' Sainsbury's; three BR Freightliner open 20' containers; three 20' BP Chemical tank containers and finally three 20' Harold Wood tank containers

China Clay and Cartic

Another new wagon in 1969 was the Bowater's china clay slurry tank. It provided a colourful addition to the freight yard while the Cartic car carrier provided the longest wagon to date modelled by Rovex. Examples of the model in good condition are now sought after by collectors. Initially it was sold with an assortment of 16 Minix cars which was incorrect for the prototype and later the number was dropped to 12.

The third new wagon was a large hopper which was a whisky wagon without a top and with the body moulded in grey. The first and last wagon assembly pack also arrived in 1969. This was the R127 mobile crane.

By 1969 the choice of whisky wagons had been reduced to two: Johnny Walker and Vat 69. The pulp wood car, which had been grey in the 1968 catalogue, was now in brown Canadian National livery. The Shell Lubricating Oil tank wagon joined the Shell Fuel Oil tanker in extinction after a long life and high production numbers. The original grey bulk grain wagon (made redundant by the more colourful whisky wagons) and the engineer's department wagon which shared the same tool also disappeared, although the operating hopper wagon would return in 1974 as a private owner wagon.

100 Ton Tanker and Ferry Van

A completely new wagon in 1970, which filled an important gap, was the 100 ton bogie tank wagon. Its first livery was that of Shell but, over the next twenty or more years, it was to be released in a range of more attractive and colourful liveries.

Also new was the Anglo Continental ferry van which reflected growing interest in the Chunnel and continental links in general.

Two more container assortments were released in 1970 bringing the total to eight. The first of these carried the 30' Manchester Lines and Fyffes containers and the other had the closed and open Freightliner containers and the BP Chemicals tank.

Following poor delivery, the Trix whisky wagon was replaced by a similar but larger model tooled up by Rovex. This adopted the two names that were finally shown on the Trix version and a topless version was offered in grey as a large hopper wagon, again replacing that proposed from Trix.

The brick wagon returned during the year looking quite wrong in bright red!

In addition to the Tri-ang Hornby models, the 1970 catalogue also had a page of wagons made by their associate company G & R Wrenn Ltd. These were from the previous Hornby-Dublo range but were given pre-nationalisation or private company liveries.

Losses from the British catalogue in 1970 included the side-tipping car, pulp wood car, CN oil tanker and, from the Battle Space series, the 'Q' car. The two converter wagons were also dropped.

The Giant Crane and Coke Wagon

The 1971 catalogue contained no fewer than eight new or re-liveried wagons. Another giant of the railway came in the form of the 75 ton crane and its three auxiliary wagons, the largest of which provided a cradle for the jib. The set was bright red and the small R127 crane and the track-cleaning car were re-liveried to match.

The coke wagon had tapered sides and was a five plank wagon with three extension boards. It was in NER livery and represented the wooden mineral wagon used for many years in the North East.

The former closed van which had been replaced in 1968 by the ventilated van with sliding doors, made

a reappearance in the rail blue livery of an Express Parcels van. The bolster wagon was available in green without a load and there were pre-nationalisation liveries for the WR brake van, the open wagon (now SR brown) and the drop-door wagon which was available in red-brown and marked 'LMS'. The ferry van was now offered in a private owner livery, that of Transfesa.

Two losses during the year were the Murgatroyd's bogie Chlorine tank wagon and, from the Battle Space series, the 4 rocket launcher. There were also changes in the range of Wrenn wagons being offered.

1972

During 1971 work had been undertaken to produce three completely new wagons in 1972. Two of these were to become part of the core of wagons made in vast quantities over the years and in many different liveries. These were the Hull and Barnsley van and the large mineral wagon. The third new wagon was to have much less impact and that was the plate wagon. The other five new wagons that year were new versions of old models. All will be covered in **Volume 3**.

It is difficult to know which of the 1972 models were released in Tri-ang Hornby boxes but we do know that the large mineral wagon was. The matter is complicated because the old boxes were not abandoned in favour of new ones, but were used up before a start was made on the stock of newly-printed Hornby Railways ones. This meant that models made after 1 January 1972 may be found in Tri-ang Hornby boxes. One example I have is a Bannockburn coke wagon (not made until 1973) in a Tri-ang Hornby box with silver seal tabs which themselves did not come into general use until that year.

Silouette of the small crane and match truck.

CHECK LIST OF WAGONS

The following wagons were made during the Tri-ang Hornby era, i.e. between May 1965 and December 1971.

The dates given in the tables are those when solo stock was manufactured and are not necessarily the years when the model was illustrated in the catalogue. Models sometimes remained in the catalogue so that stocks in the factory stores could be cleared. Some wagons may have remained in production for use in a train set when no longer available for solo sales. The dates shown also extend back into the Tri-ang Railways period and forwards into the Hornby Railways days in order to give a clear picture of production span.

The detail given in the tables for couplings, chassis, bogies and wheels applies only to models made during the Tri-ang Hornby period.

The colours listed are those in which each model appeared during the Tri-ang Hornby period

For ease of reference, in the tables the wagons have been separated out under the following headings:

1. Models of British Wagons
2. Made for Starter Sets – Standard Wagons
3. Made for Starter Sets – Integral Chassis
4. Transcontinental Freight Cars – General Issue
5. Made for Canada
6. Made for America
7. Made in Australia
8. Boxed in Australia
9. Made for Australia
10. Made for France
11. Made for General Export
12. Made for New Zealand
13. Boxed in New Zealand

NB – The information is provided here in a series of lists which allows the researcher to identify a model either by series or country of origin, using the check lists, or by basic model or shape, using the main directory. In both cases the wagons appear in the order in which the first of the type was introduced, starting with the open wagon and van which date from 1949.

1. MODELS OF BRITISH WAGONS

The first list contains British type wagons that were either already in production when the name changed from Tri-ang Railways to Tri-ang Hornby or were introduced during the Tri-ang Hornby period.

No.	British	Colour	Dates	Cpls	Chassis	Wheels
	GOODS TRUCK					
R10	BR	green	57-70	IIIc-IIId	A4a-A4b	SP1
R577	Converter Wagon	black	66-69	IIIc/HD	A3c	SS
R10A	SR	chocolate	71-74	IIId	A4b-A4c	SP1-SP2
	GOODS/FISH VAN					
R11	BR	green	57-67	IIIc	A4a-A4b	SP1
R14	Insulfish	white	52-65	IIIc	A4a	SP1
R14	Insulfish	pale blue	66-72	IIIc-IIId	A4a-A4b	SP1-SP2
R780	BR Parcels	rail blue	71-73	IIId	A4b	SP1-SP2
	SHORT BRAKE VAN					
R16	BR	brown	57-67	IIIc-IIId	A4a-A4b	SP1
	FLAT WAGON (new body)					
R17	Bolsters	brown	62-65	IIIc	A4a	SP1
R17C	& Car	brown	66-73	IIIc-IIId	A4a-A4b	SP1-SP2
R676	Bolsters	green	71-74	IIId	A4b	SP1-SP2
R18	& 2 Cable Drums	brown	62-71	IIIc-IIId	A4a-A4b	SP1-SP2
R561	& Tri-ang Container (dark blue)	brown	63-69	IIIc-IIId	A4a-A4b	SP1
R?	& Lyon's Maid Container (white)	brown	?	IIIc	A4b	SP1
R574	& Kelloggs Container (dark blue)	brown	68	IIIc	A4b	SP1
	BOGIE BOLSTER					
R110	BR	grey	61-66	IIIc	B3a-B3b	MP1
R563	& 3 Vans	salmon	67-69	IIIc-IIId	B3b	MP1
R565	2 Freightliner Containers	salmon	68-69	IIIc	B3b	MP1
R579	2 cti Containers	pale grey	68	IIId	B3b	MP1-MP2
R569	3 Cars	salmon	68	IIIc	B3b	MP1
	DROP-DOOR WAGON					
R112	BR	maroon	54-70	IIIc-IIId	A4a-A4b	SP1
R112A	LMS	various	71-72	IIId	A4b	SP1-SP2
	DROP-SIDE WAGON					
R113	BR	maroon	54-72	IIIc-IIId	A4a-A4b	SP1-SP2

No.	British	Colour	Dates	Cpls	Chassis	Wheels
	TANK WAGON					
R12	Shell	silv, grey	62-71	IIIc-IIId	A4a-A4b	SP1-SP2
R12	BP	silv, grey	62-71	IIIc-IIId	A4a-A4b	SP1-SP2
R15	UD Milk	white	57-75	IIIc-IIId	A4a-A4b	SP1-SP2
R210	Shell Fuel Oil	black	57-65	IIIc	A4a	SP1
R211	Shell Lubricating Oil	yellow	57-68	IIIc-IIId	A4a-A4b	SP1
	WELL WAGON					
R118	BR	grn,gry,brn	55-67	IIIc-IIId	B3a-B3b	MP1
	CATTLE WAGON					
R122	BR	brown	55-72	IIIc-IIId	C3a-C3b	SP1
	HORSE BOX					
R123	BR	maroon	55-71	IIIc-IIId	C3a-C3b	SP1-SP2
R578	Converter Wagon	maroon	67-68	IIIc/HD	C1c	SS
	WR BRAKE VAN					
R124	BR(WR)	brown	55-70	IIIc-IIId	C3A	SP1
R124A	GWR	grey	71-75	IIId	C3A	DP4
	GRAIN/ORE WAGON					
R215	BR Grain (working)	grey	58-66	IIIc	C2b	DS3
R215	BR Grain (non-working)	grey	67-68	IIIc-IIId	C2b	DS3
R347	Engineer's Ore (working)	black	63-65	IIIc	C2b	DS3
R347	Engineer's Ore (working)	dark green	66	IIIc	C2b	DS3
R347	Engineer's Ore (non-working)	dark green	67-68	IIIc-IIId	C2b	DS3
	BOGIE BRICK WAGON					
R219	BR	brown	59-67	IIIc	B3a-B3b	MP1
R219	London Brick	brown, red	70-73	IIId	B3b	MP1-MP2
	MINERAL WAGON					
R243	BR	grey	60-78	IIIc-IIId	A4a-A4b	DP1-DP4
	TRESTROL WAGON					
R242	BR	grey	65	IIIc	D1b	DS3
	SMALL CRANE TRUCK					
R127	BR	brown	62-70	IIIc-IIId	E	DS3
R385	BR assembly pack	brown	66-69	IIIc-IIId	E	DS3
R127	BR	red	71-79	IIId	E	DS3

No.	British	Colour	Dates	Cpls	Chassis	Wheels
	BOGIE TANK WAGON					
R247	ICI Caustic	red	62-67	IIIc	B3a-B3b	MP1
R247	ICI Caustic	blue	68	IIId	B3b	MP1
R349	Murgatroyd's Chlorine	white	63-69	IIIc-IIId	B3a-B3b	MP1
	CONFLAT					
R340	BR & 3 Cement Containers	grey	63-72	IIIc-IIId	A4a-A4b	SP1-SP2
	CAR-A-BELLE					
R342	& 6 cars	grey	65-74	IIIc-IIId	B3a-B3b	MP1–MP2
	CEMENT WAGON					
R564	Blue Circle	grey	66-71	IIIc-IIId	C3a-C3b	SP1-SP2
	LINER TRAIN					
R633	3 x 20' Freightliner	grey & red	67-74	IIIc-IIId	F	MP1-MP2
R632	3 x 20' Open	grey	69-71	IIId	F	MP1-MP2
R634	Pickford's & Containerway	blu & red	69-70	IIId	F	MP1-MP2
R635	3 x 20' – BP	whit & grn	69-73	IIId	F	MP1-MP2
R637	3 x 20' – Harold Wood	buff	69-73	IIId	F	MP1-MP2
R719	Sainsbury & Ford	gry & gry	69-73	IIId	F	MP1-MP2
R677	Manchester & Fyffes	red & yel	69-71	IIId	F	MP1-MP2
R678	3 assorted 20'	mixed	70-71	IIId	F	MP1-MP2
	BR BRAKE VAN					
R636	BR	brown	67,70	IIIc,IIId	C3A	DP1-DP4
R16	BR	brown	68-75	IIIc-IIId	C3A	DP3-DP4
	VENTILATED VAN					
R11	BR	brown	68-75	IIId	A4b	DP3-DP4
	LARGE GRAIN/HOPPER					
R647	Grain - Dewers (Trix)	blue	68	IIId	Trix-G1a	SP1
R648	Grain - Johnny Walker (Trix)	blue	68-71	IIId	Trix-G1a	SP1-SP2
R648	Grain - Johnny Walker	blue	72-73	IIId	G1a	DP4-DP5
R649	Grain - Vat 69 (Trix)	blue	68-71	IIId	Trix-G1a	SP1-SP2
R649	Grain - Vat 69	blue	72-73	IIId	G1a	DP4-DP5

No.	British	Colour	Dates	Cpls	Chassis	Wheels
R650	Grain - Haig (Trix)	blue	68	IIId	Trix-G1a	SP1
R214	BR Large					
	Hopper Wagon	grey	69-72	IIId	G1a	DP3-DP4
	CHINA CLAY WAGON					
R668	Bowaters	pale blue	69-73	IIId	G1a	DP3-DP5
	CARTIC					
R666	BR	rail blue	69-73	IIId	F	MP1-MP2
	100 TON TANKER					
R669	Shell	buff	70-72	IIId	I	MP2
	FERRY VAN					
R738	BR Anglo					
	Continental	brown	70-71	IIId	H	DP4
R741	Transfesa	blue	71-74	IIId	H	DP4
	COKE WAGON					
R781	NER	grey	71-74	IIId	A4b	SP1-SP2
	BREAKDOWN CRANE					
R739	BR	red	71-81	IIId	multiple	DS4

The following wagons, developed during the Tri-ang Hornby era, were introduced in 1972 but early examples of them are known to have been released in Tri-ang Hornby packaging :

R19	Plate Wagon	
R23	BR Salmon Rail Wagon	
R242	Trestrol with Girder	
R742	Ferry Van – Interfrigo	

Made for Battle Space
(*See separate* Battle Space Check List)

2. MADE FOR STARTER SETS - STANDARD WAGONS

These were standard British rolling stock but without printing or painting on the body moulding. This way the production cost was kept to the minimum and starter set prices could be kept down. Dates when these were used are not known and the chassis details given below are based on examples found that fall within the Tri-ang Hornby period.

No.	Standard Wagons for Starter Sets	Colour	Cpls	Chassis	Wheels
R10NP	Open Wagon	various	IIIc-IIId	A4a-A4b	DS2-SP2
R11NP	Goods Van	red, blue, khaki	IIIc-IIId	A4a-A4b	SP1-SP2
R12NP	Petrol Tank	silver	?	?	
R14NP	Fish Van	pale blue	IIIc	A4a	SP1
R15NP	Milk Tank	white	IIIc	A4b	SP2
R16NP	Brake Van	maroon	IIIc	A4A	SP1
R17NP	Bolster Wagon	brown, green	IIIc-IIId	A4b	SP1-SP2
R110NP	Bogie				
	Bolster Wagon	salmon	IIId	B3b	MP1
R113NP	Drop-Side Wagon	maroon, red	IIIc-IIId	A4b	SP1-SP2
R211NP	Lubricating Oil				
	Tank	yellow	IIIc	A4b	SP1
R219NP	Brick Wagon				
	(no buffers)	maroon, salmon	IIIc-IIId	B3b	MP1
R219NP	Brick Wagon	red	IIIc	B3b	MP1
R243NP	Mineral Wagon	blue, yellow	IIIc-IIId	A4b	DP1-DP3
R561NP	Container Wagon	dk blue, red	IIIc-IIId	A4b	SP1

The following wagon, while having a printed side, had an unpainted chassis.

No.		Colour		Chassis	Wheels
R219	Brick Wagon	red	IIId	B3b	MP1

3. MADE FOR STARTER SETS – INTEGRAL CHASSIS

These wagons were produced in a variety of primary colours with integral chassis. As many starter sets did not appear in standard catalogues and some of the wagons were produced only for limited run sets, information about wagon variations is limited. The containers used for loads were the standard moulding but with printed or labelled sides.

No.	Integral Chassis	Colour	Dates	Cpls	Chassis	Wheels
	HIGH-SIDED WAGON					
R710(NP)	Wagon	various	69-78?	IIId	Integral	DP3-DP4
	TANK WAGON					
R711(NP)	Wagon	yellow	69-78?	IIId	Integral	DP3-DP4

No.	Integral Chassis	Colour	Dates	Cpls	Chassis	Wheels
	LOW-SIDED WAGON					
R712(NP)	Wagon	various	69-71	IIId	Integral	DP3-DP4
R713	& Car & Caravan	various	69	IIId	Integral	DP3-DP4
R878	& Car & Caravan					
	- unplated	various	70/71?	IIId	Integral	DP4
R876	& Tri-ang Toys	blue	71-72	IIId	Integral	DP4
R877	& Frog Kits					
	Container	yellow	70/71?	IIId	Integral	DP4
R887	& Car	various	70-72	IIId	Integral	DP4
R888	& Van?		70	IIId	Integral	DP4
	30' LOW-SIDED WAGON					
R875	& Coca-Cola - printed					
	container	red	71-72	IIId	Integral	DP4
R880	& Tri-ang Toys					
	- container	blue	70	IIId	Integral	DP4
	30' FLAT WAGON					
R879	& Coca-Cola - labelled					
	container	blue, red	70	IIId	Integral	DP4
R880	& Tri-ang Toys - labelled					
	container	red	70	IIId	Integral	DP4
R881	& Car+Caravan		70	IIId	Integral	DP4
R889	& Car		70	IIId	Integral	DP4

4. TRANSCONTINENTAL FREIGHT CARS - GENERAL ISSUE

These were produced by Rovex during the Tri-ang Hornby period some of them being reintroduced after being unavailable for a while.

No.	Transcontinental	Colour	Dates	Cpls	Chassis	Wheels
R1150	Caboose - TR	maroon	69	IIId	B3b	MP1
R116	Gondola - TR	?	66	IIIc	B3a	MP1
R111	Hopper Car - TR	red	55-66	IIIc	B3a	MP1
R117	Oil Tanker - Shell	blue	62-66	IIIc	B3a-B3b	MP1
R126	Stock Car - TR	yellow	57-65	IIIc	B3a-B3b	MP1
R1260	Stock Car - TR	brt yel	70	IIId	B3b	MP2
R129	Refrigerator					
	Car - TR	white	57-66,70	IIIc-IIId	B3a-B3b	MP1-MP2
R136	Long Box Car -TR	?	58-66	IIIc	B3a	MP1

No.	Transcontinental	Colour	Dates	Cpls	Chassis	Wheels
R1360	Long Box Car - TR?	?	71	IIId	B3b	MP2
R137	Cement Car - TR	grey	58-65	IIIc	B3a	MP1
R138	Snow Plough - TR	green	58-68	IIIc-IIId	B3a-B3b	MP1
R139	Pickle Car					
	- Westwood	red,white	59-67	IIIc-IIId	B3a-B3b	MP1
R236	Depressed					
	Centre Car - TR	green	69?	IIId	B3b	MP1
R235	Pulp Wood Car					
	- TR	brown	68-69?	IIIc-IIId	B3b	MP1
R?	Pulp Wood Car					
	- TC	red	70?	IIId	B3b	MP2?
R344	Track-Cleaning					
	Wagon - BR	green	65-70	IIIc-IIId	B3a-B3b	MP1-MP2
R560	TC Crane Truck					
	- TR	grn & brn	62-67,69	IIIc-IIId	B2d	MS4
R345	Side-Tip Flat Car					
	- TR	salmon	64-68	IIIc-IIId	B2d	MS4
R449	Old Time					
	Caboose - TR	red	64-65	IIIc	-	MP1
R348	Giraffe Car - TR	yellow	65-68	IIIc-IIId	B2d	MS4

5. MADE FOR CANADA

The following were Rovex models manufactured during the Tri-ang Hornby years mainly in Canadian liveries for the Canadian market. The first column shows the original 'R' number and the second column the new number applied from 1970. The 'R' numbers given in brackets were those applied in 1973 (where we know them). Further information about these models will be found in the introduction to the next section and under both the British and Transcontinental headings, as models from both series were used.

Old No.	New No.	1973 No.	Canada	Colour	Dates	Cpls	Chassis	Wheels
			CABOOSE					
R115CN	–	–	CN	orange	65-68	IIIc-IIId	B3a-B3b	MP1-MP2
R115CN	R1151	R270	CN	maroon	69-73	IIId	B3b	MP1-MP2
R115CP	R1152	–	CP	brown	69	IIId	B3b	MP1
R1153	R322	–	CPRail	yellow	70-73	IIId	B3b	MP2
			GONDOLA					
R116CN	R1161	R321	CN	brown	66-69,73	IIIc-IIId	B3a-B3b	MP1-MP2

Old No.	New No.	1973 No.	Canada	Colour	Dates	Cpls	Chassis	Wheels
			GONDOLA					
–	R1160	R320	Rock Island Line					
				grey	69-70,73	IIId	B3b	MP1-MP2
			HOPPER					
R111CN	–	–	CN (working)					
				brown	67	IIIc	B2d	MS4
R111CN	R1111	R327	CN (non-working)					
				brown	68-72	IIIc-IIId	B3b	MP1-MP2
			OIL TANKER					
R117CN	R1171	–	CN	blck, blu	65-69,72?	IIIc-IIId	B3a-B3b	MP1-MP2
R1170	R323	–	Shell	yellow	69,71,73	IIId	B3b	MP1-MP2
R1170	R323	–	BP	yellow	69,71,73	IIId	B3b	MP1-MP2
			CATTLE CAR					
R126CN	R1261	R271	CN	brown	65-69,73	IIIc-IIId	B3a-B3b	MP1-MP2
R126CP	R1262	–	CP	yellow	69	IIId	B3b	MP1
			REFRIGERATOR / NEWSPRINT CAR					
R129CN	R1291	R272	CN	silv, whi	65-70,73	IIIc-IIId	B3a-B3b	MP1-MP2
–	R1292	–	CP	?	70	IIId	B3b	MP2
–	R1290	–	C&O	white	70-71	IIId	B3b	MP2
–	R1293	R273	CPRail (Refrigerator)					
				silver	70-73	IIId	B3b	MP2
–	R1353	R276	CPRail (Newsprint)					
				green	70-71,73	IIId	B3b	MP2
–	R1350	R275	Newsprint Box Car					
				white	73	IIId	B3b	MP2
–	R1330	R324	Beef Car					
				white	73	IIId	B3b	MP2
			LONG BOX CAR					
R136CN	–	–	CN	brown	65-68	IIIc-IIId	B3a-B3b	MP1
–	R1361	R277	CN	red+yell	69,73	IIId	B3b	MP1-MP2
–	R1361	–	CN	mrn+yell	69	IIId	B3b	MP1
R136CP	R1362	–	CP	brown	69	IIId	B3b	MP1
–	R1363	R278	CPRail	red	70	IIId	B3b	MP2
–	R1363	R278	CPRail	maroon	71	IIId	B3b	MP2
–	R1343	R274	CPRail	yellow	71,73	IIId	B3b	MP2
			CEMENT CAR					
R137CN	R1371	–	CN	grey	66-69	IIIc-IIId	B2d	MS4
R1370	R329	–	Wabash	grey	69,73	IIId	B3b	MP1-MP2

Old No.	New No.	1973 No.	Canada	Colour	Dates	Cpls	Chassis	Wheels
			SNOW PLOUGH					
R138CN	–	–	CN	brown	66-68	IIIc-IIId	B3a-B3b	MP1
–	R1381	R279	CN	green	69	IIId	B3a-B3b	MP1
			PICKLE CAR					
–	R1390	–	Heinz	green	69-71	IIId	B3b	MP1-MP2
			FLAT CAR					
R234CN	–	–	CN	brown	67-68	IIIc-IIId	B3b	MP1
–	R1201	–	Flat car with					
			Logs	black	71	IIId	B3b	MP2
–	R1201	–	Flat Car with					
			Logs	grey	71	IIId	B3b	MP2
			PULP WOOD CAR					
R235CN	R2351	–	CN	brown	66-69	IIIc-IIId	B3a-B3b	MP1
–	R2351	–	CN	black	69,71	IIId	B3b	MP1-MP2
R235CP	–	–	CP	black	69	IIId	B3b	MP1
–	R2353	–	CPRail	red	71	IIId	B3b	MP2
–	R2350	R372	International					
			Logging Co.	red	73	IIId	B3b	MP2
			TRACK CLEANER					
R344CN	R3341	–	CN	green	69	IIId	B3b	MP1
			CRANE CAR					
–	R5600	–	(not made) yell+blk					
			AUTOTRANSPORTER					
R342CN	R3421	–	CN	black	65-69,71	IIIc-IIId	B3a-B3b	MP1-MP2
–	R3423	–	CPRail	black	70-71,73	IIId	B3b	MP2
			CONTAINER CAR					
R633CN	R6331	–	3 x CN	silver	68-70	IIId	F	MP1-MP2
R734	R7340	–	3 x cti	red	69-71	IIId	F	MP1-MP2
R735	R7352	–	2 x CP	yellow	69-71	IIId	F	MP1-MP2
R736	R7360	–	Sea Containers					
			Inc.	yellow	69,71	IIId	F	MP1-MP2
R737	R7370	–	Manchester Liners					
				red	69-71	IIId	F	MP1-MP2
–	R7353	–	CP Ships (not made)					
				grey				
			INDUSTRIAL TANK CAR					
R247CN	R2470	–	ICI Caustic					
				blue	69	IIId	B3b	MP1
–	R3490	–	Polysar	wht+blu	70-71	IIId	B3b	MP2

Old No.	New No.	1973 No.	Canada Colour	Dates	Cpls	Chassis	Wheels
	R1210	–	**BOGIE BOLSTER** & Steel load				
–			grey	71	IIId	B3b	MP2
	R2193	–	**BRICK WAGON** CPRail red	71	IIId	B3b	MP2
–							
	R6693		**100T TANKER** CPRail (not made)				
–			stone				
			FOR STARTER SETS (no print)				
R713CN	–	–	Flat & car & caravan				
			various	69	IIId	Integral	DP3
R115TR	–	–	Caboose TR?				
			maroon	69	IIId	B3b	MP1
–	R1150	–	Caboose				
			yell/mrn?	69	IIId	B3b	MP1
–	R1153NP	–	Caboose				
			yellow	71	IIId	B3b	MP2
			M-T EXPRESS				
R219NP	–	–	Brick Wagon				
			black	67	IIIc	B3b	MP1
R115NP	–	–	Caboose red/blk	67	IIIc	B3b	MP1
R344NP	–	–	Track Cleaning Car				
			brown	67	IIIc	B3b	MP1

(Other non-printed wagons from the British standard range were also sent)

6. MADE FOR AMERICA

These were all based on standard Rovex models but with an American livery (except starter set wagons) and NMRA couplings. They were ordered by ATT of America. 800 of the Flexi-van Container wagon were left over and may have been sold off elsewhere in Tri-ang Hornby boxes. Residue ATT stock was later bought and re-boxed by Model Power.

No.	For America	Colour	Dates	Cpls	Chassis	Wheels
R10NP	Open Wagon		68	NMRA	A4b	DP3
R14NP	Fish Van		68	NMRA	A4b	DP3
R820	Container Car - Sante Fe	silver	68	NMRA	F	MP1

No.	For America	Colour	Dates	Cpls	Chassis	Wheels
R821	Container Car - Flexi-van	silver	68	NMRA	F	MP1
R822	Pulp Wood - Northern Pacific	black	68	NMRA	B3b	MP1
R823	Pulp Wood Car - Southern	black	68	NMRA	B3b	MP1
R824	Pulp Wood Car - L&N	black	68	NMRA	B3b	MP1
R710T	High-Sided Wagon		69	NMRA	Integral	DP3
R711T	Tank Wagon		69	NMRA	Integral	DP3
R712T	Low-Sided Wagon		69	NMRA	Integral	DP3

7. MADE FOR AUSTRALIA

The following were made at the Moldex factory to be sold in Australia, alongside British-made models. Most Australian production was during the Tri-ang Railways period but it did continue until 1968. It is not known how many of the following items were still being made when the production lines closed but it is most likely to have been those used for sets. Dates and chassis details are unavailable. Further details will be found in the following section dealing with British outline models.

No.	Made for Australia	Colour
R10	Open Wagon	green, blue, maroon
R11	Goods Van	brown, blue
R12	Shell Petrol Tank	silver
R13	Coal Truck	green, maroon
R14	Fish Van	white
R15	Milk Tank Wagon	white
R16	Short Brake Van	maroon
R17	Bolster Wagon	grey
R17C	Bolster Wagon + Car	grey
R18	Cable Drum Wagon	grey
R210	Shell Fuel Tank	black
R211	Shell Lubricating Oil Tank	yellow
R114	Short Box Car	orange, white
R115	Caboose	maroon, blue
R116	Gondola	green, maroon

8. BOXED IN AUSTRALIA

Moldex bought partly-finished models from Rovex during the Tri-ang Railways period and the practice probably continued after 1 May 1965. Thus Moldex finished models may appear in Tri-ang Hornby boxes but we do not have records to show us which ones these were or when they was boxed. On the other hand we do know that, for a brief period from 1968, finished models were bulk packed to Australia and boxed up by Moldex in their own Tri-ang Hornby boxes. The following were bulk packed to Australia and may have ended up as solo models in Moldex boxes and the dates given are those when they were shipped. These models are also referred to in the section on British Outline wagons which follows.

No.	Boxed in Australia	Colour	Dates	Cpls	Chassis	wheels
R10A	Open Wagon	green	68-70	IIIc-IIId	A4b	SP1
R11A	Ventilated Van - BR	brown	68-70	IIIc-IIId	A4b	SP1
R12A	Petrol Tank Wagon	silver	70	IIId	A4b	SP1
R13A	Open Wagon with Coal Load	green	68	IIIc-IIId	A4b	SP1
R14A	Fish Van	blue	68-70	IIIc-IIId	A4b	SP1
R15A	Milk Wagon	white	68-70	IIIc-IIId	A4b	SP1
R16A	Guards Van - BR(ER)	brown	68-70	IIIc-IIId	C3A	DP3
R17CA	Flat Wagon & Car	brown	70	IIId	A4b	SP1
R113A	Drop-Side Wagon	maroon	68-70	IIIc-IIId	A4b	SP1
R122A	Cattle Wagon	brown	68-70	IIIc-IIId	C3b	SP1
R123A	Horse Box	maroon	68	IIIc-IIId	C3b	DP3
R124A	WR Brake Van	grey	70	IIId	C3A	DP3
R211A	Lubricating Oil Wagon	yellow	68	IIIc-IIId	A4b	SP1
R247A	Bogie Tanker - ICI	red	70	IIId	B3b	MP2
R340A	3 Containers Wagon	brown	68-70	IIIc-IIId	A4b	SP1
R561A	Container Wagon - Tri-ang	brown	68-70	IIIc-IIId	A4b	SP1
R569A	Bogie Bolster & 3 Cars	salmon	68	IIIc-IIId	B3b	MP1
R734A	Liner Train & 3 cti Containers	red	70	IIId	F	MP2

No.	Boxed in Australia	Colour	Dates	Cpls	Chassis	Wheels
R1260A	Stock Car	yellow	70	IIId	B3b	MP2
R2350A	Pulpwood Car	red	70	IIId	B3b	MP2
R710A	Integral High-Sided Wagon	?	69-70	IIId	Integral	DP3
R711A	Integral Tank Wagon	?	69-70	IIId	Integral	DP3
R712A	Integral Low-Sided Wagon	?	69-70	IIId	Integral	DP3

9. MADE FOR AUSTRALIA

The ACT container was made for a set being assembled in Australia at the time and two of the other wagons are listed as having been produced for Australian sets although the standard models (in the previous list) may have been sent instead.

No.	For Australia	Colour	Dates	Coupling	Chassis	Wheels
R839	Freightliner Wagon ACT	white	70	IIId	F	MP1
R2354	Pulp Wood Car VR (probably not made)	?	69-72?	IIId	B3b	MP1-MP2
R1264	Stock Car VR (probably not made)	?	69-70?	IIId	B3b	MP1-MP2
R2194	Brick Wagon VR (not made)	brown	71	IIId	B3b	MP2

10. MADE FOR FRANCE

These were standard Tri-ang Hornby wagons fitted with Hornby Acho couplings for sale in France by Meccano Tri-ang. Standard Tri-ang Hornby packaging was used but labels in French were attached.

No.	For France	Colour	Dates	Coupling	Chassis	Wheels
R128F	Helicopter Car	green	65-68	Acho	B2d	MS4
R216F	Rocket Launch Wagon	green	65-68	Acho	B2d	MS4
R249F	Exploding Car	red	65-68	Acho	B2d	MS4
R348F	Giraffe Car	yellow	65	Acho	B2d	MS4
R342F	Car Transporter	grey	67-68	Acho	B3b	MP1

11. MADE FOR GENERAL EXPORT

No.	For Export	Dates	Coupling	Chassis	Wheels
R344E	Track-Cleaning Car	66-69	IIIc-IIId	B3b	MP1

12. MADE FOR NEW ZEALAND

The following wagons remained in production in the New Zealand factory during the Tri-ang Hornby years, having been previously made for Tri-ang Railways. Only the plastic bodies and roofs were made locally, the complete chassis being imported from Rovex. The dates indicate when they were available and not necessarily when they were made.

No.	New Zealand	Colour	Dates	Cpls	Chassis	Wheels
R10	Open Wagon	various	59-72	IIIc-IIId	A4a-A4b	SP1
R11	Goods Van	various	59-72	IIIc-IIId	A4a-A4b	SP1
R12	Petrol Tank - Shell BP	grey	59-72	IIIc-IIId	A4a-A4b	SP1
R13	Coal Truck	various	60-72	IIIc-IIId	A4a-A4b	SP1
R14	Fish Van	white	60-72	IIIc-IIId	A4a-A4b	SP1
R15	Milk Tank	white	59-72	IIIc-IIId	A4a-A4b	SP1
R16	Short Brake Van	red	59-72	IIIc-IIId	A3A?	DS4
R18	Cable Drum Wagon	grey	59-72	IIIc-IIId	A4a-A4b	SP1
R19	Flat Wagon & Tarpaulin	grey	59-72	IIIc-IIId	A4a-A4b	SP1
NZ20	Flat Wagon & Timber	grey	60-72	IIIc-IIId	A4a-A4b	SP1
NZ21	Flat Wagon & Container	grey	60-72	IIIc-IIId	A4a-A4b	SP1
NZ22	Flat Wagon & Cement	grey	60-72	IIIc-IIId	A4a-A4b	SP1
NZ23	Flat Wagon & Log Load	grey	60-72	IIIc-IIId	A4a-A4b	SP1
NZ24	Flat Wagon & AA Gun	grey	60-72	IIIc-IIId	A4a-A4b	SP1
NZ25	Refrigerator Van	white	60-72	IIIc-IIId	A4a-A4b	SP1
R210	Tank Wagon - Shell BP Fuel	black	59-72	IIIc-IIId	A4a-A4b	SP1

No.	New Zealand	Colour	Dates	Cpls	Chassis	Wheels
R211	Tank Wagon - Shell Lubricating Oil	yellow	59-72	IIIc-IIId	A4a-A4b	SP1
R114	Short Box Car	yell, ornge	56-72	IIIc-IIId	B3a	MP1
R115	Caboose	red/mrn	56-72	IIIc-IIId	B3a	MP1
R116	Gondola	blu/grn/mrn	56-72	IIIc-IIId	B3a	MP1

13. BOXED IN NEW ZEALAND

In 1971, and for several more years, British-made stock was boxed up in New Zealand packaging for sale locally. These may have included the following that were sent out to New Zealand for set assembly:

No.	Boxed in New Zealand	Colour
R563	Bogie Bolster with Vans	salmon
R565	Bogie Bolster with Containers	salmon
R349	Murgatroyd Bogie Tank Wagon	white
R3490	Polysar Bogie Tank Wagon	blue & white
R117	Shell Oil Tanker	blue

DIRECTORY

Wagons are a huge subject in their own right and consideration has been given as to how best to present them in a directory so that they are easy to trace. At the same time I was anxious to ensure that there was a clear link between the listing in this volume and that provided in **Volume 1**. In the belief that most users will be unsure of the wagon's 'R' number, listing in numerical order was felt not to be ideal. Instead, the models are laid out in approximate chronological order as they were in the last volume but grouped together where they share a common body tool. This will place the photographs of models of similar appearance together and thus make it easier to find what the reader is looking for.

The directory is also presented in three sections to make a distinction not only between the British outline models and those of the Transcontinental series but also to distinguish those models specially produced for starter sets. The three sections are therefore :

     A. British Outline Wagons
     B. Starter Set Wagons
     C. Transcontinental Freight Cars

Battle Space
Battle Space wagons have been grouped together in a section of their own at the end of this chapter together with locos, carriages and Battle Space sets etc.

Australian Wagons
A number of British outline models were made at the Moldex factory to be sold in Australia, alongside British-made models. Most Australian production was during the Tri-ang Railways period but it did continue until 1968. These have been listed below under the Margate equivalent. What models were still being made before the production lines closed

is not known but it is most likely to have been those used for sets. Unfortunately dates and chassis details are also unavailable.

As mentioned previously, Moldex also bought partly-finished models from Rovex during the Tri-ang Railways period. It is not recorded as to which Moldex-finished models used Australian Tri-ang Hornby boxes. Those finished models bulk packed to Australia and reboxed by Moldex in their own Tri-ang Hornby boxes for a brief period are included. The only way to be sure which models were released separately boxed and sold to Australian stores is to find examples of them.

The ACT container was made for a set being assembled in Australia in 1971 and may be the only wagon to have been designed and made at Margate, during the Tri-ang Hornby period, specifically for the Australian market. Two Transcontinental freight cars were to have been made in an Victorian Railways livery for use in the same set but are thought to have been abandoned as only 1,100 sets were ordered. Standard TC models are believed to have been sent instead. Australia wanted 400 Shell lubricating wagons in 1970 but they had been out of production at Margate since 1968 and the batch would have been too small to be economical. They had also wanted the R247 chlorine wagon in Shell livery and this was refused for the same reason.

New Zealand Wagons
Some wagons remained in production in the New Zealand factory during the Tri-ang Hornby period, having previously been made under the Tri-ang Railways name. Only the plastic bodies and roofs were made locally, the complete chassis being imported from Rovex.

New Zealand also purchased ready-made wagons unboxed from the UK, for set assembly, and some of these may have found their way into local packaging as solo models.

French Issues
These were standard Tri-ang Hornby wagons fitted with universal couplings for sale in France by Meccano Tri-ang. Alternative Hornby-Dublo couplings were provided with the model which was sold in a standard Tri-ang Hornby box but with blue end labels inscribed in French.

General Issue Transcontinental Freight Cars
With the exception of a few Transcontinental freight cars which continued to be listed in the British catalogue and models produced for specific overseas markets such as Canada, little is known about what was available to retailers overseas. It seems certain that, after the abandonment of a comprehensive Transcontinental system in the mid 1960s, many models continued to be available in the stores at Margate and could be supplied on request. Transcontinental models would turn up in stocks sent to overseas dealers five or six years after they had been withdrawn from the catalogue much to the delight of some Tri-ang enthusiasts who thought that they had missed a chance of obtaining them.

Canadian Freight Cars
The chief production of the Transcontinental models during this period was for Canada. The Canadian market was demanding authentic liveries in the early 1960s and by 1965 a range of models in Canadian National livery, including six freight cars, was available for sale in Canadian shops. Eventually nearly 50 freight cars appeared in this range; many of them reflecting the colourfulness of the North American railroads. Some, such as the CN oil tanker, were internationally available but most could only be bought in Canada. Some of the models were sent bulk packed to Lines Bros. (Canada) for use in sets assembled and boxed by them. None of the models were made in Canada.

The Canadian range was made between 1965 and 1973, the year the last Canadian catalogue was published. With the break-up of the Lines Group, Louis Marx Industries (Canada) Limited bought Lines Bros. interests in Canada. They initially decided to continue the range and even ordered a number of new models from Rovex which have been included here in order to complete the story. The company must have had second thoughts for, after 1973, no more Canadian catalogues were produced and it is assumed that no further models were ordered from Rovex.

Marx sold off their surplus stocks of Tri-ang Hornby models locally, but around 1980 a large stock of Canadian freight cars and other models was uncovered in Canada and initially offered for sale locally. Canadian collectors were able to buy models very cheaply at that time and what did not sell was shipped to the UK where it was sold by the Zodiac chain of toy shops.

Canadian Packaging
All of the freight cars which were sold solo would have been in some form of window box. Those sent out ready boxed from Margate up until 1972 would have been in standard Tri-ang Hornby window boxes; mostly the multi-lingual version of the later 1960s.

In 1970 the Canadian company, finding it had more loose freight cars than it needed for sets, decided to package some of them and, from January 1971, sold them as solo models.

For this purpose they had a bubble pack quite different from anything produced by any of the other companies in the group. Most of these packs had a printed bottom card giving details of 'Canada's Model Railway System'. The cards were numbered according to size – TR-1, TR-2, TR-3 and TR-4 and were stapled to a pale green plastic moulded plinth which held the wheels of the freight car. A strip of printed card, with the number of the model rubber

stamped onto it was wrapped like a belt around the plinth and a clear plastic dome was stapled over model and plinth.

Some models had no strip card and only a plain base card with the model number rubber stamped beneath. To make it possible to stack the packaged models in a shop, two strengthening strips moulded into the bubble fitted neatly into cut-out slots in the base card of the model above. Some of the last models packaged in Canada left the Canadian plant incomplete. I have found two examples of unopened bubble packs where the freight car is missing one set of wheels. This would not have happened at Margate!

Batches of boxed models sent out from Margate in 1973 (and possibly 1972) were in standard window boxes carrying the 'Hornby Railways' name.

Canadian Printing and Logos
Early batches of the models were detailed with an off-white paint that could be described as pale grey-green in colour but, by 1969, the paint used was a clean white. The CP Rail refrigerator, paper and box cars were lightly printed by a process other than heat printing

The Canadian National models mostly carried the 'wet noodle' logo while the Canadian Pacific models experienced a change of livery from 'Canadian Pacific' to 'CP Rail' during the life of the series. The wet noodle varied in size, the largest being found on the box and refrigerator cars. An intermediate size was provided on the gondola, hopper, cement and track-cleaning cars and a smaller version was used on the oil tanker and orange caboose. The smallest CN logo was saved for the maroon caboose, snow plough and container.

'M-T Express'
These freight cars appear to have been used in a cheap battery set thought to have been put together by Rowland G Hornby in Canada in 1967. The 'M-T'

presumably stood for Meccano Tri-ang. 2,500 of each car were sent out from Margate unprinted and received stickers on arrival. The stickers were inscribed 'M-T' (stylised 0.4" square and set on a yellow maple leaf) and a separate sticker (1.14" x 0.25") had the word 'EXPRESS'. It has been suggested that the labels were printed by Fasson of Toronto who supplied Tri-ang Railways banners for Canadian stores. For more information see B. Starter Sets (2. Miniville Sets) in the Directory of Sets.

American Freight Cars
When American Train & Track Corp. (ATT) persuaded Rovex to supply them with models for the American market they ordered 10 different wagons. Five of them were for use in beginners' sets and five were for solo sales as serious models. All of them were fitted with the standard American NMRA couplings (except perhaps the open wagon and fish van) and were supplied bulk packed for boxing in ATT's own pale grey packaging. Residue ATT stock was later bought and re-boxed by Model Power in blue boxes. This included some of the Transcontinental wagons with American road names.

Non-Printed Wagons
Non-printed wagons were used for low cost starter sets and, although mainly British outline models, are listed under B. Wagons Made for Starter Sets later in this chapter.

Between 1965 and 1971 over 1 million unprinted wagons in more than 30 different starter sets left the Margate factory. This included some sets made specially for overseas customers.

Many different varieties of standard British rolling stock were used in these sets and the only common denominator was the lack of any detail printed on their sides. This way the production cost was kept to the minimum and starter set prices could be kept lower than those of competitors.

Records of these wagons are very limited as many of the sets, in which they were used, did not appear in the general catalogue. The following sets are known to have had non-printed wagons but in many cases it is not possible to say what wagons each contained. Where known, the quantity of sets manufactured is given. The country receiving the sets is also given, where they resulted from a special order:

SET	DATES	PROD.	WAGONS USED
RS12	1969-70	6,100?	(Australia and Canada)
RS13	1969	8,900?	(Sweden and Australia)
RS18	1966-68	28,000	rocket launcher
RS19	1966-68	29,000	bogie bolster with 3 cars
RS20	1966-68	14,000	flat wagon & car, brk van
RS43	1962-65	100,000	open wagon, brake van
RS70	1964-67	313,000	open wagon, fish van
RS71	1964-65	23,000	open wagon, brake van
RS72	1965	2,400	bogie bolster & 3 cars, brake van
RS73	?	?	?
RS74	1965	2,500	2 wagons
RS75	1965	2,800	3 wagons
RS76	1965	2,650	3 wagons
RS77	1965	2,540	3 wagons
RS78	1965	2,370	2 wagons
RS82	1967	150	bogie bolster & 3 vans (Canada)
RS85	1967	24,800	open wagon, brake van
RS86	1968	104,000	mineral wagon, van
RS87	1968	35,000	brick wagon, flat wagon & car, container wagon
RS88	1968	82,000	bogie bolster & 3 cars, tank wagon, open wagon, drop side wagon
RS91	1968?	?	?
RS96	1968	?	?
RS97	1970	?	?
RS100	1969-71	11,000	?
W1	1968-70	36,000	2 open wagons, gds van
W2	1968-70	28,000	2 open wagons
W3	1968	33,000	?
W4	1969-70	57,000	open wagon, goods van
W10	1969-71	26,900	flat, Minic car, flat & orange container, red brick wagon
RS691	1972	44,000	open wagon, goods van
RS692	1972	34,000	open wagon, goods van, milk tank wagon

A. BRITISH OUTLINE

Many wagons had originally been introduced to the Tri-ang Railways system and survived into, and in some cases beyond, the Tri-ang Hornby era. Those that have already been described in detail in **Volume 1** are dealt with more briefly here.

The wagons are arranged in chronological order except that they have been grouped together where they are based on a common design as described above.

GOODSTRUCK

BR Open Wagon R10

Dating from 1949 when it was first tooled up by Pyramid Toys as part of the Trackmaster clockwork train set, the goods truck, or 'open wagon' as it was originally called, entered the Tri-ang Hornby era with a green body and numbered W1005. With the exception of slight chassis, wheel and coupling changes and occasional colour variations including platform grey, it remained in this form until 1971 when it was replaced by a Southern Railway open wagon. By June 1970, 2,680 of the original boxed wagons remained in the Finished Goods Store and

The British, New Zealand (1966/67) and Australian versions of the R10 goods truck as sold during the Tri-ang Hornby years. Colours in Australia and New Zealand varied but those sold in Britain tended to be green.

these were directed to the assembly of RS606 train sets with the idea that no stock should be left over at the end of the year.

During the six years it was available as a Tri-ang Hornby model, over 100,000 were sold as solo models and another 30,000 were sold in the RS24 Pick Up Goods set. Many more than this were produced in an unprinted version for starter sets and are described in the second section of this directory.

New Zealand Goods Truck R10NZ

The New Zealand version of the R10 Open Wagon and R13 Coal Truck, described in **Volume 1**, remained in production at Auckland during the Tri-ang Hornby years. It is not known whether R10 and R13 both used whatever colour plastic was in use at the time but the New Zealand open wagons were amongst the most colourful to be found on Tri-ang Hornby Railways.

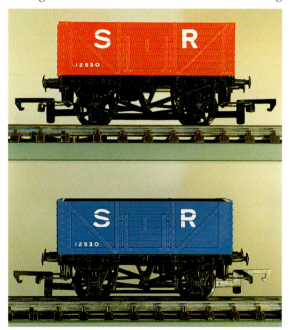

The red and blue versions of the 1971 R10A SR goods truck. Other colours probably belong to the Hornby Railways period.

The following is a list of some of the Tri-ang Railways and Tri-ang Hornby variations seen by collectors:

BODY COLOUR	NUMBER	PRINT COLOUR	CHASSIS	CPLING	WHEELS
brt red	W1005	white	A2d	IIIa/b	DS2
brt blue	W1005	white	A2d	IIIa	DS2
maroon	W1005?	white?	A2d?	IIIa/b	DS2?
pale gry	W1005	white	A2d	IIId	DS2
dk gry	W1005	white	A2d?	IIIa/b	DS2?
brt blue	W1005	white	A3a	IIIa	DS2
orange	W1005?	white?	A3a?	IIIa/b	DS2?
brt blue	W1005	black	A3a	IIIa	DS2
yellow	W1005	black	A3a	IIIb	DS2
pale gry	W1005?	black	A3a?	IIIa/b	DS2?
green	W1005	white	A4?	IIIc?	SP1?
green	W1005	black	A4a	IIIc	SP1
brt grn	W1005	yellow	A4b	IIIc	SP1
pale gry	W1005	black	A4a	IIIb	SP1
yellow	none	–	A5a	IIIc	SP2
green	none	–	A5?	IIIc?	SP2?

It is interesting to note that the only two open wagons that date from the Tri-ang Hornby period were sold without printing on them.

Australian Goods Truck R10A

During the Tri-ang Hornby period in Australia the colours most commonly made were maroon for the coal wagon and mid green for the open wagon, but bottle green and VR blue may also be found. The model had a metal chassis and was still available in 1968.

After this Moldex were supplied from the UK. About 20,000 were sent bulk packed to Australia for making into sets and some of these were probably boxed up in Moldex packaging and sold as solo models. Batches of unboxed models went out to Australia in 1968 (12,900), 1969 (1,000) and 1970 (7,000).

Converter Wagon R577

The wagon was produced for the sole purpose of enabling Hornby-Dublo operators to run Tri-ang locos and rolling stock in their trains. It formed a centre point of the amalgamation proposals which received so much publicity in 1965.

The wagon body was that of the standard open wagon but in black plastic. It used the discarded A3 wagon chassis modified to take a Hornby-Dublo coupling. In our classification, found in the 'Components' chapter, it is the A3c chassis and this was the only wagon to use it.

The first converter wagons were released in 1966 and the last year it was available was 1970. About 58,500 were made, the last batch being in 1969.

SR Open Wagon R10A

In 1970, as plans were made for the 1971 catalogue, it was decided to change the livery of four existing wagons to the livery of pre-war railway companies. The open wagon was to become an LMS vehicle while the cattle and drop door wagons were to be in Southern Railway livery. By May 1970 it had been decided to change the proposed printing for the open wagon to SR and that on the drop door wagon from SR to LMS. Each was to keep the original 'R' number but be given an 'A' suffix to distinguish it from the former issue to avoid confusion in the Dispatch Department.

At the time the company was concentrating on bright colours for its models and in keeping with this policy the SR open wagon was made of bright red plastic principally for use in train sets. A bright blue version was also available as a solo model but, as yet, the writer has not found the grey version illustrated on page 36 of the 1973 Hornby Railways catalogue and seen at the factory. Later ones were in a more accurate chocolate brown. From its first appearance in the shops in January 1971 and throughout its life the model carried the number 12530 and fine scale wheels. It was eventually replaced in 1975 by the Ocean private owner wagon.

It is not known how many were made of each colour or even how many were made in total but more than 22,000 were manufactured in 1971 alone and would have been sold in Tri-ang Hornby boxes; none being used in sets. About 42,000 were made for solo sales over the following two years and some of these would have also gone out in Tri-ang Hornby boxes, until all stocks of these cartons had been used up.

Open Wagon with No Markings
See B. Wagons Made for Starter Sets in this chapter below.

COAL TRUCK

The R13 Coal Truck, which consisted of an R10 Open Wagon with a moulded coal load, was dropped from the British range in 1962 but contin-

New Zealand and Australian R13 coal wagons. Other colours will be found especially in New Zealand.

ued in production in Australia and New Zealand into the Tri-ang Hornby period.

New Zealand Coal Wagon R13(NZ)
This remained in production until the end and has been covered in **Volume 1**. Late examples were probably non-printed.

Australian Coal Wagon R13A
The Australian-made open wagon was also used for a coal wagon which was normally russet red but did appear in other colours including green and blue. It was still available in 1968 and throughout had a metal chassis.

Although dropped from the range of models available in the UK a batch of coal wagons was made in Margate for Australia in 1968. 2,200 were sent out unboxed for use in the assembly of the R4FA set. Some of these were boxed up in Moldex window boxes for sale solo. Both grey and green versions have been found and the only part that was Australian made was the box. No further batches were sent.

GOODS VAN

BR Goods Van R11
Another model from pre-Tri-ang days, the Goods Van, had a shorter life than the goods truck, disappearing from the range in 1968 with the introduction of the ventilated van with sliding doors. The tool continued to be used for the Insulfish van and later for the BR parcels van.

The goods van had also entered the Tri-ang Hornby era with a green body and was numbered W8755, a guise it retained until its retirement. During these three years 46,000 wagons were sold, all of them solo as it was not used in a set.

New Zealand Goods Van R11(NZ)
The Auckland factory continued to manufacture

their own vans but we have no record of how many were made during the period of this book. Green, grey and white have been seen and the only late 60s example I have seen had no markings. Standard UK chassis were used.

New Zealand Refrigerator Van NZ25
Lines Bros. (NZ) Ltd continued to manufacture their own fish van but also the NZ25 Refrigerator Van which was all white with just W8755 and 10T on its sides or without any printing at all. The dates and quantities are unknown.

Australian Goods Van R11A
Australian goods vans were either VR blue or chocolate brown, in both cases with a white roof. White vans may also be found. The number carried by the van varied as did the colour in which it was printed. The following are some of the variations to be found :

VAN COLOUR	NUMBER	COLOUR
blue	W8759	white
white	W1005	black
chocolate	W 8759	yellow
blue	W8759	yellow
chocolate	M59015	yellow
blue	M59015	yellow
chocolate	M59015	white
dark green	M59015	yellow

It appears that both the blue and brown vans survived into the Tri-ang Hornby era, although it is thought likely that the blue model was dropped before the brown. All had metal chassis and the model was still available as late as 1968.

They were replaced by imported models in 1968 when Rovex shipped out 2,200 ventilated vans.

Goods Van with No Markings
See B. Wagons Made for Starter Sets in this chapter.

British and Australian R11 goods vans.

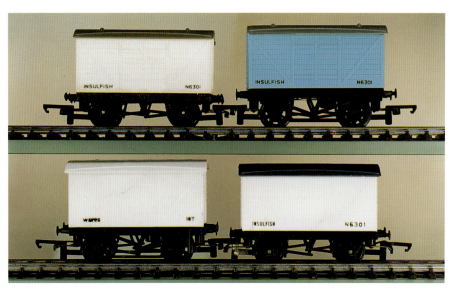

White and blue R14 Insulfish vans in the top row and a New Zealand refrigerator van and Australian Insulfish in the bottom row.

FISH VAN

BR Insulfish R14

The Insulfish van entered the new era in white but changed in 1966 to the pale blue livery that was in use on the real railways at the time. Initially the roof was grey but after the change to Hornby Railways a white roof was fitted and at the very end a slightly warmer shade of blue was used. The number it carried also changed from N6301 to N6307 at about this time. In June 1972 the Insulfish wagon was dropped in favour of a private owner fish van using the newly-introduced Hull & Barnsley van body.

About 150,000 of the blue Insulfish vans were made. Two thirds of these were sold solo and the rest were used in a variety of sets.

LNER Fish Van

As we have seen, at a meeting in February 1970 it was decided to re-livery four wagons, one in each of the liveries of the big four pre-war railway compa-

nies. The choice at that time for the LNER was the fish van but the model did not materialise as the coke wagon was chosen instead. In 1972, however, an LNER refrigerator van did arrive but made use of the newly-introduced Hull & Barnsley van body.

New Zealand Fish Van R14(NZ)

Lines Bros. (NZ) Ltd continued to manufacture their own fish van which is believed to have resembled the Australian Insulfish model. It should not be confused with the white R11 Goods Van and the NZ25 Refrigerator Van described above.

Australian Fish Van R14A

This was similar to the British wagon but had a blue roof and metal chassis and had been heat printed with a different tool. It was numbered N6301 and was available still in 1968. After the end of Australian production, 8,450 of the British-made model were sent unboxed to Australia for set assembly and it seems likely that some of these were sold

as solo models in Moldex boxes. 6,000 were sent out to Australia unboxed in 1968 and another batch of 2,450 was sent in 1970.

White and Pale Blue Vans with No Markings

See B. Wagons Made for Starter Sets in this chapter.

PARCELS VAN

BR Express Parcels Van R780

The original Trackmaster van of 1949 saw a new lease of life in 1971 when it was released in January as a BR Express Parcels Van in rail blue livery. This, however, was to be its swan song as, when the model was withdrawn at the end of 1973, the tool was retired.

The van had a dark grey roof but may be found with a lighter coloured roof from the Insulfish van or the white roof of the Hull & Barnsley van. It is questionable whether they left the factory in this state. Detail was heat printed onto the van in white

R780 Express Parcels van provided a late use for the old tool.

and it carried the number E12080. All should be fitted with fine-scale wheels. Nearly 30,000 were made including 2,500 used in the RS612 goods set in 1971.

SHORT BRAKE VAN

BR Brake Van R16

This was the original brake van tooled up by Rovex in 1953 and updated in 1957. It was to be made for three years under the Tri-ang Hornby label and throughout this time it remained brown and carried the number M73031. In 1968 it was replaced by a BR(ER) standard brake van which took the same 'R' number. As a Tri-ang Hornby model, 44,000 were sold as solo models and many more left the factory in sets.

New Zealand Brake Van R16(NZ)

The short brake van was also made in New Zealand where it stayed in production longer than its British or Australian counterparts. The model may be found in a variety of browns and maroons with grey or white roofs (with or without weather-strips) and with or without the M73031 running number. The

latest Tri-ang Hornby version found had a white roof and no printing, which once again supports the view that the New Zealand company may have abandoned printing their wagons at some stage.

Australian Brake Van R16A

The locally-made model was very similar to the British one, the plastic used being NSWR Tuscan red or maroon and the number 'M73031'. They were replaced from 1968 by the imported long wheelbase BR(ER) version of the R16 brake van.

Short Brake Van with No Markings

See B. Wagons Made for Starter Sets in this chapter.

BOLSTER WAGON

BR Bolster Wagon R17

The bolster wagon had a chequered history. It originated as a rather crude model in 1953. In 1962 it was given a new body, numbered 'B913011', and so it remained until 1966. That year a Minix car was added and the 'R' number was given a 'C' suffix to make it R17C.

Flat Wagon with Minix Car R17C

Only 8,000 R17 bolster wagons were ordered in 1965 and so, as we have seen, in order to make the model more interesting, the following year a Minix car was added and the code changed to R17C. This meant leaving the bolster pins out. The model was a success and 17,500 were ordered in the first year.

The model remained in this form until 1973 by which time about 76,000 had been made for solo sales. The colour of the wagon varied between red brown, orange brown (uncommon) and two shades of green. No one model of car was used. A further 16,700 were used in sets in the UK.

Bolster Wagon R676

As we have seen, Rovex decided to bring back the bolster wagon from January 1971, but, as the R17 number was in use on the above model, the bolster wagon was given a new number – R676. At the time R17 (the 'C' having been dropped) was using a brown-bodied wagon and so R676 was given a green body. It had bolster pins and fine-scale wheels. It may be found with or without 'B913011' on its sides and in one of two shades of green. The model also served as a match truck for the small mobile crane (R127). Over 21,000 were made between 1971 and its final year, 1974.

Australian Flat Wagon with Minix Car R17CA

The early grey crude-bodied flat wagon was made in Australia from a duplicate tool and it was this that was used for the Australian version of the R17C from 1966, with a Minix car. It was the only solo wagon to be shown in the 1968 Australian catalogue supplement and was one of three wagons featured the following year.

Flat Wagon with Minix Car (Australia) R17CA

One small batch of just 500 wagons was sent unboxed to Australia in 1970. These were probably fitted with a Minix car and sold as R17C (or more correctly R17CA if they were in Moldex boxes).

New Zealand Bolster Wagon R17(NZ)

New Zealand-made flat wagons without bolsters do sometimes turn up. These have the old body (the duplicate tools for which were sent out to Australia and New Zealand in 1960). These may have come from the unique New Zealand wagons with Minic loads as the R17 Bolster Wagon was not listed in New Zealand.

Bolster Wagon with No Markings

See B. Wagons Made for Starter Sets in this chapter.

FLAT WAGON WITH TARPAULINED LOAD

New Zealand Tarpaulin Wagon R19(NZ)

Production of this wagon had ceased in the UK by the end of 1959 but in New Zealand it remained available throughout the Tri-ang Hornby period. A non-printed version is likely to have been made in the late 1960s.

FLAT WAGON WITH TIMBER LOAD

New Zealand Timber Wagon NZ20

Made exclusively in New Zealand and described in **Volume 1**, the wagon remained in the New Zealand illustrated price list until its last edition in 1972. This has been found with a green load of timber instead of the usual yellow.

FLAT WAGON WITH MINIC CONTAINER

New Zealand Container Wagon NZ21

Made exclusively in New Zealand and described in **Volume 1**. The model remained available to the end.

FLAT WAGON WITH CEMENT LOAD

New Zealand Cement Wagon NZ22

Made exclusively in New Zealand and described in **Volume 1**. The model remained available until about 1972.

FLAT WAGON WITH LOG LOAD

New Zealand Log Wagon NZ23

Made exclusively in New Zealand and described in **Volume 1**. The model remained available through-

out the Tri-ang Hornby period. Since the publication of **Volume 1**, this model has been found with both light and dark green loads.

FLAT WAGON WITH AA GUN

New Zealand AA Gun Wagon NZ24

Made exclusively in New Zealand, described in **Volume 1** and retained in the range until 1972.

CABLE DRUM WAGON

BR Cable Drum Wagon R18

By 1965 the cable drum wagon was looking very different from the model that first appeared 12 years before. It used the 1962 bolster wagon body and carried 'closed' drums held in place by a rubber band. AEI, J&P or Pirelli transfers were used on the new

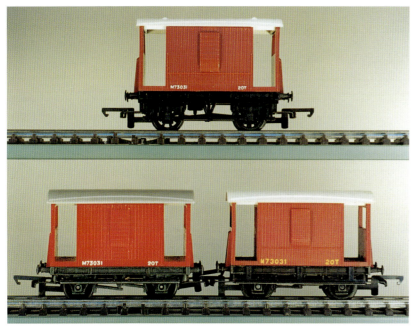

British (top), Australian and New Zealand R16 short brake vans of the Tri-ang Hornby period.

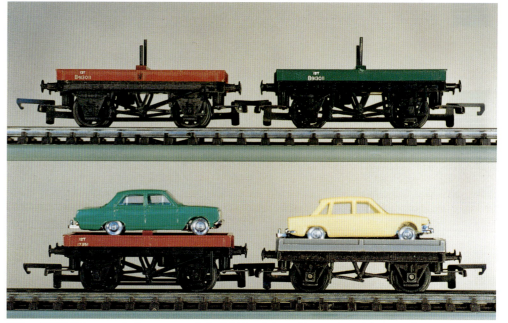

R17 and R676 flat wagons in the top row and the British and Australian versions of the R17C flat wagon with car below.

A New Zealand (top) and two Australian R18 cable drum wagons.

New Zealand wagons made during the Tri-ang Hornby period.

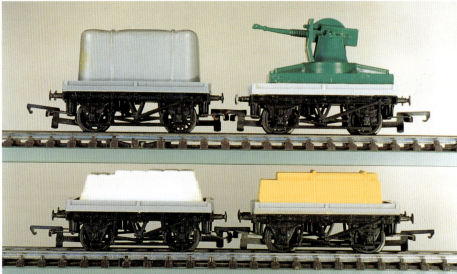

body but we can not be sure which were used during the Tri-ang Hornby period besides AEI which was illustrated in the catalogue. The cable drum mouldings could have been green or brown.

The wagon remained in the catalogue until 1972 when it became a victim of the rationalisation of models. It had been very popular with over 83,000 being packaged in Tri-ang Hornby boxes. None was used in sets during this period and the last batch was made in 1971.

New Zealand Cable Drum Wagon R18(NZ)

Production in Auckland of the former Tri-ang Railways model continued through the late 1960s with the R18 featuring in the 1972 New Zealand illustrated price list. The wagon carried the number 'M59015' and '10T' in black and Pirelli and J&P

Cables versions have been seen.

The wagon and the drums have both been found without printing and transfers and it is thought likely that these belong to the late Tri-ang Hornby period. Although shades of green were normally used for the drums, they do sometimes turn up in grey and even a silver grey body has been seen.

Unlike the British version of the wagon, in New Zealand the body and drums were not retooled and so the original body with plug-in drums remained until the end.

Australian Cable Drum Wagon R18

The Australian version of this wagon remained in production into Tri-ang Hornby days but had been replaced by the Margate version before 1968. Different shades of green can be found among the

Australian cable drums. 'Liverpool Cables', 'Pirelli General' and 'J&P Cables' may be found. There is no record of any being sent from Margate to Australia, unboxed.

FLAT WAGON WITH CONTAINER

Tri-ang Container Wagon R561

This attractive wagon remained in the catalogue until 1971 but was included the final year only to clear the remaining stocks from the stores. The last batch had been made in 1969. With the exception of an unprinted version (R561NP) which is dealt with in a later section of this chapter, the model changed very little. Variations in the number carried on the container were covered in **Volume 1** and it is not known which if any of those changes occurred dur-

Boxed New Zealand container wagon.

ing the period covered by this volume. We do, however, know that 136,000 were made during the Tri-ang Hornby period, 48,000 of them being sold as solo models.

Australian Tri-ang Container Wagon R561(A)
2,600 were bulk packed to Australia between 1968 and 1970, some of which may have been boxed up and sold solo in Moldex packaging.

Lyons Maid Container R?
The Lyons Maid container was white with just a Lyons Maid Ice Cream transfer on each side. The transfer was the same one used on the first series station buildings. The model is believed to have appeared in a set in the late 1960s. It is a very rare model.

Kellogg's Rice Krispies Container R574
The Kellogg's Rice Krispies container used the standard Tri-ang Toys & Pedigree Prams blue black container to which specially-printed stickers were attached in place of the normal ones. The number

carried by the container at the time was BK8900. It is understood that the model was used in a special set which was given away as prizes in a competition held in 1968. Just 20 were available to win. The set has been described as having contained a mixture of Tri-ang Hornby and Minic Motorways parts. The locomotive was a B12.

Some of the Rice Krispies containers appear to have a white stripe in the design but this was due to inaccurate cutting of the stickers leaving a white unprinted strip at the side. This is a very rare model.

Flat Wagon and Container with No Markings
See B. Wagons Made for Starter Sets in this chapter.

BOGIE BOLSTER WAGON

BR Bogie Bolster Wagon R110
The bogie bolster wagon had also changed substantially in 1961 and was now a very attractive model based on a 30 ton Bogie Bolster C. In 1965 the model was grey and carried the number B940052. The fol-

R561 Tri-ang Toys container wagon. A black container on the scarce orange brown wagon (most wagons were maroon).

The very rare Lyon's Maid and Kelloggs Rice Krispies (R574) container wagons.

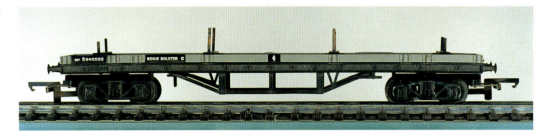

Grey R110 bogie bolster wagon.

R563 bogie bolster wagon with 3 Minix vans, R569 bogie bolster wagon with cars and R579 bogie bolster wagon with CTI containers.

lowing year was to be its last in this form. About 8,000 were released during this time.

After this the body was used as a base for various loads. *See below.*

Canadian Bogie Bolster Wagons R110(C)

It has been reported that the salmon-coloured bogie bolster was sold in Canada without buffers. There are no records of any special batches having been made for Canada and it is quite possible that standard models were sent and the buffers removed on arrival. Likewise, it is possible that those found had the buffers removed by the purchaser.

Bogie Bolster with No Markings

See B. Wagons Made for Starter Sets in this chapter.

BOGIE BOLSTER WAGON & THREE VANS

BR Bogie Bolster with Vans R563

This model first appeared in 1967 using the standard salmon-coloured bogie bolster wagon without bolster pins. A version was made for starter sets with the number missing from the side of the wagon. The van was the Ford Thames from the Minix range and a mixture of colours may be found. The most common were red, white and blue (one of each). 51,400 were sold as boxed solo models. A further 16,000 were sold in the RS89 Rail Freight train set. The wagon was not available after 1972.

New Zealand Bogie Bolster with Vans R563(NZ)

New Zealand received 5,150 bogie bolster wagons with vans bulk packed for use in the NE6 train set which contained two of these wagons. The wagons went out in small batches over a period of five years between 1968 and 1972.

Canadian Bogie Bolster with Vans R563(C)

In 1968, Canada received a batch of 1,000 bogie bolster wagons with three Ford vans and a further batch of 800 in 1970. In both cases they were sent unboxed.

BOGIE BOLSTER WAGON WITH TWO CONTAINERS

BR Bogie Bolster with Freightliner Containers R565

Between 1967 and 1969, 16,000 of this wagon were made up for use in the RS89 set. It too used the salmon-coloured bogie bolster wagon without bolster pins but instead of three vans it carried two Freightliner containers. The wagon was not officially sold in Britain as a solo model.

New Zealand Bogie Bolster with Containers R565(NZ)

Some 1,900 of the model were bulk packed to New Zealand for use in assembling the NE6 Freightmaster set.

Bogie Bolster with cti Containers R579

The wagon was a very pale grey with black printing and carried two standard red 20' cti containers. As far as I know, it was not sold solo and I do not know in which set or sets it was used.

BOGIE BOLSTER WAGON & THREE CARS

BR Bogie Bolster with Cars R569

This was another wagon assembled for use in sets in 1968. It was the same as R563 but with cars instead of vans. It is not known in what starter sets it was used but it was illustrated in the 1968 Australian catalogue supplement in their RS88A Clockwork Train Set.

Canadian Bogie Bolster with Cars R569CN

This was listed as being available to Canada in 1965 but no production or dispatch records have been found.

Bogie Bolster Wagon & 3 Cars (Australia) R569A

5,300 were bulk packed to Moldex Ltd in 1968 and it is likely that some of these ended up in Moldex boxes being sold as solo models. It was featured in the 1969/70 Australian catalogue supplement as a solo model.

BOGIE BOLSTER WAGON & STEEL LOAD

Canadian Bogie Bolster with Steel Load R1210

The only batch of these cars was made in the second quarter of 1971 when 1,770 were assembled. Most of these were sent to Canada that year, the remainder following in 1972.

The model consisted of a standard bogie bolster wagon in light grey on a black chassis with the metal pin units reversed to form retaining hoops to hold in place a load consisting of 12 lengths of System 6 rail from the R600 straight track. The lengths of rail had

Maroon BR R112 drop door and R113 drop side wagons.

Three LMS versions (R112A) of the drop door wagon.

R12 BP tank wagon.

British and Australian R15 milk tanks.

New Zealand and Australian R210 black tank wagons.

Four versions of the British Shell R12 tank wagon and two overseas versions: row 1 - small transfers and large transfers; row 2 - dark stickers on wrongly numbered wagon and stickers on a grey bodied tank; row 3 - New Zealand and Australian wagons.

not been notched for fishplates but were taken from the track production line before this stage in the process. It had been planned to use rail left over from TT production.

The model illustrated in the 1971 Canadian catalogue had no printing on the wagon sides and still had its buffers, but, in fact, the models delivered to Canada carried the standard printing used on the British bogie bolster wagon.

BOGIE BOLSTER WAGON & RAIL LOAD

BR Salmon Bogie Rail Wagon R23

This was planned for the British range for January 1972 and will be covered in **Volume 3**.

DROP-DOOR WAGON

BR Drop-Door Wagon R112

The drop-door wagon entered the Tri-ang Hornby period with a maroon body and numbered M2313. It remained in this colour until 1970 and the last remaining 2,950 boxed wagons were used up in the assembly of the RS606 train set for Christmas that year. In 1971 it was replaced by an LMS liveried version numbered R112A. Between 1965 and 1970 in excess of 57,000 were made of the BR wagon.

LMS Drop-Door Wagon R112A

As we have seen, the R112 drop-door open wagon was changed to bright red with LMS on its sides in 1971. It was originally intended that this should be printed with 'SR' but by May 1970 it had been changed to 'LMS'. This was to be the wagon's last guise and as such it survived for only two years. It was made in bright red, orange brown and maroon and carried the number 12527 and LMS heat printed in white. From January 1971, when it first appeared in the shops, it had fine-scale wheels fitted.

It would seem that 17,000 wagons were sold, all separately boxed as solo models and at least 11,000

of them on Tri-ang Hornby boxes. Although the photographs on pages 5, 6 and 7 of the 1973 catalogue depict an SR drop-door wagon in sets, this was probably only a test model. These sets were in fact issued with red SR open wagons without drop doors as the accompanying inscriptions imply.

It also seems that thought was given to replacing R112 with a five-plank wagon with drop doors. Page 26 of the 1973 catalogue shows two versions of a five-plank wagon and a lime wagon with what appear to be doors that open. These were of course pre-production models.

DROP-SIDE WAGON

BR Drop-Side Wagon R113

Although based on an LMS wagon, it seems that there were thoughts of bringing it out in GWR livery during the early 1970s craze for pre-nationalisation liveries. This, however, did not happen and instead it was dropped from the range in 1974.

It had remained a maroon BR wagon throughout the Tri-ang Hornby years during which time 82,000 were made as solo models. This however was not the sum total made during those seven years as it featured in one of the most successful sets manufactured by Rovex – the RS51/RS651 Freightmaster set. This and the RS612 set accounted for just under 100,000 drop-side wagons.

Australian Drop-Side Wagon R113A

Between 1968 and 1970, another 10,000 wagons were bulk packed and sent out to Australia where they were used to make sets in the Moldex factory but it is likely that some of these were boxed up in Moldex packaging and sold solo.

Drop-Side Wagon with No Markings

See B. Wagons Made for Starter Sets in this chapter.

PETROL TANK WAGON

Shell and BP Tank Wagons R12

The petrol tank wagon remained in production throughout the Tri-ang Hornby period. While previously the wagon had carried the Shell name on one side and BP on the other, from 1961, when the companies separated, the wagons carried either one name or the other. Whether it was a Shell or a BP tank, it had the same 'R' number.

British and Australian R211 yellow tank wagons.

R118 green well wagon of the Tri-ang Hornby period.

The BP version seems to have remained consistent. It carried the number 5057 heat printed in black on the tank sides and had two BP emblems transferred onto each side of the silver tank.

The Shell tank on the other hand usually carried the number 5056 but could be found with the number from the BP tank instead. It too had a pair of Shell emblems on each side of the tank but these could be either transfers or on gummed paper. The latter were of two distinct types distinguishable by the shade of red used. One was a dull, almost brick red while the other was a bright red. The tank colour also varied with the Shell version, being either silver or grey. The variations described above do not seem to follow a pattern but appear to have changed at random. The later tank wagons mostly had unpainted solebars.

In 1972 the model had been dropped from the catalogue not because it was unpopular, because it sold well, but because the Monoblock tank wagon had been designed to replace it. R12 was not used in any standard sets in Britain during this period, although 900 were sent to Australia for sets in 1970. Over 100,000 solo models were sold during the Tri-ang Hornby era.

New Zealand Shell Tank Wagon R12NZ

This is covered in **Volume 1** but continued in production in New Zealand using the large Minic Shell transfers. Pale grey plastic was used for the body but at times it may also have been made in silver.

Australian Shell Tank Wagon R12A

In the early 1960s the heat-printed tank wagon had been dropped to be replaced by one with either one central or a pair of square Shell stickers or transfers on each side. It could also be found fitted with the milk tank ladders with either one offset sticker or a pair of stickers per side.

An unusual version of this model has a single large Shell logo in the centre of each side, the transfer used being that from the earlier TC oil tankers. The period when this was released is not known.

By 1968 the Australian model had been replaced with the British-made version but only one batch was sent unboxed for use in sets. This went out in 1970 and consisted of 900 models some of which may have been sold solo in Moldex boxes.

MILK TANK WAGON

UD Tank Wagon R15

The milk tank wagon was a very popular model and of all the early tank wagons it was to survive the longest in its original livery. The plain 'UD' version was to have been dropped in June 1972 but it remained in the range until 1973. After this the letters were given a black outline making the wagon look similar to the models of the early 1950s. It was not until 1976 that the design was changed completely to carry the name 'United Dairies'.

During the seven years, 214,000 were sold, half of them in sets.

New Zealand UD Tank Wagon R15NZ

The R15 Milk Tank Wagon was made in the Auckland factory throughout the Tri-ang Hornby period, although its original 'UD' transfers may not have been used in later years.

Australian UD Tank Wagon R15A

This appears not to have changed much. It was similar to the British-made version but had a metal chassis throughout its life. It was eventually replaced by a supply from Margate. Over 7,000 were exported unboxed to Australia between 1968 and 1970 some of which probably ended up as solo models in Moldex boxes.

White Tank Wagon with No Markings

See B. Wagons Made for Starter Sets in this chapter.

FUEL OIL TANK WAGON

Shell Fuel Oil R210

About 7,000 black tank wagons were made in 1965 and almost all had left the factory by the end of the year. Models from this final batch would have still had 'Tri-ang' on the chassis (A4a), IIIc couplings, SP1 wheels and may or may not have had painted solebars.

New Zealand Shell Black Tank Wagon R210NZ

A black Shell tank wagon survived into the Tri-ang Hornby era. This had the early tank wagon body, abandoned in the UK in 1957, and used the large 'SHELL' transfers from the Minic Push-and-Go range. In New Zealand imported plastic chassis were used. In later years it seems likely that the wagon was sold without transfers.

Australian Shell Black Tank Wagon R210A

The black Shell tank wagon with single central square transfers continued to be available in Australia until 1967. All had metal chassis and the early tank body. Some were issued with the UD wagon ladder.

LUBRICATING OIL TANK WAGON

Shell Lubricating Oil R211

The yellow tank wagon survived until the end of 1968, the last year of manufacture, and almost 48,000 were made over the period. These later models were made in a paler shade of yellow and the latest of all had A4b chassis.

New Zealand Shell Tank Wagon R211NZ

A yellow version of the early bodied tank wagon was made in New Zealand and was available from 1959 until 1972.

Australian Shell Tank Wagon R211A

The yellow Shell tank wagon remained available from Moldex after the black version had been dropped and the silver one had been replaced by imported ones from the UK. It had a small square Shell logo applied either as a transfer or a sticker. All the wagons had metal chassis and the early tank body. It survived at least until 1967 and during this time various shades of yellow were used for the body.

Shell Lubricating Oil (Australia) R211A

In 1968 a batch of 6,000 was sent out to Australia from Margate for use in sets but almost certainly some of these were sold solo in Moldex boxes.

Indeed, it was illustrated in the 1969/70 Australian catalogue supplement as a solo wagon. The 'A' suffix in this case suggests Australian packaging. A further order of 400 wagons was placed in 1970 but these were not sent as it was not economical to make such a small number.

Yellow Tank Wagon with No Markings

See B. Wagons Made for Starter Sets in this chapter.

WELL WAGON

BR Well Wagon R118

The well wagon had gone through many colour changes in its early years but, like the open wagon and van, by 1965 had settled down to green. It was not to survive long after this. It did not appear in the 1968 catalogue and stocks at the factory had been

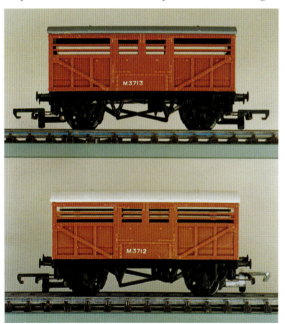

R122 cattle wagons of the mid and late '60s with both number variations including the white roofed version.

R123 late horse boxes including the white roofed version with glazed windows.

R124 BR Toad made up until 1970 and the R124A GWR version of 1971.

cleared before the end of 1967. 13,000 were made during this period, the last ones probably in 1967.

The body mould was also used that year for the TC crane car (R560), Battle Space assault tank transporter (R568) and tank recovery wagon (R631) but saw little use after that.

CATTLE / SHEEP WAGON

BR Cattle Wagon R122

The stock wagon entered service on the Tri-ang Railways system in 1955 and remained in production in some form or other until the end of 1994. With a life-span of 40 years it was Rovex's longest surviving model. Unlike most of the early wagons it underwent few changes in the 1950s and 1960s but was to blossom in the 80s and 90s when it was released in SR, LMS, GWR and private ownership liveries. Throughout the period covered by this volume it remained unchanged with its orange brown body, grey roof and the number M3713 (although W6297, from the WR brake van, may be found on some post 1968 examples of the model).

50,000 solo models were made over the seven-year period and a further 96,000 were used in the R51/R651 Freightmaster set. Together with those sent bulk packed to Australia, some 150,000 were made during the Tri-ang Hornby years.

Australian Cattle Wagon R122A

Some of the 3,900 bulk-packed models sent out to Australia between 1968 and 1970 are likely to have been boxed up in Moldex boxes and sold as solo models in Australia.

HORSE BOX

BR Horse Box R123

Until 1971 this model underwent only minor changes. By 1965 it had a C3 chassis, pin-point axles, IIIc couplings, maroon body, grey roof and B547 in small yellow characters heat printed onto the sides. The only significant change it saw in the next seven years was when it was briefly given a white roof possibly for the last batch or two. These seem very hard to find and could have resulted from a post sales roof switch but I think that this does not account for them all. The only examples I have seen have been very late production models. One of the examples in my own collection also has the windows 'glazed' and it is not clear whether this was the form in which it left the factory or if the glazing has been fitted by a previous owner. I suspect the latter.

The last batch was made in 1971 when about 1,000 came off the production line but in all 130,000 models were made over the Tri-ang Hornby period of which 96,000 were used in sets. There were thoughts of issuing the model redecorated and marked with 'Newmarket' or 'Return to Epsom' for 1972 but instead it was given a GWR livery and released as R123A.

Australian Horse Box R123A

In 1968, 2,000 of the standard BR horse box (R123) were bulk packed to Australia and some of these were sold in Moldex boxes.

Converter Horse Box R578

Following the introduction of the converter wagon based on the open wagon, it was soon realised that it would not be suitable for joining together Hornby-Dublo and Tri-ang coaches in passenger trains and so in 1967 the R578 Converter Horse Box was introduced. This had the normal maroon plastic body carrying the usual B547. For the chassis the discarded C1 chassis tool was found in the tool store and cleaned up and converted to take a Hornby-Dublo coupling at one end. This we call the C1c in our classification and no other wagon used it. It was made for the first two years only during which time 25,000 were produced. There were sufficient stocks after this for it to remain in the catalogue in 1969 after which it was dropped.

WR BRAKE VAN

BR Livery R124

The third of the trio of long wheelbase wagons introduced at the end of 1955 was the Toad or ex-GWR brake van. This had also retained its original orange-brown livery and continued to do so until 1970 after which it changed to a pseudo-GWR livery and took the number R124A. The latter is dealt with below.

Over 65,000 of the brown WR brake van were made during the period, none of them being used in sets in the UK. In June 1970, with the GWR version planned for the following year, it was decided to use up the remaining stock of 5,000 in the RS606 train set which was to go on sale that Christmas. For this the

R215 grain wagon with and without mechanism.

Below: The R347 engineer's wagon was available in dark green instead of black from 1966.

Right: R219 brick wagon showing the original BR version made up until 1967 and the London Brick version made from 1970.

brake vans and other stock were packed in their window boxes inside a set box. The model was to have been used in the range of train sets made for Australia in 1971 but had to be replaced by R16 as stocks of the BR Toad ran out.

Australian WR Brake Van R124(A)

1,250 were sent unboxed to Australia in 1970 for use in sets and some of these may have been sold solo in Moldex boxes. It had been intended to send the GWR version to Australia in 1971 but there is no evidence of any having been bulk packed. This was the last year any were made.

GWR Brake Van R124A

The R124 Toad had a new livery in January 1971, and fine-scale wheels. The body from this date was platform grey and carried 'GWR' and the number 'GW57740' heat printed in white on its sides. Initially it retained the plain white roof with a chimney at the train end but, with the introduction of the LMS brake van in 1974, a roof with rain strips and a chimney in the centre was tooled up and used on

both models. While the rain strips were authentic enough, the central chimney spoilt the look of the WR brake. In 1976 the model was re-launched with a new 'R' number, a new body and later still an authentic GWR livery.

The model was withdrawn from the range in 1992, making it one of the longest-surviving models. While we do not know how many were made in GWR livery we do know that over 8,000 were made in 1971 and sold solo in Tri-ang Hornby boxes.

GRAIN WAGON

BR Grain Wagon R215

This should not be confused with the so-called 'whisky' wagon which replaced it and was a larger model. R215 was the earlier model introduced by Rovex in 1958 and had not changed after it had settled down with a grey body in 1960. In this form it lasted until the end of 1968 with the exception of losing its gravity unloading mechanism for the last two years. 18,000 were made without the mechanism compared with only 12,000 made during the two

previous years when an unloading mechanism was still included.

ORE WAGON

Engineer's Wagon R347

This was the lower half of the grain wagon. From 1966 the body colour of this model changed from black to dark green. The model was made in this colour for just three years (1966-1968) during which time 19,000 passed into the stores. 12,200 of these were made in 1966 after which the unloading mechanism was deleted.

BRICK WAGON

BR Brick Wagon R219

The brick wagon was modelled on a Great Northern Railway wagon built to carry bricks from the London Brick Company's brick works at Fletton near Peterborough in Cambridgeshire to London. It was one of a batch of only 25 built by the Leeds Forge Co. Ltd, each designed to carry 50 tons of bricks. The LNER later increased the number to 50. No. 451004 was one of the batch made by the GNR and would have had oval buffers.

Despite its small numbers on the real railways it was modelled by all the leading British model railway manufacturers and some of the minor ones too. They all wanted a bogie open wagon and found the choice very limited. Of all the models made, the Tri-ang version was supreme with its well moulded body and, in the early years, its brick load.

Although 1967 was the last year the BR version of this attractive wagon was listed, the last batch was made the year before when about 7,000 passed through the stores. In all 14,000 were made in 1965 and 1966. There are no variations worthy of note.

London Brick Company Wagon R219

While the wagon did not run in private ownership, in 1970 Rovex decided to bring out the model in London Brick Company livery. The new wagon was the bright red of London Brick Company lorries with the detail heat printed in white. A brown version also exists but is not common. It seems a shame that the authentic LNER bauxite livery was not chosen instead.

The model was dropped after 1973 but not until 21,000 solo models had passed through the stores and a further 25,000 had been used in train sets RS601 and RS613.

CP Rail Brick Wagon R2193

The model was numbered 'CP 342826' in black with other coded detail on a bright red body which also carried 'CP Rail' in white and the CP Rail logo in black and white at one end. The red was the same colour as that used on the London Brick Company wagon in the British series and may well have been made at the same time. The underframe was painted flat black.

Only 1,200 were made in the second quarter of 1971 and these were sent to Canada in 1971 and 1972. All are thought to have been in Tri-ang Hornby boxes.

M-T Express Brick Wagon R2190 (R219NP)
See B. Wagons Made for Starter Sets.

Victorian Railways Brick Wagon R2194 (R2190)

This was planned in 1971 for the RS103 train set for Australia to be released the following year. It was to be a red wagon with 'an Australasian road name'. Port Kembla was suggested as a name for it but a mock-up of a brown brick wagon marked with a large 'VR' in white was made and is now in a private collection. It was also to have been boxed up as a solo wagon for sale in Australia and South Africa and offered bulk packed to New Zealand but none was made.

Brick Wagon with No Markings
See B. Wagons Made for Starter Sets in this chapter.

MINERAL WAGON

BR Mineral Wagon R243

The mineral wagon was to remain in the range until 1978. Although too short, it was a very good model. It survived virtually unchanged and the principal differences one notices, after eliminating those of the chassis, concern the shade of grey plastic used. Over its 19 year life the model was made in a wide range of greys according to what was in use in the factory at any time.

During the seven years covered by this volume, 132,000 mineral wagons were produced; 45,000 of them for use in train sets. It is interesting to note that none were ordered for sets made overseas.

Mineral Wagon with No Markings
See B. Wagons Made for Starter Sets in this chapter.

Proposed R2194 Victorian Railways brick wagon of 1971.

R2193 CPRail brick wagon catalogue illustration.

TRESTROL

BR Trestrol Wagon R242

A batch of 1,300 was made in 1965 and the next year the model was withdrawn. It remained out of production until returned with a girder load (*see below*). It seems likely that the version made in 1965 was either platform grey or pale greeny grey and was inscribed 'TRESTROL E C' at the left-hand end of the lower section and '55T B901600' at the right-hand end of the same section.

Trestrol with Girder R242

The model was first seen in the 1972 catalogue. It originally appeared in a Tri-ang Hornby box but will be covered in **Volume 3**.

Canadian Trestrol Wagon R242(C)

This was another wagon that is reputed to have been sold in Canada without buffers and again there is no record in the census of a batch having been specially made for Canada.

SMALL CRANE TRUCK

Crane Truck R127

The small crane of 1962 vintage was typical of many to be found throughout the country on signalling and civil engineering duties which required a crane of 5 to 10 tons capacity. During this period it remained red brown until 1971 when it changed to bright red to match the 75 ton crane introduced that year. 60,000 were made as Tri-ang Hornby models with 3,200 being produced in that final year. The model remained popular, providing a cheaper alternative to the larger wagon. In 1980 it was reliveried in yellow.

Crane Kit R385

A kit version of the wagon was sold in an assembly pack between 1966 and 1969. The parts came in a

Colour variations of the R243 BR mineral wagon as it appeared during the Tri-ang Hornby years.

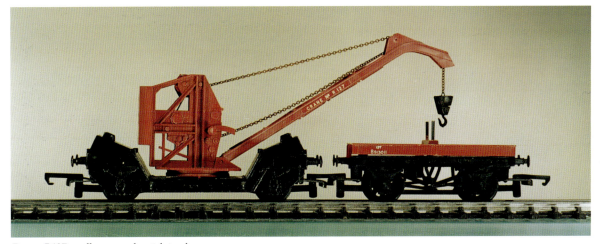

Brown R127 small crane and match truck.

pre-formed plastic tray inside a red cardboard box covered with cellophane.

The heavy metal chassis of the crane itself came with its wheels already fixed in place. Tapped holes were provided for the crane pivot and couplings. The match truck had spoked plastic wheels on pinpoint axles (SP1) and the body was brown to match the crane. The set came with an instruction sheet and was easy to assemble. About 13,000 were made and all were brown. The price was 15/9d.

Design not Adopted

This was not the only design of small crane considered by Rovex. One of their regular model makers made a white plastic model of a mobile crane on a shorter wheelbase. The model has survived in the ownership of its maker and appears not to have used any parts from the R127. It had a larger cabin and it is quite possible that, while it was a more attractive model, due to its small chassis it was hard to lower its centre of gravity thus making it less stable when in use.

BOGIE TANK WAGON

ICI Caustic Tank R247

The red ICI bogie caustic tank wagon was first introduced to the Tri-ang range in 1962. With the exception of the bogies and colour shades the wagon went through no changes until the late 1960s. In 1967 the last batch of 1,100 red tanks was made at Margate. This was despite the fact that it had been dropped from the catalogue the year before. As few as 4,000 may have been sold in Tri-ang Hornby boxes. There was a proposal to reintroduce the model in 1977 but this was not to be.

The pre-production model has survived intact having been kept by the model maker when he left the company. In 1995 it was bought by a collector. It is painted red and carries the ICI logo painted on.

Australian Caustic Tank R247(A)

Some of a batch of 1,600 red ICI tank wagons, bulk packed to Australia in 1970, were sold solo in Moldex boxes.

Blue ICI Tank Wagon R2470 (R247CN)

Collectors have always been fascinated by the blue version of this wagon and it is perhaps worth recording here what I know of it.

It seems likely that in 1968 Canada expressed an interest in selling the ICI tank wagon in blue livery as this would explain why they included it in their 1969 catalogue and ordered a batch of 2,500. For some reason the order was not dispatched although the wagons were made.

Early in 1969, both Canada and Australia wanted a Shell version of this same wagon and as both were prepared to take substantial quantities Rovex agreed to manufacture it. The Canadian company then changed its mind asking for Polysar tank wagons instead. This left the Australian order too small on its own and so in March 1970 they were sent 1,600 of the blue ICI caustic tank wagons instead. The remaining wagons were probably released onto the British market, possibly unboxed through Beatties stores although a few may have leaked through to Canada. Canada received a Shell bogie tank wagon in 1970 but it was the Transcontinental model.

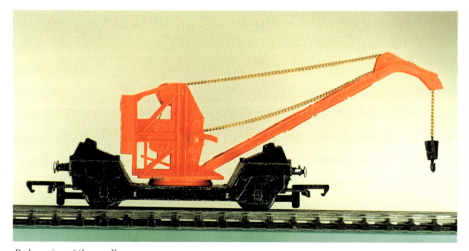

Red version of the small crane.

R385 small crane assembly pack.

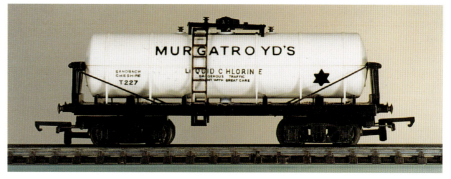

We also know that 200 of the blue ICI tankers were dispatched abroad in 1973 and Canada seems to have been a likely place for these. A further 100 were released in Britain that year and 300 remained in store at the year end.

Murgatroyd's Chlorine Tank R349

This was the white bogie tank wagon which dates from 1963. It outlived the red ICI tanker, surviving until 1970. No significant variations have been noticed although one with a buff upper half (supposedly not as a result of exposure to light) has been found. 20,000 were made during the Tri-ang Hornby era and all were sold solo.

Canadian Murgatroyd's R349(C)

The Canadian company requested their wagon to be in Polysar livery instead but even so 1,000

Above: Red and blue R247 ICI tank wagons.

Top right: R349 Murgatroyd's Chlorine tank wagon.

Right middle: An unusual variety of Murgatroyd's tank wagon.

Right: R3490 Polysar Industrial tank car for Canada.

Murgatroyd's Chlorine tank wagons were bulk packed to Canada in 1968.

New Zealand R349(NZ)

About 1,000 went unboxed to New Zealand in 1969/70.

Polysar Industrial Tank Car R3490

As we have seen above, this was supplied to Canada as a replacement for the blue ICI version and instead of the proposed Shell version of this model which the Canadian company had cancelled before the production stage. It used the same tools as the British bogie tank wagons but had a white moulded body heat printed in blue, including the whole of the central part of the tank. The chassis was made of the bright blue plastic used for the starter set wagons known as rainbow blue but an example may be found with a BR blue chassis. The ladders and platform were black.

The Polysar Rubber Corporation was, and is, one of the world's major producers of latex rubber but now forms part of Bayer AG of Germany. Although an American company, it has a major production facility in Sarnia, Ontario, an area well known for its petrochemical industry.

The Polysar bogie tank wagon was first illustrated in the 1970 Canadian catalogue but was not shown as a new model possibly because R247 had featured in the catalogue the previous year. It was first ready for shipping in May 1970 and it remained in the range to the end, appearing in the 1973 catalogue. About 6,000 were made. It was also used in Australian sets and was to have been offered boxed solo to Australia and South Africa in 1972. However this appears not to have happened.

New Zealand Polysar R3490(NZ)

It seems that about 600 Polysar wagons were sent out, bulk packed, to New Zealand in 1972. These may have been sold in New Zealand packaging.

CONFLAT L

BR 3 Container Wagon R340

The Conflat L with three type L containers was another 1962 introduction. It remained in the range until the end of 1972 and changed little. It was to have been withdrawn at the end of 1970 but a reprieve was granted. The most significant variations were in the shade of plastic used especially for the grey containers. The wagon itself remained red brown but at least one batch was produced in the orange brown of the cattle wagon. 158,000 were made during this period, three fifths of them being used in sets.

Australian 3 Container Wagon R340A

Australia took delivery of a further 1,600 unboxed wagons in 1968 and 2,000 spread over 1969 and 1970. While these were supplied for set assembly, it is known that some of them were boxed up in Moldex packaging and sold individually on the Australian market.

TIERWAG CAR TRANSPORTER

Car-a-Belle R342

Tierwag was the code name for an early design of car transporter for the Motorail train traffic. It weighed 12.5 ton and only six were built for use on BR, although there may have been a further batch for continental working.

The model was a highly successful judging by the number sold. 70,000 went through the works in 1965 alone and a further 123,000 during the next six years. The Car-a-Belle train set accounted for about 80,000 of all those made. The model remained in the catalogue until 1974.

One variation to look for has the 'No Hump Shunting' printed to the right of the centre post instead of the left. This appears to have been an early example. Models were made in two different shades of plastic. The early colour was a pale greeny grey used for the grain and ore wagons, while the plastic of later models was a pale grey with a tinge of red. It may be found in a red and yellow wraparound window box.

The R340 Three Container Wagon showing both the common red-brown and rarer orange-brown bodies.

Prototypes of the containers in use.

French Car-a-Belle R342F

This was one of the few wagons included in the Hornby Acho catalogue. 5,000 were supplied with Acho couplings and sent to France in two batches, one in 1967 and the other in 1968. The standard British model was used in grey.

CN Auto Transporter R3421 (R342CN)

This was one of a number of Canadian National models introduced in 1965 and it survived to the end of the series. The 1965 Canadian catalogue showed a grey model which was almost certainly the British version. The actual model produced for Canada was moulded in black plastic and had 'CN 700184' and 'CANADIAN NATIONAL' heat printed in white along the side of the lower deck. The cars were standard Minix issues.

The 1970 catalogue shows the model with the 'CN' logo on side panels set high on the side of the wagon. Examples of this variation have not been reported.

Over 27,000 CN auto transporters were made, making it one of the most popular models in the Canadian range. Up until 1968 these were all boxed but in 1969 a batch of almost 11,600 CN auto transporters were sent bulk packed from Margate.

CP Rail Auto Transporter R3423

This model dated from May 1970 and looked the same as the CN version, R3421, but carried 'CPRail' and the logo on two panels, specially fitted for the purpose, low on each side of the car. It too carried standard Minix cars. The panels came from the tool used to make advert panels for the whiskey wagons.

5,750 were made in 1970 and a further 3,000 in 1973. 4,200 were sold boxed. The model was amongst the stocks returned to the UK in 1980 for sale through Zodiac toy shops.

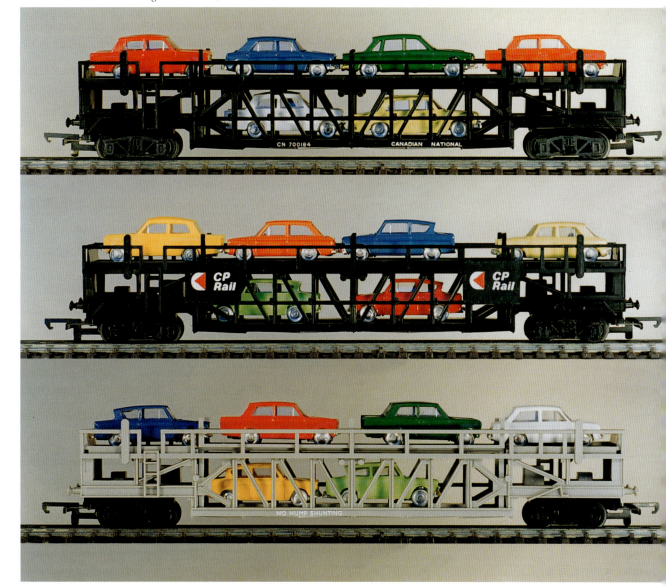

Car-a-Belle car carriers showing: R342 British, R3422 CPRail version and R3421 CN.

CEMENT WAGON

Blue Circle R564

With the exception of Battle Space models the R564 Cement Wagon was the only new wagon in the Tri-ang Hornby range in 1966.

The model was based on a wagon built by Metropolitan Cammell for Associated Portland Cement in the early 1960s which was itself a high-capacity development of the Presflo wagons. Grey was the correct colour for these tanks and that was the colour they remained for the Tri-ang Hornby period. The bright yellow version did not arrive until 1972. Perhaps the most attractive version was the Readymix cement wagon produced by Hornby in 1974 for the Australian market.

The model used the 13ft chassis developed originally for the cattle van and horse box in 1956 and the body fitted over the top, concealing the original solebars and making the wagon too wide. It had spoked pin-point wheels (SP1) and the top half of the body lifted off. The body carried Blue Circle stickers and was heat printed 'Non Pool'

The 1967 catalogue shows the model with the number LA200 on its side. This may have been a picture of the pre-production model. It is thought unlikely that any were made in this form. The actu-

Grey R564 Cement Wagon.

al number chosen for production was LA211 and this appeared on the samples photographed for the model railway press reviews. The yellow and Readymix versions were not numbered and both will be covered in **Volume 3**.

Nearly 5,000 were ordered in 1965 but the first batch was not made until 1966. Batches were made every year to 1971 with a total of around 45,000 of the grey version being made in all. The largest batch was of 13,500 in 1967 and the smallest was of 2,500 in 1971. It seems that the grey version was not used in any sets.

Rugby Portland Cement R28

This was a proposal for January 1972 but sadly did not come to fruition. It was to have been a re-livery of R564.

LINER TRAIN and 3 CONTAINERS

In **Volume 1** we looked at the Roadrailer modelled by Minic for operation with Tri-ang Railways. The prototype of this appeared in Britain in 1961 as a move to thwart competition from road transport but this was the time of the Beeching Report which amongst other things backed the ISO (International Standards Organisation) container as the only valid system for road/rail transportation.

By the late 1960s a container revolution was taking place which turned the shipping industry on its head. The major shipping companies came together to form new container handling companies. This was the only way the large amounts of capital required to invest in the new container ports could be found. Standardisation was controlled by the International Standards Organisation (ISO) who determined the size of containers so that they could be fitted to wagons and lorries the world over. The logos and liveries of the container-handling companies soon became a familiar sight on our railways.

The first revenue-earning liner train in Britain

The pre-production mock-up of the Freightliner container showing both front and undetailed back.

ran on 15 November 1965 between London and Glasgow. This carried 13 Freightliner wagons with 39 containers. Over a period of five years the government spent £25 million developing the system.

Container wagons were made up into fixed rakes and only the end vehicles had conventional drawgear at the outer ends. Rovex were the first to produce a container wagon for the British market and chose a centre car, i.e. one without buffers.

In keeping with prototype practice, Rovex made both 20' and 30' ISO containers which fitted onto a common wagon frame with reproduction four-wheeled Ridemaster roller-bearing bogies at either end. These had a scale wheelbase of 6' 6" and the scale length of the wagon was 62' 6". The bogies had smaller than usual pin-point wheels to allow 8' high containers to be carried, but they were faithful replicas of the Ridemaster roller-bearing bogies used on the actual wagons.

R633 Freightliner container wagons showing early and later livery.

When first introduced in 1967 the Tri-ang Hornby wagon frame was black but by 1969 it was being made in rail blue. The prototypes had a special type of coupling which eliminated the need for buffers. They also had lettering and markings on the solebar which were not reproduced on the model.

The box containers were assembled from four mouldings – walls/floor, a roof and two doors. This produced a simple model that could be turned out in a wide range of liveries.

Three versions of the main moulding of the 20' container exist:

1. panelled sides with a large and a small 'flat' onto which name and logo were printed,
2. panelled sides with a long 'flat' onto which 'Freightliner Limited' could be printed and
3. corrugated sides with no 'flats'.

Besides the two sizes of box container, Rovex also made open and liquid/powder tank containers which were not so often seen on the real railways. The containers were held in place by 12 locating pegs protruding from the top of the wagon frame.

3 x 20' Freightliner Containers R633
The three 20' containers on the R633 model made by Tri-ang Hornby carried the rail grey and red Freightliner livery used on the containers on that first ran to Glasgow. Each of the three containers carried a different number on a small self-adhesive label. The numbers were: 05B41, 05B71 and 05B17.

Under the provisions of the 1968 Transport Act, the enterprise was vested in a company, set up for the purpose, called Freightliner Limited. This was put under the control of the National Freight Corporation and became the largest overland container transporter in the world. With the move

came a new livery. The containers now carried 'Freightliner Limited' on their sides. Hornby picked up this change on their model in 1970.

It became the feature of the Tri-ang Hornby advertisements in the model railway press in May 1967 under the title 'HELP GET LINER TRAINS MOVING HERE'. It was priced 17/11d.

The model proved to be an immediate success. Over 20,000 were ordered by retailers in the first year of production but Rovex could only produce half that number. From January the next year they offered a presentation set (R645) containing the Hymek and three Freightliner container wagons. During the year they doubled their production figures for solo models and all sold before the end of the year. Then, with the introduction of a larger range of container wagons, the demand for R633 dropped.

It had been planned to drop the model in 1972 but it remained in continuous production until 1974. During the Tri-ang Hornby period some 52,000 solo units were made and a further 25,000 were sold through sets. It seems that of the solo units, about 39,000 had the containers with the original Freightliner livery and the remaining 13,000 carried containers with 'Freightliner Limited' on their sides.

Hornby changed the livery again in 1982 to the Freightliner flagship livery of blued-white and red (first introduced on the prototype in 1978). Then, after several years absence from the range, it reappeared in 1992 with the containers in the red and yellow of Railfreight Distribution, a fusion of Speedlink and Freightliner.

The January 1968 edition of *Model Railway Constructor* contained an article showing how the wagon could be detailed and the container converted to a range of variations.

3 x 20' Open Containers R632
This model, which was released at the end of 1969, used the standard blue liner train wagon frame as its

R632 Open Containers.

R365 BP tank containers and R637 Harold Wood Containers.

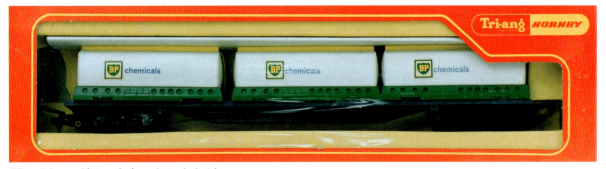

BP containers with transfer logos instead of stickers.

base but carried three open 20' containers. These were in grey with one red end. They are believed to have been based on the 4' high Type H container with detachable load-bearing sides of which some 600 were made by both St Rollox and Cowlairs for Freightliner between 1964 and 1968 for use on British Railways. The stackable containers, which had holes cut in their sides to reduce their overall weight, carried crates, barrels, cases and steel bars

The model was made between 1969 and 1973 during which time about 14,000 were sold. It was to have been dropped in 1972 but survived a while longer.

3 x 20' BP Liquid Containers R635

Another 1969 addition was the liner train wagon fitted with three 20' BR liquid or powder gravity discharge tank containers in BP livery. The container was based on a demonstration tank (No2) built by Darham Industries in 1965 for the launch of the Freightliner service. It was a 15 ton general purpose tank designed to carry 4,000 gallons and called a 'Demtank 4000'. The BP production model was slightly different in construction.

The model containers were finished in green, yellow and white livery. The BP symbol was a self-adhesive label (smaller ones at the end of the tank) but the word 'chemicals' was heat printed in pale blue. Examples may be found with transfers instead of stickers. These are the type with the BP shield in a square which were used on the smaller R12 petrol tanks. The example I have came in a Tri-ang Hornby box.

Records indicate that about 13,000 were made and the model was withdrawn at the end of 1973, two years later than originally planned. Some were sent bulk packed to Canada for set production and some of these ended up as solo models in bubble packs numbered '635'.

3 x 20' Harold Wood Containers R637

This model used the liquid or powder containers from R635 above but in the very attractive ochre and brown livery of Harold Wood. This time all of the livery detail was provided with self-adhesive stickers. It was also available between 1969 and 1973 during which time about 8,700 were sold. At the time it was withdrawn there were about 1,000 still left in the stores.

To select the names to be used on the containers, Rovex obtained a lot of photographs which are still in the company's files. Harold Wood happened to also be the name of the Chief Buyer at Universal Stores and the name was chosen to amuse him.

3 Assorted 20' Containers R678

This was also added to the range in February 1970 and consisted of one each of the type B Freightliner, type H open and BP Chemicals containers. It survived for only two years during which time about 24,700 were used in the RS600 and RS602 sets and only 1,750 sold solo.

3 x 20' Canadian National Containers R6331 (R633CN)

This was only the second version of the container wagon to be produced. The original Freightliner version had come into production in 1967 and was ordered by Canada. It was decided instead to produce a Canadian National version and 2,000 were made in 1968, although for some reason they were not shipped out that year. These were unboxed and intended for the TS905 set.

The model had grey containers (as for the Freightliner version) printed in black with the CN logo and 'CANADIEN NATIONAL'. Earliest examples had paper stickers with the number '05B17' applied to both sides. The containers were from the first mould design with all-over detail except for two small panels on each side to take the printing. The wagon frame was black.

Overseas containers: R6331 CN (Canada), locally packaged R7340 CTI container wagon (Canada) and R839 ACT (Australia).

A further batch of 4,550, this time all boxed, was made in 1970. These had the container sprayed silver and had no number stickers. They may be found on either black or blue wagons. The model was dropped from the range after 1971.

3 x 20' cti Containers R734 (R7340)

The liner train or container car is a truly international sight and it was therefore natural for Rovex to sell both standard issue models and models in local liveries to the Canadians.

This model was principally made for Canada although it was also sent to Australia (*see below*) and the single container was used later in a standard British set. The car carried three red 20' containers each heat printed in white on both sides with a 'T' and 'cti'. Both type 1 and type 2 containers may be found. As we have seen, type 1 have two small panels on each side to take the printing and type 2 containers have a long panel for printing as required for

R820 Santa Fe and R821 Flexi-Van made for ATT of America.

the updated Freightliners Limited livery on the standard issue wagon (R633). Cars used for the cti containers were either black or blue.

Batches were sent out in 1969, 1970 and 1971. These totalled 6,300 models. They included 2,100 that were sent unboxed for the R468 set and the remnants of these were packaged in Canadian bubble packs for solo sales and simply numbered '7340'.

Australian cti Containers Wagon R734A
This was sent out to Australia bulk packed in 1970 and may be found in Moldex boxes. A total of 1,250 were sent.

3 x 20' Santa Fe Containers R820
This was made to order for ATT of America. The flat car was the black standard liner train wagon with NMRA couplings and carrying three silver 20' containers printed in red with 'Santa Fe Transportation

Co.' and the Santa Fe logo. 10,000 had been ordered but only 7,400 were delivered, all of them in 1968.

3 x 20' Flexi-Van Containers R821
This was another model for ATT and had the same silver containers as R820 but they were printed with 'FLEXI-VAN SERVICE', 'THE MILWAUKEE ROAD' and the number '450', all in red. 10,000 were ordered but only 5,800 were bulk packed to America and 800 were left in the stores at the end of 1968. These may have been sold off elsewhere in Tri-ang Hornby boxes. This model also had NMRA couplings for use on American model railroads.

3 x 20' ACT Containers R839
This was released in May 1970 and consisted of a standard liner train wagon with three white 20' containers printed in blue with the ACT logo. It was made for Australia for set production in 1971 but the

wagons went over in 1970 and at times they were referred to under the number R8390. Only 500 wagons were sent but there were an additional 5,100 containers made that year that appear to have been sold in the UK during the year.

In 1971 and '72 the wagon appeared in the RS102 Express Freighter Set which accounted for the manufacture of a further 3,500 wagons. The model had also been scheduled for Canada, Australia and South Africa and bulk packed to New Zealand as a solo wagon that year but it appears that none were sent.

LINER TRAIN and 2 CONTAINERS

The wagons used for the 30' containers were identical to those of the 20' ones but by now they were moulded in rail blue plastic. The containers correctly had smooth sides but showing the lifting recesses and the opening doors in the ends of the containers were also correctly detailed.

Pickfords and Containerway Containers R634
R634 had two 30' ISO private owner containers. The Pickfords was in the traditional navy blue associated with that company and the Containerway unit was in red. The names were heat printed onto the sides. The model was featured in the Tri-ang Hornby advertisement in August 1969 and was reviewed in *Model Railway Constructor* in October 1969.

Only one batch of 15,000 models was made and that was in 1969. Two thirds of these sold in the first year but after that, sales dwindled until R634 was dropped from the 1972 catalogue. A further 4,700 wagons were assembled for use in the 1970 RS600 and RS600A sets.

Sainsbury's and Ford Containers R719
This model also featured in the August advertisement and was identical to R634 but with a different

A page from the 1970 catalogue.

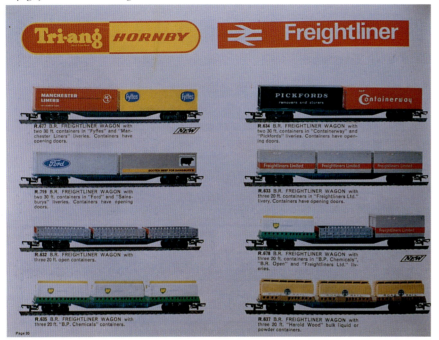

R634 Pickfords and Containerway, R719 Sainsbury's and Ford and R677 Manchester Liners and Fyffes.

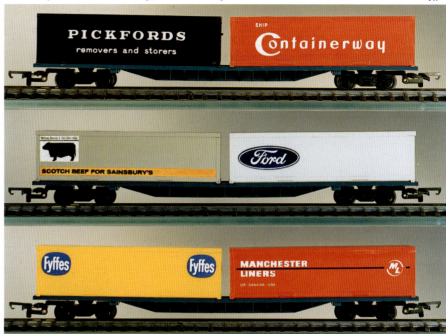

Part of a Tri-ang Hornby advertisement for August 1969.

assortment of ISO 30′ containers. The Ford container was pale grey with the traditional Ford symbol on its sides while the other carried the message 'Scotch Beef for Sainsbury's'. All livery detail was carried on self-adhesive stickers. The latter container, which was also grey, was really that of William Donald & Sons of Aberdeen whose stickers it carried.

A single batch of 20,000 was made in 1969 and these lasted until 1973, the last year it was listed. Half the models sold in the first year. It had been planned to drop the model in 1972 but stocks were too high by the end of June 1971 and it was decided to keep it in the catalogue for a further year.

Manchester Liners and Fyffes Containers R677
The range of container wagons was further extended in 1970 with the ML and Fyffes version. The

Manchester Liners container used the same red plastic mouldings as the Containerway model. The detail was all printed onto the sides including the countries of use – 'UK CANADA USA'. This provides a hint of why it was chosen. A pair of containers were used on a Canadian version of the wagon (R737).

The Fyffes container was bright yellow with two Fyffes stickers on each side and one at the door-less end.

The model remained in the range for only two years during which time about 20,000 were used in the RS602 set and a further 8,700 were sold solo.

2 x 30′ Canadian Pacific Containers R7352 (R735)
The standard container car carried two 30′ yellow containers with a large white panel printed onto

their sides with the name 'Canadian Pacific' in script reversed out. The model was made exclusively for Canada.

Nearly 7,000 were made of which 1,500 were produced in 1969 for sets but were not sent out to Canada until 1970. These were for the R468 set of 1969 which was not listed in 1970 but is known to have been made.

2 x 30' Sea Containers Inc. Containers R7360 (R736)
This model, also for Canada, was very similar to R7352 above but carried different printing. In red was printed 'CANADIAN PACIFIC' and beneath it in black was 'sea containers inc.'

The model was also needed for the R468 Dockside Containerisation Switcher set and so 2,500 unboxed models were made in 1969. This was the only batch recorded as going into the stores and yet, during the stock taking at the end of 1969, records show that all that could be found of the model were 1,860 boxed cars. This suggests that a small number of unboxed ones were sent and the rest boxed up at Margate. These were not sent out to Canada until 1971 and no others were made.

2 x Manchester Liners Containers R7370 (R737)
The Manchester Liners container tends to be associated with the British range but it was first seen in the Canadian catalogue in 1969. Furthermore, the combination of two of these units on a standard container car was exclusive to the Canadian market. The containers were marked 'MANCHESTER LINERS' and carried an ML logo. They also carried the names 'UK - CANADA - USA'.

Although 1,500 unboxed models were made in 1969 they were still in the Rovex stores at the end of the year. Half of a batch of 1,200 boxed wagons did, however, go to Canada that year. In 1970, 760 boxed and 1,900 bulk packed units were dispatched and a further 1,000 boxed models followed in 1971.

2 x CP Ships Containers R7353
This model was shown in the 1971 Canadian catalogue but it is believed not to have been made. It was to have been assembled at Margate in the spring of 1971 but there is no record of any going through the stores. A CP Ships container was introduced much later and made between 1982 and 1985 for container wagon R036. It was sold principally in the UK.

R7352 Canadian Pacific, R7360 Sea Containers Inc. and R7370 Manchester Liners Containers for the Canadian market.

#7353—Flat Car with two 30' C.P. Ship Containers
#7353—Wagon plat avec deux conteneurs C.P. Ship de 30'

Picture of the proposed CP Ships Containers as illustrated in the Canadian catalogue.

Tartan Arrow container from the RS602 Senior Freightliner set.

BR (ER) BRAKE VAN

BR Brake Van R16A (R636, R16)

By 1967 the R16 brake van was looking very dated. It had been in production since 1953 and children were more accustomed to seeing long wheelbase brake vans on the real railways.

For 1967 Rovex tooled up the body of a standard BR 20 ton goods brake to fit the WR brake van chassis (C3A). The model had an orange-brown body mounted on a black chassis. The roof was grey although later, for a short time, it was made in white plastic. The body was well moulded and included such details as duckets, grab irons, lamp brackets, door and side lights, footboards and a stove pipe chimney. The number it carried was B952698.

The prototype was 16' long but the model measured only 14'. The two feet were missing from the chassis, the body being accurate and an improvement on the former Hornby-Dublo version as far as moulding detail was concerned. The model had pin-point axles when first released.

The new model, priced 6/1d, was to survive as long as its predecessor. About 65,000 were sold solo between 1967 and the end of 1971, but this was only the tip of the iceberg. The model was manufactured in hundreds of thousands for use in most of the freight sets made during the period.

When illustrated in the 1967 catalogue it was given the reference number R636 and the first year's supply (10,800) was made under that number. The following year it took over the number R16A, the 'A' distinguishing it from the model it replaced and this suffix was dropped as soon as stocks of the short brake van had been used up. From 1976, when it became a silver seal model, it carried the number R218.

An attempt was made to fit a rear red light and the experimental model has survived in a private collection, but it was decided not to put this version into production.

Australian BR(ER) Brake Van R16A

An additional 13,000 went unboxed to Australia for use in its home-made sets and some of these were sold off solo in Moldex boxes.

VENTILATED VAN

BR Van with Sliding Doors R11A

The ventilated van was introduced in 1968 to replace the ex-Trackmaster R11 goods van of 1949 vintage. The sliding doors gave it greater play value although they always looked badly out of scale. The prototype design was chosen to allow for loading by fork-lift truck and the failure to provide a model of one with miniature pallets and loads may have been a lost opportunity. The wagon was also 1'9" underscale.

While the colour of the roof varied it seems that the colour of the body plastic remained a constant orange brown throughout its life. It always carried the number 'B784287' and the inscription 'empty to Marazion WR' but a variety of chassis and wheels were used.

The model was in production until 1978. By the end of 1971 about 72,000 had been made, 7000 being sold in the RS24A Pick-up Goods Set. The model took the number R11 from the original goods van which it replaced and until the stocks of the earlier model had been used up it was given an 'A' suffix.

Australian BR Ventilated Van R11A

With the ending of Australian production in 1968, 6,800 ventilated vans were bulk packed to Australia over the period 1968 to 1970 for use in sets. In 1968 Rovex shipped out 2,200 unboxed vans and further batches of 1,500 in 1969 and 3,100 in 1970. Some of these probably found their way onto local shop shelves as solo models in Moldex boxes.

BULK GRAIN WAGON

Whisky Wagons R647-R650

In 1968, quite a stir of speculation was started when the Trix 'Whisky' bulk grain wagon appeared in the Tri-ang Hornby catalogue. Rumours were rife and it was speculated that Trix was about to join the list of toy companies in the Lines Bros. Group. The mystery deepened when Trix wagons began to appear in the shops packed in Tri-ang Hornby boxes. Even the pre-formed plastic tray that held the model rigid in the carton was purpose made to take the wagon.

British Trix had introduced the 'Whisky' wagon to their range in 1967. It was based on the BRT 35 ton bulk grain wagon built by Pressed Steel Ltd to carry malt barley grain from East Anglia to distilleries in Scotland. The Trix range included vehicles with advert boards for Johnny Walker, Haig, Vat 69, King George IV, Dewers, Crawfords, White Horse, Jamie Stuart and Abbots Choice all in blue livery, and The Maltsters Association in yellow. On the real wagons the two boards did not necessarily carry the same advertisement, the boards bearing no relationship to the company using the wagon.

So how did Trix wagons end up in Tri-ang Hornby boxes?

The official story, supported by correspondence in my possession, is that Mr Rozsa of British Trix approached Richard Lines at the 1967 Brighton Toy Fair with the suggestion that Tri-ang couplings should be fitted to Trix rolling stock. He also offered

to supply Rovex with their new bulk grain wagon which could be fitted with couplings supplied by Rovex for sale in the Tri-ang Hornby range.

Six types of the wagon were available at this stage and a sample of each was sent to Rovex; one of them fitted with couplings and with wheels taken from a Tri-ang Hornby R564 cement wagon. Rovex estimated a list price of 8/- and a trade price of 4/6d. As they would be looking for 40% profit themselves and would have a packaging cost of 5d they could afford to buy the Trix wagons in at 2/3d each. This offer was made to Trix for a possible order of 5,000 models. Trix felt this was too low a price and countered with 4/2d each. It was finally agreed in July that Rovex would order 4,000 wagons (1,000 each of 4 designs) at 3/11d each, with Rovex providing the couplings free to be sent with the order. At the same time Richard Lines chose the four names that he felt would be most familiar. These were included in the 1968 catalogue as follows :

R647 Dewers
R648 Johnny Walker
R649 Vat 69
R650 Haig

The order was placed in August and the wagons were promised in September with possible repeat orders each month after that. The models were being manufactured in Austria which posed a slight problem because Tri-ang Hornby boxes bore the words 'built in Britain'. In July two of the models were sent to Fleet Illustrated to be included in the photography for the 1968 catalogue and they duly appeared on page 20 of that edition complete with Tri-ang couplings. Arrangements were also made for each of the 12 salesmen working for Rovex Industries to have a set of the four wagons fitted with tension-lock couplings.

Unfortunately Trix experienced production difficulties. No wagons arrived in 1967 and during 1968,

when Rovex had ordered 10,000 wagons, only 129 of the Dewers wagon, 159 of Johnny Walker, 54 of Vat 69 and 191 of the Haig wagon were delivered, most arriving in May and the residue in June. They were sold in DR size boxes. In the meantime, due to devaluation of the pound, the cost of the wagon increased. Trix also had to meet the cost of altering the buffers to meet requirements set by Rovex.

It was probably during 1969 that Rovex decided to make their own body tool. That year Rovex listed just two varieties – Johnny Walker and Vat 69 – and 5,408 of the former and 7,371 of the latter were delivered. How many came through Trix and how many were produced at Margate we may never know. It should be noted that untypically Rovex did not allot fresh 'R' numbers to their own models to distinguish them from the Trix versions.

Rovex were still unable to meet orders from their retailers which totalled more than 18,000 during the year. Their own version came out late in the year and at the same time Rovex used the body mould to make a batch of grey hopper wagons (R214). Both were photographed by *Model Railway Constructor* in March 1970.

A further 16,500 bulk grain wagons went through the stores in 1970.

Part of the problem at Trix in 1969 had been the delivery of the chassis and so Rovex asked them to send bodies in the meantime. In the Margate factory these were fitted to the Tri-ang Hornby 15' chassis (G1a), produced that year for the new china clay tank wagon. As the 15' chassis was slightly longer than the Trix chassis these wagons would not fit into the pre-formed plastic trays. This meant that Rovex had to produce a special tray for these few bastard wagons which they sold in DQ size boxes. I know that this happened only because I have a boxed R650 Haig, fitting this description, in my collection. The clue to its authenticity is the pre-formed tray which was clearly made to fit the oversize chassis and will fit no other wagon.

Left below: Extract from a memorandum dated 21 July 1967 from Richard Lines to H W Henden regarding the British Trix whisky wagons.
Left: Trix bulk grain wagons supplied to Rovex: R650 Haig and R649 Vat 69.
Below: Trix bulk grain wagon (R650) body on Tri-ang Hornby chassis.
Right: R648 Johnny Walker and R649 Vat 69 Triang Hornby bulk grain wagons in both bright blue and rail blue.

"I have agreed, in principle, with Trix to order from them 1,000 each of Dewers, Johnny Walker, Vat 69 and Haig, as I consider these to be the best-known brands. They will, in fact, be manufactured in Austria which may present a slight problem if our boxes state made in England but, on the other hand, if we do not export these models I do not think this really matters.

"Would Margate, therefore, raise an order on British Trix Limited, Wrexham, Denbighshire, for 1,000 each of the four brands mentioned and enclose with the order 8,000 free issue Tri-ang Hornby Coupling units. If this order is placed with Trix by the middle of August, they say they can effect delivery by September 15th, which date should be quoted on the order.

"I believe at this stage it may be possible to incorporate two of these models in the 1968 Catalogue."

In 1972 production of Trix Trains, as they were now called, ceased and the tools remained in the ownership of Liliput until they were transferred to Liliput UK in 1974. Meanwhile Rovex continued to produce their choice of two – Johnny Walker and Vat 69 – from their own tools. The Tri-ang Hornby wagon was, of course, to 00 scale and 3.5mm taller and 5mm longer than the Trix model which was in a bastard scale somewhere between 00 and H0.

Also, unlike the Trix wagons which were tampo printed, the Tri-ang Hornby adverts were printed on self-adhesive labels. They also carried the numbers 5833 on the Johnny Walker version, 5820 on the Vat 69 and BRT symbols on self-adhesive labels. They were fitted with disc wheels and were priced 10/6d.

Vat 69 was withdrawn from the Hornby Railways catalogue in 1972 and Johnny Walker in 1974, but the model reappeared in the Hornby range in 1987

in Heygates livery and later still in a plain blue livery without advertisement boards.

Later in 1992, through a middle man, the Liliput tools ended up in the hands of Bachmann Industries (Europe). The grain wagon subsequently appeared as part of Bachmann's British outline range in a variety of finishes, having been manufactured in the Far East. This meant that more than 30 years after the tale began, both models were once again competing on the British model market.

LARGE HOPPER WAGON

BR Large Hopper Wagon R214

When the original grain wagon (R215) was introduced in 1958 it was accompanied by an ore wagon (R214) which used the same moulding tool for the body. This happened again in 1969. By the end of the year the tool for the 'whisky' wagon body was used to make over 9,400 large hopper wagons. The model took the same number as its 1958 predecessor – R214. Between 1969 and 1971, the last year of production, about 19,000 were made. In 1973 a private owner version, Robert's (R102), was introduced in black livery and this will be found in **Volume 3**.

The picture in the 1969 catalogue showed the model on a Trix chassis. This was because the Tri-ang 15' chassis was not ready at the time the photographs were taken for the catalogue. It is thought unlikely that any Tri-ang grey large hopper wagon bodies were sold on Trix chassis.

SLURRY TANK WAGON

Bowater's China Clay R668

This model was also photographed on the wrong chassis for the 1969 catalogue and for the same reason as above. In its case the C3b chassis of the engineer's wagon was used for the picture but the production model had a new 15' chassis with a 60mm wheelbase (type G), also to be used on the 'whisky' wagon.

The prototype wagons were introduced in 1966 to carry china clay in liquid form from Cornwall in block trains to Bowaters paper mills in Kent. No doubt the subject was chosen for a model both because it was a familiar sight in Kent where Tri-ang

R214 BR Large Hopper Wagon.

Railways were designed and made and because of its bright and (for its time) unusual colour.

The model probably first appeared in the shops late in 1969 with only about 1,000 being made that year. About 15,000 wagons were made in all between 1969 and 1971 and a further 7,700 in 1972-73.

The pale blue wagon body carried the number 1025 on its sides and, unlike the heat-printed white star, the number was applied as self-adhesive labels which have a tendency to fall off when they dry out. It would surely have been easier for Rovex to have also heat printed the number or were they intending to offer alternative numbers? Variations worth looking out for are models with either white or blue ladders as well as two different sizes of ladder, the smaller ones being later. It is thought that the model as originally released had a white ladder.

Dubonnet R674

Thought was given, in 1970, to the possibility of releasing this model as a Dubonnet tank wagon with a view to it being available in 1971. This would have meant moulding the body in a different colour and a change of lettering. The idea was dropped at the end of June and I am not aware of a mock-up surviving or even being made.

This was not the only occasion that other uses of the slurry tank were considered. Amongst a number of pre-production models sold by Hornby Hobbies Ltd to a local retailer in the early 1990s was a standard pale blue slurry tank with an Italian WW2 air force roundel stuck onto each side. This model is now in a private collection.

CARTIC CAR CARRIER

The prototypes were first seen in September 1964 and their principal use was for transporting cars from the factory around the country or to the docks for export.

BR Cartic Car Carrier R666

The BR version of the prototype, with the Motorail emblem, first came into use in 1966. As the Motorail business was largely seasonal to meet the needs of the holiday trade, car manufacturers used the vehicles during the off-season when they could be hired more cheaply. The wagons were in rakes of four with the outside vehicles being different to the inside ones; the latter being shorter and sharing both bogies. The Tri-ang Hornby model consisted of two outside wagons.

The articulated wagon is thought by many to be one of the most attractive of the period. It consisted of two double-storey car carriers joined by a shared bogie. At the time it was an expensive model costing the equivalent of eight open wagons or two coaches and this may be the reason why its load was cut from 16 to 12 cars in 1971. On the other hand the reduction may have been made because the proto-

R666 Cartic car carrier as supplied with 16 Minix cars individually boxed.

types carried only 12 cars! The assortment of cars on each wagon had no pattern to it.

The *Model Railway Constructor* review in March 1970 pointed out that while the bogies were accurate and the length correct, the body rode the equivalent of 18" too high above the track. It also pointed out that the insignia boards should be carried on top of the top deck and not be mounted on its side.

The model first appeared in 1970 in BR blue when 12,500 were made. It seems that this was the only year the BR version was made as it took until the end of 1973 for these to sell after which the model was withdrawn from the catalogue. The tools then lay in store until 1989 when they were cleaned up for reuse, this time to produce the model in the orange livery of Silcock Express. The Cartic used the liner train bogies.

100T TANKER

Early in 1966, two bogie tankers appeared on the railways. One was made by Alcan Industries Ltd and the other by Metro-Cammell. They were turned out in the liveries of Esso and Shell Mex & BP, the latter being from Metro-Cammell. After a period of trials, these 100 ton wagons went into full production. By the end of 1969 about 1,200 were in use in Britain. While they were also made by a number of other companies they all looked very much alike.

Most of them built by Metro-Cammell and two thirds of them in the Shell Mex & BP Ltd livery.

Shell R669

May 1970 saw their introduction to the Tri-ang Hornby system. A pre-production mock-up was on display at the Tri-ang House Toy Fair early in the year and the model was to be the source of many interesting and colourful liveries over the years, surviving well into the 1990s. A new bogie moulding was made especially for the tank

The picture in the 1970 catalogue appears to be that of the pre-production model. This is even clearer in the Canadian catalogue and presumably this was all that was available when the catalogue was being put together.

The body was moulded in pale grey plastic, representing the colour reserved for petroleum loads, and had a mid grey walkway fitted along the top. The lack of yellow and white bands along the sides of the tank, as seen on the 1982 version of the model, indicates that this model was in the revised colour scheme introduced to the railways after 1967. Black plastic ladders at each end were mounted onto the tank body rather than the walkway. Heat printed onto the sides were '2309' in red and two stars in black indicating 'fast traffic'. The Shell symbols were applied as self-adhesive stickers and should correctly have been accompanied by BP stickers. The chas-

Variations of the R688 Bowater's china clay wagon.

A strange version of the china clay tank wagon found in the Hornby factory.

R669 Shell 100T tanker.

The catalogue picture of the proposed CPRail 100 ton tanker.

#6693—C.P. Rail 100 Ton Oil Tanker.
#6693—Wagon-citerne C.P. Rail de 100 tonnes

sis and bogies were black.

The Shell bogie tank remained in the range until the end of 1973 and 17,000 solo models were sold. In fact all stocks had been sold by the end of 1972 but it seems that a few Shell tankers were made in 1973 at the time the British Oxygen tank (R669) was in production. This was no doubt in response to orders generated by its continued presence in the catalogue. These, however, do not show in the company's records but can be identified because they had a white body, the same as the bulk oxygen tank.

According to notes made in 1971, the company was considering how it could brighten up the wagon for 1972. A brighter colour tank and a red chassis were suggested but the idea was not adopted. It was to reappear in 1981 as a mail order model.

CP Rail 100T Tanker R6693

Although advertised in the 1971 Canadian catalogue the model was not delivered. It was to have been made in the spring of 1971 but none were recorded on the factory census sheets.

FERRY VAN

BR Anglo Continental R738

Another modern railway vehicle to arrive in 1970 (autumn) was the long-overdue ferry van which had been planned five years before. It was based on the prototype series built by Pressed Steel Co Ltd in 1962/63, the initial batch being 150. They originally worked the Harwich-Zeebrugge ferry but were later to be found in all parts of the British mainland. Three braking systems were fitted in order to make

the wagon suitable for different overseas railway systems.

It was the longest four-wheel vehicle so far modelled by Rovex and to ease it round the tight curves the pairs of wheels were mounted on bogies. The inward end of each bogie carried a hook to which a rubber band could be attached to self-centre the wheels.

A pre-production model was on display at Tri-ang House for the 1970 Tri-ang Toy Fair and attracted a lot of interest.

The production model had a single red-brown roof and body moulding with detail heat printed onto the sides in white and black. The body clip fitted to the chassis and faithfully recreated the steel bracings, rivets and ventilators found on the prototype. According to the review in the November 1970 edition of *Model Railway Constructor* it was exactly scale length (41' 9.5") and had the equivalent of a 26'3" wheelbase.

It had arrived in the shops in May 1970 and remained in the catalogue for two years before being replaced by the Interfrigo version (R742). It was to return to the range again in 1978 under the same 'R' number and in 1980 was given a paint finish. It was to go on to enjoy modern BR liveries. About 13,500 were originally made and sold under the Tri-ang Hornby label, the last of this stock leaving the factory in 1972.

Transfesa R741

Transfesa was a Spanish company whose international vans could be recognised on railways in Britain by their striking blue livery. The model used the same moulding as R738 but in bright blue plastic. The red and white detail was applied by heat printing. It first appeared in the shops in January 1971 and remained available until 1974 during which time 9,000 were made.

The Transfesa Van may be found with a rail blue instead of electric blue body. Closer inspection may

A review of the ferry van in the November 1970 edition of Model Railway Constructor.

R738 BR Anglo Continental ferry van.

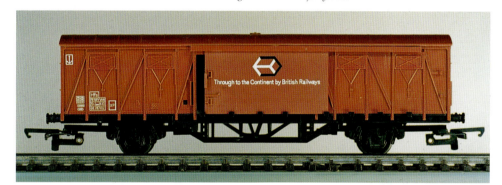

TRI-ANG OO 4mm
BR 20ton Continental Ferry van

This vehicle was announced in the 1970 Tri-ang Hornby catalogue and shown at the Tri-ang Toy Fair in January. It was scheduled for release in the summer and here it is—well worth waiting for!

The prototype is the series built by the Pressed Steel Co Ltd in 1962/3 when 150 were produced (it is possible that more were subsequently ordered), and were originally put to work on the Harwich-Zeebrugge Ferry. They are a common sight on all parts of BR. As the vehicles are designed to run over the Continental lines, three braking systems are employed vacuum, compressed air and hand lever.

The model is exact 4mm scale as regards length (41ft 9½in), wheelbase (26ft 3in), width (7ft 3in) and sliding doorway (13ft long). It is only a bare fraction out on the height and doorway height, a matter of 3mm (9in) at the most and this very slight difference is not noticeable.

The body is a one-piece moulding in brown plastic, which is clip fitted to the chassis by spigots at the ends and on each side near the door. The body moulding faithfully portrays the steel sides, the bracings, rivets and ventilators of the prototype, while the door actually slides to the full width opening. The chassis is correct and accurately moulded in black plastic with correct axleboxes, and solebar attachments. The only thing missing is one slinging hook attachment on one side, but this is carrying criticism a bit far for something the size of a pin-head! Nevertheless we mention it to show how accurate and faithful this model is to the prototype.

This is the longest 4-wheeled vehicle in the Tri-ang range (possibly the longest 4-wheel British outline model in any range), as it is 7⅛in over buffers with a wheelbase of 4⅛in. To allow the vehicle to negotiate sharp radius curves the axle units are pivoted at their centre and held in line with an elastic band. This sounds to be a crude system but in fact it works and the axles automatically self-centre. However, on curves above 2ft 10in radius (maybe less) this is not necessary and our tests proved that the elastic band in self-centring the axle put a slight friction on the van, so that with the band removed it is very free running and the vehicle holds the track admirably.

The vehicle is of all-plastic construction with the exception of metal pin-pointed axles and the Tri-ang tension lock coupling. We welcome this addition to the range of modern image models and we are sure it will be popular with "toy" and scale layouts alike, particularly if the latter fit metal wheels.

The R741 Transfesa ferry van in bright blue and rail blue.

reveal that this is electric blue plastic that has faded in the sun.

Interfrigo R742

This model, planned in 1971, seems to have reached the shops after the name-change deadline and therefore may be available only in Hornby Railways packaging. The model will be covered in **Volume 3**.

COKE WAGON

NER R781

In 1970 it was decided that the range for the 1971 catalogue should include a low-cost coal wagon. The result was the coke wagon whose body moulding tool cost about £800. This compares with £4,500 to tool up the large breakdown crane. It is not known whether the model was based on a actual prototype but the slightly tapered sides gave it a North Eastern Railway look.

At the formation of the LNER the NER contributed over 10,000 12T mineral wagons; by far the highest number of any of the constituent companies.

As it first appeared in March 1971 the model had a dark grey body with 'NER 52220' heat printed in white on its sides. It was fitted with fine-scale (SP2) wheels from the start. The model remained in production until 1974 and during this time was produced also in platform grey and rail blue. About 13,000 were made in 1971 but much greater quantities were to go out in Hornby Railways boxes.

Other Liveries

The coke wagon in Bannockburn livery arrived in 1973. The Roberts Davy and Barrow liveries followed later.

LARGE BREAKDOWN CRANE

Red R739

The small R127 mobile crane had been a good model with plenty of play value but it could not compare with the giant cranes now being offered by continental modellers. In 1971, the time had come for a 75 ton railway crane on the Tri-ang Hornby system. The mock-up was seen at the 1971 trade toy fair at

Tri-ang House at the end of January and the Editor of the *Model Railway Constructor* commented that it looked excellent, probably surpassing the standard of the old Hornby-Dublo one of 1959 which had long been, and still is, a firm favourite amongst collectors.

The crane was based on a design by Cowans Sheldon Co. Ltd of Carlisle and 12 (ten steam and two diesel) were ordered by BR in 1960. They had the greatest lifting power of any railway cranes in Britain at that time.

The model was mounted on two four-wheeled bogies. Two weight-relieving bogies and a match truck to support the jib when in transit and to carry the side jacks, completed the set. The jacks fitted into slots in the side of the crane and provided lateral support when the crane was in use. Metal weights were also screwed to the underside of the weight-relieving bogies in order to lower the centre of gravity.

The crane unit was fitted with the old split-axle wheels to enable it to take the tight curves found on Tri-ang Hornby layouts and the chassis was made in two halves that were pushed together to trap the axles. There should have been a covering, extending from the cab roof to cover the winding drums, which was pivoted to lift as the jib was raised but this was missing on the model. A removable chimney was, however, supplied so that it could be taken down when the model was in motion on the track.

The drums holding the chain rotated by means of large wheels on the side of the crane body. The prototype crane was 58'1" long over the buffers and, with the couplings compressed, the model was exactly to scale length. The match truck was made from a integral low-sided wagon (R712) normally found in the toy train sets and as such was much too short. Consequently the jib of the crane was shortened to 175mm instead of 196mm.

The model was scheduled for release in September 1971 and both the *Railway Modeller* and

Hand-painted pre-production sample for the NER coke wagon.

Boxed example of the NER coke wagon.

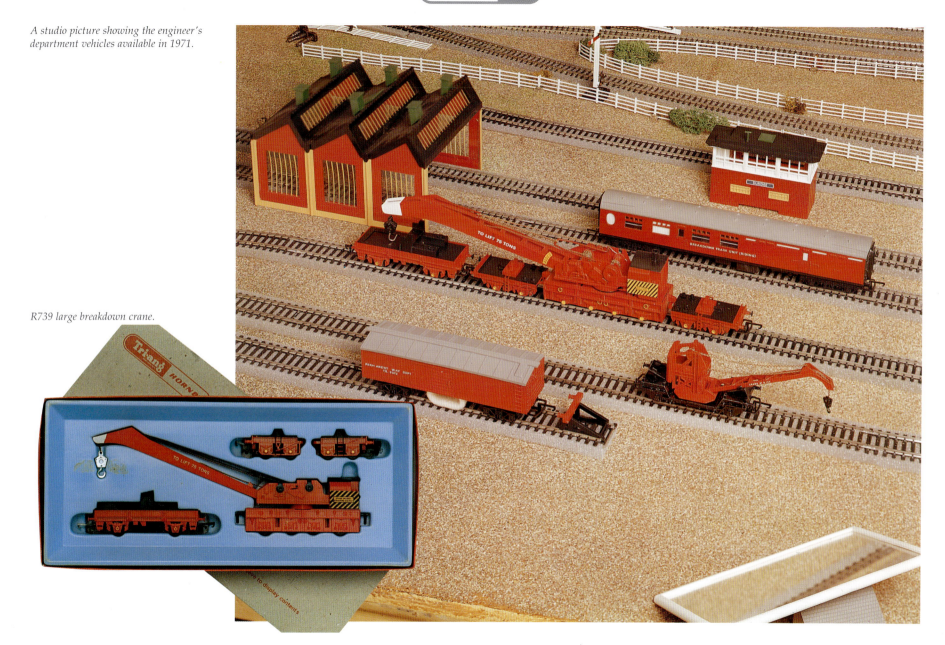

A studio picture showing the engineer's department vehicles available in 1971.

R739 large breakdown crane.

A model railway press review of the new crane and riding car.

Above: The crane ready for travelling. Right: Jib raised and chimney in place.

TRI-ANG   OO 4mm
Breakdown crane   £2·75

In the 1971 catalogue Tri-ang announced this new model for release in Mid-71. We are especially pleased to announce that it is now available. The prototype of this model is the 75ton crane manufactured by Cowans, Sheldon & Co Ltd of Carlisle, one of which is based at the Stratford mpd. Twelve such cranes (ten steam and two diesel) were ordered by BR in 1960 and they possess greater lifting power than any other railway cranes in Britain.

The set is made up of the crane itself, together with two weight relieving bogies and a match truck to support the jib. In the match truck are four jacks (right angled pieces of plastic), which fit into slots on either side of the crane to stabilise it when slewing. This is very necessary for the model will topple over unless these packs are in position; we understand the prototype will tend to do the same. All plastic construction is used except for metal pin-pointed axles and the fine metal chain used for raising the jib and hook. Metal weights are also screwed to the underside of the weight relieving bogies. The drums holding the chains rotate by means of large black wheels on one side as shown in the photograph—the other side merely has the pivots. The finish is all red, except for the black drums, cab rear and roof and top surface of the bogie and match trucks. The yellow "zebra" marking on the sides of the cab are paper stickers.

The total length of the prototype crane and bogies is 58ft 1in over buffers and the Tri-ang model is exact (providing the couplings are compressed). However, the length of the prototype match truck shown on our dimensioned drawing is 35ft 1in over buffers with a wheelbase of 22ft 6in. Unfortunately, the model only measures 112mm with a wheelbase of 55mm, so this part of the set is much shorter than it should be; although it is possible that Tri-ang made it shorter as such a length of rigid truck would have difficulty in negotiating sharp radius curves. The jib has likewise been shortened to 175mm (instead of 196mm-49ft), and the height from rail level to top of cab is 1mm too much, but from our draw-

ing and photographs there should be an extension of the cab roof covering the winding drums, which is pivotted at the cab end and raises with the jib. This could easily be made from Plastikard. A removable boiler chimney is supplied, which is always removed when travelling to clear the loading gauge. In spite of the small errors mentioned above, the crane looks right as far as its size is concerned and will be a welcome addition to the range of available rolling stock. We did not like the red finish, but of course this can easily be painted where necessary, such as the underframes etc, and the rest generally "dirtied" up a bit to make it look a little more realistic. In fact there is a lot of super-detailing that can be done to make this model ideal for scale layouts.

TRI-ANG   OO 4mm
Breakdown Riding coach   £1·07

With the introduction of the breakdown crane, above, comes a riding coach to provide accommodation and eating facilities for the crew. This in fact is the standard Tri-ang Thompson brake 3rd coach as reviewed in the April '70 issue and is now finished in red with a white roof, black underframe and ends. Some of the windows have been rendered opaque by having white plastic inserted behind them. This vehicle gives a good impression of the breakdown unit, but would look better if "dirtied" before being employed on a scale layout.

Model Railway Constructor reviewed it in December that year.

To go with the red crane, to make up a breakdown train, the Thompson brake coach was turned out at the same time in bright red as a crew coach. This is reviewed in the chapter on 'Coaches'.

It is not known how many large cranes have been made by Rovex but, as a guide to its popularity, over 8,000 were sold in 1972. A little over 1,000 had left the factory the previous year against orders in excess of 4,000, suggesting that it was late reaching the shops.

The red version remained in production until 1982 when it was reliveried in yellow and received better detailing. In this form it survived well into the 1990s. In 1989 a 'weathered' version was introduced to the Thomas the Tank Engine series and in 1995 it returned to red livery.

PLATE WAGON

BR Plate Wagon R19

The model was planned for January 1972 and was intended to replace the R18 Cable drum wagon. From the start it had fine-scale wheels. Possibly this model came too late to be found in Tri-ang Hornby packaging. It will be covered in **Volume 3**.

HULL & BARNSLEY VAN

Refrigerated Van R21

Planned for January 1972 and dealt with in **Volume 3**.

LARGE MINERAL WAGON

Cory R22

This was planned for January 1972 and from the start it was seen as an opportunity for a variety of private liveries. It will be covered in **Volume 3**.

MONOBLOCK TANK WAGON

Shell R20

This was to have been released in January 1972 but was delayed to the following year and will be dealt with in **Volume 3**.

B. WAGONS MADE FOR STARTER SETS

1. Non-Printed Wagons

Between 1965 and 1971 over 1 million unprinted wagons left the Margate factory but they were sold only in low-priced sets. Their 'R' number usually carried the suffix 'NP'. Those I have come across are listed here:

OPEN WAGON

Various Colours R10NP

This is probably the most common of the NP wagons. Colours that may be found include blue, maroon, orange brown, red and green. The earliest in my own collection dates from 1960 and the latest known use of the wagon was in the 1972 RS691 and R692 clockwork sets.

Open Wagon for America R10NP

These were almost certainly supplied for a Miniville Train and Station Starter Set which were probably fitted with NMRA couplings. 120,000 were made in 1968 and sent out to America bulk packed. They would have had no printing and were probably green; although with such a large order they could have been made in any colour the customer required.

A selection of wagons from the Tri-ang Hornby period which were made for use in starter sets.

GOODS VAN

Various Colours R11NP

This is believed to have been used between 1965 and 1972 and may be found in blue, vermilion and khaki. The most common seems to be the red version.

Fish Van R14NP

This was used between 1964 and 1967 and was the same as the above, but the van was in the pale blue used for the Insulfish van. It is not known whether a white one exists.

Fish Van for America R14NP

In 1968, ATT of America ordered 60,000 fish vans fitted with NMRA couplings. These were made the same year as the unprinted open wagon but in only half the quantity as the set they were intended for contained two open wagons and a van. They were most likely light blue in colour and had no printing.

TANK WAGON

Petrol Tank Wagon R12NP

Silver coloured but without printing or stickers. I have not seen an example yet but it is thought to exist.

Milk Tank Wagon R15NP

This was definitely used in the RS692 set of 1972 and may well have been in other earlier sets. It was white and carried the normal ladders but without any printing.

Lubricating Oil Tank Wagon R211NP

This was just a yellow tank wagon identical to the R211 but without printing. It is not known when it was used but the example I have seen dates from the mid to late 1960s.

BRAKE VAN

Brake Van R16NP

This is a fairly common wagon as it was used in a number of sets. It always seemed to be in the normal red-brown plastic.

BOLSTER WAGON

Bolster Wagon R17CNP

This wagon was either red-brown or green and carried a Minix car. It is known to have been in use between 1966 and 1968.

DROP-SIDE WAGON

Drop-Side Wagon R113NP

Both vermilion and maroon examples of this wagon have been found. The only set known to have used this model was RS88 in 1968.

BRICK WAGON

No Markings R219NP

This may be found in either red brown, salmon-brown or bright red, the last being hardest to find. Some of the R219NPs were released without buffers.

London Brick Company R219NP

The model was without a blackened chassis but with the normal 'LONDON BRICK COMPANY' printing on the sides of a bright red body.

M-T Brick Wagon with Log Load R219NP

In 1967, 2,500 of the brick wagon in black plastic were sent to Canada for a battery set. The logs it carried were almost certainly those from the side-tipping car R345. The wagons appear to have been used in a set thought to have been put together by Rowland G Hornby in Canada, details of which are given above in the introduction to the Directory of Wagons. For more information *see* B. Starter Sets (2. Miniville Sets) in the 'Sets' chapter under the heading – 'Directory'.

MINERAL WAGON

Mineral Wagon R243NP

The mineral wagon in its non-printed form may be found in yellow or bright blue and was used in the RS86 set of 1968 amongst others.

CONTAINER WAGON

Container Wagon R561NP

All examples I have seen of this model have had the red-brown flat wagon and either a navy blue or vermilion container. Neither wagon nor container carried any markings. It is known to have been used in RS87 of 1968.

BOGIE BOLSTER

Bogie Bolster Wagon and three Cars R563NP

The 1961 version of the bogie bolster wagon often appeared on its own with a clockwork tank engine. The wagon body was the normal salmon colour and the cars were standard Minix models. The wagon was missing its buffers but had bolster pins.

CABOOSE

Yellow Caboose R1153NP

This was a plain yellow model identical in every way to the CP Rail caboose (R1153) but without any printing on the sides. It was used in 1971 in the Canadian-assembled No.918 Chugga set with standard integral starter set wagons. Only 2,750 of the cars were made of which 1,400 went to Canada in 1971 and the remainder appear to have been boxed up in Tri-ang Hornby boxes, without a printed flap label but with the number '1150' written on by hand.

The caboose was exported to Canada in non-printed form as R1153NP. A similar version of the oil tanker is not recorded but would presumably be numbered R711NP.

The rocket-launching wagon from the RS74 set.

M-T Express Caboose R115NP

This was orange and black like R115CN but without printing. Instead it carried stickers on its sides inscribed 'M-T'. There were 2,500 made of this model and supplied for the 1967 battery set assembled in Canada. Further details of the set may be found in the introduction to this Directory of Wagons.

TRACK-CLEANING CAR

M-T Express Track-Cleaning Wagon R344NP

These freight cars appear to have been used in a cheap battery set thought to have been put together by Rowland G Hornby in Canada, details of which are given in the introduction to this Directory of Wagons. A single batch of 2,500 cars was made for Canada in 1967. These had a black roof and an orange-brown body without any printing but with stickers inscribed 'M-T' (stylised 0.4" square and set on a yellow maple leaf) and a separate sticker (1.14" x 0.25") with the word 'EXPRESS'.

CANADIAN MILITARY WAGONS

Canadian Rocket Launcher R570CN?

This was a one-off wagon used in the RS74 export train set which is believed to have gone to Canada in 1965. The wagon was probably unique to this set and consisted of the short wagon chassis with the turret of the 4 rocket-launching wagon mounted on it. The turret carried no markings but there were four rockets packed in the box. 2,500 were made, all of which were sold abroad.

Canadian Radar Tracking Car R567CN?

It is not known whether this was made but if it was it may well have consisted of the radar-tracking car body mounted on a wagon chassis for a cheap starter set to be sold in Canada.

2. WAGONS MADE WITH INTEGRAL CHASSIS

'Integral' was a term used in the factory when referring to rolling stock which had a body and chassis incorporated in a single moulding. This, of course, was done for cheapness of manufacture and while it was usually used in connection with the starter set rolling stock, introduced at the end of the 1960s, the first time this money-saving technique was practised was in the 1950s when wagons for the Primary series were made.

There were five basic models introduced in 1968 for the toy train sets:

High-sided Wagon
Low-sided Wagon
Tank Wagon
Long Flat Wagon
Coach

The coach in actual fact had a separate chassis!

There were also a number of additional models created by adding loads to the low-sided wagon and the long flat wagon. The latter was longer than the rest of the models so that it could carry a standard 30' container. All except the coach and tank wagon were produced in a variety of bright colours known in the factory as 'toy red', 'rainbow blue', 'BP yellow' etc. The coach and tank wagon were always yellow, although there is evidence that Rovex did contemplate a green version.

As many starter sets did not appear in standard catalogues and some of the wagons were produced only for limited run sets, information about them is difficult to obtain. We do, however, know that the sample of the low-sided wagon, the first to be made, was ready by July 1968 and the other two wagons were ready in September.

The high- and low-sided wagons and the tanker were to survive for more than 20 years. From around 1973 the tension-lock coupling was replaced by an integral hook-and-loop coupling but not before the coupling mounting had been strengthened resulting in a bulge in the ends of the wagon body. When the body tools were altered in 1973 to take the integral coupling these bulges disappeared.

In 1986, after the introduction of Thomas the Tank Engine and Friends, the wagons were adapted to take flangeless wheels to run inside plastic moulded track provided in some of the cheaper Playtrains sets.

The wagons in the following list have been arranged in numerical order :

HIGH-SIDED WAGON

Standard Issue R710 (R710NP)

This model tends to be in shades of red, maroon, blue or green, although yellow was later used in the Playtrains series. Between 1968 and 1971 well in excess of half a million integrated high-sided wagons were made, making it the most numerous of these cheap wagons during the Tri-ang Hornby period. Also listed as R710NP. Many were bulk packed to Australia, New Zealand, Canada and the United States for use in sets assembled over there.

High-sided Wagon for America R710T

17,150 of the high-sided wagon were made with NMRA couplings for the States in 1969 but were still in the store at the end of the year. These had been ordered by ATT of America and there is no record of them after the end of 1969.

TANK WAGON

Standard Issue R711 (R711NP)

Until recent times it seems that the tank wagon was always yellow and usually heat printed with 'Shell' and the Shell logo. This detail was later tampo printed onto the wagon at which time the Shell symbol became much larger. The original integral tanks from the late 1960s had no markings and these may be found in both the common yellow and a bright canary yellow. These were listed as R711NP.

A quarter of a million integral tanks were made between 1969 and 1971. In 1986 a white milk version was made for the R183 Thomas the Tank engine set. The model was also exported for overseas set production.

Tank Wagon for America R711T

This was a standard integral tank wagon with NMRA couplings fitted. 17,500 were made for ATT in 1969 but were still in the stores at the end of the year. There is no record of their having been dispatched.

LOW-SIDED WAGON

Standard Issue R712 (R712NP)

The low-sided wagon was withdrawn in the early 1970s and possibly did not reappear until 1986 for use in Playtrains sets. Consequently most examples found have tension-lock couplings. It was also listed as R712NP. During the Tri-ang Hornby period 360,000 were made for use in sets. Common colours were green, blue, and both deep and canary yellow (known as BP yellow), red (shades) and black (scarce) may also be found. Once again the model was bulk packed to Australia, New Zealand, Canada and the USA for use in locally-made sets.

Low-sided Wagon for America R712T

Those made for the USA were fitted with NMRA couplings. They had been ordered by ATT in 1969 but no records have been found of them having been dispatched. 24,750 were made and 24,300 remained in the stores at the end of the year.

Low-sided Wagon and Car & Caravan R713 CN

2,500 of these went to Canada in 1969 for use in their TS906 Chugga switcher freight set planned that year and the No.918 set of 1970. They consisted of a Minix car and caravan (or 'auto and house trailer' in

The four basic integral freight vehicles used in starter sets from 1969.

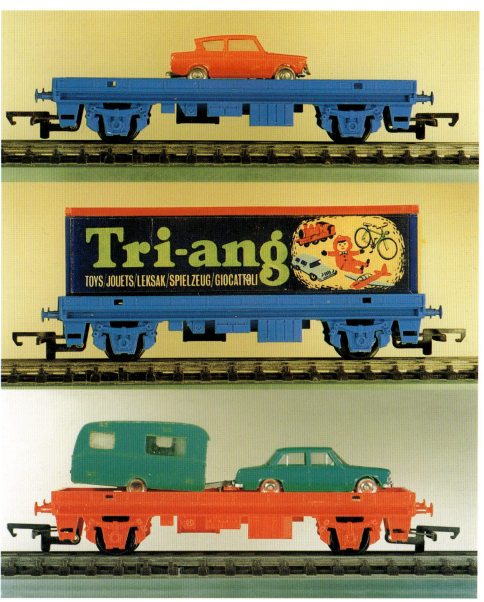

A range of integral wagon loads used in starter sets and thought to date between 1969 and 1971.

Canadian) on an R712 low-sided wagon.

Despite the 'CN' suffix it was almost certainly a standard issue integral low-sided wagon. No doubt the CN referred to the fact that with this particular load it had been ordered by the Canadian company and was not available to anyone else.

Low-sided Wagon and Car & Caravan R878

The low-sided wagon was fitted with a Minix car, which was made with an non-plated chassis, and caravan. Leaving out the plating process is unlikely to have cut the cost by much but it was obviously thought to be worthwhile. There was a choice of six Minix cars made for this purpose. The model is believed to have been used only in UK sets in 1971.

Low-sided Wagon & Car R887

The wagon was rainbow blue and the car was any one out of the current Minix range. The model was used in sets during 1971/72 (although it may have been available from 1968) and it is thought that 216,000 were made.

Low-sided Wagon & Van R888

This variation on the R887 was probably made in 1970 but I have not found evidence of it.

Low-sided Wagon & 20' Tri-ang Toys Container R876

The low-sided wagon was used with a yellow 20' Tri-ang Toys container. The container has red doors and a deep blue roof and the inscription was in red. It was available between 1971 and 1972. There may also be a blue version of the container.

Low-sided Wagon & 20' Frog Kits Container R877

This time a 20' yellow container, carrying a Frog Kits advertisement, was mounted on a low-sided wagon. The wagon appears to have been made in 1971.

LWB LOW-SIDED WAGON (30')

The long wheelbase version of the low-sided wagon was long enough to carry a 30' container and the sides were lower than those of the low-sided wagon (R712). It was used as the basis for several other models that follow and was in manufacture around 1970. It is slightly better detailed than the other integral wagons in having strut and planking detail on its underside. All examples seen by the writer had tension-lock couplings and colours used included blue, deep blue, red and yellow. Early examples have flat ends and later ones have a reinforcing bulge. Although at least 160,000 were made they are not easy to find today.

The wagon should not be confused with the long wheelbase flat wagon which had no sides but a flat bed with upright pegs for locating containers.

Long Low-sided Wagon & 30' Coca-Cola Container R875

This is believed to have been the l.w.b. low-sided wagon with a different Coca-Cola label on the sides of a 30' container. The container was red and was used between 1971 and 1972. 11,000 were made.

Long Low-sided Wagon & 30' Tri-ang Toys Container R880

This was a blue Tri-ang Toys container with a red roof and red doors mounted on a l.w.b. low-sided wagon for use in the RB2 set of 1970.

LWB FLAT WAGON

This was a completely flat wagon except for four locating pegs for containers. It was illustrated in the 1970 Trade Catalogue and all examples shown were yellow. I have never seen this wagon and it is possible that it was replaced by the long wheelbase low-sided wagon before going into production. The uses to which it could have been put are as follows:

Long Flat Wagon & 30' Coca-Cola Container R879

It was introduced into sets in 1970 and was probably replaced by R875 the following year. It used the standard integral l.w.b. flat wagon in yellow and standard 30' container with a red roof but printed paper stuck on the sides. The paper was blue with four large white discs per side onto which was printed by offset litho the Coca-Cola logo in red. The advert was supplied by Coca-Cola at the factory's request. 24,000 are believed to have been made.

Long Flat Wagon & 30' Tri-ang Toys Container R880

This had the detail printed onto paper stuck to the sides of a standard 30' container. The container had a blue roof and the paper sides were red with Tri-ang Toys printed in yellow. In white were the products in foreign languages and beneath these toy icons in blue. The model appeared in 1970, probably for one year only, and 63,000 are believed to have been made.

Long Flat Wagon and Car & Caravan R881

This model used the longer wheelbase flat wagon which was more suitable than the low-sided wagon for the Minix car and caravan load but there were not any sides to secure the load. 24,000 were probably made for sale in RB sets in 1970.

Long Flat Wagon & Car R889

Almost 40,000 of this model were made for use in RB sets in 1970. It consisted of a Minix car on the 30' integral flat wagon.

C. TRANSCONTINENTAL FREIGHT CARS

Only those models in production during the Tri-ang Hornby years are listed here. Some of the freight cars have been well covered in **Volume 1** and of

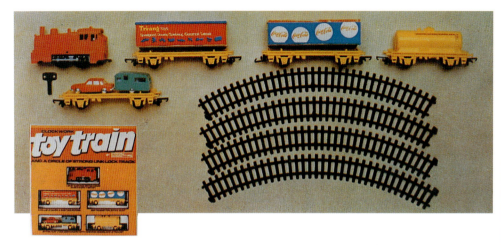

An illustration from the 1970 Trade catalogue showing the R880 and R879 container wagons, a 20' Tri-ang Toys container and a 30' Scalextric container. All were produced for cheap starter sets and are now rare and much sought after.

R344 track-cleaning cars.

these little further detail is given here. All are displayed here in the order in which the original model was introduced and all derivatives are listed with the original model. If the reader wishes to see a list of, for example, only those models made for Canada, a check list of these has been provided above for this purpose.

SHORT BOX CAR

New Zealand TC Short Box Car R114

While the small box car in the UK became the track-cleaning wagon, in New Zealand and Australia it survived for much longer. The New Zealand version was still in the catalogue in 1972 and showing little change in appearance. Colours seen have been shades of yellow or orange, although a dark green version has been found. While normally printed in black, red has been found and the roofs and doors vary between light and dark grey. Later models were made in a more translucent plastic and sometimes had the doors stuck in place instead of sliding.

Australian TC Short Box Car R114

The model remained in production in Australia until 1967 after which it was dropped. Those made during the later years were white but there is no guarantee that other colours, and especially orange or yellow, were not used. It is also thought likely that the 'Tri-ang Railways' inscription and 'TR22831' were used on the sides of the model until the end, despite the illustrations in the 1965 Australian catalogue supplement of a Trans-Australia version being planned. Later models had plastic bogies and some had the doors stuck in place and no door runners.

TRACK-CLEANING CAR

BR Track-cleaning Car R344

For many years this was the only track-cleaning vehicle available and it tended to take its colour from the BR engineering liveries. This meant that up until 1964 it was black. Its placement in the Transcontinental section recognises that despite its livery it was still a Transcontinental freight car using the body of the former R114 short box car described above.

It entered the new era in dark green livery, a colour retained until the last batch was made in 1970. The model was very popular with about 85,000 green ones produced between 1965 and 1970.

Red Version R344

It was shown in the 1971 catalogue in bright red livery but none was made that year and so the red version is strictly a Hornby Railways model. It was made in that colour between 1972 and 1976.

Export Track-cleaning Car R344E

Around 1965 it had been necessary to withdraw from sale in the UK stocks of the track-cleaning car which were supplied with carbon tetrachloride as the cleaning fluid. There was felt to be no reason why these should not still be supplied to overseas stockists and so the stocks of earlier models were renumbered R344E and steadily disposed of overseas over the next four or five years. The reason for this and the destination of this Rovex-made model are not known. It is quite possible that the change of cleaning fluid occurred at about the time the colour of the car changed from black to green.

CN Track-cleaning Car R3341 (R344CN)

The Canadian model had the same dark green body as the British track-cleaning car and had either a grey or black roof. It carried a large white CN logo on each side which was applied using the same tool as that used on the gondola and hopper car.

In 1969, 561 boxed and 2,750 unboxed models were made for the Canadian R469 presentation set but only 1,250 of the unboxed together with all of the 560 boxed models were sent that year. It is not known what happened to the 1,500 unboxed models that remained in the Margate stores at the end of the year. It seems likely that they were sent out with a batch of the standard issue track-cleaning wagons the following year to use up the stock.

M-T Express Track Cleaning Car R344NP

See B. Wagons Made for Starter Sets.

CABOOSE

TC Caboose R115 (R115TR)

The caboose, TR 7482, was perhaps the most important of the Transcontinental freight cars, being needed wherever a set was required. It consequently appeared in many guises. The basic model remained widely available in 1965 and 1966 but was not made in either year. After this, the only listing of it to be found was in 1969 and 1970.

TC Caboose for Canada R115TR, R1150

Both maroon and brown versions may be found in Canada. Towards the end of 1969 about 2,500 were made, of which more than 2,000 were sent unboxed to Canada for the TS906 Steam Switcher Freight Set. They were not in fact dispatched until 1970 and the remainder were sent to Canada boxed the same year. The number '1150' was hand written on one end of the box.

1969 had also seen the production and dispatch of a batch of 1,200 cabooses for Canada which were numbered R115TR some of them unboxed.

New Zealand TC Caboose R115

The model was made at the Auckland factory throughout the 1960s and was inscribed 'TRI-ANG RAILWAYS' to the end! Later models were made in a translucent plastic and without a running number. I have seen only red or maroon examples. The roof colour varied between light and dark grey.

Canadian cabooses: R115CN, R1151, R1152 and R1153. Note the shorter steps on the two later models.

R115 New Zealand and Australian cabooses.

R114 Australian short box car.

R3441 Canadian National track-cleaning car.

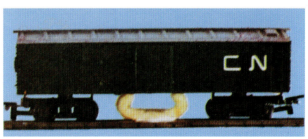

Australian TC Caboose R115

Australia continued to make its own caboose until 1967 after which it was dropped. Later ones were blue with plastic bogies. These would have been heat printed with the original 'TRI-ANG RAILWAYS 7482' inscription and, it would appear, not the 'TRANSAUSTRALIA' illustrated in the 1965 Australian catalogue supplement.

CN Caboose R1151 (R115CN) (R270)

The model first appeared in the catalogue in 1965 with a body in a soft orange-red, black roof and a large, light greyish-green 'CN' with the number beneath it on each side. The first shade of orange seems to have been available in 1965 and 1966 but a brighter orange plastic was used for those sold in 1967 and 1968.

The pre-production model for this livery has turned up in Canada, suggesting that it may have been sent out to the Canadian company for approval before going into production. The body had been moulded in black plastic and spray-painted a dark red. The detail was correctly in white and the number it carried was 'CN3470'. This was the model used for the illustration in the 1965 Canadian catalogue and one of the pictures in the 1967 catalogue.

By the time the 1969 Canadian catalogue was illustrated the model had assumed a different livery. Judging by the picture on page 3 it was still orange but the CN logo (now smaller) and the number (now 7482) had been separated to opposite ends of the wagon side. The roof was now grey. No evidence of this version has come to my notice.

A maroon version, with these new markings, took over in 1969. This had a grey roof and a grey or black chimney. MP2 wheels were fitted in 1971 and in 1972 the steps were modified to avoid System 6 point motors. The model may be found in Hornby Railways boxes either as R1151 or R270, the latter being the number allocated to the model in 1973 when it was found that four digit numbers could not be handled by the works computer.

A paler version is believed to date from around 1970.

8,000 Canadian National cabooses were made in 1965 and a further 15,000 over the next three years. In addition, in 1968, a batch of 7,500 was bulk packed to Canada for use in sets. A further 9,000 boxed and 14,000 unboxed models were sent out to Canada between 1969 and 1971. A final batch of 8,600 was sent out in Hornby Railways boxes in 1973. Surplus stocks of the wagon were returned to the UK for sale in 1981.

M-T Express Caboose R115NP

See B. Wagons Made for Starter Sets.

CP Caboose R1152 (R115CP)

The Canadian Pacific caboose arrived in 1969 when 5,700 were shipped unboxed for set production. It was also illustrated in the Canadian catalogue that year. In 1970 the first boxed models went to Canada, but the batch contained only 600 and so must have left a number of people disappointed. A further 6,300 unboxed models had been made at the end of 1969 to meet 1970 orders but only 3,800 of these were sent out in 1970, the remainder going to Canada the following year after having been put in boxes for solo sales.

The model had an orange-brown body and grey roof with either a grey or black chimney. They carried the inscription 'CANADIAN PACIFIC' and '346346' in white on each side. After 1971 it was dropped from the range.

CP Rail Caboose R1153

The CP Rail caboose was first shown in the 1970 Canadian catalogue and stocks of it were still held by Louis Marx Industries (Canada) Ltd in 1974.

The model was bright yellow with the black and white logo and 'CP Rail' in black on the sides. It also carried the number 'CP 35644' and other detail printed in black. The roof was yellow but the smoke stack was black. The yellow plastic used in later batches was more translucent than earlier ones. The last examples had modified steps to clear System 6 point motors.

About 4,800 were made in 1970 of which only 800 were sent out boxed. A further 4,750 boxed models were dispatched between 1971 and 1972. A final batch of 5,200 was ordered by Louis Marx in 1973 and these cars will have been sent out in Hornby Railways packaging.

Canadian Starter Set Caboose R1153NP

See B. Wagons Made for Starter Sets.

Proposed New-Bodied Caboose R114

Rovex realised that their caboose left a lot to be desired and in 1971 were planning a new body. This was primarily for Canada and was to have been ready for January 1972. Presumably it did not reach a high enough priority and was dropped from the Programme.

GONDOLA

TC Gondola R116

The last stores recording of Transcontinental gondolas was in 1966 when about 750 were available. As these were not recorded the previous year it mightbe argued that they were made in 1966 but it seems a very small number to produce and could have been old stock uncovered. The R116 number was taken over by the Rock Island Lines version supplied to Canada and detailed below.

New Zealand TC Gondola R116

In New Zealand the Tri-ang Railways version survived to the end in 1972. Over the years it appeared in many different colours a few of which are listed opposite:

The version pictured in the 1972 illustrated price list had 'TRI-ANG RAILWAYS' on its sides and no number. Examples may be found with the couplings bolted onto the stocks instead of being attached with eyelets.

R1111 Canadian National hopper car.

Australian TC Gondola R116

The Australian made gondola was dropped after 1967. The late ones carried the inscription 'Express Delivery' under marked with a wavy line. They had plastic bogies and were normally Tuscan red or mid green.

*Right: Gondolas –
R116 New Zealand made,
R116 Australian made,
R1161 Canadian National
and R1160 Rock Island Lines.*

Body colour	Printing colour	Printing	Cpl.	Bogies
greeny blue	white (transfer)	TRI-ANG RAILWAYS TR3576	II	metal
greeny blue	white	TRI-ANG RAILWAYS TR3576	II	metal
brighter blue	white	TRI-ANG RAILWAYS TR3576	II	metal
brighter blue	white	TRI-ANG RAILWAYS TR3576	III	metal
maroon	white	TRI-ANG RAILWAYS TR3576	III	metal
pale green	white	TRI-ANG RAILWAYS TR3576	III	metal
deep green	white	TRI-ANG RAILWAYS TR3576	III	plastic
deep green	yellow	TRI-ANG RAILWAYS TR3576	III	plastic
blue green	yellow	TRI-ANG RAILWAYS TR3576	III	plastic
bright green	white	TRI-ANG RAILWAYS	III	plastic
dark green	white	TRI-ANG RAILWAYS	III	plastic
deep green	white	Express Delivery	III	plastic
deep green	yellow	Express Delivery	III	plastic

CN Gondola R1161 (R116CN)

The model did not reach production until 1966 which may explain why there was no picture of one in the 1965 Canadian catalogue although it was included in the price list.

The body was moulded in brown plastic and rather crudely heat printed with white lines representing lines of technical codes. The car carried the number 'CN141101' and the usual CN logo.

The model survived to the end of the series and was among the models returned to the UK in 1980. Almost 12,000 were made of which 4,100 were dispatched unboxed, some later being packaged in Canada. The last batch of 2,400 was delivered in Hornby Railways boxes.

Rock Island Line Gondola R1160

The model was first sent out to Canada in 1970 and it survived long enough to be amongst those repatriated in 1980. Only two batches were made, the first, of 3,750 (mostly made at the end of 1969) was sent out to Canada in 1970 and a further 3,000 in 1973. Half of the first batch were unboxed for use in the No. 913 Cross Canada Diesel Freight Set and all of the last batch were in Hornby Railways boxes.

The body was pale grey (it can also be found in the lighter BR rail grey) and 'NORTH AMERICAN' and 'ROCK ISLAND LINE' were heat printed in black on the sides.

HOPPER CAR

TC Hopper Car R111

About 1,800 were made in 1965 and a further 250 in 1966. After this all hopper cars made were in Canadian National livery.

CN Hopper Car R1111 (R111CN)

Like many of the Canadian National freight cars R1111 was moulded in orange-brown plastic but, arriving in 1967, it was late in being introduced to

the series. The sides were heat printed with 'CANADIAN NATIONAL CN9 8103' and the large CN logo.

The first batch of 1,400 models sent over in 1967 were all boxed and had the gravity unloading mechanism and metal B2d bogies and MS4 wheels. Many of them would have been used in the PS267 Work Train Presentation Set and are quite scarce today.

The following year, 1,700 unboxed models were sent for set production together with a further 1,200 in boxes. These and subsequent batches would have had the unloading mechanism replaced by a plastic plate. In 1970, 5,700 were sent for the production of the TS901 Cross Canada Diesel Freight Set. Remnants of these were used in the No.926 Canadian National set of 1971. In all 21,300 were made, the last of them returning to Britain in 1980.

OIL TANKER

Shell Blue Oil Tanker R117

A batch of 2,500 oil tankers was made in 1965. These were probably the dark blue tankers with the small Shell stickers. A further batch of 300 seem to have been made in 1966 possibly from existing mouldings in the factory, as it does not seem likely that the mould would have been got out of the store and cleaned up for such a small number. Some of the blue bodies ended up in the R117CN production line and came out as CN tankers. As by then production had already changed to the CN livery for the oil tanker, all of the 300 1966 models may have ended up with this livery.

In 1969 and 1971 the tanker was moulded in yellow plastic and finished either with red shell stickers or with BP transfers. These were produced principally for the Canadian market and are dealt with below.

New Zealand Shell Oil Tanker R117

The blue shell oil tanker had been used in New

Zealand sets from 1961 and these must have been delivered bulk packed in the early years in sufficient quantity to last many years. There were no records of any being supplied during the Tri-ang Hornby period until 1970 and possibly 1971 by which time the model may have been yellow. Even as late as 1969 the illustrated price list showed the oil tanker as having the 1955 body finish. Some blue oil tankers were boxed and sold as solo models. These had Shell transfers but no other markings.

CN Oil Tanker R1171 (R117CN)

In 1965 the blue Shell oil tanker gave way to a black Canadian National model on the production lines at Margate. The moulding carried the CN symbol and the number 'CGTX 20044' in white.

For some unknown reason a batch of blue CN oil tankers was made in 1968 and heat printed in off-white. Possibly the factory ran short of black plastic and so blue was used to finish off a batch of freight cars. Alternatively, some blue bodies left over from the Shell oil tanker production were used up to finish a batch rather than put the tool back on the moulding machine for such a small number. It seems unlikely that Rovex would have run short of black plastic in view of the large quantity they used and so the second explanation seems more likely. The blue version is known to have turned up in the TS.673 set which was being assembled in Canada at that time.

As one might expect, blue CN tankers are something of a rarity today. In contrast, the conventional black model is probably the most common of the Canadian range as it was one of the few freight cars, in Canadian livery, that was featured in the UK range (1968-69). During this period 8,500 UK orders were received.

Another uncommon version with a black body has the 'CN' logo on each side, both at the same end. On normal cars the 'CN' was always on the right as one looked at the side of the car. This variation

*Oil tankers:
R117 Shell and
BP tankers
and R1171 CN
tankers in black
and blue, the
latter being quite
scarce.*

*Stock cars: R1260A made for Australia, R126CN and R1261 both CN versions and R1262
Canadian Pacific.*

belongs to 1972 or 1973.

A little over 20,000 of the CN oil tanker were made between 1965 and 1969 but only 800 were bulk packed to Canada for use in sets, these going out in 1968. The rest went out in boxes.

Shell/BP Oil Tanker R1170 (R323)

The yellow Shell bogie tanker had first appeared in the 1969 Canadian catalogue, both in the TS903 set and as a solo model and 2,600 had been made probably towards the end of the year, the last 1,200 being boxed. It was not until 1970 that these were sent to Canada. The following year a further 4,140 went to Canada, all of them boxed. A final batch of 3,600 went out in Hornby Railways packaging in 1973.

It seems likely that all these models were moulded in yellow plastic and finished either with red shell stickers or with BP transfers. They carried no other detail. The model was always advertised as a Shell tanker and so it is fair to assume that the BP version resulted from a shortage of Shell stickers and a good supply of BP transfers from the R12 petrol tank wagon production line. How many of each were made is not known. The Shell version was amongst the surplus stocks brought back to the UK in 1981 and sold by Zodiac.

The model was also listed as R1171 in 1972 but this is thought to have been an error.

It would appear that a small quantity of the Shell (or BP) oil tankers ended up in New Zealand. These went out in 1970 and were boxed. A batch of unboxed cars was ordered in 1971 but it is questionable whether they were dispatched.

STOCK CAR

TC Stock Car R126

Nearly 3,000 stock cars were made in 1965 and 500 were left in stores at the end of the year. An example of this version has appeared as late as 1971 in a Canadian bubble pack. The 'TR 2742' printing on it was described as smaller than usual. This is an example of how Canadian bubble packs serve as time capsules. So long as the seal has not been broken, you know that the contents date from the period that this type of packaging was in use.

Victorian Railways TC Stock Car R1264

There was a proposal to produce the stock car in VR livery for use in the RS102 set assembled at the Moldex factory in Melbourne but the numbers required were too small to make it worthwhile and the Australian plant was issued with a TC version instead (see below). It is likely that a sample model was made in the proposed livery.

Australian TC 1970 Version R1260A

In 1970 a batch was made for sets being assembled in Australia. About 1,550 were sent and these may have been the bright yellow model inscribed 'TRANSCONTINENTAL' and 'TC 1260' in black. This had a yellow body and roof but black underframe. The model was later seen in the RS 102 set of 1972 and some 250 boxed models were released from the stores in 1972.

Australian 1972 Version R1296 (R1330)

A special version of the wagon was considered for the Australasian and South African market for 1972 but dropped; not before it had been allocated the number R1296, later changed to R1330.

CN Cattle Car R1261 (R126CN) (R271)

This was another of the original CN freight cars introduced in 1965. It was shown in the catalogue as yellow but in fact it had an orange brown body and roof and 'CANADIEN NATIONAL' and 'CN 172350' in off-white on the side panel. The 1967 Canadian catalogue showed both yellow and brown versions of the model but I have not yet received any evidence of a yellow version.

In 1969 a change occurred in the printing when 'CANADIEN NATIONAL' gave way to the 'CN' wet spaghetti logo printed in white. This version is not very common as the model seems to have reverted back to the original design, but printed in white rather than off-white, in 1970. It was among the stocks of unsold models returned to the UK in 1981.

Between 1965 and 1967 11,000 boxed models went out to Canada. In 1968 a batch of 2,550 unboxed cattle cars was dispatched followed by 2,200 the next year and another 2,200 in 1970. Between 1970 and 1971 a further 1,500 boxed cars were sent and a further 2,600 in Hornby Railways packaging were shipped out to Canada in 1973 bearing the number R271 on the boxes. Some of the latter returned to Britain in 1981.

CP Cattle Car R1262 (R126CP)

Another of the new Canadian Pacific freight cars announced in 1969 was the cattle car. This, like the CN version, was a standard stock car moulding. It was finished in all over yellow with 'CANADIAN PACIFIC' and 'CP 503588' in black on the side boards. It remained unchanged until the series was wound up with stocks left for disposal on the 1974 price list.

They were made only in 1969 and around 4,300 unboxed models were sent out to Canada for the TS903 train set followed in 1970 with a further 1,500 for the No.915 set. Amongst unsold Canadian stock packaged in bubble packs by the Canadian company were some R126 TC stock cars but the packaging carried the number '1262' for the CN version. To add to the confusion, there is no record of R126 being bulk packed and sent to Canada. This suggests that not all the R1262 models bulk packed to Canada were in the correct CP livery. Probably there were insufficient of the CP model to meet the order and so numbers were made up with the TC version but given the CP cattle car 'R' number.

Between 1970 and 1971 2,150 boxed models were supplied, the first batch of which had been made at Margate late in 1969.

REFRIGERATOR CAR

TC Refrigerator Car R129

It is possible that none was made in 1965 but a batch of 8,000 was made in 1966. These would have been the standard Transcontinental version with a white body and black roof and the TR shield on the sides.

At the end of 1968 there were about 300 unboxed models in store being held for New Zealand and between 1970 and 1973, about 1,000 boxed models passed through the stores.

CN Refrigerator Car R1291 (R129CN) (R272)

Another of the original six cars in Canadian National livery, R129CN was first manufactured in all over grey-silver. By 1969 this had been changed to a white body and black roof and in this livery it survived into Hornby Railways boxes. This, however, does not tie in with a comment in a memorandum sent by Stafford to Richard Lines in December 1970:

"Canada have requested that the 1291 Refrigerator Car be issued with silver sprayed body and a change note should be issued accordingly"

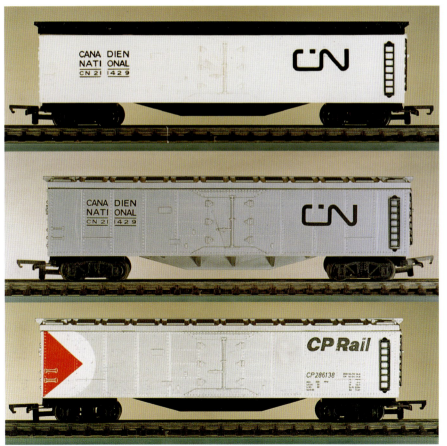

Refrigerator cars: Canadian National in both white (R1291) and grey-silver (R129CN) and R1293 CPRail in silver.

This suggests that the model returned to silver livery in 1971, assuming that Canada's request was fulfilled.

The model carried a large CN logo and the words 'CANADIEN NATIONAL' and 'CN 21 1429' all in black. In the 1966 catalogue supplement it featured in the No2 presentation set.

It seems that about 9,600 of the silver version were made of which 1,500 were sent to Canada unboxed for use in sets. About 16,000 of the white version were sold of which some 7,700 were unboxed. In 1973 about 3,900 were sold in Hornby Railways boxes by now renumbered R272 and some of the bulk packed models found their way into bubble packs – strangely carrying the 'R' number '1290'.

CP Refrigerator Car R1292

Not much is known about this model except that it

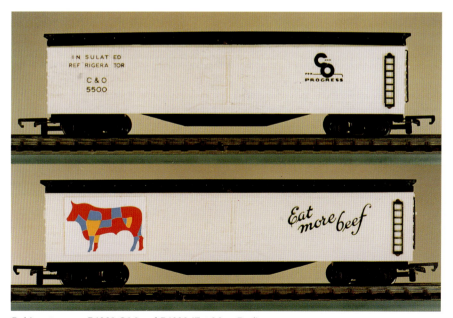

Refrigerator cars: R1290 C&O and R1330 'Eat More Beef'.

was made only in 1970 when just 900 went through the stores. These sold as solo models in boxes.

C & O Refrigerator Car R1290

This model was manufactured at the end of 1969 and 2,200 unboxed cars went to Canada in 1970. A further batch of just 600 were boxed and sold slowly over the next three years. Some of the unboxed cars were later packaged in bubble packs for sale as solo models. It seems likely that in 1973 a further batch of 300 or 400 went out to Canada in Hornby Railways boxes.

The model had a white body and either black or white roof and underframe. On the sides it was heat printed in black with 'INSULATED REFRIGERA-TOR' and 'C&O 5500' and at the other end 'C AND O FOR PROGRESS'.

CP Rail Refrigerator Car R1293 (R273)

This was moulded in grey and sprayed silver. It carried a large CP Rail logo in metallic red and white at one end of each side and 'CP Rail' and 'CP288138' and a mass of other coded detail in black at the other end. The underframe was black. It arrived with the other CP Rail equipment around May 1970 and was amongst the surplus Canadian models being sold by Zodiac in 1981. The 1970 catalogue contained a picture of a pre-production model.

The first batch to reach Canada in 1970 was bulk packed and contained a little over 7,000 models. The remaining 8-9,000 dispatched over the next two or three years were boxed, the last ones probably in Hornby Railways boxes and numbered R273. No more than 4,000 would have carried this number.

The model was wrongly labelled 'R1353 CP Rail Newsprint Car' in the 1973 Canadian Hornby Railways catalogue. In the CP Rail colour coding, yellow is used on certain refrigeration cars.

CP Rail Newsprint Car R1353 (R276)

The first version of this model was made in May 1970 from the refrigerator car (R129) tools. A pre-production model was illustrated in the 1970 catalogue.

Green was the CP Rail livery for freight cars carrying wood products and the model had a green body, roof and underframe and carried the CP Rail logo in black and white on the right hand side and 'CP Rail' and 'CP81030' and other detail in white at the other end.

A little over 4,500 of the model were sent to Canada that year of which 3,800 were sent bulk packed for use in sets. When production was repeated in 1971 the Long Box Car (R136) tool was used and this version is dealt with below. In 1973 a further 3,000 were made, presumably with the refrigerator body, and these were dispatched in Hornby Railways boxes and were probably marked 'R276'.

Beef Car R1330

Another freight car ordered by Louis Marx and issued in 1973 in Hornby Railways boxes. 3,000 were made and residue stocks came to Britain in 1981.

The model was white with a black roof and underframe and carried a large sticker on each side showing a segmented bovine in red, blue and yellow. At the other end, heat printed in black, was 'Eat more beef'.

Australian Beef Wagon R1296

This was proposed for the 1972 RS103 set for the Australian market. It was originally shown as R1296 suggesting that an Australian name was to be used on the wagon, and then it was changed to R1290 suggesting a standard refrigerator car. Yet later still the set was to have the R1330 Beef Car described above. It is quite possible that a pre-production model of R1296 exists carrying an Australian road name. The model was to have also been boxed up as a solo model for Australia and South Africa and be offered bulk packed to New Zealand.

Newsprint Car R1350 (R275)

This was a very late addition to the range being planned for 1972 but coming in 1973. It used the refrigerator car body in white plastic and carried on its sides the names of newspapers. The model had a black roof and underframe. 3,000 were made and sent out to Canada boxed, probably as R275.

LONG BOX CAR

TC Long Box Car R136 (R1360)

About 2,150 were in the stores in 1965 but these had been carried over from the previous year. In 1966, however, it seems that a further 870 were made. There is no record of anymore being made but in 1971 about 1,150 were recorded in the stores at Margate.

CN Long Box Car R1361 (R136CN) (R277)

In its original form the body roof and underframe were moulded in brown plastic and 'CANADIAN NATIONAL CN523976' was heat printed in white at one end and the CN logo at the other. This first appeared in 1965 and the following year was shown in the No2 presentation set.

By 1969 the model had changed to a vermilion red body and yellow roof and underframe but retained the same heat printing in white on its sides. In this form it survived until the end being amongst the stock returned to the UK in bubble packs. 1973 models, of which 3,000 were boxed and sent to Canada, were in a darker red with more translucent yellow fittings.

Although the model was shown in sets quite early on it is interesting to note that none was sent out unboxed until 1968. About 8,000 CN box cars were made in brown livery and a further 14,000 in various versions of red and yellow. 400 of the former and 8,500 of the latter were unboxed. The 3,000 boxed models that went out to Canada in 1973 probably carried the number R277 on their boxes.

An unusual version of the model has a maroon body and yellow doors, the latter having been moulded in maroon but painted in a semi-matt yellow.

The roof and underframe were in the same yellow as the Canadian Pacific Stock Car and they had MP1 wheels. This variation is thought to have originated in 1969 and it has been found in both DO window boxes and Canadian bubble packs.

CP Long Box Car R1362 (R136CP)

The model first appeared in the catalogue in 1969 but remained in the series until 1973. It had a brown body and roof and was heat printed in white with 'CP 258599' and 'CANADIAN PACIFIC RAILWAY'.

Only 2,600 models were made, all of them in 1969 and all but 93 of them were sent out to Canada unboxed for use in the TS902 Canadian Pacific diesel freight set. Any models that were not used up by the end of the year were packaged in bubble packs for solo sales.

CP Rail Box Car R1363 (R278)

The freight car made its debut in May 1970 as a red box car numbered 'CP202199'. It had a vermilion red body, roof and underframe and carried a large black and white CP Rail logo at one end and at the other, in white, 'CP Rail' with much technical detail beneath it. 5,000 were made of which only 1000 were boxed.

In 1971 a maroon version of this same model was made which also differed in having a black underframe although some were moulded in yellow and painted black. 3,000 maroon cars were sent out to Canada that year, all boxed.

Vermilion cars on CP Rail are used for general purpose conveyance but it is not known what maroon ones are used for if they exist.

Although the model was allocated a new 'R' number (R278) in 1973, there is no record of any having been made that year.

CP Rail Box Car R1343 (R274)

For some reason another version of the CP Rail box

Below:
Newsprint cars: R1353
CPRail and R1350 carrying
the names of North American
papers. Both were made using
the refrigerator car body.

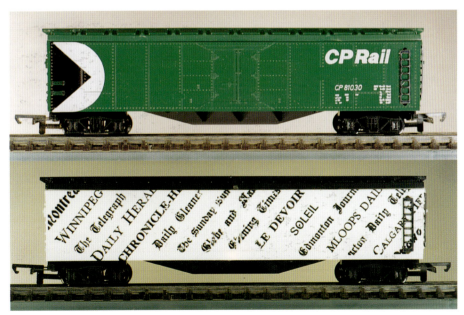

Long box cars: R136CN with a brown plastic body and R1361 with a red body and yellow roof. The catalogue illustration of the R1362 Canadian Pacific long box car shows that it was the same brown as R136CN.

car was sent out to Canada in 1971 with a different 'R' number. This was yellow with a yellow roof and black printing and underframe (or yellow painted black). It carried less technical detail printed on the sides and was given the running number 'CP 35644' from the R1153 CP Rail caboose.

A batch of 3,000 was made in the spring of 1971 but only 1,900 were dispatched that year to match the Canadian order. The remainder were dispatched in 1972. Another 3,000 were made and sent in 1973 and these should have had R274 on the boxes.

CP Rail Newsprint Car R1353

The CP Rail Newsprint Car had been made with the refrigerator car body in 1970 but when production was repeated in 1971 the long box car (R136) tool was used instead. The model had a green body, roof and underframe and carried the CP Rail logo in black and white on the right hand side and 'CP Rail' and 'CP81030' and other detail in white at the other end.

All models were boxed. Only 2,000 were made initially but a further batch of 3,000 was sent out in 1973 in Hornby Railways boxes. Some of them returned to Britain in 1980 for sale by Zodiac Toys.

CEMENT CAR

TC Cement Car R137

Some 2,000 were made in 1965 but of these between 400 and 600 remained unsold in the stores until 1968 when most of them were disposed of.

CN Cement Car R1371 (R137CN)

The model was first listed in the 1966 Canadian price list but not illustrated until the following year.

The model had a greenish-grey plastic body which carried a large CN symbol and 'CANADIAN NATIONAL' and 'CN11 3000' in brown. The underframe was black. The first batch of the model carried the gravity unloading mechanism and so it was fitted with B2d metal bogies and MS4 wheels. Later models, instead of the unloading mechanism, had a body coloured filler plate which had the tendency to fall out.

Long box cars: the three colourful CP Rail cars - yellow (R1343), maroon (R1363) and vermilion (R1363).

Below: Cement cars: the rare R 1371 CN cement car, this one without the gravity unloading mechanism, and the more common R1370 Wabash version

For the first two years, only boxed models were sent to Canada but, in 1968 and 1969, unboxed batches were dispatched. Some 4,400 boxed and 2,400 unboxed models left the factory during the four years it was in production.

Its place was taken by the Wabash cement car as early as 1969 but sufficient unsold stocks of the Canadian National model ensured that it remained listed until 1971.

Wabash Cement Car R1370

The Wabash cement wagon appears to have been introduced in 1969 when just 500 were made. These were not sent out to Canada until 1970. In all, records show that no more than 5,500 were made but when this is compared with almost 8,000 of the CN version having been recorded going through the store, something appears to be wrong. The Wabash model is not hard to find, unlike the CN model. One possible answer is that some of the later batches of models listed as CN cement cars where in fact the Wabash version. This would also fit with the CN version being withdrawn from the catalogue so early on.

The Wabash cement car had a grey body and roof and black underframe. In white on its sides it had heat printed 'WABASH' and 'WAB 31627'. It had no unloading mechanism. From 1970 the roof was charcoal coloured.

Australian Iron Ore Wagon R1374 (R1375)

This was proposed for the RS103 set of 1972 and was initially allocated the number R1374 and, later, R1375. While a mock-up was probably prepared, the wagon almost certainly did not go into production. It was to have been the standard grey but carrying an Australasian iron ore mining insignia.

Besides being used in the set it was also to have been packaged as a solo model for Australia and South Africa and be offered to New Zealand bulk packed.

Snow ploughs: R138 TC and R1381 CN versions together with a photograph of the catalogue illustration showing the CN version in dark green.

SNOW PLOUGH

TC Snow Plough R138

The standard livery snow plough remained in production with batches being made up to 1968. After the end of 1965 a total of 5,500 were made, the largest batch being 3,600 in 1968.

CN Snow Plough R1381 (R138CN) (R279)

The first CN snow ploughs were made in 1966 and that year more than 1,000 crossed the Atlantic. They were similar to the standard Transcontinental model but had a black roof and brown body and carried a small CN symbol and the inscription 'Warning –

Close Wings Before Passing Through Shed' and 'Built in Britain by Rovex' other small writing heat printed in white near the front of the plough. This version was to form part of the PS267 Work train Presentation Set.

By 1969 the model had a new green and grey livery with the same small version of the CN symbol (unlike the pre-production model illustrated in the catalogue) heat printed in white on its sides. 'Built in Britain by Rovex' had been deleted.

In all about 8,000 were made of which possibly half were the green version. It also seems that about 2,600 of the green ones were bulk packed to Canada. This was one of the surplus models sold off in bubble packs. None was made after 1971 even though

R1390 Heinz Pickle car.

R1200 Canadian National flat car with logs.

the model was allocated a new 'R' number (R279) in 1973.

Due to its popularity with Canadian scale modellers, many were 'improved' and consequently both versions in 'as manufactured' form are quite scarce today.

PICKLE CAR

Westwood Pickles R139

The Westwood pickle car remained in production during 1966 and 1967 and the version shown in the Canadian catalogues was the one with a red head board and white tanks. Approximately 1,000 were made during this period. The model was replaced in 1970 by the Heinz pickle car.

Heinz Pickle Car R1390 (R139CN)

After the fictitious Westwood Pickles, when the Heinz Pickles car arrived in 1969, it was nice to see a name with which people could associate. The car had a grey roof and green body with white tanks. Each of the four tanks carried World famous '57' heat printed in red, two facing left and two facing right. Along the side of the car the name 'HEINZ PICKLES' was printed in white.

Three batches, totalling in all 10,500 models, were made between 1969 and 1971. All were sent boxed

and so it seems unlikely that any would have been packaged in bubble packs in Canada. The model remained in the catalogue until 1973 but with no production in 1972 or 1973 it also seems unlikely that it was ever packed in Hornby Railways boxes. No doubt someone will prove me wrong on both points!

DEPRESSED CENTRE CAR

TC Depressed Centre Car R236CN

This model was also listed as R236CN suggesting that it was in CN livery but such an example is not known to me. In 1968 2,400 R236CN models were recorded as having entered the stores that year, 800 of them for dispatch to Canada. Little more is known of any of these and it was dropped from the Canadian catalogue in 1969. The last illustration of it showed it as a green bodied wagon.

FLAT CAR

CN Flat Car R234CN

This is a rare model. Only 2,000 were made, most of them in 1967 and a few in 1968. About 600 were sent to Canada unboxed for use in the PS267 Work Train Presentation Set. The model did not survive long enough to receive a 4 figure 'R' number.

The CN flat car had a brown body and come off

the same production line as the CN pulp wood car using the same printing tool. The main differences were the provision of eight metal stanchions and the absence of a log load. It was packaged in the same 'Tri-ang Railways' box used for its Transcontinental predecessor.

Flat Car with Logs R1200 (or R1201)

This is another rare freight car as only 800 were made. These passed into the stores in the second quarter of 1971 but were not dispatched to Canada, boxed, until the following year.

The model consisted of the grey or black Transcontinental flat wagon with its eight load retaining stanchions and three logs. The logs were stained dowels similar to those used for the R345 Side Tipping Car, but longer. Why such a small number of models was available, in a choice of colours, can be explained in that the pulp wood wagon was made in these colours and they shared the same body.

A grey example has turned up in a bubble pack and a black one in a 1972/3 Hornby Railways box. The latter was heat printed with 'CN655309 CANADIAN NATIONAL' printed in white on its sides. This is the same inscription as that carried by the R235CN Canadian National Pulp Wood Car.

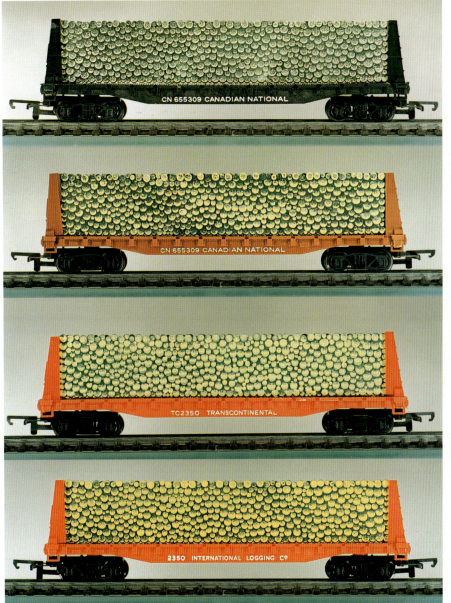

Pulp wood cars:
Canadian National black
(R2351) and the more
common brown
(R235CN), R2350
Transcontinental and
R2354 International
Logging Co.
Below: R822 Northern
Pacific, R823 Southern
and R824 L&N; all made
for ATT of America.

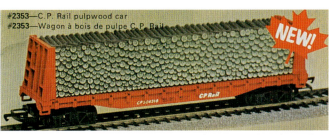

Catalogue illustration of the rare R2353 CPRail version of the pulp wood car.

PULP WOOD CAR

TC Pulp Wood Car R235

In 1968 a stock of 588 was recorded. Whether these were old stock or a small batch assembled to meet a customer order we do not know but that year the grey version of the model was once again featured in the British catalogue.

In 1969 a batch of 5,675 was made and most of them were sold that year. Although the illustration in the 1969 catalogue was of the brown CN-liveried model these are thought to have been brown-bodied pulp wood cars with 'TR4415' instead of the CN markings.

A further batch of 1,270 unboxed models was made in 1970 for dispatch to Australia for assembly of sets.

Victorian Railways Pulp Wood Car R2354

This was to have been produced for the RS102 set but was, instead, replaced by the standard TC Pulp Wood Car in stock at the time. Had it been made, it was also to have been boxed up as a solo model for the Australian and South African markets and offered bulk packed to New Zealand in 1972. While a pre-production model may exist, it seems very unlikely that it went into production.

CN Pulp Wood Car R2351 (R235CN)

The model was principally moulded in brown plastic but black examples, probably dating from 1969, may be found. They both carried the same 'CN655309 CANADIAN NATIONAL' heat printed in white on their sides. It was first listed in 1966 and illustrated in the 1967 catalogue, where it is shown fitted with side irons which were absent in later pictures and have not been included on examples seen by me.

R2351 must have been a very popular model. Just over 10,000 CN pulp wood cars were sent boxed to Canada between 1966 and 1972. In addition, a further 3,600 were sent out unboxed for use in set assembly.

CP Pulp Wood Car R235CP

In 1969 just 2,800 black CP pulp wood cars were made and sent to Canada unboxed for use in sets. As no others were made, this model is quite scarce even though it could still be found on dealers' shelves in Canada in 1970.

It carried the inscription 'CANADIAN PACIFIC' and 'C.P.520047' on both sides. Both body and ends were black. The model did not survive long enough to receive a four-figure number.

CP Rail Pulp Wood Car R2353

The model was another of the new freight cars shown in the 1971 Canadian catalogue. It was vermilion with 'CP 304318' and 'CP Rail' in white on the sides and the usual load of greenish logs. An example has been found with black retainers which may date from the last days of production. In 1971 a single batch of 2,200 was made around May and supplied boxed to Canada.

TC2350 Transcontinental Pulp Wood Car R2350

The final version of the model was in vermilion plastic with black bulkheads and with the inscription 'TC2350 TRANSCONTINENTAL'. Neither date nor quantity are known, although it was most likely made when the CP Rail version was in production.

Australian TC Pulp Wood Car R2350A

This would have been the above model sent out to Australia bulk packed in 1970 of which some possibly found their way into Moldex boxes.

International Logging Co Pulp Wood Car R2354

This was one of the models produced at the request of Louis Marx & Co. of Canada Ltd after 1971 and is therefore a Hornby Railways model. The name 'Tri-ang' was removed from the moulds and 100 were made in 1972 and a further 3,000 in 1973. It is likely that all were issued in Hornby Railways boxes.

There was some confusion over numbering with the model sometimes being referred to as R2354 and

sometimes as R2350. Private owner versions of wagons normally took the number ending in '0' but in the case of the pulp wood car this was already being used as the standard wagon was still in production.

Northern Pacific Pulp Wood Car R822

Made exclusively for ATT of America and fitted with NMRA couplings, the wagon was moulded in black plastic and had the road name heat printed in white on its sides. 4,950 of the cars were supplied in 1968. The models were sold in grey ATT boxes having been sent to the States bulk packed. Some of the models were acquired by Model Power and re-boxed in their blue boxes.

Southern Pulp Wood Car R823

The details given for R822 above also apply to this model. 5,000 were made and dispatched in 1968.

L & N Pulp Wood Car R824

The details given for R822 above also apply to this model. 5,220 were made and dispatched in 1968 but, in 1970, 'frustrated' stock was found in the stores at Margate and offered to retailers at 2/7d per model.

TC CRANE CAR

TC Crane Car R560 (R5600)

The model had been added to the TC range in 1962. In 1966 and 1967 two further batches of this model were made totalling 2,750 models for the two years. It is thought that these were brown and green.

The model was listed in the Canadian part of the production census having been supplied to Canada for some time. The last batch of crane cars sent to Canada had been in 1969 when 1,000 were sent for the R469 presentation set with a residue of 300 being boxed up and sent the following year. In 1971, however, it was shown in the Canadian catalogue with a black car and a yellow crane but there is no record of any being made either that year or, indeed, at any time after.

R5600 TC crane car.

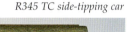

R345 TC side-tipping car

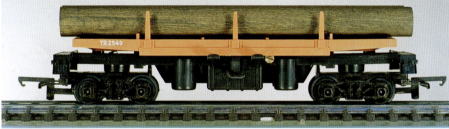

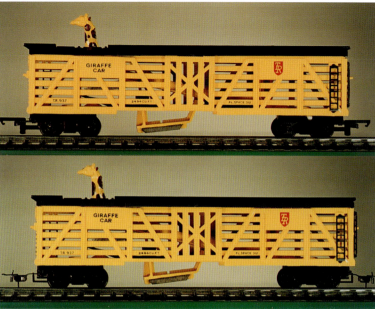

Giraffe car from the R348 set showing it both as sold in Britain and sold through Hornby Acho in France - note the couplings.

SIDE-TIPPING CAR

TC Side-Tipping Car R345

Production of the boxed sets continued into the late 60s. 4,500 were made in 1965, 1,500 the following year and then a last batch of 3,500 in 1968. If the 1965 sets were packed in Tri-ang Hornby boxes there were probably almost as many Tri-ang Hornby sets manufactured as Tri-ang Railways. It was last listed in 1969.

The bright red version with a load of black pipes belongs to the Hornby Railways period and will be covered in **Volume 3**.

OLD TIME CABOOSE

TC Old Time Caboose R449

3,000 were made in 1965 and sold slowly over the next four years. For further details of this car, *see* **Volume 1**.

GIRAFFE CAR

TC Giraffe Car R348

While designed and initially made before the change to Tri-ang Hornby, the giraffe car set reached the shops in the spring of 1965 and the box carried the name Tri-ang Railways. Most sets would have been sold during the Tri-ang Hornby years. 8,500 were made in 1965 and a further batch of 5,500 the following year. Numbers in the census for 1968 suggest that a further small batch of about 1,000 were made that year. It was last offered for sale in 1970 and the final stocks left the Margate factory in 1971.

The model came as part of a set in a very attractive box. Fuller details of the set are given in **Volume 1**. A study of the lid illustration shows that the artist must have had an example of the model to copy which did not have a magnet fitted to the trigger bar. The ladder was also missing.

Model for France R348F

In 1965, 1,500 models went out to France in Tri-ang Railways illustrated boxes but with supplementary French instruction leaflets. They were provided with Hornby Acho couplings and the ducking post rail clip was different. This appeared to have been adapted to Acho power clips (7940) but with the spring terminals removed and with an extension provided to the rear to take the post socket. This was necessary because the original Tri-ang ones would not have fitted Acho track. In the French factory circular stickers were stuck over the red 'R348' on the ends of the boxes. The new stickers appear to have been superfluous as they merely renumbered the model 'R348' but in black!

BATTLE SPACE

INTRODUCTION

Early Military Models

There had been military models in the Tri-ang Railways range since the R216 Rocket Launcher was introduced in 1957.

The Amalgamation Leaflet showed, on the back page, that two sets without track and with different 'R' numbers were originally planned.

Lines Bros. had learnt early on that to have appeal toys needed gimmicks and the military range was full of models that 'did things'. They tried to make them simple to use and, wherever possible, mechanically, rather than electrically, operated so that they could be sold as self-contained models, ready for use, without the need to buy additional power units.

I vividly recall evenings spent with my children fighting intense battles with a whole range of Battle Space military hardware. Missiles crossed in the skies above the living room carpet to plough into rows of Airfix military vehicles and Battle Space commandos lined up at either end of the room. Anything hit was removed from the battleground until one side had nothing left. I recall that one of the most devastating missiles was the Honest John that could wipe out three or four tanks at one go.

At Margate it was Richard Wansworth who galvanised everybody into making Battle Space. The rocket launcher had been selling well for many years and the helicopter car had become one of the companies most successful products. It was therefore natural to assume that children were crying out for more of the same.

By the end of the Tri-ang Railways era there were seven military vehicles available, most of them in a mid green and carrying NATO markings.

These were as follows:

No.	Model	Colour	Marking
R216	Rocket Launching Wagon	grey	M3971
R128	Helicopter Car	green	NATO TR301
R239	Bomb Transporter	green	NATO TR7190
R341	Searchlight Wagon	green	NATO TR7192
R249	Exploding Car	red	9841
R343	Four Rocket Launcher	green	NATO TR191
R248	Ambulance Car	white & green	RAMC

The Launch of Battle Space

There had been no new models in 1964 or 1965. The following year saw the start of Battle Space with three whole pages of the catalogue devoted to it. The first two were a spread displaying trains of khaki wagons and a matching Jinty tank engine. It soon became clear that the latter was available only in the RS16 Strike Force 10 set which besides the Battle Space Jinty also contained two more new items: the R562 Catapult Plane Launching Car and the R568 Assault Tank Transporter.

The rest of the wagons in the display were the pre-Battle Space military wagons now in khaki with the exception of the Ambulance Car which was unchanged. The helicopter had changed from red to yellow but the missiles and bomb had retained their original colours.

While pages 16 and 17 were devoted to battle, page 18 was devoted to space. Here was another train set, this time in red and blue and called RS17 The Satellite Set. It contained a red 0-4-0DS pulling an R566 Spy Satellite Launching Car and an R567 Radar Tracking Command Car.

Also on this page was a strange looking object called a Propeller-Driven Battle Space Car. The text said that by varying skilfully the speed of the propeller both forwards and backwards, the Space Car could be driven at a hair-raising speed and could negotiate inclines and declines safely. My personal experience of the car was that it was capable of embedding itself in the scenery, by its nose spike, if thrown into curves too fast! It was also capable of embedding itself into the odd unwary spectator on occasions, the most susceptible being those who like to view the action 'at eye level'.

1966 also saw the appearance of a set of six Battle Space Commandos. They could be bought separately boxed as R164 or later as additions to boost sales of Battle Space wagons. Meanwhile the large stocks of NATO green military wagons left in the stores at Margate were bulk purchased by Hattons of

The Satellite set as illustrated in the 1966 catalogue.

Pre-production models of the gun turret and rocket-firing tank, the former showing design changes.

Liverpool who then proceeded to sell them off at bargain prices through the pages of the model railway press.

The Range Expands

1967 saw the peak of Battle Space development. Coverage of the range spread to four pages in the catalogue and another eight pieces were added to the series. In order to keep costs to the minimum all models were based on existing tools.

The rarest of all Battle Space cars, the R571 'Q' Car, was made from an exploding car. The R525 Command Car was a Transcontinental Mail Car set in a new livery. The R630 POW Car was a TC Stock Car and the R639 Sniper Car was the Giraffe Car which itself was a development of the Stock Car. The TC Crane Car had become the R631 Tank Recovery Car.

The R671 Multiple Ground-to-Air Missile Site and the R670 Twin Ground-to-Air Missile Site were both ground formations made from expanded polystyrene and fitted with missile firing units borrowed from other models while the R672 Honest John Pad was the rocket launcher mounted on a yard crane base and fitted with a Dinky Toy rocket.

The End

Large numbers of the models were sold in 1966 and the management at Rovex must have felt content with the action they had taken. Ambitious production targets were set for 1967 but the repeat orders failed to arrive. Only half the stock left the stores during the year. Large stocks of the models sat in the stores at Margate for a number of years gathering dust. Almost over night the market had collapsed. Only one new piece was added after 1967 although some of the items used in the sets and the tank were offered separately. In 1970 the decision was taken to run the series down. The range was listed in the 1971 catalogue in order to clear existing stocks. By 1972 Battle Space had gone. Some time during 1972 large quantities of the models were removed from the stores but the fate of them is not recorded.

While modellers were unenthusiastic, the Battle Space series is fondly remembered by many thousands of men who grew up in that period. Today the series attracts special attention from collectors with good boxed sets and scarce models fetching high prices. The diorama boxes, that a lot of the wagons were packed in, are particularly sought. The stagnation of stock at Margate was also experienced by retailers and many of the diorama boxes survived in shop stores to be disposed of to hungry collectors years later. It's an ill wind that blows nobody any good.

Was the series a success?

The ability to produce such an extensive range of models for such a small tooling cost at a time of recession in the model railway industry must have been a considerable comfort to Rovex and showed Tri-ang ingenuity at its best. It also meant that when the series failed in 1967, it was not a major financial calamity. The military wagons had sold well when they were in bright colours but, despite high sales in the first year, sales fell off within two years of the new Battle Space livery being introduced. Khaki was just too dull for children.

When Hornby Hobbies tested the market again in the early 1980s the models used for their R580 and R579 Task Force Action sets were brightly coloured. The models included the tank transporter, the exploding car and, of course, the ever popular helicopter car.

Overseas Sales

It was not only in Britain that Battle Space found a market. The models were exported all over the world. Of the five wagons chosen for the Hornby Acho catalogue and provided with Acho couplings, three were from the military range. These were: R216F rocket launch wagon, R249F exploding car and R128F helicopter car. France also received the Defender set.

Types 5 and 2 packaging for Battle Space models – the shrink-wrapped tray and panorama tray.

Early military wagons were well represented in the 1962-63 Australian catalogue but as this publication concentrated on Australian-made or assembled items they were not to be included in subsequent editions. It must be remembered that the Australian publication was a supplement for the British catalogue rather than a catalogue in its own right.

The 1965 Canadian catalogue had offered the full range available at the time and in 1967 the Canadian company was offering the PS367 All-Action Presentation Set which contained the "Battle Rescue Helicopter Car, operating Multiple Missile Launcher, Exploding Car and Anti-aircraft Searchlight Wagon with electric Searchlight". The models illustrated were all standard ones from the Battle Space range. Thirteen of the Battle Space series were also available as solo models. Two years later the All-Action Presentation set was back in the Canadian catalogue as R367 with the same selection of wagons. Canada dropped Battle Space in 1971.

PACKAGING

Six types of packaging were used. These were:

1. End flap box.

At least six sizes of box were used and they were numbered on the inside flap – BS1, BS2 etc. The main body of the box was red and the end flaps yellow. They carried the name 'Battle Space!' and logo. On the sides of the box were pictures of some of the series with descriptions of them.

2. Diorama box

The two Battle Space sets and several of the vehicles were sold in a diorama tray box which was shrink wrapped. The vehicle was mounted on a card printed track with a battle scene card as a back cloth. These came with a set of seven commandos, one of which was displayed in a special display alcove in the frame of the box. Once again the 'Battle Space!' name and the logo were prominent. So too was the inscription 'Rail Based Combat Units'.

3. Lift-off lid box

These were used for the two ground-to-air missile sites. The top of the box carried a picture of the contents in use and the name 'Battle Space!'. Two sides carried pictures of models in the range.

4. Cellophane bag and stapled card

These were used for the set of six Commandos. The top card, stapled to the bag, was a standard one used for sundries and small accessories and was one of the only examples of standard packaging used with this series.

5. Shrink-wrapped tray

These were rather crude. They looked like a type 1 box but with one side cut away to expose the model within and the whole shrink wrapped and slid into a grey outer card sleeve.

6. Standard Tri-ang Hornby window boxes

Both the space models may be found in standard packaging as well as Battle Space.

CHECK LIST OF BATTLE SPACE MODELS

Dates

The dates given in the following tables are those when models were made. They remained in the catalogue well beyond these dates.

Detail

The couplings, wheels and bogies indicated are those carried by the model during the Tri-ang Hornby period.

Boxes

The number in the box column refers to that used above under the heading 'Packaging'.

A. Sets

No.	Sets	Colour	Dates	Box
RS50	Defender Set	green	64-65	-
RS16	Strike Force 10 Set	khaki	66-67	2
RS17	The Satellite Set	red & blue	66-67	2
RS92	Battle Space Turbo Car Race	not made		
R397	Satellite Presentation Set	not made		
R398	Battle Space Presentation Set	not made		

B. Motive Power

No.	Locomotives	Colour	Dates	Cpling	Box
R588	Jinty (in sets only)	khaki	66-67	IIIc	set
R756	0-4-0DS (in sets only)	red	66-67	IIIc	set
R752	Battle Space Turbo Car	red	67-68,70	none	1

C. Coaches

No.	Coaches	Colour	Dates	Cpling	Bogie	Wheels	Box
R248	Ambulance Car	wht+grn	63-67	IIIc	t3b	MS4-MP1	1, 2
R725	Command Car	khaki	67	IIIc	t3b	MP1	1

D. Wagons

No.	Wagons	Colour	Dates	Cpling	Bogie	Wheels	Box
R216	Rocket Launch Wagon	grey	57-65	IIIc	B2d	MS4	-
R216K	Rocket Launch Wagon	khaki	66-67	IIIc	B2d	MS4	1
R128	Helicopter Car	green	62-65	IIIc	B2d	MS4	-
R128K	Helicopter Car	khaki	66-69	IIIc	B2d	MS4	?
R239	Bomb Transporter	green	62-65	IIIc	D1b	DS3	-
R239K	Bomb Transporter	khaki	66-67	IIIc	D1b	DS3	2
R341	Searchlight Wagon	green	63-65	IIIc	D1b	DS3	-
R341K	Searchlight Wagon	khaki	66-67	IIIc	D1b	DS3	1
R249	Exploding Car	red	63-65	IIIc	B2d	MS4	-
R249K	Exploding Car	khaki	66-67	IIIc	B2d	MS4	?
R343	4 Rocket Launcher	green	63-65	IIIc	D1b	DS3	-
R343K	4 Rocket Launcher	khaki	66-67	IIIc	D1b	DS3	2
R562K	Plane Launch Car	khaki	66-67	IIIc	D1b	DS3	2,6
R566	Satellite Launch Car	blue+red	66-67	IIIc	B2d	MS4	2,6
R567	Radar Tracking Car	blue+red	66-67	IIIc	C1A	DS3	1,6
R568	Assault Tank Trans.	khaki	66-67	IIIc	B3b	MP2	2
R571	'Q' Car	khaki	68	IIIc	B2d	MS4	1
R630	POW Car	khaki	67	IIIc	B3b	MP2	1,2,5
R631	Tank Recovery Car	khaki	67	IIIc	B2d	MS4	2
R639	Sniper Car	khaki	67	IIIc	B2d	MS4	1,5

WAGONS FITTED WITH FRENCH COUPLINGS

Dates given are those when the models were dispatched.

No.	Wagons for France	Colour	Dates	Cpling	Bogie	Wheels
R128F	Helicopter Car	green, khaki	65?-68	Acho	B2d	MS4
R216F	Rocket Launch Wagon	grey, khaki	65?-68	Acho	B2d	MS4
R249F	Exploding Car	red, khaki	65?-68	Acho	B2d	MS4

E. Missile Emplacements

No.	Missile Emplacements	Colour	Dates	Box
R670	Twin Missile Site	khaki	67	3
R671	Multiple Missile Site	khaki	67	3
R672	Honest John Pad	grey	67	1

F. Sundries

No.	Sundries	Colour	Dates	Box
R164	Commandos	khaki	66-67	4
R673	Assault Tank	khaki	67	1

DIRECTORY OF BATTLE SPACE MODELS

1. SETS

DEFENDER SET

RS50 Green 0-6-0DS (R152), three military wagons (R341 searchlight, R343 four rocket launcher, R249 exploding car), R487 clip, R488 uncoupling ramp, oval of Super 4 track.

This set was first made in 1964 but over 5,000 of the Defender set were made in 1965. It is not known,

however, whether those of this second batch were sold in the original Tri-ang Railways box or whether a new box bearing the Tri-ang Hornby name was printed. The set was also covered by **Volume 1**.

RS50F **The Defender set** was also offered to the French through the Hornby-Acho catalogue and called Le 'Train Militaire'.

About 775 special sets were made in 1967 which were sent over to France in two batches; one in 1967 and the other in 1968. They would have been provided with Acho track and had a French label applied to the box on arrival in France. The set

remained in the Meccano Tri-ang French catalogue until 1970.

STRIKE FORCE 10 SET

RS16 Battle Space Jinty (R558S), two wagons (R562 plane launcher, R568 assault tank transporter), uncoupler, oval of Super 4 track & 12 Commandos.

About 24,000 of this action set, called 'Strike Force 10', were made in 1966 and 1967 with, it seems, a further 1,500 sets being made in 1970. All but about 800 sold in the UK. The pre-production model for

Battle Space models still available in 1971.

this set may be seen in the R280S amalgamation leaflet of 1965.

THE SATELLITE SET

RS17   Battle Space 0-4-0DS (R756), two wagons (R566 satellite launcher, R567 radar car), uncoupler, oval of Super 4 track & 12 Commandos.

This was a less popular set and, although it was available for the same five years, only 10,000 were made in 1966 and 1967. It was dropped from the catalogue in 1971. Once again, the pre-production set may be seen on the back page of the Amalgamation leaflet of 1965. Here the wagons are seen with ochre-coloured bases instead of the red in which they were made.

Battle Space Turbo Car Race RS92

This set had been offered to retailers in 1968 but was not listed in the standard catalogue. It may have been a joint railway/motorway set. There were only 580 orders received for it and so it was probably felt to be uneconomical to produce. It is not known what the set would have contained.

Satellite Presentation Set R397

Planned but not made.

Battle Space Presentation Set R398

Planned but not made.

For further military sets *see also* – RS18, RS74, PS367, R367 in the Chapter on 'Sets', the last two being found in the section on 'Canadian Presentation Sets'.

The RS16 Strike Force 10 and RS17 Satellite Train sets.

2. MOTIVE POWER

JINTY

Battle Space Jinty R588

The locomotive was a standard Jinty but with a body moulded in khaki-coloured plastic and bearing Battle Space stickers on the tank sides. It appears not to have been sold on its own but was found in the RS16 Strike Force 10 Set. About 24,000 were made during 1966 and 1967 of which about 800 went abroad.

BR Khaki Jinty

The locomotive has been found with BRc logo transfers in place of the Battle Space stickers.

NORTH BRITISH 0-4-0DS

Battle Space Diesel Shunter R756

The diesel shunter was a version of the North British type of 0-4-0 which had previously been available in plain blue for starter sets or, until 1967, in full BR green livery as R559. This Battle Space version of the model was red and had Battle Space stickers on its sides. As we have seen, about 10,000 were made and it is thought that these all date from 1966 and 1967.

R588 Battle Space Jinty and R756 Battle Space diesel shunter.

TURBO CAR

Battle Space Turbo Car R752

It will be noted that the illustration in the 1966 catalogue is quite different from that produced and rather than being the picture of a pre-production model it is more likely to be an artist's impression. This is supported by the fact that none were manufactured that year.

About 2,500 were made in 1967. These can be identified by the plastic nose spike they were fitted with. As we have seen, the spike was capable of impaling the Turbo car in the scenery or anything else that got in its way. After some expressions of concern by distressed parents, Rovex modified the design, replacing the plastic spike with one made of rubber. This revised design is thought to have been available in 1968 when 11,400 were made. A further batch of 2,500 was made in 1970 in order to use up the stocks of motors, packaging and castings but the model was dropped from the catalogue the following year.

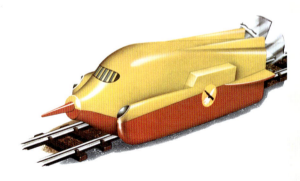

R752 Space Turbo Cars with plastic and rubber nose spikes respectively and the picture from 1966 catalogue.

3. COACHES

AMBULANCE CAR

RAMC Ambulance Car R248

This was based on the second series Transcontinental Baggage/Kitchen Car and dated from 1963. It was in white plastic with a broad green window stripe and doors. The cars carried the red cross emblem on the sides and roof. The latter was always a sticker but those on the sides could be either stickers or heat printing.

It was available in the British catalogue until 1967 and reappeared in 1971 after a break of three years. It looks as though this late appearance was to use up old stock as the last year of manufacture was 1967. During 1965-67 about 14,000 were made and these may be found in both closed Battle Space boxes and the diorama display packaging.

ANZAC Ambulance Car R248A

Up until 1967 the Australian company produced

their own version of the Ambulance Car from the tools of the first series Baggage Car. The model had a white body, grey roof and red doors. On its sides in red was heat printed 'Ambulance Car' and 'ANZAC'. It also carried two small Red Cross stickers on the sides and a large one on the roof. The model was mounted on Australian tooled bogies.

COMMAND CAR

Battle Space Command Car Set R725

This was the Transcontinental travelling mail set but made in khaki plastic with a black roof and chassis. Even the dispatch bags it picked up from a khaki gantry were khaki. The car carried the words 'COMMAND CAR' heat printed at one end and 'D778' in large characters on a self-adhesive label at the other end of each side. The sides also carried Battle Space stickers.

It was in the shops by mid 1967 and remained in the range until 1971. The set retailed at 28/2d and came complete with a set of Commandos, track trigger ramp, lineside dispatch gantry and separate collecting bin. The set was made only in 1967, when 4,400 were produced, and it was sold in a closed box.

4. WAGONS

ROCKET LAUNCHING WAGON

Grey Rocket Launcher R216

This was the oldest in the military range having been introduced to the Tri-ang Railways system in 1957. The wagon remained grey until 1966 when it changed to khaki without going through the NATO green stage. In 1965, its last year of production as a grey model, over 18,000 were made. The earliest models had rockets made in clear plastic but, with the exception of a few silver ones, after that they were always red with a black rubber tip.

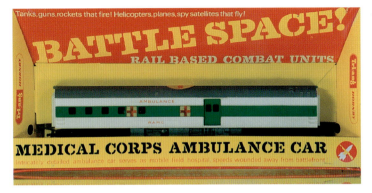

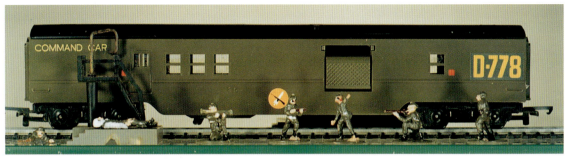

RAMC ambulance car (R248).

R725 command car.

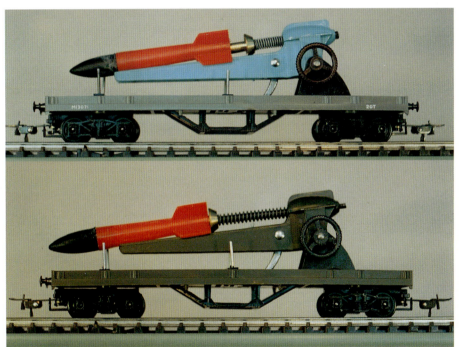

R216K rocket launching wagon.

R216F Versions of the rocket launching wagon fitted with couplings for the French market.

Lionel version of helcopter car which inspired the Tri-ang model and the Tri-ang Battle Space version.

R128F helicopter cars for France.

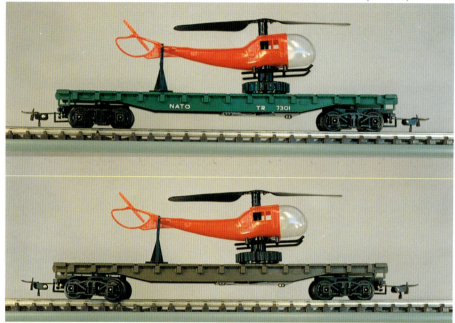

French Grey Version R216F

The first batch of 1,530 grey rocket launchers went to France in 1965 in Tri-ang Railways window boxes. These were the standard model in standard packaging but a pair of Hornby Acho couplings were supplied in the box and the boxes had the contents label inscribed 'R216 Wagon Lance Fusees' attached on arrival at the Meccano France factory.

Battle Space Rocket Launcher R216K

From 1966 until 1971 the model had a khaki body and launcher. It carried a Battle Space sticker on the top of the decking and had a red rocket. Two bolsters were provided on the decking and, although it was not clear what these were supposed to be for, they made the model look more detailed.

A total of 20,000 were made in this colour scheme in 1966 and 1967 after which no more were pro-

duced and the remaining stocks took until 1971 to clear.

French Battle Space Version R216F

Further batches of 1,900 in 1967 and 1,800 in 1968 were sent to the Meccano France factory and these were the khaki Battle Space version. These also went in standard boxes but had just 'Ref R216' printed in blue on the end labels which were attached on arrival at the Calais factory. There were no further shippings.

Once again, the models had a pair of Hornby Acho couplings supplied in the box for the purchaser to use if required.

Canadian Rocket Wagon R570CN

There is a record of a Canadian Rocket Wagon but no record of manufacture or of orders received or

dispatches made but the date given is 1965. It is strange that it was allotted an 'R' number of its own that bears no relationship to any other military model. A possible explanation is that this number belongs to the strange wagon which appeared in the R74 train set which is described and illustrated on page 84. This would also explain why production figures are unavailable as items used in sets were not counted separately.

HELICOPTER CAR

NATO Helicopter Car R128

The early version of this model, added to the Tri-ang Railways range in 1962, had a grey body but by 1965 it was mid green with NATO markings and the number TR7301. It carried a red helicopter and in 1965, over 23,000 passed through the stores.

Tri-ang HORNBY
Model Railways

BATTLE RESCUE HELICOPTER CAR R.128

The Tri-ang Hornby Battle Rescue Helicopter Car sends the Helicopter soaring into the air. The mechanism is designed for operation by manual or trackside control. Launching power is supplied by a spring inside the winding drum. The teeth on top of the drum engage with corresponding teeth at the base of the Helicopter rotor shaft. Upon release of the spring by a movement of the trigger, the drum spins the rotor shaft causing the rotor blades to lift the Helicopter into flight.

Before operating the Helicopter it is advisable to push the car round the track by hand, with the Helicopter in position, to ensure that it does not encounter any obstructions (the Trigger, in the operating position, will strike the ramps on a Hopper Unloading Bridge).

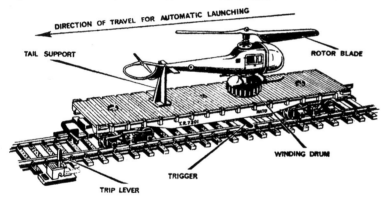

DIRECTION OF TRAVEL FOR AUTOMATIC LAUNCHING

TAIL SUPPORT

ROTOR BLADE

WINDING DRUM

TRIGGER

TRIP LEVER

OPERATING PROCEDURE

Place the Car (without Helicopter) on the track facing in the right direction.

Set the trigger to the operating position as illustrated, at right angles to the track, before winding the spring.

Turn the winding drum clockwise until the spring is fully wound.

Position the Helicopter on the Car with the rotor shaft in the centre of the winding drum and the tail skid in the centre of the tail support.

Turn the rotor blade slightly to permit the driving teeth of the Helicopter and the winding drum to engage, and ensure that the blades do ot project over the sides of the Car (so that they do not foul tunnels. etc

TRAINS HO Tri-ang

WAGON LANCE-HÉLICOPTÈRE - réf. R.128
avec Attelages Universels

L'hélicoptère du 'wagon porte-hélicoptère' peut réellement s'élever dans les airs.

Lorsqu'une pression sur le levier de détente détend le ressort, le remontoir en tournant entraîne l'arbre du rotor et les pales, et l'hélicoptère se trouve lancé dans les airs.

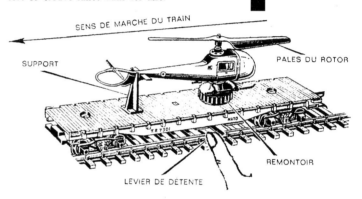

SENS DE MARCHE DU TRAIN

SUPPORT

PALES DU ROTOR

REMONTOIR

LEVIER DE DÉTENTE

MODE D'EMPLOI

Placez le wagon (sans l'hélicoptère) sur les rails, dans le sens de la marche. Placez le levier de détente comme indiqué sur l'illustration, avant de procéder au remontage qui doit être fait dans le sens des aiguilles d'une montre (le levier doit former un angle droit avec les rails)

Placez ensuite l'hélicoptère sur le wagon en insérant l'arbre carré commandant le rotor au centre du remontoir, et en posant la queue de l'appareil sur le support prévu à cet effet. Tournez légèrement le rotor pour permettre aux dents de l'hélicoptère de s'engager dans celles du remontoir.

Il vous suffit maintenant, pour lancer l'hélicoptère, de pousser légèrement le levier de détente vers la droite, soit avec un doigt, ou encore une butée amovible placée le long de la voie.

NOTE.

Lorsque votre train est en marche, assurez-vous que les pales du rotor ne dépassent pas la largeur du wagon, afin d'éviter toute collision avec des obstacles, tunnels par exemple.

MECCANO-Tri-ang 70-88, av. Henri-Barbusse, BOBIGNY (Seine)

British and French instructions for the R128 helicopter launch car showing that as the trackside trigger would not fit onto the Hornby-Acho track the instructions were changed to show how to trigger it by hand.

French NATO Version R128F

A batch of 1,530 of this NATO green model, with its red helicopter, was sent to the Meccano factory in France in 1965 in normal Tri-ang Railways window type boxes complete with a pair of Hornby Acho couplings in the box for the purchaser to fix. On arrival, a white label, inscribed 'R128 Wagon Lance Helicoptere', was stuck on each end.

It is interesting to note that the wagon was not supplied with the usual lineside trip as this would not have fitted Acho track. The instruction sheet, which was printed in France in the style of the British ones but in French, does not mention the lineside trip but instead indicates that a suitable block placed beside the track will release the helicopter. Indeed the instructions explain how to operate it by hand and the illustration, taken from the British leaflet, has the trip lever removed from the drawing and a thumb drawn in instead!

Battle Space Helicopter Car R128K

The khaki version replaced the NATO one in 1966 and was made in 1966, 1967 and a further batch in 1969. In total 35,000 were made in this form, most of them in the first two years. The body of the wagon was khaki and carried a Battle Space sticker on the decking. The helicopter was a pale yellow.

French Battle Space Version R128F

Further batches of the helicopter car were sent to France in 1967 and 1968 but this time the model was the khaki version. They too were supplied with Hornby Acho couplings and sold through the Hornby Acho catalogue. About 4,200 were sent, all in their original boxes with identification stickers, marked 'Ref R128', fixed on arrival to the end flaps of the box.

1982 Version

This had a white wagon and yellow helicopter and will be reviewed in **Volume 3**.

BOMB TRANSPORTER

NATO Bomb Transporter R239

This was introduced in 1962 and survived until the end of 1965 after which it was replaced by the Battle Space version. In 1965 over 7,600 were made of which 2,500 remained at the end of the year. These were not cleared until 1967.

The body of the wagon was mid green and the sides were marked 'NATO' and 'TR7190'. The bomb carried on two black frames was red and the metal nose was held in place by a spring. When this was stretched a cap could be inserted. The origins of the bomb are described in **Volume 1**.

Battle Space Red Arrow Bomb Transporter R239K

The Battle Space version was described in the 1966 catalogue as the Red Arrow Bomb Transporter. The body of the wagon was now khaki. The bomb remained red but now sported Battle Space stickers.

7,600 of the new version were made in 1966 and another 5,400 the following year. After an initial run

on the models, stock stagnated, the last 2,000 still residing in the store at Margate in 1971.

SEARCHLIGHT WAGON

NATO Searchlight Wagon R341

This was another model planned for introduction in 1962 which did not arrive until the following year. The model had a mid green body and superstructure with a working searchlight and a moving radar scanner mounted on the latter. It carried a blue and red military symbol on the sides and the NATO and TR7192 markings in white. In 1965, its last year, 8,400 were made in this livery.

Battle Space Anti-aircraft Searchlight Wagon R341K

1966 saw a new livery and a new name. The model was now khaki and carried Battle Space stickers on the sides of the superstructure. Only two batches were made of this version and these were in 1966 and 1967. The initial batch of 12,160 almost sold out during the year but only half of the 1967 batch of 12,100 sold during the year and the remainder took until 1971 before being disposed of.

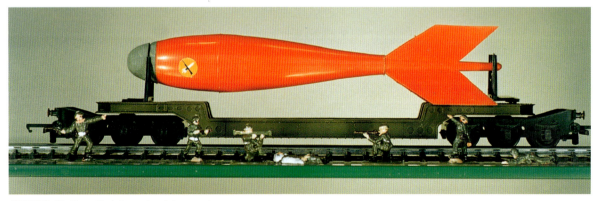
R239K Battle Space Red Arrow bomb transporter.

EXPLODING CAR

NATO Exploding Car R249

This is one of the best remembered of the military wagons. Besides the surprise element of its operation, its bright red colour was appealing and sadly lost when it became a Battle Space wagon in 1966. The model was almost a direct copy of the Lionel exploding car even down to the heat printing on its sides and it had first appeared in the Tri-ang Railways range in the autumn of 1963. In its final year, almost 15,000 were made putting it in third place after the helicopter car and the rocket launching wagon in terms of production volume.

French NATO Version R249F

In 1965 a batch of 1,500 red exploding cars was sent to the French factory for distribution through the Hornby Acho catalogue. These probably had Tri-ang Railways window type boxes with French identification labels attached on arrival. A pair of Hornby Acho couplings were provided in the box which the purchaser could fit themselves if they wished.

Battle Space Exploding Car R249K

The Battle Space exploding car had a khaki body and Battle Space stickers on the sides but otherwise was the same as its predecessor. Batches totalling 22,000 models were made in 1966 and 1967. No more were made after this and the last of the stock survived until 1971.

French Battle Space Version R249F

In 1967 and 1968, two more batches of the exploding car were sent to France with Hornby Acho couplings in the box. This time they were in Battle Space khaki and in Tri-ang Hornby boxes. 1,000 went out in 1967 and 1,300 in 1968.

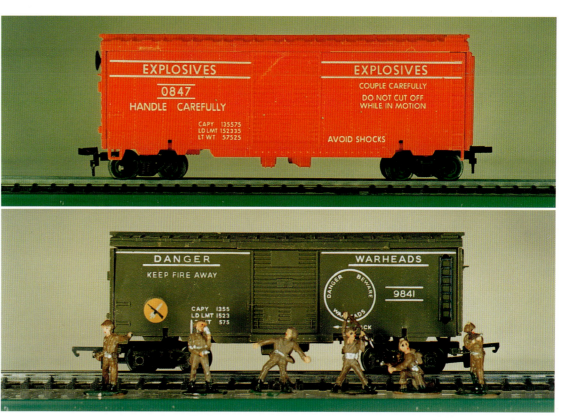

Lionel and Tri-ang exploding cars

R341K Battle Space Anti-aircraft searchlight wagon.

R249F exploding car with Hornby Acho couplings for sale in France.

1982 Version

This was red with a grey roof and had different markings. It will be reviewed in **Volume 3**.

4 ROCKET LAUNCHER

NATO Four Rocket Launcher R343

This model also joined the Tri-ang Railways range in the autumn of 1963. Like the NATO bomb carrier and searchlight wagon, it had a mid green Trestrol body. Onto this a cabin and a revolving missile launcher had been mounted. The launcher could fire four missiles at once each mounted on its own spring-loaded launch rod. In 1965, 14,500 were made and all but a very few had been sold by the end of the year.

Battle Space Multi-Missile Launcher R343K

This replaced the NATO version in 1966 and like the other Battle Space wagons had a khaki body and superstructure and Battle Space stickers. The missiles were now called Red-Eye Rockets. As a sign of the model's popularity, over 30,000 were made in 1966. Only 2,500 of these remained in the stores at the end of the year.

PLANE LAUNCH CAR

NATO Plane Launch Car R562

Records indicate that 750 Plane Launch Cars were made in 1966 numbered R562 (without the 'K' suffix). It could be they were in NATO green livery but the pre-production model is clearly finished in Battle Space livery. A sample needs to be found to confirm what form wagons, from this first batch, took.

Battle Space Plane Launch Car R562K

The model consisted of a Trestrol wagon mounted with a girder built launch ramp. The power was provided by a rubber band which launched a cardboard aircraft when a catch on the side of the wagon hit a lineside trip. The plane flew quite well but did not look very realistic. It could be made to perform aerobatics by adjusting the elevators and rudder. It came with a spare rubber band and a set of commandos. A profound observation in the *Model Railway Constructor* is worthy of repetition:

"We are sure that these exciting operating accessories will keep many youngsters faithful to their railways long after the thrill of seeing a two coach express chasing its tail round an oval track has palled, and it must be faced

that it is during this period that so many budding enthusiasts are lost to the hobby."

This was certainly the belief at Lines Bros. and lead to, amongst other things, the introduction of gimmick vehicles to the newly-acquired Dinky range.

The pre-production model of the plane launcher may be seen on the back page of the amalgamation leaflet. It was shown with a solid launch ramp fitted to a NATO green Trestrol wagon. Later another pre-production model was made with a 'girder' ramp as eventually manufactured. The latter, which was in Battle Space livery, was sold by Hornby Hobbies to a private collector in 1994.

In 1966, almost 11,000 of the model were made in Battle Space livery for solo sales and a further 24,000 for use in the Strike Force 10 Set where it was referred to as a 'Close Support Plane'. The following year, 4,700 more were made for solo sales making a production total of about 40,000.

SATELLITE LAUNCH CAR

Satellite Launch Car R566

This is another model that may be seen on the back page of the amalgamation leaflet in its pre-production state. It was shown with an ochre-coloured wagon and blue superstructure. The model first appeared in 1966 both as part of the RS17 The Satellite Set and as a solo wagon priced 16/-. As part of the advance publicity, towards the end of 1965, the company must have sent out photographs of the impending model together with a press release. The reports in the rival model railway magazines at the time had matching reviews suggesting that they had been lifted from an official circular.

The wagon was the TC flat wagon as converted for use as the helicopter launch car. The same launch mechanism was used to spin the satellite so that it was thrust into the air. It was red with a blue superstructure (control room), the latter having a caboose chimney stuck in the roof. The satellite was red with yellow wings and a 'chromed' nose. The reviewer in the *Model Railway Constructor* remarked that:

" *Younger enthusiasts should have great fun terrorising the rest of the family playing '1984'* ".

15,500 solo models were made and 10,000 for use in the Satellite Set. All production took place in 1966 and 1967 after which stocks stagnated. As late as 1971 there were still 6,700 solo models in the Margate stores. What became of these is unknown but they appear to have been disposed of during the year.

One strange fact emerges from the production records, however. Of the 6,400 solo models made in 1966, 1,400 were listed as 'R566G'. I have no explanation for this and can only guess that the first batch made were in some way different but still considered suitable for sale. I have one that was sold in a Tri-ang Railways window box rubber stamped on the end with the contents but it is numbered 'R566' without a 'G'. This was many months after the packaging should have changed to Tri-ang Hornby.

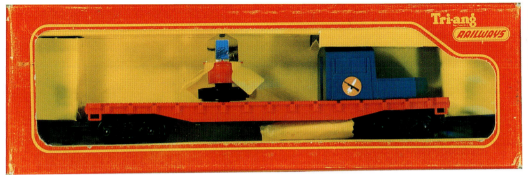

R566 satellite launching car as sold in a standard Tri-ang Railways window box.

RADAR TRACKING CAR

Radar Tracking Car R567

Here again the pre-production model is shown in the amalgamation leaflet in blue with an ochre-coloured base plate which was mounted on a standard 12' wagon chassis. What can not be seen from the picture is the original method adopted for turning the satellite tracking dish. This consisted of a rubber wheel mounted mid way between the two chassis wheels on one side of the wagon and set at the right height to make contact with the rail. Through the centre of the wheel was a splined shaft which rubbed on a small drum suspended flat beneath the centre of the wagon.

I bought this pre-production model in 1994 and found that it did not work at all well which explains why this method of traction was abandoned. The baseplate on the model had been repainted red instead of ochre and the square dish shown in the leaflet picture had been replaced by a small segment out of the side of a disc covered in sticky foil. The

R343K four-rocket launcher.

R562K Battle Space plane launch car.

R567 radar tracking car and pre-production model.

Proposed and actual mechanism used on the radar tracking car.

Battle Space logo was hand painted onto both sides of the wagon, slightly larger than the stickers eventually used. The model shows no means by which the light dome could be lit.

Traction on the production model was supplied by a rubber band fixed around the two wheel axles and passing around the sides of a large drum suspended flat beneath the centre of the wagon. This version worked well until the rubber band perished. The top of the drum acted as a rotating switch which caused the light in the dome, on the upper surface of the wagon, to switch on and off as the wagon was moved along the track; power pickup being through two of the wheels. The dish used on the production model was a plastic moulding which had been plated to give it a silvery shine and the chassis used was the old cast metal long-wheelbase brake van chassis (C1B) that had been replaced in 1963, for general use, by a plastic one. The treads of the wheels were milled for extra traction.

The model seems to have been delayed due, no doubt, to the problem of getting it to work properly. Only 1,260 solo models were made in 1966. 12,300 were made the following year but no more were produced after that. A further 10,000 were made for assembly of the Satellite Set in 1966 and 1967.

Almost 7,000 solo models had to be disposed of in 1972. The only boxed example I have is not in a Battle Space box but a standard Tri-ang Hornby window box with printed labels stuck on each end. As the label indicated, it came with Battle Space Commandos.

Canadian Radar Tracking Car R567CN
This appears to have been a radar tracking car planned for the Canadian market but records indicate that none were made or that all produced were used in set production. Could this have been a similar case to the Canadian Rocket Launching Wagon R570CN described above? If so, it may have consisted of the radar tracking car body mounted on a wagon chassis for a cheap starter set.

ASSAULT TANK TRANSPORTER

Assault Tank Transporter R568
This too was illustrated in pre-production form on the merger leaflet with a green NATO well wagon and a khaki tank. There is no evidence in the records of any having been made in the lighter green. The production model was illustrated in the November issue of *Railway Modeller* and the December 1966 edition of *Model Railway Constructor*. The model shown did not have Battle Space logos.

It seems that Rovex spent some time getting the revolving gun turret on the tank correct. Pre-production attempts at designing a working turret have survived and are in a private collection. The tank turret, eventually developed, carried and fired two

R568 assault tank transporter.

of the Red-Eye rockets we had previously seen on the R343 4-Rocket Launcher and which the *Railway Modeller* told us fired 400-500 scale feet. They must have had fun working that out!

The tank had flangeless spoked wheels and came with seven Battle Space Commandos described by *Railway Modeller* as:

"This hand-picked corps of superb fighting men stand fully 7 scale feet in their plastic boots"

From the start the model was dispatched in the display pack which make these popular to collect. The price was £1-0s-5d. It was also one of the vehicles in the Strike Force 10 Set for which 24,000 were made in 1966 and 1967. 5,700 solo models were made in 1966 all of which sold during the year. A final batch of 15,000 was made in 1967 but 10,000 of these remained in the stores at the end of the year. There were still 4,800 to dispose of in 1972.

1982 Version

The model returned in 1982. Both the wagon and the tank were sand coloured and will be reviewed in **Volume 3**.

POW CAR

POW Car R630

Priced 12/10d, the Prisoner of War Car made its appearance in mid-1967. It was made from the stock car tools with a khaki body and black roof and under frame. The two side panels carried self-adhesive labels printed with '334' and 'POW' in large yellow characters. The model, which had sliding doors and a set of Commandos, was available in the range until 1971.

Only one batch of 4,800 was made in 1967. About 2,000 sold during the year and the rest gradually disappeared with the last 30 leaving the stores in 1971.

SNIPER CAR

Sniper Car R639

This made use of the Giraffe Car tooling but instead of a giraffe, a sniper ducked as the model passed the trigger rail. The khaki-coloured car carried the number D453 heat stamped in yellow onto a side panel on each side and Battle Space stickers on the other panel. The roof was black and the sniper, khaki to match the body of the wagon. The doors were moulded, as part of the body, in a closed position. My own example came in a closed Battle Space box complete with a polythene bag of seven Commandos and two actuating rails, with brown plastic clips, strapped to the side of the wagon with

R630 POW and sniper cars.

a rubber band. There was also a Sniper Car instruction sheet numbered R639/44640/1.

The model arrived in 1967 when a total of 4,510 were made. 2,000 of these were left in the stores at the end of the year and these hung around until the last 200 were disposed of in 1972.

TANK RECOVERY CAR

Tank Recovery Car R631

This used the standard Tri-ang working crane originally made for the R127 crane wagon but now in khaki and mounted on a khaki well wagon. One platform of the wagon carried a large '106' in yellow on a self-adhesive label and a two-pin bolster and the other had a Battle Space sticker. The number label was also stuck on the back of the crane. The wagon, with its set of Commandos, was available by mid 1967 and remained in the range until 1971. It retailed at 17/11d.

A little over 4,800 were made in 1967 of which 1,800 remained at the year end. No more were made after this and the final 250 models left the store in 1971.

'Q' CAR

G-10 'Q' Car R571

A late attempt was made to revive the Battle Space series with the introduction of a new model in 1968. This was the 'Q' Car which was a variation of the exploding car but as the sides and roof were cast off a twin Red-Eye rocket launcher was revealed. The khaki body of the car was plain except for a large 'G-10' in yellow on self adhesive labels on each side.

Whether its lateness in introduction to the range was due to technical difficulties or because it was a late attempt to inject new life into the series I do not know but I suspect the latter. The model arrived in 1968 and was given pride of place in the four pages of the catalogue devoted to the Battle Space series.

Only 2,586 models were made and all but 60 of these sold during the year in BS4 boxes. It seems only 2,700 had been ordered by retailers and no doubt it was felt that this was insufficient interest to warrant continuing with the model. As a result, the G-10 Q Car is the rarest of the Battle Space series and consequently is much sought after by collectors. Its inclusion in the 1969 catalogue was to help retailers sell stocks bought the previous year.

5. MISSILE EMPLACEMENTS

TWIN MISSILE SITE

Twin Missile Site R670

This model came in a large square box with a lift-off lid. The body of the model was a moulded piece of expanded polystyrene painted green. It represented a hill top built in with a gun emplacement and steps. The gun in this case was the turret from the Battle Space Assault Tank with two Red-Eye rockets. The

R631 tank recovery car.

The much sought-after R571 'Q' car.

R670 and R671 missile sites.

set also included a packet of Commandos. To fit the gun onto the moulding, a black plastic boss had to be made.

It reached the shops in 1967 and this was the only year any were made. 4,900 were produced of which 2,400 remained in the stores at Margate at the year end. There were still 1,000 left in 1971.

MULTIPLE MISSILE SITE

Multiple Missile Site R671

This was very similar to the last model but had a taller box. The base moulding and black boss were the same but this time the revolving turret from the 4-Rocket Launcher was fitted. It was supplied with four Red-Eye rockets and a set of commandos.

It also appeared in 1967 but only 3,800 were made. 2,300 remained at the end of the year and these gradually sold until the stock ran out in 1972.

HONEST JOHN PAD

Honest John Pad R672

This used the base of the yard crane as the pad and to this was fitted the rocket launcher from the R216 Rocket Launching Wagon. On one side of the rocket guide a red sticker carried the number '5261' in white followed by a black square. On the opposite side of the model, a Battle Space sticker had been attached to the anchor block. The Honest John rocket was a long white moulding with a black rubber

The 6 and 7 character sets of R164 Commandos, packaged as sold and the catalogue illustration which shows a completely different set of figures!

tip and is believed to have come from the Dinky production line at Liverpool.

4,600 were made, all in 1967. 1,300 remained at the end of the first year.

6. SUNDRIES

COMMANDOS

It is thought that these were originally supplied by another company in the Lines Bros. Group called Scale Figures Ltd of Tunbridge Wells which was acquired in 1966.

R.164 Battle Space Men. Specially designed to man the Battle Space Units.

R672 Honest John rocket launcher.

Set of Six R164

Most Commandos found today are of this type. They were moulded in brown plastic and had the upper side of the large bases hand painted green. Pink, brown, silver red and black paint was used to add detail to the figures. All had bases to which at some time 'Hong Kong' was added on the underside; or was it originally there and later removed?

The figures could either be bought as a set in their own right in which case they were packed in a cellophane bag with a Tri-ang Hornby card folded over and stapled at the top, or they were packed in a polythene bag and stuffed out of sight into the wagon diorama boxes. To show what was inside, a seventh Commando was slotted into an open compartment beneath the wagon. This seventh figure was any one of the six designs and it was pot luck which one you got. The packaging indicated that seven Commandos came with the wagon.

The figures were:
1. Section leader pointing ahead and holding a pistol
2. Soldier throwing a grenade
3. Soldier walking while using a radio on his back
4. Soldier kneeling with a gun
5. Soldier bayoneting
6. Soldier standing aiming a rifle

Set of Seven R164

The later sets of figures, made from a different tool, came in a small polythene bag stuck down with sellotape and pushed into the box with the model they were to accompany. There were seven different figures in all possibly because all literature referred to seven because of the diorama method of display described above. They included five with large circular bases and two lying down without bases. They were moulded in a grey-green plastic thus avoiding the need to paint the bases. Faces and hands were crudely painted as were guns, webbing etc. They had no inscription on the underside of the bases.

R673 assault tank.

The set consisted of the following:
1. Section leader pointing ahead and holding a pistol
2. Soldier throwing a grenade
3. Soldier kneeling with a bazooka on his shoulder
4. Soldier kneeling firing a rifle
5. Soldier walking forwards with a gun at the ready
6. Soldier crawling with a rifle
7. Injured soldier in bandages, lying down

ASSAULT TANK

Assault Tank R673

This was the tank from the R568 Assault Tank Transporter which could be bought separately. It came with two Red-Eye rockets and a set of Commandos. Only one batch was made. These passed through the stores in 1967 and totalled 4,840 in number. It was sold in a BS1 box.

The tank had started life in the Minic Push-and-Go series and was featured on the Tri-ang Railways R241 Bogie Well Wagon with Tank Load. For the Battle Space series the turret was replaced, after quite a bit of experimenting, with a rocket firing turret. A pressed steel chassis was fitted with TT driving wheel centres (no tyres) running on rolling-stock type axles.

OPERATING ACCESSORIES

By 1965, operating accessories were not listed as a range in their own right but were dispersed amongst other models in the price list. The following were the models carried over into 1966 from the former Tri-ang Railways range:

RT266	Set of 2 Station Lamps
RT267	Fog Signal
RT268	Bell Signal Set
R345	Side Tipping Wagon Set (see 'Wagons')
R348	Giraffe Car Set (see 'Wagons')
R402	TPO Royal Mail Set
R405	Colour Light Signal Set
R406	Automatic Train Control
R408	Electric Turntable
R573	Double Track Colour Light Gantry

Those with an 'RT' prefix had also been available in the TT range. The 'T' was finally dropped in 1969.

Deletions and Additions

The first operating accessories to be dropped from the catalogue were the RT268 Bell Signal Set, RT267 Fog Signal and R406 Automatic Train Control which survived to the end of 1967.

In 1966, the R263L and R236R Point Operated Signal (for left and right points) were added to the range but were available for only three years.

There was only one other new non-wagon operating accessory associated with the Tri-ang Hornby period and that was the R675 Freightliner Depot which was first seen in the 1969 catalogue. This was available both as a solo model and in the RS602 Senior Freightliner Set. The lorry shown in the picture, although planned, was not made.

The R345 Side Tipping Car Set was withdrawn in 1970 but was to return later as a Hornby Railways model.

The remainder of the range of operating accessories survived the seven years of the Tri-ang Hornby label. These included the R348 Giraffe Car Set, R402 TPO Royal Mail Set, RT266 2 Station Lamp Set, R573 Double Track Colour Light Gantry and R408 Electric Turntable.

As we have seen, in 1970 thought was being given to a ticket operated train set but this was not to materialise until 1972 and is outside the scope of this volume.

CHECK LIST OF OPERATING ACCESSORIES

All operating accessories made during the Tri-ang Hornby period are listed here together with the years they were made. Some were still listed in the catalogue after these dates as old stock was cleared. The number in brackets refers to the type of track the model was made to fit.

No.	OPERATING ACCESSORY	Dates
R402	Travelling Post Office - Royal Mail Set - maroon	62-68
R402	Travelling Post Office - Royal Mail Set - blue (4)	69
R402M	Travelling Post Office - Royal Mail Set - blue (6)	70-72
RT405	Colour Light Signal Set (4)	62-69
R405M	Colour Light Signal Set (6)	70-71
R406	Automatic Train Control Set	62-68
R408	Electric Turntable - grey & brown (4)	63-69
R408U	Electric Turntable - grey & black (6)	70-74
RT266	Set of 2 Station Yard Lights	64-68,70
RT267	Fog Signal	64-66
RT268	Bell Signal Set	64-65,67

No.	OPERATING ACCESSORY	Dates
R573	Double Track Colour Light Gantry (4)	66-69
R573M	Double Track Colour Light Gantry (6)	70-71
R263L	Point Operating Signal (colour light) Left Hand	66-67
R263R	Point Operating Signal (colour light) Right Hand	66-67
R572	Operating Lift Bridge	not made
R675	Freightliner Depot Crane & Container	69-74
R575	AEC Articulated Tractor and Trailer	not made
	for the following - see 'WAGONS'	
R127	Crane Truck	-
R344	Track Cleaning Wagon	-
R345	Side Tipping Car Set	-
R348	Giraffe Car Set	-
R560	Crane Car	-
R739	Operating 75Ton Crane	-
	for the following - see 'BATTLE SPACE'	
R128	Helicopter Launch Wagon	-
R216	Rocket Launching Wagon	-
R239	Bomb Transporter	-
R249	Exploding Van	-
R341	Searchlight Wagon	-
R343	4 Rocket Wagon	-
R562	Catapult Plane Launching Car	-
R566	Spy Satellite Launching Car	-
R567	Radar wagon	-
R568	Assault Tank Transporter	-
R571	Q Car	-
R631	Tank Recovery Wagon	-
R639	Sniper Car	-
R670	Twin Ground to Air Missile Site	-
R671	Multiple Ground to Air Missile Site	-
R672	Honest John Pad	-
R725	Command Car	-
	for the following - see 'TRACK'	
R663	Point Control Pack	-
	for the following - see 'BUILDINGS'	
R475	Platform Crane	-
	for the following - see 'SUNDRIES'	
RT297	Power Clean Brush	-

DIRECTORY OF OPERATING ACCESSORIES

ROYAL MAIL SET

Maroon Short Coach R402

This remained in production until the end of 1968. The appearance and operation of the set was described in detail in **Volume 1** and will not be repeated here. The mail bags and tops of the receiving bins were now bright red and the receiving bin and pickup arm were separate units that clipped to the rail instead of incorporating the track as earlier ones had done. The earlier sets had had only two bags but with the R402, four bags were supplied. More than 28,000 sets were made during the Tri-ang Hornby period and the maroon version was withdrawn as the blue and grey one took over after 1968.

A feature of this period was the attractive illustrated box in which the set came. The box was DN size and carried an illustration on one side of a mail train passing the collection hook and receiving the bag while a railway worker stood on the platform. The sides of the box showed pictures of other models that could be bought.

Blue and Grey Short Coach – Super 4 R402

This first appeared in the catalogue in 1969 and was really an updating of the maroon model to take account of the livery changes on the real railways. In 1969 it was illustrated in the catalogue with red mail bags but by the following year they were shown as grey.

In 1969 the box was still DN size but was now the standard red and white type relieved only by a sticker on one broad side carrying the message 'Now in NEW B.R. Livery'. 8,200 were made in 1969 for solo sales.

Blue and Grey Short Coach - System 6 R402M

For 1970, it was slightly redesigned to fit the new

R402 operating mail coach sets, an advertisement promoting the set and unloading sorted mail from a real travelling post office.

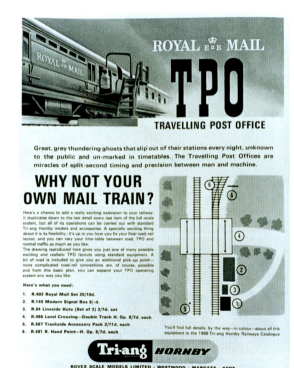

The instructions that came with the operating mail coach set.

Tri-ang HORNBY
Regd Trade Mark

Model Railways

[SYSTEM] **6** [TRACK]

OPERATING MAIL COACH SET—R.402M

(British Patent No. 749657)

R.402M set consists of a Special Mail Coach, four "mail bags" S.5077, two press-fit operating ramps X.757, one "pick-up hook" and base X.755, and one "receiving bin" and base X.756. These parts clip-fit to System 6 track only.

When fitting operating ramps X.757 to the track, they should be placed in the correct position and the ends pressed down firmly so that they clip home between the underside of the rail and the moulded track base.

With the addition of this set to a Tri-ang Hornby layout, it is possible to reproduce the method used on full-sized railways for collecting mail bags from isolated points and delivering them without the train having to stop.

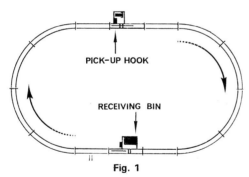

PICK-UP HOOK

RECEIVING BIN

Fig. 1

Initially it is recommended that the parts be incorporated in a simple oval to accustom the user to their function. Later they may be used in more complicated layouts in conjunction with buildings, etc.

To ensure successful working the following instructions should be carried out carefully.

Decide in which direction it is required to run the train and fit one operating ramp and the "pick-up hook" to the left of the approaching train.

The relationship between the "pick-up hook" base and the ramp is illustrated in Fig. 2.

It is advisable to have a straight length of track immediately before and after the "pick-up hook".

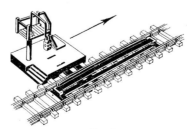

Fig. 2

System 6 track and at this stage the set picked up an 'M' suffix to its number. This version was actually due for release in March that year but due to tooling difficulties was delayed a further month.

As a solo model, batches were made each year between 1970 and 1972 and these totalled about 10,000. In 1970 and 1971 it was also used in the RS604 Night Mail Set and for this and its variants, nearly 14,000 TPO sets were made.

TURNTABLES

Super 4 R408

The last year that the R407 hand-operated turntable was available was 1965. From 1966 onwards, only the R408 electric version could be bought. The X319 conversion kit was withdrawn at the same time.

The R408 is well described in **Volume 1**. Between 1965 and 1969, about 11,000 electric turntables were made for the Super 4 track system and presumably packed in Tri-ang Hornby boxes.

System 6 R408U

The turntable stayed in production until 1974 but was converted to System 6 in 1970 as a result of pressure from the public. At this time it picked up a 'U' suffix to the 'R' number. While the Super 4 version had had a brown rotating gantry and motor housing, on the System 6 version these were black.

The system 6 version of the turntable arrived in November 1970 but even as it was being made, consideration was being given to a new version to be numbered R602. The tool costs were to be investigated before proceeding further. Subsequently the existing R408U was retained. In 1970 and 1971, about 4,000 of the System 6 version of the electric turntable passed through the stores.

FOG SIGNAL SET RT267

Available from the spring of 1964, the Fog Signal was to be made for only three years during which time about 17,000 were produced. Of these, some 10,000 fell into the Tri-ang Hornby period and yet all I have seen were in Tri-ang Railways boxes. The packaging was attractive with an illustrated lift-off lid.

BELL SIGNAL SET RT268

Only about 3,850 bell signal sets were made and these in 1964, 1965 and 1967. The latter consisted of only 500 units. They do not often turn up and the only ones I have seen have still been in their packaging which carried the Tri-ang Railways name. The model is reviewed in **Volume 1**.

TWO ASPECT COLOURED LIGHT SIGNAL SET

Super 4 RT405

These were released in 1962 and are described in **Volume 1**. Judging by the numbers made they sold quite well. Between 1965 and 1969 some 42,000 were made and dispatched in standard Tri-ang Hornby AA boxes.

System 6 R405M

In 1970 the model was revised so that it would clip to system 6 track and these sets were numbered R405M. The revised set was made for just two years before being withdrawn. During this time 12,000 were produced.

DOUBLE TRACK COLOUR LIGHT GANTRY

Super 4 R573

This set consisted of two signal heads, the same as those used on the point-operated signal (R263), mounted on a white plastic gantry which spanned two tracks. The gantry was a white version of that used for the catenary system and carried clips for the catenary wires. The base of the mast one side had five sockets for the necessary operating wires covered by a flat casing with a quite realistic safety tread moulded on. From the box two thick three-core cables ascended the gantry mast and fed the two signals which faced in opposite directions.

Although the bases were designed to clip into

Right: R408U electric turntable box label for the later version suitable for both Super 4 and System 6 track.

Far right: R348F giraffe car distributed through Hornby-Acho in France.

Below: RT405 colour light signals.

Super 4 track they also had screw holes so that they could be fixed down to the baseboard if used with any other make of track. The set also contained the power leads, a yellow two-lever switch frame and an instruction leaflet. It was priced at 29/6d.

The first batch of signal gantries was made in 1966 and the Super 4 version remained in production until 1969. During this time 10,000 were made and sold in an illustrated box which depicted two units in use on a real railway. Some boxes were originally printed in 1965 or 1966 and another batch was printed in 1967.

System 6 Version R373M

In 1970 the set was revised so that the gantry could clip to System 6 track and at this time it was renumbered R373M. In this form it was made for two years with a total production of 4,000 units.

Some of the boxes originally printed in 1965 or 1966 lasted through to 1970 when a label was stuck over the end panels to indicate change to R373M and System 6 track.

POINT OPERATING SIGNAL

Super 4 R263L, R263R

This was the subject of the official Tri-ang Hornby model railway press advertisement for June 1966. It operated by a simple mechanical switch which clipped to the sleeper base of Super 4 points and fitted over the point lever. From the switch three wires lead to the signal and to the power supply. The red aspect was shown when the points were not set for the road on which the train was running. When the point was changed, the slide of the switch made contact and the signal changed accordingly. The signals could only be used to protect the points from the trailing direction although it should have been possible to wire the switch to a splitting junction signal for the facing direction so that a green light was shown for whichever route was set up.

The set cost 15/6d and came with an illustrated leaflet to show the correct wiring. All the parts in the set could be bought separately.

The signals looked quite good, although the bulbs were too large for realism and viewed from the front

the light from the bulbs looked white and only faintly tinged with colour. They were reviewed in the *Model Railway Constructor* in May 1966.

Left- and right-hand versions were available, each being identified by an 'L' or 'R' suffix. In 1966, almost 7,000 left-hand and 5,800 right-hand sets were made and dispatched in standard red AA boxes. Further batches totalling 7,700 were made in 1967 and these may have been the last except for a special supply made in 1969 for Canada. The latter consisted of 1,540 of each type.

Other Signals

See the chapter on 'Buildings and Trackside Accessories'.

AUTOMATIC TRAIN CONTROL SET

Super 4 R406

The automatic Train Control Set was in the shops in 1962 and remained in production until 1968. It is thought that a total of 25,000 were made, of which

Studio picture showing the colour light signal gantry in use.

R573 double track colour light gantry.

R.406 AUTOMATIC TRAIN CONTROL SET. One train can be made to start and stop another automatically or to control Colour Light Signals. Operates on 12 Volts DC.

Left: R406 automatic train control and R263 point operating signal.

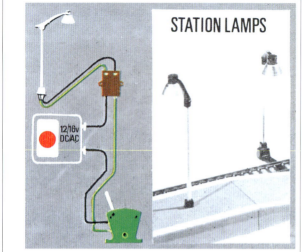

STATION LAMPS

R.266U PAIR OF STATION LAMPS. Operating from 12/16 volts A.C./D.C. the lamps are packed complete with wire and junction box. They may be clipped to Super 4 or System 6 Track or attached to Island Platform Units. The wiring diagram shows how a lamp may be simply wired to your power controller using a green Lever Frame Section R.47 as a switch.

Station lamps illustrated in the 1970 catalogue.

Left: RT266 Yard lamps.

8,800 were in the Tri-ang Hornby period. A description of how it worked will be found in **Volume 1**. The set continued to be sold in AE size boxes which had been reprinted in 1965 or 1966 with the Tri-ang Hornby logo.

System 6 R406

It was reissued in 1972 adapted for System 6 track.

PAIR OF STATION YARD LIGHTS RT266, R266U

In 1963 the station yard lights came out as a set of four (RT265) and a year later they were also available in a pair. The set of four was not made after 1964 and yet at the end of 1965 there were still 1,600 sets in the stores. As these were not recorded the following year they may have been broken up to use in the RT266 sets.

The instruction leaflet that came in the box showed how the lamps could be plugged into the holes in the platform units, fixed to the baseboard using the S6435 adapter or clipped to the track using the standard R455 mast base track clip. Spare bulbs were available as S6379.

The RT266 pair of lamps remained in production until 1968 and were sold in standard AC boxes with the contents rubber stamped on the ends. There was then a break of one year followed by a batch of 5,440 in 1970, which may be identified by the 'U' suffix to the number on the box. This, while remaining an

AC size, now had printed labels and reference to the suitability of the lamps for both Super 4 and System 6 track.

OPERATING LIFT BRIDGE R572
Planned for 1965 but not made.

FREIGHTLINER DEPOT CRANE & CONTAINER R675
The model was due for release at the start of 1970 but, due to necessary tool modifications, it was delayed and probably arrived in the shops in February. An article in the October 1970 *Railway Modeller* described how the model could be motorised. It was the subject of a full page picture in

the 1970 catalogue displayed with the R2008 Freight Depot, the proposed AEC Tractor and Trailer with a Tartan Arrow container and an assortment of container wagons.

As the catalogue told us, the crane had three operating movements. It could grasp or release a container, lift or lower it from rail or road vehicle and transfer it across two rail tracks to a roadway, clearing road and rail wagons in its path.

Once again, Rovex had succeeded in designing a complex model that was robust and could be operated without the need for a power supply. Its simplicity of design meant that it was also comparatively cheap to produce.

More than 8,700 were made in 1970 and were sold

Instructions for the yard lamps.

Tri-ang HORNBY
Regd. Trade Mark

Model Railways

SET OF TWO STATION LAMPS R.266

This set operates from a supply of 16 volts A.C. or 12 volts D.C., and the lamps may be mounted in three different ways as described below.

PLATFORM MOUNTING

The lamps may be mounted on platform units, other than those supplied with R.5083 and R.5084A, the bases being designed to fit into the square holes in the platforms. The two leads (black and green) from each lamp must first be passed through the hole and out under the side of the platform as illustrated (Fig. 1).

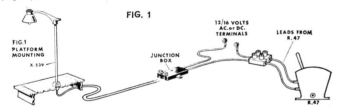

FIG. 1

FIG.1 PLATFORM MOUNTING X.539

JUNCTION BOX

12/16 VOLTS A.C. or D.C. TERMINALS

LEADS FROM R.47

R.47

BASEBOARD MOUNTING

To secure a lamp to a permanent baseboard, the wires should be pased through the square hole in Base Adaptor S.6435, into which the base of the lamp is then fitted. The small hole in the adaptor will take a track pin to hold it in the desired position on the baseboard. Make sure that the leads pass through one of the slots provided, before hammering the pin home (Fig. 2).

FIG.2 BASEBOARD MOUNTING

S.6435

CLIP-FITTING

If it is preferred, the lamps may be clipped to the track base by using the base clips into which the base adaptors slide after being fitted to the lamps (Fig. 3).

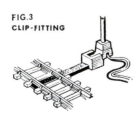

FIG.3 CLIP-FITTING

Boxed freight depot crane and container.

Picture of a freight depot crane from the 1971 catalogue.

Below: Freight depot illustrated in the 1970 catalogue showing the crane and freight depot and the proposed articulated tractor and trailer that was not made.

R.675 B.R. FREIGHTLINER DEPOT CRANE with container. The crane has three operating movements, it can grasp or release a container, lift and lower it from rail or road vehicle and transfer it across two rail tracks to a roadway, clearing loaded road or rail vehicles in its path. (Road and Rail Vehicles not included).

R.2008 FREIGHT DEPOT complete with loading bank.

before the end of the year. Another 4,500 came in 1971 and further batches up to the end of 1974.

CONTAINER FREIGHT MANAGER'SOFFICE AND LOADING BAY R2008

This has been listed in the chapter on 'Buildings and Lineside Accessories' and is also mentioned on page 476.

AEC ARTICULATED TRACTOR AND TRAILER R575

It had been planned to introduce this model in 1969 as part of the container depot scene. Indeed, a pre-production model was displayed at the Tri-ang House Trade Show early in the year. A study of the catalogue picture suggests that the cab of the Minic Motorway mechanical horse (the tools of which were now at Margate) had been used for the mock-up.

It was to have been based on the AEC articulated unit with the latest Ergomatic cab. I do not know why the model, which was planned for the autumn, was dropped before reaching production.

CONTROL EQUIPMENT

POWER UNITS

Thanks to the rules concerning Purchase Tax, we have little information about the various power units made for or by Rovex. In order to avoid tax, they could not be advertised with the rest of the model railway equipment but had to be treated as an independent range. Leaflets were produced illustrating and describing the units but these have not survived as well as the train catalogues.

The equipment was sold under the name 'Tri-ang Power Units' and later, probably from 1972, as 'Rovex Power Units'. With the marketing changes in

the mid 1960s, the 'Tri-ang Power Units' logo was updated to make it look more like that for 'Tri-ang Hornby'.

A look at the factory census sheets shows a large range of units in the stores at Margate over the Tri-ang Hornby years. Few of these were stocked for British model railways. Due to different power systems around the world and different regulations concerning design of electrical equipment, Rovex found it necessary to make many variations to their standard British designs for overseas sales. This took a lot of time consuming research and each variation carried a different reference code.

Inherited Units

In 1965, there were no sudden changes in the basic range of equipment. The Rovex-made RP3, RP4.5 and RP5.5 power controllers of 1963 remained in use along with the P40 battery connector, P41 connectors and P42 speed control unit, all of 1960 vintage. These are described and illustrated in **Volume 1**.

The company continued to service these early units up until 1970 when, in November, the company announced that replacement parts for the units were becoming scarce and that they could no longer be guaranteed repairs when sent in to the Service Department. Instead, the policy would now be to offer alternative service replacement units corresponding as nearly as possible to the specification of the damaged unit.

The 1967 'Banks' Power Controllers

The first major change in the range came in 1967 when a new series of power controllers was launched. In order to conform to the requirements of the Consumer Protection Act, Rovex had to redesign their power units. From now on the transformer windings had to be separately insulated, the cases completely sealed and the mains lead internally clamped.

The new range of power units was built for Rovex

by the Hinchley Engineering Company, based in Wiltshire. This company held the patent on the resistance mat used which automatically gave half-wave rectification on the first third of speed to provide fine control at low speed. It then went onto full wave at one third speed. With the accent on safety, Rovex hit on the idea of naming their new power units after banks; thus the RP13 was The Westminster and RP14 was The Barclay. This theme carried over into publicity material with the claim 'safe as banks'

Mail order houses who bought train sets from Rovex were demanding sets which were complete and ready to run without the need for their customers to go out and buy power units or even batteries. For this market, Rovex designed the RP15 which was available from 1969. It was a small relatively cheap unit that could be held in the hand if so wished.

All power units were to have a combined Tested/Power supply label on the flex for 1971 and a draft of the wording was sent to the NJE, in December 1970, for their approval. A new type of packaging was also planned for 1971 consisting of a plain corrugated cardboard box (suitable for mail order or bulk supply if required) with a separately printed lid for solo sale.

A one-colour leaflet in priced and non-priced versions was planned for 1970.

Keeping Up with Fashion

In 1972, it was foreseen that certain items of the 1971 range would continue to be offered as well as the following new items:

RP51 1.2 amp transformer
RP18 2.2 amp transformer
RP53 0.6 amp variable transformer controller

Following a study of power units being marketed in Britain by other manufacturers, in 1970 plans had been drawn for the production of a range of completely new power units in 1972. These appear to have been based largely on ideas borrowed from Hammant & Morgan.

The principal features of the new range would be twofold. They would firstly be modern in design and have the high quality finish associated with expensive electrical equipment; secondly, the individual units might be free standing or capable of being coupled together in rigid form to provide one control panel. As far as possible they would share the same basic case with positive power linking between units on the rear or side of the case.

This new series of power units will be described in **Volume 3**.

Overseas-Made Units

Transformers and power controllers were very heavy items of equipment to transport and it was therefore impractical to export British equipment to the major overseas markets. Instead, it was cheaper to leave the overseas companies in the Lines Bros. Group to manufacture their own or find a local company who would make units to their specification.

While technically these were the same or similar to basic Tri-ang units, they often looked different externally. The difference was more pronounced in some countries than others. Australia more or less kept to the shape of the early Tri-ang units but the Canadian equipment, which were mostly imported from the USA or Japan, bore no visual resemblance to the units made in the UK.

SWITCHES

These varied little from those originally produced for the Tri-ang Railways system. Australia and New Zealand had the tools to make their own and these were almost identical to those supplied from the UK.

CHECK LIST OF CONTROL EQUIPMENT

The following tables list the power units that passed through Margate between 1965 and 1971, some albeit briefly and in small quantities. The other tables include overseas models, switches etc.

No.	GENERAL RELEASE POWER UNITS	Dates
RP3	power controller	63-66
RP4.5	power controller	63-66
RP5.5	power controller	63-66
RP13	The Westminster power controller	67-69
RP14	The Barclay power controller	67-70
RP15	mail order power controller	69-71
RP18	The Midland transformer	70-71
RP40	battery connector	60-71
RP41	connect for leads	60-69
RP42	circuit control unit	60-71
RP51	transformer rectifier (not made)	-
RP53	power unit	71

No.	OTHER POWER UNITS STOCKED	Dates
RP5.S3		66-67
P5.34	power controller	65
RP5.5SI	110V power controller for export	65-67
RP8	transformer	65
RP8SI	110V transformer	65
RP9	transformer for Scandinavia	70
RP9D	transformer for Denmark	65-67
RP9S	transformer for Sweden	66-68
RP14SI	110V power unit for export	68-70
RP15SI	110v power unit	70-71
RP16	power controller	71
RP17	power pack	71
RP18SI	110V Transformer	71
RP20	power controller	70-71
RP20SI	power controller	71
RP30	power unit	69
RP40D	battery control unit	70
RP43	battery box controller	69

No.	OTHER POWER UNITS STOCKED	Dates
RP50	power controller	71
RP50D	transformer	70
RP52	power unit pack	70-71
RP1770	power unit	68-70
RP1770SI	110V power unit	68-70
WN7	power controller (see Wrenn)	71
WN8	circuit control unit (see Wrenn)	71
J11	power unit for Jump Jockey	71
J24	power pack for Jump Jockey	69-71
J300	110V for Jump Jockey	71

No.	NEW ZEALAND MADE POWER UNITS	Dates
P15	power controller	62-66
P15A	power controller	67-71?
P42	circuit controller	59-71?
P149	power controller	68-71?

No.	AUSTRALIAN MADE POWER UNITS	Dates
P2A	transformer	62-71
P5A	power controller	59-66
P5TA	power controller	66-71
P6A	power controller	60-69
P8A	transformer	63-71
RP42	circuit control unit	58-67
RT42A	circuit control unit	67-69
P40	battery connector	59-67

No.	CANADIAN MADE POWER UNITS	Dates
PA102C	power controller	63-68?
PA200C	power controller	63-68?
2200	power controller	67-68?
PA102C	power controller (Meccano Tri-ang)	68?-70
PA200C	power controller (M Dual Electro)	68?-70
PA2200	power controller (Meccano Tri-ang)	68?-70
1000	power pack (Tri-ang Railways)	70-71
1005	power pack	70
1000	power pack (Hornby Railways)	72-73

No	UNITS BULK PACKED FOR OVERSEAS	Dates
RP40C	battery box (Canada)	67
RP43	battery box controller (Australia)	69
RP43	battery box controller (New Zealand)	69
S5083	leads (Canada)	69-71

no	SWITCHES ETC.	dates
RT44	lever points	65-71
RT46	lever signal	65
RT47	lever track	65-71
RT144	lever frame base	65-71
R663	point remote control pack (see track)	68-71

DIRECTORY OF CONTROL EQUIPMENT

A number of the Tri-ang power units remained available into the Tri-ang Hornby period. These are listed here and are described in **Volume 1**.

N.B. All power units have been listed in numerical order, rather than the usual chronological order, in order to make it easier to trace a unit.

Australian Transformer P2A

This locally-made power controller was first introduced in 1962 and could operate three trains when used with an RT42 Circuit Controller. Its outputs were 12V DC and 15V DC giving a total of 2 amps. The unit remained in production until 1971.

Power Controller RP3

This had only a 12V DC output of 1 amp and was contained in a moulded black plastic case with a red control knob. It was made from 1963 to 1966, with more than 10,000 being made in 1965 and again in 1966. It had a thermal resetting cut-out button.

Power Controller RP4.5

This had two 1 amp 12V DC outputs, only one of which was controlled. A High-Low resistance control switch was fitted to allow for variations in the current consumption characteristics of different types of motor. It was made from 1963 until 1966 and had a metal casing with a sloping face as the control panel. Over 61,000 were made in 1965 and a further 6,000 in 1966.

Australian Power Controller P5A

This replaced the imported P5 in 1959 and looked very much like the original. It remained in production in Australia until 1966. The P5A was the principal power controller made for model railway enthusiasts in Australia. It had the following outputs: 12V DC controlled, 12V DC uncontrolled and 15V AC for operating electrical accessories. It had a total output of 2 amps.

Australian Power Controller P5TA

The P5TA was a transistorised version of the P5A which it replaced in 1966. It remained in production until 1971.

Power Controller RP5.5

This shared the same case as the RP4.5 but besides the two DC outputs it also had a 15V AC output for control of electrical accessories. It too was made between 1963 and 1966 and production during the last two years was 36,000 and 28,000 respectively.

RP5.S3

The unit was made between 1966 and 1967 and only 167 were recorded through the stores during that time.

Power Controller P5.34

This was made in 1965 but only 162 were recorded.

Australian Power Controller P6A

The P6A was introduced in 1960 as a replacement for the earlier P3A. It was designed primarily for overhead railway systems and had two completely independent DC outlets. There was a 12V DC controlled, 12V DC uncontrolled and a 15V AC outlet. These provided 2.5 amps continuous (AC 1.5 amps intermittent). Production of this unit ceased sometime in 1969.

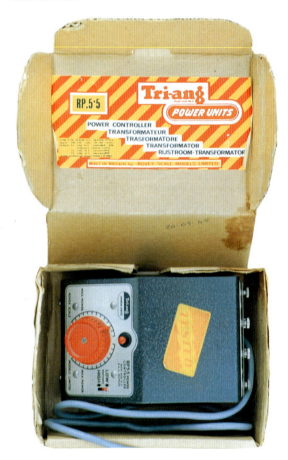

RP5.5 power unit.

Any correspondence with the factory regarding this unit should quote **No. RP.4·5**

Tri-ang *POWER UNITS*
Regd Trade Mark

RP.4·5 POWER CONTROLLER

FOR USE WITH 200/250 VOLTS ALTERNATING CURRENT 50-100 CYCLES MAINS ELECTRICITY SUPPLY ONLY

READ THESE INSTRUCTIONS CAREFULLY

(1) TRI-ANG RP.4·5 Power Controller is a general purpose unit which converts A.C. (Alternating Current) mains to low voltage D.C. (Direct Current). One Direct Current output can be controlled from zero to maximum output, and the polarity reversed, through the variable resistance. A HIGH-LOW resistance control switch is fitted to this section to allow for variations in the current consumption characteristics of different types of motors. The other Direct Current output is uncontrolled.

(2) This unit converts 200/250 volts A.C. (50-100 cycles) to:
    1. 12 volts D.C. controlled.
    2. 12 volts D.C. uncontrolled.
    Total continuous rating 1 amp maximum.

(3) The two D.C. outputs, derived from a single secondary winding on the transformer, are suitable for use with separate isolated circuits but must not be used for controlling two separate supply systems with a common return having variable polarity.

(4) This Power Controller must **NEVER BE CONNECTED TO D.C. MAINS.**

(5) A suitable 3-pin plug must be affixed to the flexible mains lead on the input side. The three wires MUST be attached as follows:—
    RED   ⎫ Each to one of the two smaller pins on the plug.
    BLACK ⎭
    GREEN— to the largest (Earth) pin on the plug. NEVER to either of the other two pins.
If the power supply is fitted with 2-pin plugs and sockets, consult a competent electrician before use.

(6) Two pairs of marked terminals are fitted to the output side of the Power Controller.

(7) The Unit may become slightly warm when operating.

(8) In the event of a short-circuit or overload, in the controlled D.C. circuit, the automatic cut-out device in the Controller should operate, causing the red button to jump upwards. Upon this occurring the Control Knob must be reset to zero immediately, and the cause of the short-circuit or overload traced and remedied. The red button should then be pushed down to the normal operating position. UNDER NO CIRCUMSTANCES must the button be held down by external pressure.
Care must be taken to ensure that the uncontrolled output is not subjected to short-circuits or overloads.

(9) Do not unfasten any of the screws securing components in the Unit housing.

(10) It is inadvisable to stand a Power Controller on carpets or rugs, as the flow of air through the ventilating louvres may be restricted.
ALWAYS DISCONNECT THE UNIT FROM THE MAINS WHEN NOT IN USE.

ROVEX SCALE MODELS LIMITED, WESTWOOD, MARGATE, KENT

RP.4·5/39553/7   MADE IN ENGLAND   Printed in England

1966 instruction leaflet for the RP4.5 power controller.

Transformer RP8

This transformer was made for 1965 only when nearly 133,000 were produced. It is thought to have been a heavy duty 3.5 amp unit which was suitable for slot-cars as well as railways. It probably had 12V DC and 15V AC terminals.

Australian Transformer RP8A

Australia made their own version of the RP8 between 1963 and 1971.

Scandinavian Transformer RP9

It was offered in the Export Price List in 1965 as being approved by SEMKO and DEMKO. It appears, however, that the units that went out to Denmark and Sweden went under their own codes as shown below. It was advertised as being suitable for Scalextric and Minic Motorways as well as Tri-ang Hornby Railways. The only units recorded as having passed through the stores with the plain RP9 code were just 264 transformers in 1970.

Transformer (Denmark) RP9D

This was in production between 1965 and 1967 and over 11,000 were made.

Transformer (Sweden) RP9S

Between 1966 and 1968 a total of 6,500 of this unit passed through the stores at Margate.

Power Unit J11

This was probably a power unit for Jump Jockey of which only 74 were made in 1971.

The Westminster RP13

This was new in 1967 and introduced to conform to the Consumer Protection Act. It was made for Rovex by Hinchley Engineering and remained in production until 1969 with units still being available in 1970. In all a little over 64,000 were made.

The unit, which was named after the Westminster Bank, had Microfine Control (patent application No. 19902/66) which operated automatically from the Hinchley half-wave rectification built into the printed circuit. There was a self-setting thermal cut-out button which took about 10 seconds to reset and had a single 12V DC output (max 9.6 VA). It was housed in an updated steel case, still with a sloping panel but with rounded edges, a 'hammered' finish and rubber feet. The red control knob rotated through 300° providing fine speed adjustment.

A review in the *Model Railway Constructor* told us that the unit measured 7.5" x 4.5" and was 3" deep sloping to 1.5".

"It is extremely neat in appearance, having the one large control knob and a press toggle switch for reversing, but the front is marred by having the output terminals just above the reversing switch, which means that the wires trail over the front, and while they do not get in the way they look unsightly and would be far better positioned at the rear – particularly if the controller is to be fitted to a panel."

On the half-wave rectification the reviewer remarked:

"We would suspect that this is operative on control settings 1-4 as after this there is a 'kick' of approximately 1.3V increasing the output from 5V to a little under 7V. This however does not affect the running but merely increases acceleration slightly at this particular point, but otherwise the output rises evenly throughout the range."

Under test the controller was described as being 'delightful to use'. All the models tested could 'creep along the track at a scale walking speed' whether light or with a train. Acceleration was found to be smooth and despite the low price the power unit was 'unbeatable'.

This and the Barclay power unit were the sole feature of official Tri-ang Hornby advertisement during the autumn of 1967.

The Westminster (Export) RP13E
No!, not an extra strong beer but an export version of the Westminster offered to overseas buyers in 1968/69.

The Barclay RP14
This power controller, named after Barclay's Bank, and also made by Hinchley Engineering, was to the same specification as the Westminster but had three outputs. Besides the controlled 12V DC (9.6VA) there was an uncontrolled 12V DC and a 16V AC. It too was introduced in 1967 and was deleted for 1971, by which time 124,000 had been produced. The maximum continuous output rating was 18VA. It had a Microfine Control running device incorporated in the main control (patent application No. 145/67).

The Barclay (Export) RP14E
Export version of the Barclay offered to overseas buyers in 1968/69.

The Martin RP15
This power controller was named after Martin's Bank, later taken over by Barclay's, and had an output of 6 VA at 12 volts DC (0.5 amps) with a variable controlled DC output. It was made between 1969 and 1971 as the principal unit to be offered for mail order. It had been designed to a specification prepared by Rovex and during these three years, over 47,000 were made. The plastic body was a buff colour and the two halves were riveted together to form a wedge shape. The control knob, direction rocker switch and plastic terminal nuts were red and it was protected by an automatic cut-out. It was designed to be held comfortably in the hand.

In 1970, discussions were being held with the management in Australia concerning the suitability of the power unit for the Australian market and

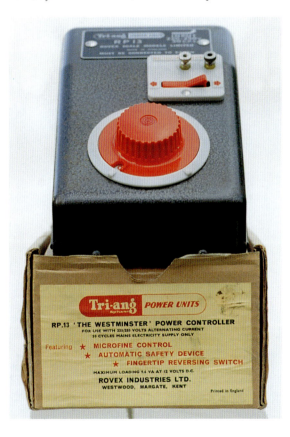

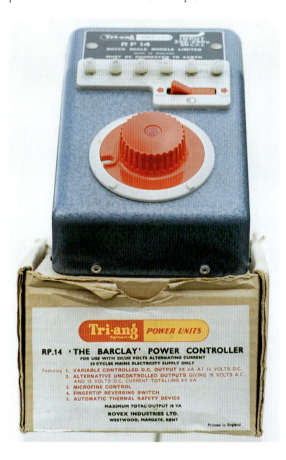

Banks power units RP13 (Westminster), RP14 (Barclay) and RP15 (Martin) for mail order sets.

whether it should be made there. It seems likely that the demise of the company occurred before any were produced.

In 1972, the unit became R915 in the Hornby Power Units range.

New Zealand Power Controller P15 (P15A)

This was a power controller made in New Zealand for local sales between 1962 and 1971. In 1967 it became P15A presumably due to some alteration in design. For a long time it was the only power controller available from the New Zealand factory.

Advertisement for the RP13 and RP14.

Power Controller RP16

This dates from 1971 and offered one 12V DC controlled output and one 15V AC output for electrical accessories. It came in a flat rectangular case which claimed to be double insulated. It carried the 'ROVEX Tri-ang' trademark as well as that of 'Tri-ang POWER UNITS'. It was made for one year only and about 2,800 were produced.

Power Pack RP17

This appears to have been planned for 1969 but none were made until 1971 when about 2,500 were manufactured. Was this the Lloyd Power Pack?

The Midland RP18

The RP18 was to become the Midland Power Unit, named after the Midland Bank. It had an output of 26 VA at 12 volts DC (2.16 amps) with a surge output of 2.5 amps. It was being purchased from Hinchley and in November 1970 it was decided that an alternative source of supply would be investigated. 500 were bought in 1970 and 32,000 the following year.

Work was being undertaken on the RP27 in 1970 as a replacement for the RP18 transformer introduced that same year. However, it remained available in 1971 under the name 'Smoothflow'.

Power Controller RP20

This was a new power controller being introduced in 1970 as an improved RP14. Its output was 30 VA at 12 volts DC (2.5) with variable controlled DC output and separate uncontrolled DC and AC outputs. 15,000 were made for 1971.

Power Pack J24

This was a Scalextric unit which in 1971 was replaced by the RP51 transformer.

Power Unit RP30

One batch of 1,500 passed through the stores in 1969.

Any correspondence with the factory should quote **No.** 270668

The RP.13 and RP.14 POWER CONTROLLERS have been specially designed and constructed to conform to the requirements of the CONSUMER PROTECTION ACT. The Transformer windings are separately insulated, the Cases completely sealed and the Mains leads internally clamped.

Each unit has the following outstanding features:—

★ **Microfine Control** operates automatically from **Half Wave Rectification** built into the **Printed Circuit** (Patent Pending).

★ **All Outputs** protected from **Overload** by a **Self-Resetting Thermal Cut-Out**.

★ **Easy Access Output Terminals** and **Reversing Switch**.

★ **Long Rubber Feet** provide **stability** for **One Hand Operation** and prevent **Scratching**.

★ **Deep Drawn Steel Case** for **Strength**, rivetted to prevent dangerous **Tampering**.

RP.13 'THE WESTMINSTER' Power Controller

INPUT	TYPE OF OUTPUT	MAX. OUTPUT
220/250 Volts A.C. 50 Cycles	Variable Controlled Direct Current	9·6 VA at 12 VOLTS D.C.

RP.14 'THE BARCLAY' Power Controller

INPUT		TYPE OF OUTPUT	MAX. OUTPUT
220/250 Volts A.C. 50 Cycles	1.	Variable Controlled Direct Current	9·6 VA at 12 VOLTS D.C.
	2.	Uncontrolled 12 volts Direct Current	8·4 VA
	3.	Uncontrolled 16 volts Alternating Current	

MAXIMUM TOTAL OUTPUT 18 VA

Power Controllers must NEVER BE CONNECTED TO D.C. MAINS.
A suitable 3-pin plug must be connected to the mains lead. The three wires MUST be attached as follows:—
RED
BLACK } Each to one of the two smaller pins on the plug.
GREEN — to the largest (Earth) pin on the plug. NEVER to either of the other two pins. If the power supply is fitted with 2-pin plugs and sockets, consult a competent electrician before use.
THE THERMAL CUT-OUT DEVICE fitted inside these units protects the outputs in the event of overloading by cutting off the supply to the output leads. Upon this occurring the Control Knob must be set to zero immediately, and the cause of the overload traced and remedied. The Thermal Cut-Out will reset itself automatically, after a period, and normal operation may be resumed. If the unit continually cuts out check (a) that there are no short circuit conditions, and (b) that the maximum total output of the unit is not being exceeded. The Unit will become warm when operating.
RP.14 ONLY
The two D.C. outputs, derived from a single secondary winding on the transformer, are suitable for use with separate isolated circuits but must not be used for controlling two separate supply systems with a common return having variable polarity.
ALWAYS DISCONNECT THE UNIT FROM THE MAINS WHEN NOT IN USE

Made in Great Britain by
ROVEX INDUSTRIES LTD.
WESTWOOD, MARGATE, KENT

RP.13/14/49961/4

Printed in England

Instruction leaflet for the Westminster and Barclay power controllers.

Battery Connector RP40

The battery connector, which arrived in 1960 and was described in **Volume 1**, was made in large quantities over the years and various versions may be found. They were usually black but yellow plastic was used later. The wording was embossed on the surface of the plastic on some, and printed on others.

This became R940 in the Hornby Power Units range in 1972 but during the Tri-ang Hornby days some 56,500 were made for solo sales. They were mostly sold in sets but were packaged in boxes of three to be separated from their packaging and sold singly in the shop.

Battery Connector (Australia) RP40A

Sometimes Moldex imported their battery connectors and sometimes they made them locally. No records have survived to indicate accurately which years they were locally made or how many.

Battery Connector (Denmark) RP40D

In 1970 a batch of 2,500 was made. They must have been a variation on the RP40 for some special purpose and may have been left over from a much larger batch made for export sets for Denmark. There were none left at the end of the year.

Battery Connector (Canada) RP40C

This was a version of the Battery Connector made for Canada in 1967. 2,500 were sent bulk-packed for use in Canadian set assembly.

Circuit Control Unit RP42

The RP42 was the standard circuit control unit. It had a black plastic body and red control knob, cut-out button and resistance slide. The latter gave a choice between high or low power. In 1970 work was being undertaken on the RP44 which was intended to replace it.

In 1972, the RP42 became R942 in the Hornby

Power Units range but not before 137,000 of the unit had been made during the Tri-ang Hornby years.

Australian Circuit Control Unit RT42A

Australia was using the R42 Circuit Unit from 1958 until 1969. These were made from the tools sent out in the late 1950s for train set production. In later years it was redesigned to give it a stove-enamelled metal box with a sloping face. Both input and output terminals were positioned at the back of the box.

New Zealand Circuit Controller P42

Originally known as the RT42 but changed to P42 in 1962, the casing moulding was made from the same tools as RT42A above and the unit served the same purpose. It remained in production until 1971.

Battery Box Controller RP43

The battery box was made for use in certain starter sets sold in 1969 but none was made for solo sales.

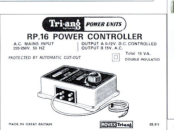

Wrenn power units WN7 and WN8 with a not suprising similarity to the Tri-ang RP15 and RP42.

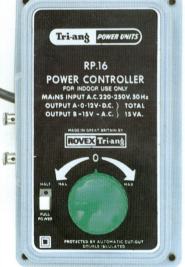

RP16 and box label.

Australian control unit.

New Zealand power controller P15A.

Battery Box Controller (Australia) RP43

A quantity of battery boxes was made specially for Australian set production. 9,000 were sent out bulk packed in 1969.

Battery Box Controller (New Zealand) RP43

New Zealand was also sent a supply of battery boxes for set assembly in 1969. About 1,500 were sent bulk packed. They were also available solo as well as in sets for that year only.

Transformer RP50

This was a 1971 model of which 2,200 were made.

Transformer RP50D

This may have been a version of the RP50 which was made specially for sale to Denmark. Nearly 4,600 went through the stores, all of them in 1970.

Double Insulated Power Unit RP51

This double insulated power unit had a single uncontrolled output of 14 VA at 12 volts DC (1.16 amps). The unit had a black plastic case with two terminals. It was protected by an automatic thermal safety cut-out. Production was supposed to have started at Margate late in 1970 but none was recorded as passing through the stores. It first appeared in the 1971 price list and was supposed to replace the Scalextric transformer J24.

Power Unit Pack RP52

This was a dual pack containing an RP42 circuit control unit paired with either a J24 or an RP51 transformer. It was available in 1970 and 1971 and came in a plain brown box with a black and white label on one end. Over 3,200 were made.

Power Unit RP53

This was a new unit for 1971 to be made at Margate and released at the provincial shows. It was a transformer/controller based on the old Hornby No1

RP52 containing the RP51 and RP42 circuit controller.

Canadian 2200, PA102C and PA200C power units.

unit. The controller had a slightly jerky four-stage speed control. Its output was 7.2 VA (0.6 amps) at 12 volts DC. It was seen as a possible replacement for the RP15 as a mail order transformer/controller by 1972. In November 1970 it was decided to look for the original Hornby tools rather than borrow the moulds from Calais where it was presumably being made for the Hornby Acho system. It became R953 in the Hornby Power Units range in 1972 although the output quoted for this unit was different. 7,000 were made in 1971.

Canadian Power Controller PA102C

This was called a train pack and between 1963 and 1968 was made in the USA for Lines Bros. (Canada)

Ltd. It was designed to CSA standards for use in Canada and had a golden finish with black printing. It was for a 120V 60Hz power supply and provided a 1 amp single 12V DC controlled output for train operation and a 16V AC output for operation of electrical accessories. Instead of the usual rotating knob for speed control, this unit had a pivoted lever mounted on the side of the case. Direction was controlled by a simple switch calibrated '15...30...45...120 full' presumably representing miles per hour!

From 1968 or 1969, they were made in Canada to a different design. They now carried the Meccano Tri-ang name but 'Tri-ang Railways' remained. The speed control was now on the underside with a half exposed wheel with a projection mounted on it. The direction switch was also now on the underside. This unit was to survive only until around 1970 and was priced $11.

Both units were fitted with automatic circuit breakers.

New Zealand Power Controller P149
This was offered new to New Zealand modellers in 1968 and was locally made. It was a unit with a sloping face in the Tri-ang tradition but was a lot taller and squarer. The terminals were mounted along the narrow top strip. It was described as 'electronic' presumably referring to the fact that it made use of transistors.

Canadian Power Controller PA200C
This was another American-made power controller for Lines Bros. (Canada) Ltd to CSA requirements but this time referred to as a 'Twinpower Unit'. It had a similar gold finish to the PA102 referred to above but, instead of a side lever for speed control, it had two rotating knobs on the face of the unit. These gave two controlled 12V DC outputs but there were also a fixed 12V DC connection for a third train and a 15V AC output connection. The unit cost $20.

It was made in this form between 1963 and 1968 after which it was replaced by a unit which carried the name 'M Dual Electro'. It was built to the same specification but with a 2.5 amp output instead of 2 amp and it retained the PA200C code. The revised model survived only to 1970.

Canadian Power Pack 1000
The '1000' was a development of the PA2200 described below. It was made in Japan for Meccano Tri-ang and carried the name 'Tri-ang Railways' for the first year it was in use but 'Hornby Railways' after that. The box was black and printed in yellow.

It had a speed control and a direction switch and one unit was required for each train. The unit had an open circuit output of 17.5 V AC and 15.5 V DC to a maximum of 10 VA 1 amp.

Canadian Power Pack 1005
This was included on the 1970 and 1971 price lists but not referred to in the catalogue. Whether it was made is not known.

Power Unit RP1770
Probably a transformer for Minic Motorway more than 7,000 of which were taken into the stores in 1968, possibly from the Canterbury factory when Minic Motorway production was transferred to Margate.

Canadian Power Controller PA2200 (2200)
This unit was introduced to the Canadian market in 1967 and was made in Japan. It was made for Lines Bros. (Canada) to CSA standards probably as a cheaper version of the PA 102. Initially it was referred to as the '2200' but later as the 'PA2200'.

The flat box was red and carried the Tri-ang Railways name. It provided a controlled 12V DC and a 15V AC output. Speed control was by way of a pinned lever on the underside of the casing and here too was located a direction switch. The unit had an automatic circuit breaker. A useful feature was a keyhole slotted baseplate which allowed the user to hang the unit onto two round-headed screws.

Around 1968 or 1969 it was replaced by a similar looking unit which was made in Canada for Meccano Tri-ang but still carrying the Tri-ang Railways name. This was available only until 1970 and cost $10.

110V Units
110 Volt versions of power units carried an 'SI' suffix. These included the following:

Type	Code	Notes
power unit	RP5.5SI	800 made between 1965 and 1967
power controller	RP8SI	940 made in 1965
power unit	RP14SI	480 made in 1968
power unit	RP15SI	660 made in 1970 & 1971
power pack	RP18SI	2,260 made in 1971
power controller	RP20SI	Planned but none made before the end of 1971
power unit	RP1770SI	300 in store for Minic Motorways in 1969
power unit	J300	This is thought to have been made for overseas sales of Jump Jockey. 27,400 were in the Margate stores in 1971

SWITCHES

Lever (Points) RT44
These switches had a black casing and were used for changing electric points. They gave a brief charge of electricity as the lever passed an electrical contact and that tripped the solenoid. They were first made in 1955 and remained in production for many years. About 220,000 were made during the Tri-ang Hornby period.

Lever (Signal) RT46
The yellow switches were for changing colour light signals and switched between two contacts – one for red and one for green. The last batch of 2,600 solo switches was made in 1965 after which they were supplied only with the signals.

Lever (Track) RT47
The green levers were used for isolating track and

were a simple on/off switch. They were made throughout the Tri-ang Hornby period, a total of about 110,000 being produced.

Lever Frame Base RT144

This was the base onto which the lever switches could be mounted and connected together as required. It was a small black plastic moulding which was supplied to shops in boxes of six. Batches were made every year during this period and a total of 40,000 passed through the stores.

LEADS

Canada and Australia S5083

These were described in literature as 'plug and tag' and were 7.5" long. Large quantities were sent to Canada in 1969, 1970 and 1971 totalling 115,000.

POINT REMOTE CONTROL PACK R663

(*See* 'Track')

RP42 circuit controller, switches RT44 (black) for points, RT46 (yellow) for signals, RT47 (green) for isolating track, box of 4 RT144 lever frames and S5083 leads.

CATENARY

In 1961 the catenary system had been modified. The tin-plated steel pressings had given way to wire. The system underwent consolidation in 1962 but this principally meant that fewer parts were sold individually but had to be bought in sets instead. Despite this, the system continued to grow. The final addition, the double track gantry, arrived in 1965 and so is technically a Tri-ang Hornby model.

The catenary system remained available throughout the Tri-ang Hornby years but slowly diminished. With the change of name to Hornby Railways in 1972 it finally disappeared. In 1988 the overhead power supply system returned but was never as highly developed as it had been in the mid 1960s.

CHECK LIST OF CATENARY

No	Model	Dates
R314	Mast Base	65-68
R415	Catenary Wire and Clip	65-71
R416	15' Catenary Set	65-69
R416U	15' Catenary Set	70
R417	10' Catenary Set	65-67
R418	Catenary Extension Set	65-69
R418U	5' Catenary Extension Set	70
R419	Mast with Link	65-68
R419C	2 Single Masts and Bases	69
R419U	3 Single Masts and Bases	69-71
R455	Mast Base Clip	65-68
R580	2 Double Track Gantries	65-69
R580U	3 Double Track Gantries	70

DIRECTORY OF CATENARY

Mast Base R314

This was the screw fixing type first seen in 1960. It

Any correspondence with the factory regarding this model should quote **No.**

Tri-ang HORNBY Model Railways — OVERHEAD POWER SUPPLY SYSTEM

INSTRUCTION LEAFLET

READ INSTRUCTIONS BEFORE OPERATING TRAIN

CATENARY, MASTS, ETC.

Overhead Power Supply equipment for use with Tri-ang Hornby Model Railways Track consists of the following items:—

R.415 Catenary Wire with Clip 15″ (38·1 cms.)
R.416 15′ Catenary Set
R.417 10′ Catenary Set
R.418 Catenary Extension Set
R.419 Single Track Mast with Link
R.455 Mast Base Clip
R.314 Mast Base (Screw fixing type)
R.580 Pair of Double Track Gantries

Catenary Set R.416 contains sufficient equipment to fit catenary to 15 feet of track.

Catenary Set R.417 sufficient for 10 feet of track.

Catenary Extension Set R.418 contains wire and masts, etc., for 5 feet of track plus one Overhead Diamond (X.351) and 2 Junction Links with clips (X.352) and thus enables catenary to be fitted to a Diamond Crossing and two Points as explained below.

ASSEMBLY

Each Single Track Mast R.419 requires a Mast Base Clip R.455. The foot of the mast slides into the base which clips to the track as illustrated in Fig. 1.

Fig. 1

R.314 Mast Base (Screw fixing type) may be secured to a baseboard in positions where Mast Base Clips R.455 cannot be fitted. The hole provided will take a No. 2 countersunk wood screw.

One X.353 Phase 2 Power Mast will be required for each circuit of catenary. The Power Mast fits to a Power Mast Base X.354.

For a simple oval of track, one mast will be needed for each section of track. One of the masts must be a Power Mast.

R.415 Catenary wires with clip are 15″ in length and may be cut to any shorter length which may be required. They may be used with straight and curved track but in the latter case should be suitably curved by hand, to follow the radius of the centre of the track, before fitting them to the masts.

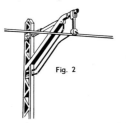

Fig. 2

The method of fixing catenary wires to R.419 Masts is illustrated in Fig. 2. It should be noted that a Mast Link S.5866 is supplied already fitted to each mast.

In the case of X.353 Power Mast (Fig. 3), Mast Link S.5866 must be fitted. It is first looped over the boom. The loop in the spring is then passed up over the mast link and held there while the catenary wire is clipped up into the slot in the base of the mast link. The spring is then released and presses down on top of the catenary wire to make electrical contact.

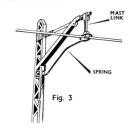

MAST LINK

SPRING

Fig. 3

X.351 Overhead Diamond is used, as illustrated in Fig. 4, to fit catenaries above both tracks in a diamond crossing. One catenary wire passes right through one arm of the fitting and the ends of two wires are clipped into the other arm.

Fig. 4

To fit Overhead equipment to a siding, a Junction Link with clip X.352 should be used on the mast situated before the point, in place of the mast link (Fig. 5).

The Phase 2 Catenary wires can be connected to the earlier pattern Catenaries (R.306 to R.309 now discontinued) by means of Catenary Adaptor S.3358.

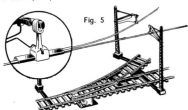

Fig. 5

LOCOMOTIVES

Tri-ang Hornby locomotives designed to operate with the Overhead Power Supply System collect current from the overhead Catenaries through Operating Pantographs or may be powered from the track. Each model is fitted with a three position switch permitting instant selection of overhead or track control. The central position isolates the locomotive from both supplies.

Lubrication and Maintenance of these models is similar to the description given in the OO/HO Gauge Instruction leaflet.

B.R., ELECTRIC LOCOMOTIVE R.351 (ALSO MODELS BUILT FROM CKD KIT R.388)

The three position switch on this model is disguised as the centre section of the long ventilator on the roof.

To remove power bogie from this model, first set the switch on the roof to the Neutral position and remove roof screw. Move the sliding contact right forward until the wide part of the slot coincides with the pivot nut. The body can then be lifted off. Before refitting the roof the sliding contact on the body should be set to the Neutral position in line with the switch in the roof.

This model may be lubricated from below (see Fig. 6).

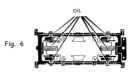

OIL

Fig. 6

Fig. 7

Fig. 7 shows the position on one type of power bogie of the letter "I", on other bogies it is positioned in the centre of the sideframe, this side must be nearest the power mast for operation from overhead Catenary.

ELECTRIC LOCOMOTIVE R.753

The changeover switch on this locomotive is situated in the centre of the roof. In keeping with British Railway practice, later models are fitted with one pantograph only.

To lubricate the power bogie in this model partially remove the body by withdrawing the securing screw located on the underside. The body remains attached to the chassis by one lead which should not be strained. Lubricate model at points indicated in Fig. 8.

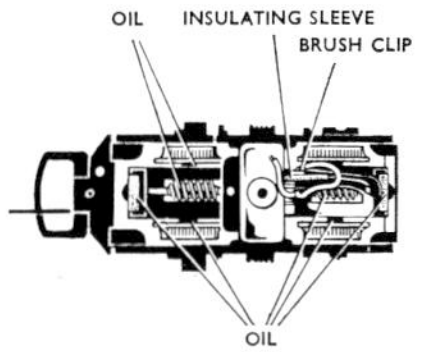

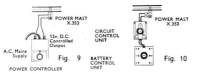

OIL   INSULATING SLEEVE   BRUSH CLIP

Fig. 8

OIL

POWER SUPPLY AND CONTROL SYSTEM (OVERHEAD)

Tri-ang Hornby electric locomotives are designed to operate from a supply of 12/15 volts Direct Current (D.C.). The methods of obtaining such a supply are detailed in the OO/HO Gauge Instruction leaflet.

When the pantographs are lowered and the switch on the locomotive is in position (TK) for track control the operation is as described for other Tri-ang Hornby locomotives.

To provide power to the catenaries the leads from the 12/15 volts D.C. supply are connected to the base of the Power Mast X.353.

Alternative Power Supply arrangements are shown in Figs. 9 and 10.

WIRING FOR OVERHEAD OPERATION

POWER MAST X.353

12v. D.C. Controlled Output

A.C. Mains Supply

CIRCUIT CONTROL UNIT

POWER MAST X.353

POWER CONTROLLER

Fig. 9

BATTERY CONTROL UNIT

Fig. 10

It is possible with certain limitations, to operate two locomotives on the same track, under independent control, if one is powered by current from overhead and the other from the track.

To do this, it is essential that the two 12/15 volts D.C. supplies are independent and some suggested arrangements are illustrated in Figs. 11, 12 and 13.

Do not attempt to operate the two locomotives on the same track from one battery control unit or from a single power unit. A power unit with two D.C. outputs can be used only if the design is such that the two D.C. outputs are derived from separate secondary windings of the transformer and are therefore suitable for controlling two separate supply systems having a common return.

WIRING FOR TWO TRAINS ON ONE TRACK

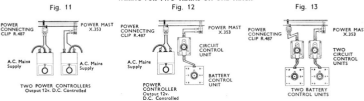

Fig. 11

POWER CONNECTING CLIP R.487

POWER MAST X.353

A.C. Mains Supply

TWO POWER CONTROLLERS
Output 12v. D.C. Controlled

Fig. 12

POWER CONNECTING CLIP R.487

POWER MAST X.353

CIRCUIT CONTROL UNIT

A.C. Mains Supply

POWER CONTROLLER
Output 12v. D.C. Controlled

BATTERY CONTROL UNIT

Fig. 13

POWER CONNECTING CLIP R.487

POWER MAST X.353

TWO CIRCUIT CONTROL UNITS

TWO BATTERY CONTROL UNITS

HINTS ON OVERHEAD OPERATION

If a locomotive operating from overhead is coupled to a locomotive under track control, it may be found necessary to reverse the direction in which the track controlled locomotive is facing to avoid creating a short circuit through the couplings.

When operating two trains on the same track — one overhead and one track controlled — it is essential that each has an independent 12/15 volts Direct Current (D.C.) supply as the two supply systems have a common return through one of the rails.

Remember that two trains can be operated in opposite directions on the same track — therefore 'drive carefully' to avoid damage from collisions.

If difficulty is experienced in operating a train, a further study of the instructions is suggested.

GENERAL

Additional information regarding track operated and overhead supply locomotive models is given in the publication R.282 "Tri-ang Hornby Model Railways Operating Manual" priced at 9d. direct from the factory.

Details of track geometry and the elements of layout construction is contained in the booklet R.166 "HO/OO Gauge Track Plans" at 4/- plus 6d. postage.

Built in Britain by

ROVEX INDUSTRIES LTD, WESTWOOD, MARGATE, KENT, ENGLAND

OPSS/10/46964/8   Printed in England

Overhead power supply system leaflet.

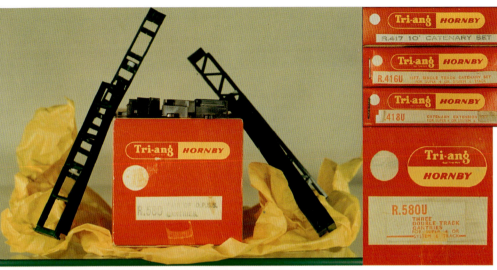

A selection of catenary equipment.

remained in production until 1968 and during the last four years about 22,000 were produced.

Catenary Wire and Clip R415

The last year of production for the catenary wire and clip was 1971 and during the Tri-ang Hornby years over 100,000 were made. It consisted of a single wire with a tubular joiner on one end. Although supplied to shops in packets of 12, they were sold singly.

15' Catenary Set R416, R416U

Up until 1969 some 18,000 of these sets were sold in Tri-ang Hornby boxes. A further batch of 1,000 suitable for both Super 4 and System 6 track was made in 1970 and sold as R416U. It was in the shops by March but by the end of June 1970 stocks had nearly been cleared and it was deleted from the 1971 catalogue. The set contained 17 masts with links and universal bases. There were 12 catenary wires with clips, a power mast and base and four mast links and overhead wire clips.

10' Catenary Set R417

Like the R416, the shorter set had been introduced in 1962 but it did not survive long after the name change; probably to about 3,000 sets. It was last made in 1967. It contained 12 masts (one power) and eight wires.

Catenary Extension Set R418, R418U

The 5' Catenary Extension Set, also introduced in 1962, remained in production until 1970; 11,000 as R418 and 2,000 in 1970 as R418U (suitable for Super 4 and System 6). The latter type arrived in the shops in March 1970. The set had five masts (one power), four wires and contained junctions for two points and a diamond crossing.

Mast with Link R419

This was simply the green mast without a base but with the plastic clip to which the catenary wire was attached. The last batch was made in 1968 and about 35,000 were made in the Tri-ang Hornby period.

2 Single Masts and Bases R419C

In 1969, instead of buying masts singly they were sold in pairs and with bases. 7,500 were made like this.

3 Single Masts and Bases R419, R419U

Also in 1969 one could buy the masts in sets of three with bases and these remained in production until

1971 with a total of 6,500 sets being made. Those available in the shops from March 1970 onwards carried the 'U' suffix to the 'R' number as an indication that they would also fit the new System 6 track.

Mast Base Clip R455

During the period that the catenary masts were sold on their own, so too were the mast bases with a clip to fix them to the track. These had first arrived in 1962 and the last batch was made in 1968. About 50,000 were made in the Tri-ang Hornby years.

TRACK GANTRY

2 Double Track Gantries R580

Designed in the Tri-ang Railways period, the double track gantry did not arrive until 1965. It was green like the single track masts and came complete with two base clips and two catenary clips. They were sold in pairs in DD size boxes, printed around 1965/66, until 1969. They were initially priced 4/11d and over 12,000 boxes were sold.

3 Double Track Gantries R580U

In 1970, about 2,000 DC size boxes were assembled with three double track gantries to a box. The six track clips came in a separate brown envelope inside the box. These were sold as R580U and had a printed label on one end of the box describing the contents and indicating their suitability for System 6 track. It was included in the 1971 catalogue even though only 230 boxes remained in the store.

Illustrations of catenary from the 1967 catalogue.

INTRODUCTION

Earlier Systems

By 1965, the Tri-ang Railway system was based solidly on Super 4 track and, as its name implied, it was the fourth track system to be made by Rovex.

The first had been introduced for the original Rovex train set in 1950. This was non-universal in that each piece had to be the correct way round to connect to its neighbour. This original track was of a poor geometry which made further expansion of the system difficult. In order to correct this, but more importantly to provide universal coupling, a second track system known as 'Universal' or 'Standard' track was introduced in 1952.

Series 3

In the search for lower production costs and greater realism, Series 3 track was launched in 1958. This abandoned the all-in-one grey plastic base of the first two systems and, like modern track, had only a black plastic sleeper web into which the rails were slid. It kept to the geometry of the Standard track which was very restrictive to imaginative layouts.

Rovex did not find it worthwhile drawing their own rail section as this would have required steel wire to be tin coated through a special die. Also, the malleability of the rail was critical since if it was too hard, curved rails would not retain their correct curvature. Instead, the rail was bought ready drawn from John Rigby & Sons Ltd in the north of England.

Although manufacture of Series 3 had ceased by 1965, stocks remained in the stores at Margate until 1968 when some 50,000 pieces (including 1,800 point motors) were disposed of. The Track Bumper (buffer stop), made for overseas sales, remained in produc-tion after the demise of Series 3 track; the last production being recorded as 1969 to meet an order from Canada.

Despite the gradual replacement of Series 3 in Britain and Australia and in Canadian set production, it continued to be made in the New Zealand factory, which retained the tools they had previously had to share with the Australian factory. Indeed, New Zealand continued with Series 3 track right up to 1971 despite a very late start in replacing it with System 6.

Super 4

Series 3 was much criticised for the unrealistic gaps between the sleepers and was compared unfavourably with the finer quality 2-rail track now made for Hornby-Dublo, Trix and other manufacturers. This led to the most comprehensive track system yet produced by Rovex which was launched in 1962 and christened Super 4. Despite the demand for finer scale rails, Super 4 was coarse scale. The excuse given at the time was that the high cost of modifying all locomotives and rolling stock to run on finer scale rails made it uneconomical to produce.

The full range of parts produced for the new track was described in some detail in **Volume 1** and will not be repeated here. Four items were added during the Tri-ang Hornby years. These were the R394 Hydraulic Buffer Stop, which was really a Tri-ang Railways model but was late in arriving, the R642 and R646 third radius curves which arrived in 1968 and the R663 Point Remote Control Pack of 1969.

On a Question of Scale

On the subject of scale, a reader of *Model Railway Constructor* responded in January 1967 to an earlier explanation by Henden, the then Managing Director of Rovex Scale Models Ltd, that Super 4 track and over-scale wheel flanges were persisted with by Rovex because it was easier for children to handle:

"It is claimed that Tri-ang wheel standards are designed for ease of re-railing and to allow wide track tolerances in the interests of cheaper production. If Tri-ang track was markedly cheaper than others this would be acceptable"

The writer went on to point out that HO track to a much finer scale was now widely used on the Continent and, in some cases (e.g. Playcraft), it was cheaper than that made by Rovex. Moreover, the wheels used on some Continental models were very crude and yet they run without difficulty on the scale track.

In his response in the March edition, Henden defended Super 4 on the grounds that the thicker base allowed sufficient clearance under the rail for a whole assortment of lineside accessories to be clipped and held rigidly in place. He went on to say:

"There were, however, two disadvantages. The track sections needed to be somewhat over-scale so that the tongues on the accessories could be sufficiently thick to attach the accessory securely. The other disadvantage was that the method of track manufacture was necessarily more complicated than on other types which did not have this special feature. Nevertheless, we feel that our track has a very special advantage for our younger users and we would not wish to sacrifice this on the alter of pure scale."

Fine Scale

By 1969, the men at Rovex had second thoughts on this point. They were well aware that their track was no longer of an acceptable standard and that adhering to coarse scale could severely affect future sales. They were not going to be hurried into a decision, however, and went to some trouble to look around the world at track other model railway manufacturers were making before deciding to develop their fifth track system.

The problems they faced were numerous.

Besides the obvious costs involved, there was the problem that existing locomotives and rolling-stock would not happily operate on fine scale track and many accessories designed to fit Super 4 would need to be modified. It was also recognised that fine scale track was less universal than the existing Super 4 which could happily be used for a wide range of British and foreign models. The principal problem was the pointwork which, to retain its universal nature would require a moving frog, but this was unacceptable to the scale devotees.

Then there was the fact that dimensions would be more critical and tolerances would have to be rigidly held throughout the production changes, resulting in greater tooling and production costs. At the time there was a variation of up to 0.02" in back-to-back wheel measurements but to run on fine scale track the tolerance would have to be within + or - 0.005".

Series 5 Investigation

A 'Series 5 Investigation' was implemented early in 1968. The track of 27 different model manufacturers was subjected to tests at the Margate factory with a view to establishing its interchangability. The tests proved that the dimensions of the heavy Tri-ang Rovex wheels were not compatible with those of other makes and that design changes were therefore unavoidable.

It was found that, if Tri-ang Hornby locomotives and rolling stock were fitted with Hornby-Dublo type wheels, both Tri-ang Hornby and Hornby-Dublo (2 rail) models would be able to run on Super 4, Atlas Snap Track, Fleischmann, Peco Streamline, Lima, Playcraft, Trix, Hornby-Dublo and the new Series 5 when produced. Harold Hendon concluded that the use of existing Hornby-Dublo wheels would probably result in a considerable saving of development and production time.

It was proposed that the new Series 5 track should look like Atlas Snap Track but, in order to take Playcraft, Jouef, Wrenn and Graham Farish wheels and the existing Tri-ang Hornby locomotives and rolling stock, would need to be to a slightly different specification. The flat-bottomed rail used by Peco and Welkut for some years was identified as being the most suitable for universal use. It was further thought that Fleischmann and Peco Streamline had similar specifications to those required for Series 5.

System 6

At some stage the name was changed from 'Series 5' to 'System 6' and samples of the track were sent out to retailers on 18 December 1969. The first pieces of track were to have been in the shops in January 1970, but, due to delays in the supply of materials and particularly fishplates, it was mid-February before the new track could be bought. Over 950,000 pieces of System 6 track sold in that first year, together with over 33,000 copies of the new Track Plan Book which had been rewritten for the new track system.

System 6 was used in all train sets made from 1 January 1970 with the exception of clockwork and battery-powered sets. The latter continued to receive Super 4 track, the company arguing that this was easier for small children to put together and was tougher. It meant that a lot of clip-fit accessories had to be modified and these were identified with a 'U' (for universal) suffix to their 'R' number. These could be fitted to both types of track and were also available from 1 January 1970.

The arrival of System 6 meant flexible yard lengths in a choice of steel or nickel-silver which were manufactured for Rovex by their subsidiary, G&R Wrenn Ltd. System 6 retained the geometry of the Super 4 track and the feature of the half width sleeper at each end which provided a realistic looking joint that was very strong. The sleepers were to scale size and spacing and moulded in black plastic with a wood-grain finish. The solid drawn steel rail was also of the correct scale section. Most important, the new track was compatible with many other makes of track by then available.

In 1965 Rovex had introduced the R476 Converter Track to allow Hornby-Dublo operators to buy and connect to Tri-ang Hornby Super 4 track which had been adopted as the standard of the 'amalgamated' system. The Converter track had been a good seller and, with the exception of 1969, large quantities had been made every year. Ironically it was now advertised to encourage modellers to convert back to scale track. 87,000 pieces were sold in 1970.

The first nickel silver System 6 was available in 1973.

The boxes of System 6 had the yellow colour left out when they were printed. Thus they were red and white. This colour scheme was also used for later Super 4 track packaging. It was felt that this would not matter as track boxes were rarely displayed in shops, the contents being sold loose. Moldex in Australia had their System 6 boxes printed to match the UK ones.

In March 1970 the management declared its intention to cease production of Super 4 track in 1972 but this was not made public at the time. Instruction was given that no more solo boxing for Super 4 track was to be ordered.

The old Super 4 system was last illustrated in the catalogue in 1971 and in the retail price list in 1973. It is interesting to note that it was dropped from the trade price list after 1972 and must have remained in

The photographic studio where System 6 track is being laid ready for a session shooting pictures for the 1971 catalogue and other publications.

the retail list a further year in order to help retailers to clear their stocks.

System 6 track became the British standard and underwent several changes over the years. From 1976 it was not promoted as 'System 6' but just as 'Hornby Track' although the catalogues still used the name in connection with the converter track which continued to be available until 1983.

CHECK LIST OF TRACK

The dates given in this check list are the years in which items were actually manufactured (except in the case of Australia and New Zealand where they are the dates when the items were identified in the price lists). They sometimes appeared in the catalogue a year earlier due to a delay in the development work and sometimes a year or two after production had ceased in order to clear outstanding stocks from the stores. As dates used in **Volume 1** were based on availability rather than production, the dates quoted sometimes vary.

As the System 6 track system was to be available far into the foreseeable future no termination dates are given here. If any of the pieces were subsequently dropped, the termination dates of them will be given in **Volume 3**.

The following list includes track sent bulk packed to Canada in 1972. It should be noted that as this fell within the Hornby Railways period, it is strictly speaking outside the limits of this volume. It is, however, included here as Canadian set production ceased in 1973 and for the sake of completeness, all matters concerning Canadian production have been kept together in this volume.

SERIES 3		
No.	AUSTRALIAN MADE	Dates
R83A	Track Bumper	58-67
X156A	Point Motor LH	60-68
X157A	Point Motor RH	60-68
R190A	Straight	58-68
R191A	Quarter Straight	59-68
R192A	Eighth Straight	59-68
R193A	Curve Small Radius	58-68
R194A	Half Curve Small Radius	59-68
R195A	Curve Large Radius	62?-68
R196A	Half Curve Large Radius	62?-68
R290A	Diamond	59-68
R291A	Hand Point LH	59-68
R292A	Hand Point RH	59-68

SERIES 3		
No.	NEW ZEALAND MADE	Dates
R83	Track Bumper	64-72
R103	Buffer Stop	59-72
X156	Point Solenoid LH	59-69
X157	Point Solenoid RH	59-69
R169	Railer	65-70?
R190	Straight	59-72
R191	Quarter Straight	59-72
R192	Eighth Straight	59-72
R193	Curve Small Radius	59-72
R194	Half Curve Small Radius	59-72
R195	Curve Large Radius	68-72
R196	Half Curve Large Radius	68-72
R199	Yard Straight	63-72
R290	Diamond	59-70?
R291	Hand Point LH	59-70?
R292	Hand Point RH	59-70?
R293	Electric Point LH	59-69
R294	Electric Point RH	59-69
R487	Power Connecting Clip	63-72
R488	Uncoupling Unit	65-72
NEW ZEALAND MADE TRACK PACKS		
R429	Twin Line Conversion Set	69
R439	Passing Loop Set	69
R454	Siding Set	69

SUPER 4		
No.	SOLO TRACK	Dates
X404	Point Motor	62-68
R394	Hydraulic Buffer Stop	65-69
R394U	Hydraulic Buffer Stop	69->
R410	Underlay 16'6"	63-70
R411	Double Track Spacers	63-65
R433	Underlay for R490	63-70
R434	Underlay for R491	63-70
R435	Underlay for R492	63-66,68-69
R436	Underlay for R493	63-69
R437	Y Point	63-70
R438	Underlay for R437	63-64,66,69
R476	Converter Track	66-83
R480	Double Straight	62-72
R481	Straight	62-72
R482	Quarter Straight	62-72
R483	Double Curve 1st Radius	62-72
R484	Curve 1st Radius	62-72
R485	Double Curve 2nd Radius	62-72
R486	Curve 2nd Radius	62-72
R487	Power Connecting Clip	62-72
R488	Uncoupling Ramp	62-72
R489	Long Straight	63-72
R490	LH Point	62-72
R491	RH Point	62-72
R492	LH Diamond Crossing	62-71
R493	RH Diamond Crossing	62-70
R494	Buffer Stop	62-69
R494U	Buffer Stop	70-71
R497	Isolating Track	62-70
R642	Curve 3rd Radius	68-69
R646	Double Curve 3rd Radius	68-69
R663	Point Remote Control Pack	69
R663U	Point Remote Control Pack	69-70
TRACK PACKS		
R7	Track Pack	63-65
R8	Track Pack	63-65
R166	Track Plan Book	63-69
R167	Track Pack	65-67
R429	Twin Line Conversion Set	65-70
R439	Passing Loop Set	65-70
R454	Siding Set	66-70
R680	Oval & Siding	68-70
R681	Oval, Siding & By Pass	68-70
R682	2 Ovals & 2 Sidings	68-70

No.	BULK PACKED FOR CANADA	Dates
R480	Double Straight	69
R481C	Straight	67
R483C	Double Curve Small Radius	67
R487C	Power Connecting Clip	67
R488C	Uncoupling Ramp	67
R642	Curve 3rd Radius	69
R646	Double Curve 3rd Radius	69
	SPECIALLY BOXED FOR CANADA	
R83	Track Bumper	68-69
R167	Track Pack	68-69
R167A	Track Pack	69
R439	Passing Loop Set	70
R680	Oval & Siding	70

No.	BULK PACKED FOR AUSTRALIA	Dates
R494A	Buffer Stop	68

No.	AUSTRALIAN MADE	Dates
X404A	Point Motor	62-69
R480A	Double Straight	62-69
R481A	Straight	62-69
R482A	Quarter Straight	62-69
R483A	Double Curve 1st Radius	62-69
R484A	Curve 1st Radius	62-69
R485A	Double Curve 2nd Radius	62-69
R486A	Curve 2nd Radius	62-69
R487A	Power Connecting Clip	62-69
R488A	Uncoupling Ramp	62-69
R489A	Long Straight	63-69
R490A	LH Point	62-69
R491A	RH Point	62-69
R492A	LH Diamond	62-69
R493A	RH Diamond	62-69
R497A	Isolating Rail	62-66?
	AUSTRALIAN MADE TRACK PACKS	
R7A	Track Pack	64-67
R8A	Track Pack	64-66
R429A	Twin Line Conversion Set	65-69
R439A	Passing Loop Set	65-69
R454A	Siding Set	66-69

No.	BULK PACKED FOR NEW ZEALAND	Dates
R476	Converter Track	71
R488	Uncoupling Ramp	68-71
R494	Buffer Stop	70-71

	SERIES 5	
No.	TRACK FOR SETS	Dates
R500	Straight	not made
R505	Double Curve Small Radius	not made
R524	Power Connecting Clip	not made

	SYSTEM 6	
No.	SOLO TRACK	Dates
R600	Straight	70-
R601	Double Straight	70-
R603	Long Straight	70-
R604	Curve 1st Radius	70-
R605	Double Curve 1st Radius	70-
R606	Curve 2nd Radius	70-
R607	Double Curve 2nd Radius	70-
R608	Curve 3rd Radius	70-
R609	Double Curve 3rd Radius	70-
R610	Quarter Straight	70-
R611	Pack of Fishplates (see Sundries)	71-
R612	LH Point	70-
R613	RH Point	70-
R616	Isolating Track	70-
R617	Uncoupling Ramp & Marker	70-
R618	Power Connecting Clip	70-
R619	Flexible Steel Track	71-
R620	Flexible NS Track	71-
R624	Underlay for R612	71-
R625	Underlay for R613	71-
R638	Underlay Roll	70-
R6166	Track Plans Book	70-
	TRACK PACKS	
R686	Passing Loop Set	71
R687	Siding Set	71
R695	Twin Line Conversion Set	71
R680-6	Oval & Siding	71
R681-6	Oval, Loop & Siding	71
R682-6	Twin Ovals & 2 Sidings	71

No.	BULK PACKED FOR CANADA	Dates
R600	Straight	72
R618	Power Connecting Clip	72
	SPECIALLY BOXED FOR CANADA	
R681	Track Pack	69
R686	Siding Track Pack	not made
R1670	Track Pack	70
R1671	Track Pack	70
R1672	Track Pack	70
R1673	Track Pack	70
R6670	Track Pack	71
R6671	Track Pack	71
R6672	Track Pack	71
R6673	Track Pack	71
R6674	Track Pack	71

No.	AUSTRALIAN MADE	Dates
R600	Straight	70-72
R601	Double Straight	70-72
R605	Double Curve 1st Radius	70-72
R606	Half Curve 2nd Radius	70-72
R607	Double Curve 2nd Radius	70-72
R617	Uncoupling Ramp & Marker	70-72
R618	Power Connecting Clip	70-72

No.	NEW ZEALAND (BULK PACKED)	Dates
R612	Left Hand Point	70-71
R613	Right Hand Point	70-71
R617	Uncoupling Ramp	70-72
R618	Power Connecting Clip	70-71

No.	NEW ZEALAND MADE	Dates
R600	Straight	70?-72?
R601	Double Straight	70?-72?
R605	Double Curve 1st Radius	70?-72?
R606	Half Curve 2nd Radius	70?-72?

DIRECTORY OF TRACK

SERIES 3 TRACK

This remained in production in New Zealand until 1972. The 12 items of Series 3 track still in production in Australia at the start of the Tri-ang Hornby years remained in the price lists until 1968. It was not listed in the UK after 1965 although sizeable stocks of the track remained in the stores at Margate at that time.

For further detail of the individual pieces of track and the various moulds used, see **Volume 1**.

AUSTRALIAN SERIES 3

Series 3 production had started in Australia in 1958 using a duplicate set of tools which they were to share with the New Zealand and South African factories. In fact, they never reached South Africa.

By 1965 there were 12 items available. As they continued to be listed up to 1968 it is assumed that they were still being made in the Tri-ang Hornby period. Certainly some were packaged in Tri-ang

Hornby boxes. At first glance all pieces looked identical to that made at Margate but closer inspection of the underside reveals different inscriptions. Some pieces were marked 'Tri-ang MADE IN AUSTRALIA (or AUST) N.Z. & S. AFRICA' while items added to the range later were marked 'MADE IN AUSTRALIA' or 'BUILT IN AUSTRALIA'. This was done by dropping a piece, bearing the inscription, into the tool before use.

All Australian Series 3 track that I have seen carries the name 'Tri-ang' on its own and not 'Tri-ang Railways' or 'Tri-ang Hornby'.

Track Bumper R83A

The tool for this had been sent out to Australia and New Zealand in 1958 and the Track Bumper was available in the Australian price list until 1967. For some reason it was listed under stations and lineside accessories while the British style of buffer stop was listed under track.

Point Motors X156A, X157A

There were left- and right-handed versions of the

solenoid type motor. They were first listed in 1960 and were inscribed 'BUILT IN AUSTRALIA'. They were sold three to an MXT15 size Moldex box but with instruction slips from Margate

Straight R190A

Any straights made during the Tri-ang Hornby period would have been inscribed 'Tri-ang R190 MADE IN AUSTRALIA & N.Z. & S. AFRICA'

Quarter Straight R191A

Introduced in 1959, this was inscribed 'Tri-ang R191 MADE IN AUSTRALIA'.

Eighth Straight R192A

The eighth straight rail was so small that it was difficult to add the lengthy inscription on the underside. It was marked: 'Tri-ang R192 MADE IN AUSTRALIA' and sold 24 to an MXT7 size Moldex box

Curve Small Radius R193A

Any small radius curves made during the Tri-ang Hornby period would have been inscribed 'Tri-ang R193 MADE IN AUSTRALIA & N.Z. & S. AFRICA'. They were sold 12 to an MXT18 size Moldex box.

Half Curve Small Radius R194A

This was inscribed: 'Tri-ang R194 MADE IN AUSTRALIA'. They were packed 12 to an MXT17 size box wedged in with a sheet of white tissue paper.

Curve Large Radius R195A

No information available.

Half Curve Large Radius R196A

No information available.

Diamond R290A

The Diamond Crossing was first made in Australia in 1959 and carried the inscription 'Tri-ang MADE IN AUSTRALIA R290'. They came wrapped in white tissue paper inside an MTX21 box.

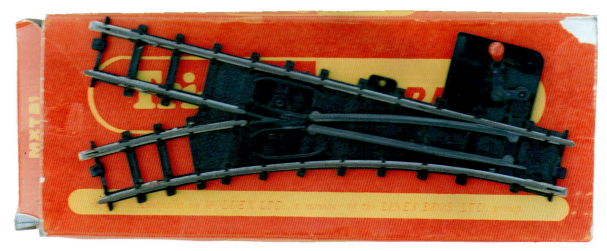

Australian left-hand Series 3 point with its different colour lever.

New Zealand-made Series 3 track showing that the tool was made to be shared with the Australian and South African factories.

Boxes of Super 4 track.

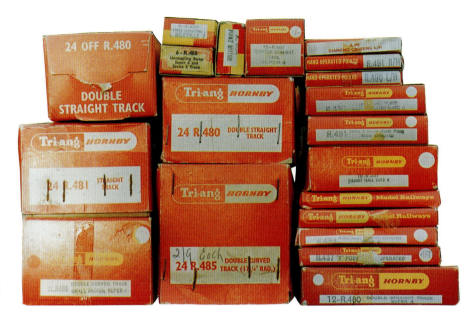

Hand Points R291A, R292A

A few years ago I acquired some brand new Series 3 points from Australia. Each was wrapped in white tissue paper and slipped into an MXT21 size Tri-ang Railways box but with the British points instruction leaflet, which was for Super 4 points and not Series 3. The points carried the inscription 'Tri-ang R292/294 (or R291/293) BUILT IN AUSTRALIA'. Unlike the British version, the left-hand point had a red lever.

NEW ZEALAND SERIES 3

As we have seen, the New Zealand company had to share tools with Moldex in Melbourne, later slotting into each tool the panel that indicated the items were made in New Zealand. As the track looked exactly

the same as that made in Australia, space is not wasted here repeating the information given above.

The packaging, on the other hand, was different. Besides the fact that the New Zealand company had their own boxes (albeit similar in design to those used in Australia and Britain), the track was packed in different quantities. The R194 half curves, for example, were packed six to a box the size of the AA used by Margate and large radius curves were packed six to a coach box.

NEW ZEALAND TRACK PACKS

In 1969, the New Zealand company were persuaded to manufacture track packs using their Series 3 track but based on the Super 4 sets made at Margate and using similar packaging. They were listed for one

year only, adopting the Super 4 numbers, and included the following:

Siding Set R454

This contained three straights, one large radius curve, one point, an uncoupling unit and a buffer stop. The set cost $2.95.

Passing Loop Set R439

The set contained two points, seven straights, two quarter straights and two large radius curves. The set cost $5.95.

Twin Line Conversion Set R429

This had 10 straights, 12 large radius curves, two-points and a power clip.

SUPER 4 TRACK

As we saw in **Volume 1**, with the exception of underlay, most of the Super 4 track range arrived in 1962 and remained more or less intact until 1972. This means that it was made for more years under the Tri-ang Hornby label than that of Tri-ang Railways. 31 individual items were manufactured during the Tri-ang Hornby years and these are listed below. In addition there were 10 different track packs available at various times during this period.

SUPER 4 SOLO TRACK

Packaging

Initially Tri-ang Hornby packaging was the same as that for Tri-ang Railways with just the name changed. Thus standard straights were supplied in 12s in a DJ size box while the double large radius curves and double straights were in the long flat BM size packaging more usually associated with cate-nary sets. Points and diamond crossings were sold singly, the diamonds and 'Y' point in the flat DV boxes and the left- and right-hand points in a BA, changing later to the slimmer BH. At one stage the diamond crossing was sold in this slim box with a label stuck over the end to cover up the printing indicating that the box was really intended for hand points.

Early track sets came in illustrated boxes but these were replaced by ones which showed the track layout on the box top with a list of contents.

At some time before 1968, R480 double straights were available in a new type of box which had side flaps rather than end flaps or a lift-off lid. This was also printed without the yellow colour and so looked plain red with reversed out white writing.

By 1968 more Super 4 boxes were being made without the yellow. The R490 and R491 Points were now sold in the wider DV boxes as previously used for the R437 'Y' Point.

Underlay was sold in AT, CX and DS boxes.

Double Straight R480

While the double 1st radius curves held the record for the greatest number made of any Super 4 item of track, the double straight held the record for the greatest number sold solo during the Tri-ang Hornby period. The figure was just short of 2,500,000.

The tool for the sleeper web was made by the Riverhead Tool Co. (RTC). Track with the 'Tri-ang' name erased was made from about 1967/68 onwards. Sometime two R481 webs were used instead of the single R460 web. The track was 33.5 cm long.

Originally the track was sold 12 to a BM box , then 24 to a side flap box (no yellow) from January 1967 and finally, sometime after mid 1967, in a box of 24 with a lift-off lid, also minus the yellow.

Canada received a bulk supply of double straights in 1969 when 6,000 were sent.

Straight R481

The R481 straight was 16.8 cm long. Earlier track had 'R481' on the sleeper web but this was later

erased so that the web could be used for double straights as well, as referred to above. Initially the track was sold 12 to a DJ box but, from January 1967, it was sold in boxes of 24 with the yellow left off. Without counting those used in sets, over 2,000,000 were sold during this period.

Canada received 5,000 R481 straights in 1967. These were bulk packed and used for set assembly in Canada. In factory records these were referred to as R481C.

Quarter Straight R482

These were 1.5" (3.8 cm) long and were sold in boxes of 12 using the standard AA box. Over 900,000 were made during the Tri-ang Hornby years.

Double Curve 1st Radius R483

These were initially packed 12 to a DL size box for solo sales and during the seven years, over 1,374,000 were sold this way. By far the greatest number of small radius double curves, however, were sold in train sets each of which contained eight.

A single sleeper web was used which, around 1967/68, had the 'Tri-ang' name and 'R' number erased, probably so that the track could be used in Miniville sets which did not carry the Tri-ang name on them. It seems that the original supply of DL size boxes lasted beyond this time and yet, from January 1967 the track was sold to retailers in boxes of 24. Like other boxes used for track at this time, these were printed only in red.

Canada needed 20,000 1st radius double curves in 1967 for the assembly of sets for sale locally and these were sent bulk packed that year. Records refer to them as R438C.

Curve 1st Radius R484

Just under one million first radius curves were made for sale in Tri-ang Hornby packaging. The radius of the R484 and R483 curves was 14⅝" (37.2 cm) and 16 of the former or eight of the latter made a circle.

They were initially sold in boxes of a dozen and later in a red and white box of 24.

Double Curve 2nd Radius R485

These were the length of two R486s which gave an arc of 45°, requiring eight for a circle. The spacing from the 1st radius curve allowed the straights to be correctly spaced to take two point junctions on both sides of an oval layout.

They were originally sold 12 to a BM size box but from 1967 were packed in 24s in a box with a lift-off lid and printed in red only. One and a half million were made during the Tri-ang Hornby period.

Curve 2nd Radius R486

The R486 and R485 had a radius of 17¼" (43.8 cm) and an arc of 22.5° which meant 16 were needed for a circle. They were packed 12 to a box. A little under one million were made during these years.

Long Straight R489

These was made using two R480 webs and were 67 cm long. They sold in a long plain brown cardboard box with a label stuck on mid-way along its length. It proved to be a popular piece of track and 1.2 million were made for solo sales during the Tri-ang Hornby period.

Isolating Track R497

This was a standard straight which had a break in one of the rails and sockets for single wire connectors. Almost 100,000 were sold in Tri-ang Hornby packaging but production ceased at the end of 1970. It was catalogued for 1971 in order to clear stocks.

Curve 3rd Radius R642

An advertisement in the model railway press in March 1969 extolled the virtues of the newly introduced 19.875" (sold as 20") radius track. It had been developed in response to requests by the public. This enabled the purchaser to have three tracks parallel tracks allowing more exciting layouts. The larger radius curves also reduced locomotive and carriage overhang and therefore added realism.

It was released in 1968 when 52,000 pieces were made. A further 100,000 were made in 1969 and then production ceased. Sales were not as good as expected and 53,000 were left in the stores at the end of 1969. These took several years to clear, by which time Super 4 production had ceased completely.

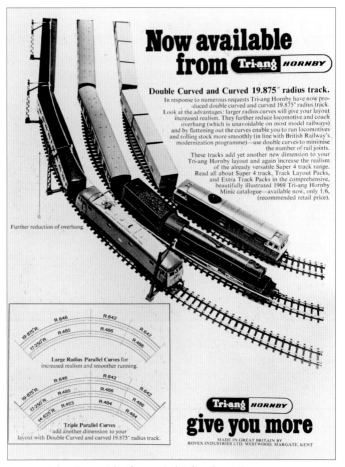

Advertisement for the new 3rd radius Super 4 curves.

The track was sold to retailers in boxes of 24 at £1 a box and retailed at 1/6d each.

Canada received a batch of 1,500 bulk packed in 1969.

Double Curve 3rd Radius R646

It was released early in 1969, a few having been made at the end of the previous year. None were made after 1969. More than 160,000 went into the stores that year and some 50,000 remained at the year end. This remaining stock took several years to clear by which time all production of Super 4 track had ceased.

Like the standard R642 curve, the R646 was sold in boxes of 24. Retailers bought them for £1-13s-6d a box and sold the track individually for 2/6d.

Canada received a batch of 2,000 bulk packed in 1969.

Diamond Crossings R492, R493

Unlike the Series 3 track system, Super 4 had both left-hand (R492) and right-hand (R493) diamond crossings. One straight was 16.8 cm long (a standard straight) and the other was 18.1 cm. The combination gave the equivalent of a 2nd radius curve in either direction.

These were always sold individually boxed, the minimum purchase quantity for retailers being three. Prior to 1967 it seems that when the Tri-ang Railways boxes ran out, BH size hand point boxes were used with a printed label stuck over the inscription on the end flaps. In 1967 a batch of DV boxes was printed for diamond crossings and these seem to have lasted until the end of production.

The last batch made was to have been in 1970 with their retention in the 1971 catalogue in order to clear stocks but supplies of the left-hand version had fallen to under 300 by the end of 1970 and so another small quantity had to be made in 1971 to meet orders. Just over 70,000 of each type were sold in Tri-ang Hornby boxes.

Hand Points R490, R491

These were also sold individually boxed as R490 Left-Hand Point and R491 Right-Hand Point. As they sold better than the diamond crossings, the minimum purchase order for retailers was six. They retailed at 7/11d in 1965 and between 1965 and 1971, 800,000 of each were made.

They had a radius of 17.25" (43.8 cm) and an arc of 22.5°. They were 16.8 cm long and so were the equivalent of a standard straight and a standard 2nd radius curve.

They went through three stages of boxes. Once the Tri-ang Railways packaging was used up, a batch of BH size boxes were printed with the new logo. These were very slim and must have been difficult to pack. They carried the inscription 'HAND OPERATED POINTS' on the small end flaps and the actual contents were rubber stamped onto a white panel. The next type of box was the larger BA container which could be used for a wider range of products including level crossings. Early ones carry the name 'ROVEX SCALE MODELS LIMITED' while those printed after 1966 were inscribed 'ROVEX INDUSTRIES LTD'. The third container was printed after mid-1969 as it carried the name 'ROVEX Tri-ang'. This was a DV size box which was wider, thinner and shorter than its predecessor. The points were also sold in blister packs that could be hung on a shop display. The backing cards for these were printed in 1967 or 1968.

Y Point R437

The 'Y' point curved both ways on a radius of 17.25" (43.8 cm) and an arc of 22.5°. It was made between 1963 and 1969 with just under 100,000 being made during the Tri-ang Hornby years. The largest batch of nearly 30,000 was made in 1967. All stocks had gone by the end of 1970.

The only packaging I have seen was a DV box marked 'ROVEX SCALE MODELS LIMITED' and it is quite possible that this was the only one used.

Point Motor X404

The point motor was made between 1962 and 1968 with over 200,000 being made in the last four years. At the end of 1968 there were about 200 motors left in the stores and these probably went for Point Remote Control Pack production (*see below*). The motor was packed in a red and yellow CU size box.

Point Remote Control Pack R663, R663U

In 1969, Point Remote Control Packs replaced loose motors. The pack contained an X404 point motor, an R44 lever frame section and a set of leads. The set was sold in a standard red AA box or as a card mounted blister pack with a hole punched in the top with which to hang it on a peg on a display stand. The back of the card carried fitting instructions in six languages. During the three years the pack was sold in Tri-ang Hornby AA boxes, at least three different labels were used on the ends of the boxes.

Over 36,000 packs were made up in 1969 and nearly all sold before the end of the year. Towards the year end it was decided to renumber the pack R663U and the first 4,000 packs with the 'U' suffix were made in 1969 although not sold until 1970. A further 33,000 were made that year bringing the total number of R663U packs to 37,000. None was made after 1970.

Double Track Sleepers R411

1965 was the last year of production of these which had first been introduced in 1963. About 1,440 were made that year and were probably bagged up 12 to a cellophane envelope with six envelopes stapled to a Tri-ang Railways display card. Very few of these double-track sleepers have survived probably because they were lost or were thrown away by people not knowing what they were.

Buffer Stop R494

The Super 4 buffer stop was made up of two mould-

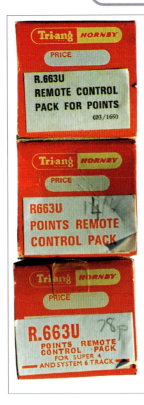

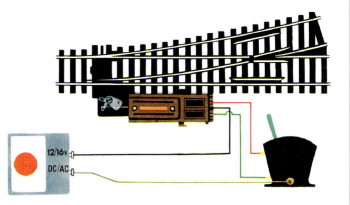

R663 point remote control pack.

R494 buffer stops and R394 hydraulic buffer stop.

ings: a dark brown frame which clipped between the sleepers and a red beam and buffers which was glued on top of the frame. In 1970 the frame moulding was altered and the colour of the frame changed to black to match the new System 6 track. The unit was unusual in fitting both straight and curved track.

It was sold as R494 up until the end of 1969 after which it became R494U for its last two years of production. About 750,000 of the Super 4 version were made from 1965 until end of production in 1969.

They were sold to shops in boxes of six using the standard red AA box and shops sold them singly.

With home production coming to an end in Australia, a supply of buffer stops was needed in 1968 to replace the R103 Series 3 type that the Australian company had been making up until then. A total of 10,000 were sent out to Melbourne, bulk packed.

In New Zealand, with the arrival of System 6 track the Auckland factory required a supply of appropriate buffer stops and 500 of the new universal type

(R494U) were sent out in 1970 and an additional 5,500 the following year. These were all bulk packed and may have been sold in local packaging.

Hydraulic Buffer Stop R394

This had a cream-coloured body and a black top carrying a red lamp, a colour scheme it retained to the end of the Tri-ang Hornby period. The buffers were metal and were spring-loaded so that they worked. The model was sold singly, originally in a AC box which was much too large. From around 1967 the packaging was a red and yellow CH size box and later ones had all-over red boxes.

It was supposed to have been available in 1964 but the first batch was made the following year. By the end of 1969 over 150,000 had been made.

Track Bumper

Although no longer available in the UK, this relic of the Standard and Series 3 track systems continued to be needed in Canada. In 1986, 3,200 were sent out to Canada and the following year a further 5,100 were sent. It appears that these were boxed when sent but I do not know what packaging was used.

Power Connecting Clip R487

While usually dark brown, some red ones were made. They could be used with both Series 3 and Super 4 track. They were packed in sixes in a CU size red and yellow box. Every electric train set contained one and so a large number were made. Over 600,000 were made for solo sales during the Tri-ang Hornby years.

For set production, Canada required a supply of unboxed power connecting clips and 2,500 were sent out in 1967. These were recorded as R487C.

Uncoupling Ramp R488

This was also dark brown and included in many sets. It could be used for both Series 3 and Super 4 track and was packed six to a CU size box. Originally the contents were rubber stamped onto the end of the box but for late batches a printed label was fixed to one end flap. Over 400,000 were made for solo sales during this period.

In addition to the power connecting clips referred to above, Canada needed 2,500 uncoupling ramps for train set assembly and these were also sent out in 1967 bulk packed. They were listed as R488C.

New Zealand also received a supply of R488 Uncoupling Ramps each year between 1968 and 1971. These were sold with the Series 3 track. A total of 2,900 were supplied bulk packed.

SUPER 4 UNDERLAY

Underlay 16'6" R410

This was a continuous piece for single track, imprinted to take the track sleepers, which could be used with both straight and curved track. It was supplied rolled in CX size boxes and was last made in 1970. It was catalogued for 1971 in order to clear stocks. During the period covered by this volume, 138,000 rolls were made and sold.

Boxes of R487 power clips and R488 uncoupling ramps.

Tri-ang Hornby track underlay.

Underlay (for R490 and R491) R433, R434

These had the imprint of the respective point and were sold six to an AT size box. R433 fitted the R490 left-hand point and R434 fitted R491. They were also manufactured up until 1970 and about 125,000 of each were made during this period.

Underlay (for R492 and R493) R435, R436

About 22,000 of each of the two diamond crossing underlays were made and sold in Tri-ang Hornby packaging. These were also supplied to retailers six of one type to an AT box.

R476 converter track.

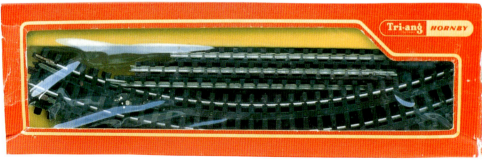

Track pack for Canadian train sets.

Underlay (for R437) R438

This was the underlay for the Y point and was also dropped after 1970 but not before over 22,000 had been made since 1965. It was also sold in boxes of six.

SUPER 4 CONVERTER TRACK

Converter Track R476

In 1966, 56,542 pieces of converter track were made to connect Super 4 to Hornby-Dublo. This was available until the end of 1968. With the introduction of System 6 track in 1970 it was decided to start making the R476 converter track again. The track was supplied to shops in an AC size box but then sold singly. It was a poor fit with the new track and alterations to solve the problem were carried out early in 1971. The track was eventually dropped from the price list at the end of 1983.

The R476 was also supplied to New Zealand in 1971 to help them convert from Series 3 track to the new System 6. For this, 5,500 pieces were sent unboxed.

Short Converter Track

A shorter version was also made for use with Hornby Acho track to help sell Tri-ang in France. Like the standard pieces they had aluminium rails but the middle of the web was cut out to shorten it while still leaving the printing visible.

The short converter track was also used in Minic Motorway sets made in France which had rails included.

SUPER 4 TRACK PACKS

With the exception of the R167 Track Pack, track sets came in two forms: extension sets for the train set oval and self-contained track layouts that were not intended to be 'add-ons' but bought to go with either of the R644 or R645 train packs. The Track Layout Packs were added to the range in 1968.

Track Pack R7

Made between 1963 and 1965 in the UK, 2,444 packs were produced in the final year. It converted your train set track to a double oval with two left-hand points.

Track Pack R8

This was also made between 1963 and 1965 and was identical to R7 but the points were right-hand. A little over 1,000 were made in the final year. *See* **Volume 1**.

Track Pack R167

This track pack was a boxed oval of track for use in assembly of train sets. It was first made in 1965 and remained in production for three years. Over 27,000 were made. *See* **Volume 1**.

Additional supplies were made for Canada where they replaced loose track in train set assembly. Ordinary DP size, pre-1967, window boxes were used. 4,000 were sent out in 1968 and a further 14,500 in 1969. 3,600 more were sent that year but these probably had different contents as they were numbered R167A.

Twin Line Conversion Set R429

This was introduced in 1965 as a replacement for R7 and R8. The set contained:

R481 Straights (10)
R485 2nd Radius Double Curves (8)
R491 Points (2)
R487 Power Connecting Clip

The illustration on the box lid showed a large open landscape dominated by long runs of straight railway track and sidings. The last but small batch of R429 sets was offered to retailers in April 1970 and this brought total production up to more than 34,000.

It was included in the 1971 catalogue in order to clear stocks. Once box stocks ran out, instruction was given that the sets were to be boxed up in plain red and yellow boxes. The box had two inserts and a base card and the two points were sometimes packed in their standard carton before being put in the pack. Two sheets of tissue paper were also used.

Passing Loop Set R439

This set also first appeared in 1965 and provided the following track with which to enlarge the basic train set oval and at the same time incorporate a passing siding:

Extra Track Packs

Extra Tracks Packs are designed to simplify the early stages of building up your Train Set Oval.

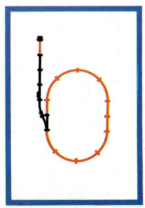

R.454 SIDING SET.
Contains Point, 3 Straights, Curve, Uncoupling Ramp, Buffer Stop and Loading Gauge. Shown added to Train Set Oval.

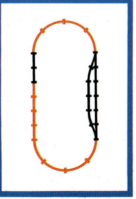

R.439 PASSING LOOP SET.
Contains 2 Points, 8 Straights and 2 Curves. Shown added to Train Set Oval.

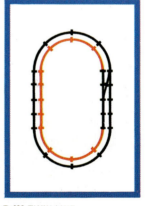

R.429 TWIN LINE CONVERSION SET.
Contains 2 Points, 10 Straights, 8 Double Curves and Power Clip. Shown added to Train Set Oval.

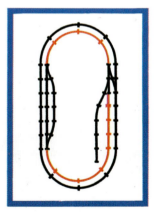

Train Set Oval to which has been added all three Extra Track Packs.

R454 and R429 track packs of 1965/66, and information about track packs from the 1968 catalogue.

Track Layout Packs

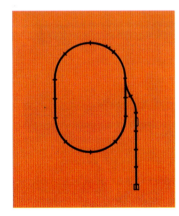

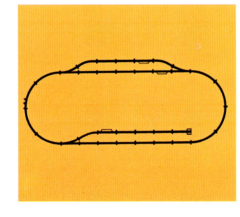

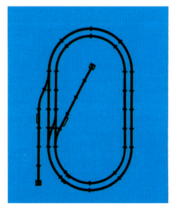

R680 and R681 track layout packs of 1968, and information about track layout packs from the 1968 catalogue.

R481 Straights (8)
R486 2nd Radius Curves (2)
R490 LH Point
R491 RH Point

The track was packed in an illustrated box with two inserts and two sheets of tissue paper. The points in this set were separately boxed in their normal cartons (same as above). Over 28,000 sets were made, the last being in 1970. The set was also supplied to Canada in 1970.

Siding Set R454

The siding set dates from 1966 and survived until 1970 during which time over 40,000 were made. This set had a simple illustrated box with no inserts, the illustration being of a pair of tracks running up to crossing gates and a siding with three wagons on the left. The set contained:

R490 LH Point
R481 Straights (3)
R486 2nd Radius Curve
R488 Uncoupling Ramp
R494 Buffer Stop
R479 Loading Gauge

Oval & Siding R680

The R680 set was first made in 1968 and last made in 1970. It had enough track to make an oval with a siding off to the left. It contained:

R481 Straights (7)
R483 1st Radius Double Curves (8)
R486 2nd Radius Curve
R487 Power Connecting Clip
R488 Uncoupling Ramp
R490 LH Point
R494 Buffer Stop

A new box design was introduced for the 1968 layout packs. These had lift-off lids with a list of contents and a layout plan on the top and pictures of lineside equipment around the sides. Over the three years the set was made 22,000 were produced.

Oval, Siding & By Pass R681

This 1968 set consisted of a large track pack containing 38 pieces. It came in a similar but larger box to R680 which contained the following:

R481 Straights (19)
R483 1st Radius Double Curves (8)
R486 2nd Radius Curves (3)
R487 Power Connecting Clip
R488 Uncoupling Ramps (3)
R490 LH Points (2)
R491 RH Point
R494 Buffer Stop

The set was also made between 1968 and 1970 and over those three years 20,000 were produced.

2 Ovals & 2 Sidings R682

This was the largest of the track layout packs of 1968 containing 46 pieces which were sufficient to build two ovals, one inside the other and linked, and a siding inside the ovals and one outside them. The contents were as follows:

R481 Straights (20)
R483 1st Radius Double Curves (8)
R485 2nd Radius Double Curves (8)
R486 2nd Radius Curve
R487 Power Connecting Clips (2)
R488 Uncoupling Ramps (2)
R490 LH Point
R491 RH Points (3)
R494 Buffer Stops (2)

During the three years of production (1968-70), over 11,000 sets were assembled.

SUPER 4 TRACK PLAN BOOK R166

In **Volume 1** we traced the range of information on track layouts, available to customers, from the Track Layout Folder (TLF/7690/1) to the first R166 Super 4 Track Plan Book. The development of this book and the changes that occurred over the years were well documented by James Day in an article in the

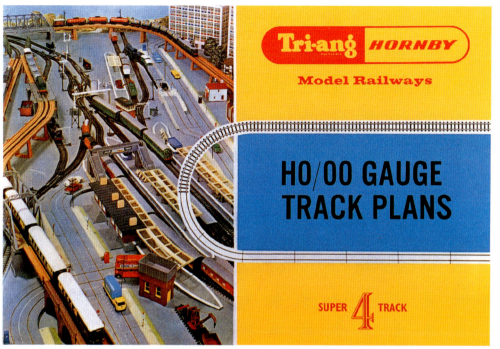

Super 4 track plan book.

Train Collectors Society News (No. 95) published in December 1995 and an update in No.97 of summer 1996. Some of the information that follows was extracted from the article with the writer's permission.

The Super 4 Track Plan Book was published in 1963 and underwent a number of changes until its replacement in 1970 by R6166 which covered both Super 4 and System 6 trackwork.

First Edition

The first edition of R166, as one might expect, carried the name 'Tri-ang Railways' in logo form in red on a yellow background, a colour scheme normally used for the TT range rather than HO/OO.

67,000 copies were printed in 1963 and, based on

orders received, it is likely that all were dispatched during the year. These first copies may be identified by the absence of the Super 4 Track logo on the front and the back page bearing the date '1963'. A further 40,000 were printed in 1964 and a final batch of 25,000 in April 1965. The latter carried the coding R/166/33078/1 but despite its late publication, still carried the Tri-ang Railways inscription.

The immediately striking feature of the book was the superb coloured illustrations of working layouts, starting at the front cover and finishing on the back. The pictures showed the railway system in use with Minic Motorway, Model-Land buildings, the Countryside Series and, finally, with Arkitex office blocks in the background – all products from the Tri-ang stable. The completeness of the model system made under the Tri-ang name was one of its major selling points and, today, is one of the features that makes it so interesting to collectors.

Pages 2 and 3 gave general information about the

Australian Super 4 track.

range and there were illustrations showing an EM2 and L1 climbing an incline with the benefit of Magnadhesion. A mosaic of small pictures showed the range of small equipment that clipped to the track. These were:

Power Clip
Single Track Level Crossing
Uncoupling ramp
Sidewalls
Buffer Stop
Inclined and High Level Piers
Water Trough
Operating Royal Mail Set
Catenary Mast
Telegraph Poles
Loading Gauge
Trackside Fencing
Colour Light Signal
Automatic Train Control Set
Pedestrian Crossing Set
Cattle Crossing Set

Pages 4 and 5 were devoted to units of track, giving lengths and radii and page 6 showed the symbols used in the layout plans to represent station parts etc. The next three pages illustrated standard track formations such as sidings and a large range of pointwork. An interesting station plan occupied

page 10 with a colour picture of the operating layout upon which it had been based (reproduced on page 355 of **Volume 1**).

The plans proper started on page 12 and there were 14 in all, not counting the railway/motorway layouts at the back. They were arranged in ascending order of complexity from a 6' x 4' oval with one siding to an exciting 18' x 5' layout spread over three pages necessitating a fold out section (one of two such sheets in the book). This final layout looked like two large layouts joined end to end.

Pages 30 and 31 provided a check list of the parts one needed for each layout. As a sign of those lovely non-inflationary times the cost of the components for each layout were given although these were updated in subsequent printings as they were affected by tax changes.

On page 32 there were two simple schemes shown for using the R406 Automatic Train Control Set and page 33 showed a variety of different junctions equipped with catenary.

The combined railway/motorway formations and layout shown on pages 34 and 35 appeared almost as an afterthought. Only one complete layout was shown and this was of a very simple type which neatly combined the twin crossing of the RMA set with a car loading ramp as contained in sets RMC and RMD.

Second Edition
With the change of name from Tri-ang Railways to Tri-ang Hornby in 1965 it became necessary to revise and reprint the track plan book and the first batch of the 2nd edition was ordered around February 1966. 52,000 were printed during the year and carried the code R/166/36926/2. Strangely, on the components pages (30/31) the prices quoted were claimed to include Purchase Tax as at 23 January 1967! Surely this should have been 1966. A further 42,000 were printed in 1967.

The standard spacing between Tri-ang Hornby Track centres is 2⅝" (6.7 cms.). In keeping with real railway practice, this allows trains to pass each other on curves without knocking into each other. This important dimension is referred to as 'Standard Track Centres'.

A passing loop merging into double track.

Double track with crossover becoming triple track which remains parallel because of the three track radii.

Tri-ang Hornby basic track geometry showing first, second and third Radii of curves.

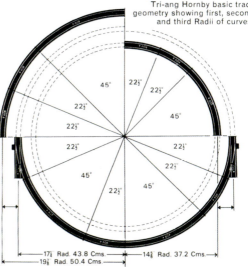

17¼" Rad. 43.8 Cms. — 14⅝" Rad. 37.2 Cms.
19⅞" Rad. 50.4 Cms.

R.6166 TRACK PLANS BOOK —
A comprehensive 32 page guide to track layouts, stations, lineside buildings and scenic effects.

the international SYSTEM 6 track...

R.600 STRAIGHT TRACK. 6⅝" (16.8 cms.) Long.

R.601 DOUBLE STRAIGHT TRACK. 13³⁄₁₆" (33.5 cms.) Long.

R.603 LONG STRAIGHT TRACK. 26³⁄₈" (67 cms.) Long.

R.604 1st RADIUS CURVED TRACK. 14⅝" (37.2 cms.) Radius.

R.605 1st RADIUS DOUBLE CURVED TRACK. 14⅝" (37.2 cms.) Radius.

R.606 2nd RADIUS CURVED TRACK. 17¼" (43.8 cms.) Radius.

R.607 2nd RADIUS DOUBLE CURVED TRACK. 17¼" (43.8 cms.) Radius.

R.608 3rd RADIUS CURVED TRACK. 19⅞" (50.4 cms.) Radius.

R.609 3rd RADIUS DOUBLE CURVED TRACK. 19⅞" (50.4 cms.) Radius.

R.610 QUARTER STRAIGHT TRACK. 1½" (3.8 cms.) Long.

R.611 PACK of 24 SPECIAL FISHPLATES for flexible track.

R.612 LEFT HAND POINT. 17¼" (43.8 cms.) Radius.

R.613 RIGHT HAND POINT. 17¼" (43.8 cms.) Radius.

R.616 ISOLATING TRACK. 6⅝" (16.8 cms.) Long.

R.617 UNCOUPLING RAMP AND MARKER SIGN.

R.618 POWER CONNECTING CLIP.

R.619 FLEXIBLE TRACK. 36" (91.4 cms.) Long.

R.620 FLEXIBLE TRACK – nickel silver rail 36" (91.4 cms.) Long. (not illustrated).

R.624 UNDERLAY for R.612.

R.625 UNDERLAY for R.613.

R.629 LEVEL CROSSING. (track not included).

R.638 ROLL OF TRACK UNDERLAY.

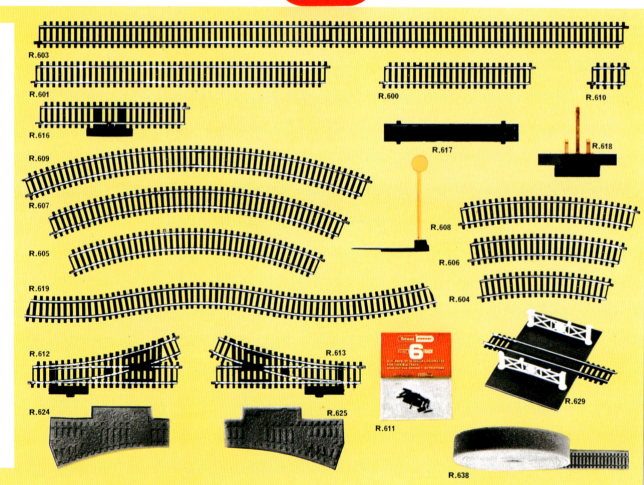

 **super** **4** **track...**

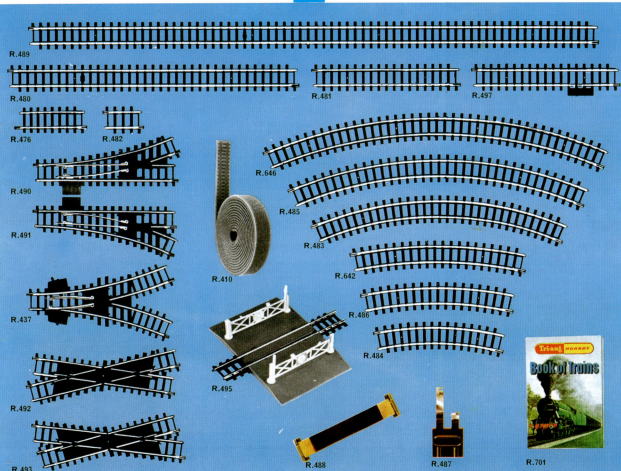

R.489

R.480

R.476   R.482

R.481   R.497

R.490

R.491

R.646

R.485

R.483

R.642

R.437

R.486

R.410

R.484

R.495

R.492

R.488   R.487   R.701

R.493

R.410 ROLL OF TRACK UNDERLAY.

R.433 UNDERLAY FOR R.490. (*not illustrated*).

R.434 UNDERLAY for R.491. (*not illustrated*).

R.437 'Y' POINT. $17\frac{1}{4}$ (43.8 cms.) Radius.

R.438 UNDERLAY for R.437. (*not illustrated*).

R.476 CONVERTER TRACK. $2\frac{7}{8}$ (7.3 cms.) Long. Connects Super 4 Track to System 6 Track.

R.480 DOUBLE STRAIGHT TRACK. $13\frac{3}{16}$ (33.5 cms.) Long.

R.481 STRAIGHT TRACK. $6\frac{5}{8}$ (16.8 cms.) Long.

R.482 QUARTER STRAIGHT TRACK. $1\frac{1}{2}$ (3.8 cms.) Long.

R.483 1st RADIUS DOUBLE CURVED TRACK. $14\frac{5}{8}$ (37.2 cms.) Radius.

R.484 1st RADIUS CURVED TRACK. $14\frac{5}{8}$ (37.2 cms.) Radius.

R.485 2nd RADIUS DOUBLE CURVED TRACK. $17\frac{1}{4}$ (43.8 cms.) Radius.

R.486 2nd RADIUS CURVED TRACK. $17\frac{1}{4}$ (43.8 cms.) Radius.

R.487 POWER CONNECTING CLIP.

R.488 UNCOUPLING RAMP.

R.489 LONG STRAIGHT TRACK. $26\frac{3}{8}$ (67 cms.) Long.

R.490 LEFT HAND POINT. $17\frac{1}{4}$ (43.8 cms.) Radius.

R.491 RIGHT HAND POINT. $17\frac{1}{4}$ (43.8 cms.) Radius.

R.492 LEFT HAND DIAMOND CROSSING.

R.493 RIGHT HAND DIAMOND CROSSING.

R.495 LEVEL CROSSING. (*Track not included*).

R.497 ISOLATING TRACK. $6\frac{5}{8}$ (16.8 cms.) Long.

R.642 3rd RADIUS CURVED TRACK. $19\frac{7}{8}$ (50.4 cms.) Radius.

R.646 3rd RADIUS DOUBLE CURVED TRACK. $19\frac{7}{8}$ (50.4 cms.) Radius.

R.701 BOOK OF TRAINS. 64 pages of interest for all Railway Fans.

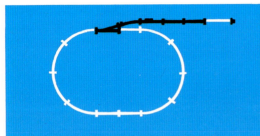

SIDING SET – R.686 System 6 Track

Contains Point, 3 Straights, Curve, Uncoupling Ramp and Buffer Stop.

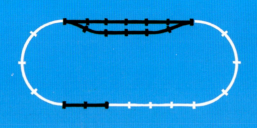

PASSING LOOP SET – R.687 System 6 Track

Contains 2 Points, 8 Straights and 2 Curves.

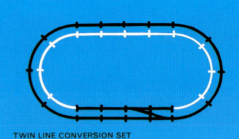

TWIN LINE CONVERSION SET

R.695 System 6 Track   **R.429** Super 4 Track
Contains 2 points, 10 Straights, 8 Double Curves and Power Clip.

Tri-ang HORNBY

Track Packs

The Track Packs shown *ABOVE* provide the easiest and most economic means of adding on extra track to that supplied with most Train Sets. The track sections in black represent the contents of the Track Pack and those in white a basic oval. Each of the Packs may be added to the oval by itself as shown or you may add two or all of the Packs to build yourself a good sized double track circuit with siding and passing loop. When ordering check whether you require System 6 or Super 4 track.

The contents of the Track Packs *BELOW* build into complete layouts for those wishing to start out with a more elaborate model railway. They are specially suitable to go with one of the Train Packs on Page 11 or, of course, with all other Tri-ang Hornby Trains.

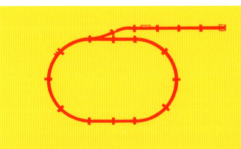

LAYOUT WITH SIDING

R.680-6 System 6 Track   **R.680** Super 4 Track
Contains Point, 7 Straights, 8 Double Curves, Curve, Power Clip, Uncoupling Ramp and Buffer Stop.

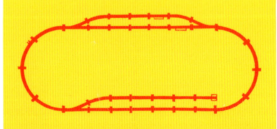

LAYOUT WITH PASSING LOOP AND SIDING

R.681-6 System 6 Track   **R.681** Super 4 Track
Contains 3 Points, 19 Straights, 8 Double Curves, 3 Curves, Power Clip, 3 Uncoupling Ramps and Buffer Stop.

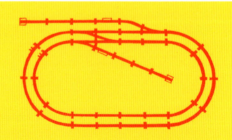

TWIN TRACK LAYOUT WITH TWO SIDINGS

R.682-6 System 6 Track   **R.682** Super 4 Track
Contains 4 Points, 20 Straights, 16 Double Curves, Curve, 2 Power Clips, 2 Uncoupling Ramps and 2 Buffer Stops.

The logos which appeared on each page were replaced with new ones. The lozenge-shaped logo was used on the front cover but inside a less common two-tier oval logo was used.

Another change that took place at this time was the removal of the water troughs, pedestrian crossing and cattle crossing from the small illustrations on page 3. A new addition, however, was the hydraulic buffer stop and the R407 hand-operated turntable had given way to the R408 electric version. Changes occurred on the components table on pages 30 and 31 where the prices had also risen.

Third Edition

This looked very much like the second edition and was probably ordered early in 1968 with a total of almost 50,000 being printed that year although it is not known whether all of these were printed in one batch and, therefore, all of this edition. The books carried the order code R/166/51053/3.

The only noticeable differences to the previous edition was the removal of the prices on page 31, the disappearance of the Lines Bros. logo from the back cover and the replacement of 'Rovex Scale Models Ltd' by 'Rovex Industries Limited' .

Fourth Edition

As this did not carry the order number on the bottom of the back page we do not know when it was printed but it was likely to have been late in 1968 or early 1969; a year in which 50,000 were again printed. 12,000 were left over at the end of the year, most of them being sold during 1970 at the same time as the new Track Plan Book R6166.

An obvious difference in this edition was the disappearance of the large oval logos and their replacement with the longer lozenge type. This was not possible on page 34 where the two-tier trademark matched better with the Minic Motorway one alongside it, or on the back cover where the trademark was superimposed on a picture and its replacement

with one of a different shape would have left an awkward hole in the picture.

The same mini-pictures were shown on page 3 but the 'R' numbers had been removed. On page 6, where the accessories are illustrated, the station building was now shown as R582 instead of the earlier R473/465/460. Another change was the removal of pictures of obsolete models from various pages. These included the Yard Crane, Continental Sleeper, Water Tower, DMU, Davy Crockett and Victorian Suspension Bridge.

The most significant change was the replacement of the whole of the feature about the R406 Automatic Train Control Set, and most of the details of the catenary system. In their place were details of how you could make your own scenery.

Correction Slip

Sometime following the publication of the 4th edition of the track plan book it was brought to the company's notice that the publication contained a number of errors. Why it should have taken six to seven years for these to come to light must be left to speculation but it resulted in the printing of a small correction slip which was carried in some of the books. The errors concerned incorrect numbering of components, items listed but not shown on plans and vice-versa and incorrect quantities of components quoted.

In August 1970 it was announced that the book was finally out of stock being replaced by the new R6166 Track Plan Book which was due for delivery in October that year but did not arrive until November.

AUSTRALIAN SUPER 4

While Lines Bros. (NZ) Ltd were happy to stick with Series 3 track, Moldex Ltd wanted the new Super 4 from its announcement. Duplicate tools were sent out to Australia in 1962 and the range was illustrat-

ed in their 1962 catalogue. By 1965, 16 items were listed as being available from Moldex as well as five track packs.

The track looked identical to the British range except that it carried the wording 'Tri-ang MADE IN AUSTRALIA (or AUST)' and the appropriate 'R' number. None of the mouldings carried an identification mark to indicate who made the tools.

The packaging was also very like that used in the UK but was inscribed beneath the Tri-ang Hornby logo: 'Made in Australia by MOLDEX LTD, a member of the LINES BROS. LTD group'.

With the exception of the R497A Isolating Rail, Australian Super 4 was available until 1969 after which System 6 was in production at the Moldex factory. Contents of the boxes were rubber stamped onto white end panels. A note in the Australian Trade Price List in 1964 indicated that Diamond Crossings were only temporarily being made in Australia and that in future they would be imported from the UK. In actual fact production remained in Australia.

R411 Double Track Spacers were advertised as coming from Australia in 1965 and 1966. They were offered in sets of 12 bagged up and mounted on a card for shop display, as in the UK. This may have been old stock no longer available from Margate or the tool could have been sent to Melbourne for the spacers to be made there once production ceased in Britain. The former seems more likely.

SYSTEM 6 TRACK

SERIES 5 TRACK

As we have seen, the new fine scale track was to have been called Series 5 but at the last minute it was changed to System 6. The change did not come fast enough for the printing of the factory census sheets for 1969 which recorded three track pieces in the R500 range with a note that they were for use in sets. The pieces were R500 Straight, R505 Double Curve Small Radius and R524 Power Connecting Clip. As no production figures were shown against these items it is assumed that no track was made with these 'R' numbers for solo sales.

SYSTEM 6 SOLO TRACK

As System 6 was available for such a short period during the Tri-ang Hornby years and yet remained the standard track system for several decades under the Hornby Railways trademark, it will be only lightly covered here but will be dealt with in detail in **Volume 3**.

Packaging

Packaging containing System 6 track or equipment specially modified to go with System 6 track carried the 'SYSTEM 6 TRACK' logo. Box sizes followed previous practice for Super 4 track but the box that once took 24 R480 Super 4 Double Straights could now carry 36 R901 System 6 Double Straights as they were much thinner.

Items of which large quantities would sell had boxes printed especially for them with the contents incorporated in the printing but for poorer selling items a printed label provided the information. Point boxes, for example, carried a printed label on the end flaps sometimes with a small 'System 6 Track' logo and sometimes without. Some had 'S/6' after the description of contents instead of a logo.

Production

During the Tri-ang Hornby period the following quantities of System 6 track were made:

Straight R600	287,000
Double Straight R601	386,000
Long Straight R603	163,000
Curve 1st Radius R604	201,000

Double Curve 1st Radius R605	226,000
Curve 2nd Radius R606	184,000
Double Curve 2nd Radius R607	114,000
Curve 3rd Radius R608	153,000
Double Curve 3rd Radius R609	144,000
Quarter Straight R610	155,000
Pack of Fishplates R611	(see Sundries)
LH Point R612	117,000
RH Point R613	120,000
LH Diamond Crossings R614	none made until 1972
RH Diamond Crossings R615	none made until 1972
Isolating Track R616	11,000
Uncoupling Ramp & Marker R617	70,000
Power Connecting Clip R618	55,000
Flexible Steel Track R619	49,000
Flexible NS Track R620	47,000
Buffer Stop R494U	170,000

Hydraulic Buffer Stop R394U

From 1970 the hydraulic buffer stop was adapted to fit both Super 4 and System 6 track and the number therefore received a 'U' suffix. 50,000 were made during this period. In later years the cream body became grey.

Points R612 and R613

These were due in the shops in May 1970 but some difficulty was experienced in getting them right. Strangely, they could get the right-hand version to work properly but were having difficulty with the left-hand ones. Rather than send out just right-hand points, they were delayed until June by which time both types were said to be working well. By November it was announced that further problems had been experienced and the factory could not keep pace with orders. In 1971 the decision was taken to fit diecast switch rails which would hold their shape better.

The pre-production model of the right-hand point

was kept by the model maker when he left the company. It was made in white plasticard with brass and steel used for the rails.

Diamond Crossings R614 and R615

Early in 1970 it was decided to introduce R614/R615 diamond crossings (subject to tool costs first being established) but by April that year it had been decided to postpone the introduction of these. In January 1971 a note was made that the design had to be updated in the light of the experience they had had with the points. This would explain why none were made during the Tri-ang Hornby period. Were boxes printed in advance carrying the Tri-ang Hornby name?

Flexible Track R619, R620

This was 36" in length and was made by G&R Wrenn Ltd. It did not go metric until 1973 (R621). The steel version of flexible track carried the number R619 but it was also available in nickel-silver as R620.

Three Way Point R640

Experiments on a three-way point were being carried out late in 1970 and early in 1971 with a view to including it in the 1973 price list but the work did not lead to production of this item.

Large Radius Points

In 1970 thought was given to producing 2ft radius points but there was concern as to how these would fit into the existing track geometry; a problem no doubt left to the Development Department. By April it had been decided to postpone the introduction of these possibly until 1972.

Power Connecting Clip R618

It appears that some early production of R618 was sent out with the two outer contacts loose. These should have been firmly anchored to the basic

Packaging used for System 6 track.

Super 4 and System 6 long straights came in long brown card boxes with a label stuck in the centre.

Super 4 point blister packs.

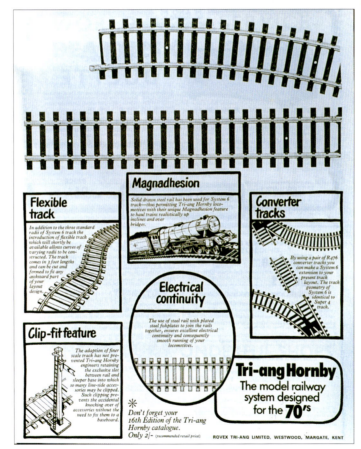

An advertisement emphasising the benefits of System 6 track.

moulding and a request was sent out to retailers in July 1970 for them to return any power connecting clips with loose outer contacts.

SYSTEM 6 TRACK UNDERLAY

The first System 6 track underlay also appeared in 1970 and used the same boxes as that for Super 4 and had the contents rubber stamped on the end flap panel. The following quantities were made under the Tri-ang Hornby name:

Underlay (for R612)	R624	22,000
Underlay (for R613)	R625	19,000
Underlay Roll	R638	37,000

SYSTEM 6 TRACK PACKS

Like the Super 4 track packs, those for System 6 came in two forms: those for adding to the oval of track provided in the train sets and those which were complete track layouts designed to be used with the trackless train packs. The System 6 track packs first appeared in 1971 and these were made for one year only. A completely new range of packs arrived in 1972.

Passing Loop Set R686

The R686 was the System 6 version of the R439 track pack and arrived in the shops in November 1970 having been delayed two months by the shortage of points. About 3,900 were made with the following contents:

R600 Straights (8)
R606 2nd Radius Curve (2)
R612 LH Point
R613 RH Point

Siding Set R687

This was the System 6 version of the R454 track pack and was also delayed by the point shortage and was not in the shops until November. About 4,700 were

made and the following were the contents:

R494U Buffer Stop
R600 Straights (3)
R606 2nd Radius Curve
R612 LH Point
R617 Uncoupling Ramp

Twin Line Conversion Set R695

The R429 Super 4 twin line conversion set was still available in 1971 but R695 was a System 6 version of it. Only a little under 1,000 were made and they contained:

R680 Straights (10)
R607 2nd Radius Double Curves (8)
R613 Points (2)
R618 Power Connecting Clip

Oval and Siding R680-6

This was System 6 track served up in the Super 4 box but with a stick-on label indicating the type of track. The set contained the equivalent pieces of track and 700 were made. The idea was to use up the stock of old boxes and when the stock had gone the set was dropped. The contents were as follows:

R494 Buffer Stop
R600 Straights (7)
R605 1st Radius Double Curves (8)
R606 2nd Radius Curve
R612 LH Point
R617 Uncoupling Ramp
R618 Power Connecting Clip

Oval, loop and Siding R681-6

This was the System 6 version of the R681 and 3,100 were made. They contained:

R494 Buffer Stop
R600 Straights (19)
R605 1st Radius Double Curves (8)
R606 2nd Radius Curves (3)
R612 LH Points (2)
R613 RH Point

R617 Uncoupling Ramps (3)
R618 Power Connecting Clip

Twin Ovals and 2 Sidings R682-6

Only 400 of this System 6 set were made. These contained:

R494 Buffer Stops (2)
R600 Straights (20)
R605 1st Radius Double Curves (8)
R606 2nd Radius Curve
R607 2nd Radius Double Curves (8)
R612 LH Point
R613 RH Points (3)
R617 Uncoupling Ramps (2)
R618 Power Connecting Clips (2)

TRACK BED

See 'Buildings and Lineside Accessories'

SYSTEM 6 TRACK PLAN BOOK R6166

System 6 track arrived in 1970 and with it the first edition of a brand new Track Plan Book. The book, which was available from the November 1970, went through many editions and is in its eighth as I write this chapter. Unlike the four editions of the Super 4 plan book, those of the System 6 version underwent substantial changes between editions and these will be covered in **Volume 3**.

First Edition

The book was the same size and format of the 1971 catalogue and contained 32 full colour pages.

The cover of the new plan book showed part of what was supposed to be an extensive layout of the period but was in fact a diorama set out in a studio for the purpose of the photographs. Dad was shown trying to 'push start' a Hall while his son was trying to re-rail wagons that Dad has knocked off with his arm! The centrepiece of the picture was a turntable on which a maroon Coronation was being turned. There were bright red station buildings and the new Tri-ang Hornby station roof units. To the right a goods yard had been made up from two ex-Minic Motorway R1008 Freight Depot buildings in maroon with black roofs. Pylons marched across the layout to join other Model-Land buildings on the left which included the three factory units.

The same diorama was featured in the 1970 catalogue but carried a different arrangement of trains. No doubt this was done so as to suggest to the public that it was a genuine layout with trains on the move and not just a studio display.

On page 3 various ideas for accommodating a model railway in a modern home were described and illustrated, showing that the company had got its priorities right where the public were concerned.

The book illustrated the range of System 6 track available and the buildings and accessories that fitted to it and around it. These were illustrated on pages 4-7 and showed the symbols used on the layouts that followed. Pages 8 and 9 provided copious notes about the use of signals and other lineside apparatus. Two pages on track radii and pointwork followed. A useful conversion table between System 6 and Super 4 was also provided.

The plans started on page 12 and a total of 13 were displayed. All of these had been tested before being included in the book. This time the list of components required was given with each plan. The first five plans in the book were designed to build up progressively starting with a simple oval with a siding and adding a passing loop, additional sidings, second circuit etc. Plans 7 and 8 demonstrated the

use of incline and high level piers to create figures of eight and plan 10 provided an attractive 'out-and-back' layout with a double-width terminus station outside the circuits. The next three plans were of yards and depots leaving 13 for the 'big one'.

This time the largest layout covered an area 15' x 6' and had just two circuits. For the first time, it was a layout with two external termini which could be operated from separate points so that two people could use the layout and send trains to each other; something my brother and I liked to do many years ago.

Pages 26 and 27 illustrated a number of station layouts and 28 and 29 showed the reader how to wire-up a layout and electrical accessories. The last two inside pages were devoted to notes on scenic modelling and flexible track usage was dealt with the back cover.

In 1970, 33,200 books were printed all of which sold during the year. A further 39,000 were printed in 1971. The second edition was released in 1972 when another 50,000 were printed but it is assumed that these all carried the Hornby Railways logo and are thus outside the scope of this volume.

CANADIAN SYSTEM 6 TRACK PACKS

As with the Super 4 track before, the Canadian company bought-in packs of track that it could place in the train sets it assembled. Thus a number of different packs were made up for them at Margate between 1970 and 1973.

Five track packs for Canada appeared in Tri-ang Hornby packaging; initially in 1970 in R1000 series of numbers and in 1971 in the R6000 series. It is known that packs were required for the following track formations but which 'R' number applies to each is not known.

40" x 32" oval 10 pieces of track
45" x 32" oval 12 pieces of track
76" x 32" oval (B shaped) 24 pieces of track

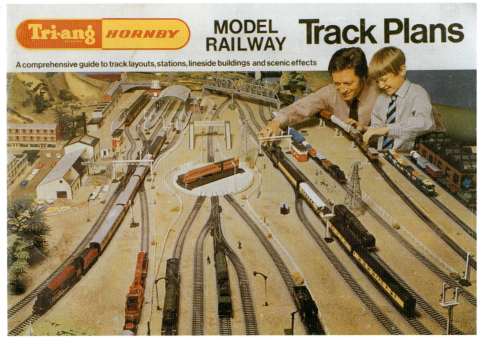

System 6 track plan book.

The following list gives the quantities of each pack that were sent out to Canada:

Pack	Code	Made
Pack A	R1670 ('70)	8,000
	R6670 ('71)	7,000
Pack B	R1671 ('70)	4,500
	R6671 ('71)	4,900
Pack C	R1672 ('70)	1,500
	R6672 ('71)	1,500
Pack D	R1673 ('70)	800
	R6673 ('71)	1,350
Pack E	R6674 ('71)	900

HORNBY RAILWAYS TRACK PACKS FOR CANADA

In 1973 four System 6 track packs were made up for sale in Canada, presumably for use in train set production. Not much is recorded about them but we know that there were four different ones and that they were all delivered. None were listed in the Louis Marx price lists for either 1973 or 1974, suggesting that there was not any surplus stock after train set production ceased.

The packs made were:

Pack A R705
This formed a circle and 10,000 packs were made.

Pack B R706
A second radius oval was formed with this pack of which 2,000 were made.

Pack C R707
A first radius oval was formed from this pack and 3,000 were ordered but records showed only 500 made that year. It is possible that a further 2,500 were sent in 1974 but there is no record of this.

Pack D R708
This pack provided track for a 2nd radius figure of '8'. Only 1,000 were made.

AUSTRALIAN SYSTEM 6 TRACK

As Moldex were both manufacturing and importing System 6 track between 1970 and 1972 it is very difficult to be sure which items they made. The following list is based on what I have seen and what I have deduced by logic:

Straight R600
Double Straight R601
Double Curve 1st Radius R605
Half Curve 2nd Radius R606
Double Curve 2nd Radius R607
Uncoupling Ramp R617
Power Connecting Clip R618

The mouldings were inscribed 'Tri-ang HORNBY R... MADE IN AUST. & N.Z.' Australian-made track looked like the British, with black sleeper webs and steel rail.

NEW ZEALAND SYSTEM 6 TRACK

We know that System 6 was listed in the New Zealand price lists in the last two years and I have assumed it was made there rather than imported from Australia.

PLASTIC TRACK FOR STARTER SETS

Curve R412
In 1970 a series of clockwork train sets were made which did not use conventional track but had eight pieces of moulded plastic track. Only the curve was made by Rovex as each set contained only a circle. The track may be found moulded in grey or black. Instead of the usual 'RS' number, these sets were given an 'RB' prefix. Over 200,000 of these sets were sold and so more than 1.6 million R412 plastic curves must have left the Margate factory during that short period.

Evidence that System 6 track was made in New Zealand.

INTRODUCTION

Second Series Buildings

Production of the original series of buildings designed in 1952, continued in New Zealand and Australia but from 1962 the second series of station buildings had become the mainstay of the range in the UK. There were over 50 buildings and accessories available to the purchaser by 1965 and many of these lines were selling surprisingly well considering the recession in the model railway industry at the time.

As we saw in **Volume 1**, the new range of buildings provided realistic looking station complexes but, because they were skilfully designed to keep manufacturing costs down, they could be bought at comparatively low prices.

Consolidation

Despite their success, the recession did require some consolidation of the range in 1966 and the following were cut from the catalogue. It is assumed that they did not make their way into Tri-ang Hornby packaging.

R148   Platform Accessories
R149   Cattle Crossing
R186   Field Fencing Set
R187   Pedestrian Crossing Set
R269   Extension to Victorian Suspension Bridge
R432   Girder Bridge Set
R472   Island Platform Waiting Room
R475   Platform with Operating Crane
R478   Set of Six Telegraph Poles

In actual fact, few of the building and lineside elements were withdrawn. The R148 Platform Accessories continued to be incorporated in station

sets and in the late 1960s a special batch was made for an American customer. While the bridge set disappeared, the girder bridge and its supports continued under their old numbers – R78 and R77. The Island Platform Waiting Room was now incorporated into the R588 station set but did reappear as a solo model in 1971. The crane base was required for the Honest John Rocket Launch Pad in the Battle Space series but the platform crane was to return many years later.

Hornby-Dublo Buildings

In 1966 the ex-Hornby-Dublo range of buildings was added to the catalogue. Initially this was to dispose of stocks of the buildings found at the Meccano factory, but later, some were manufactured at Margate. Old Hornby-Dublo stock sold officially through Rovex as Tri-ang Hornby may be identified by a Tri-ang Hornby sticker on the box.   In at least one case, the R5030 Island Platform Kit, the model was sold in a red Tri-ang Hornby box.

The 1966 catalogue featured the following:

R5005   Two Road Engine Shed Kit
R5006   Engine Shed Extension Kit
R5015   Girder Bridge
R5020   Goods Depot Kit with Working Crane
R5030   Island Platform Kit
R5083   Terminus and Through Station
        Composite Kit
R5084   Canopy Extension Kit for R5083
R5086   Platform Extension
R5087   Platform Fence for R5086 or R5089
R5089   Side Platform Extension for R5083
R5092   Double Track Tunnel

It will be noted that the original Hornby-Dublo catalogue number was retained but given an 'R' prefix.

The other new addition to the building range in 1966 also came from another company. This was the R589 Ultra Modern Station Set which was made up of Arkitex parts formerly marketed in construction kits by Spot-On Ltd, another company in the Lines Group. The assembly pack was sold in a plain red box with Tri-ang Hornby stickers or in a specially-printed box with an illustrated lid.

Deletions

During the next few years the range of buildings and accessories did not grow any further but deletions began to appear. By 1968 the Ultra Modern Station Set had been dropped along with three of the ex-Hornby-Dublo range:

R5015   Girder Bridge
R5087   Platform Fence
R5092   Double Track Tunnel

From the former Tri-ang Railways range we also lost the R586 Pair of Platform Fences with name boards and names although these continued to be available in station sets. The R474 Station Upper Floor was also dropped as a separate item – being only available in sets but it was to return in 1971 in a slightly different form.

In 1969, more of the ex-Hornby-Dublo range was dropped from the price list as stocks ran out. These included:

R5020   Goods Depot Kit
R5030   Island Platform Kit
R5086   Platform Extension
R5089   Side Platform Extension

The R5083 and R5084A Terminus and Through Station Assembly Pack and the Extension Pack were now being made at Margate and in a different colour scheme. The maroon plastic of the Tri-ang

The Tri-ang Hornby range got off to a good start with an attractive, versatile and inexpensive station system which was to last until the late 1970s.

station buildings was used for the buildings and the platforms had changed to grey to also match those in the main Tri-ang Hornby range. They left the factory in plain red boxes with a Tri-ang Hornby sticker on the end. That year also saw the withdrawal of the R414 Double Curve Level Crossing which had been an excellent idea for small layouts.

1970 saw the loss of the R587 Trackside Accessories Pack which contained mile and gradient posts and whistle signs. Also to go that year were the R456 Straight Walls for the salmon-coloured incline piers, the R461 Platform Unit with Subway, which was to return later, and the R588 Island Platform Set which would be replaced the following year by a similar set.

Short-Lived Stations

In 1970, for one year only the catalogue advertised the R689 Mainline Station Set and R688 Island Platform Station Set which were replaced by the slightly smaller R4 Suburban Station and the R1 Wayside Halt in 1971.

1970 also saw the appearance of the R71A Modern Footbridge, to replace the R71, which like other, later, models would have been cheaper to manufacture. The old footbridge had been in production since 1953.

Three of the remaining four ex-Hornby-Dublo buildings also went in 1970. These were:

R5005 Engine Shed
R5006 Engine Shed Extension Kit
R5084 Canopy Extension Kit for the
R5083 Terminus & Through Station

The maroon, Margate-manufactured R5083 Terminus and Through Station was, itself, in its last year of production.

New Station Kits

The Hornby-Dublo station had been expensive to make and so it was difficult to keep the price of it down to a level which the public could afford. It had been necessary to find a cheaper alternative. During

1970, Rovex tooled up their own design of station canopy and the opportunity was taken to completely revise the range of stations for 1971. Six new stations replaced the previous range and these were:

R1 Wayside Halt
R2 Village Station
R3 Town Station
R4 Suburban Station
R5 City Station
R6 Central Station

R5 and R6 utilised the new canopy units. The rest of the platforms, canopies, fences, seats and buildings used to make up the six new sets were already in existence as part of the second series station system.

Bright Red Buildings

In order to appeal to younger enthusiasts, the 'brick' buildings in the series changed colour from maroon to bright red. This was a controversial move amongst modellers but, as the company always insisted, children were their main market and the introduction of bright colours tended to boost sales. The company were entering a very colourful period which marked the early 1970s. Competition from new firms entering the model railway market later forced them back to realistic finishes but, for a period, the 'bright colour' advocates had their way.

The following station parts were now offered also as solo items:

R74 Station Roof Unit (over-track canopy)
R471 Platform Wall
R474 Terminus Building/Platform Unit
R582A Ticket Office/Platform Unit

The Terminus Building was in fact the reappearance of the R474 Station Upper Floor & Clock Tower which had been dropped at the end of 1967. Platform and steps had now been added to the box.

The R145 Modern Signal Box and R146 Modern Engine Shed were also released in 1971 in bright red plastic and 1971 saw the introduction of the small platform section which was used in larger station sets as a support for the terminus building or to fill

in gaps when building a terminus station layout.

In 1971 the range of former Tri-ang Railways buildings and lineside accessories was substantially condensed. This meant the loss of the following:

R88 Water Crane
R281 Set of 5 Train Staff
R383 Set of 5 Platform Figures
R284 Set of 5 Coach Figures
R413 2 Loco Crew
R458 Small Station Set
R459 Large Station Set
R479 Loading Gauge
R496 Double Track Level Crossing
R582 Ticket Office/Platform (replaced)

CHECK LIST OF BUILDINGS & LINESIDE

For ease in tracing individual models they have been listed below in numerical order. Those listed under 'General' were available in the UK and world-wide and almost all appeared in the standard catalogue. Some models were made for particular countries such as Canada or were supplied bulk packed for use in sets or for packing locally on arrival. These are listed under the name of the country receiving them and are of more importance to collectors for the packaging they were sold in than for their contents.

Dates

Wherever possible, dates refer to the years when a model was made rather than when it was available. Thus one may find a model in the catalogue after the dates quoted below. It was common practice to keep a model on sale until stocks had cleared and it was very difficult to sell models not illustrated in the catalogue.

No.	BUILDING	Colour	Dates
	GENERAL		
R1	Wayside Halt	red	71-77
R2	Village Station	red	71-77
R3	Town Station	red	71-77
R4	Suburban Station	red	71-77
R5	City Station	red	71-74
R6	Central Station	red	71-74
R43H	Home Signal - metal	white	55-67
R43H	Home Signal - plastic	white	68-75
R43D	Distant Signal - metal	white	55-67
R43D	Distant Signal - plastic	white	68-75
R71	Footbridge	grey	53-69
R71A	Double Track Footbridge	grey	70-79
R74	Station Roof Unit	green	71-72
R77	Bridge Supports	grey	54-68
R78	Girder Bridge	grey	54-68
R78C	Girder Bridge & Supports	grey	69-72
R84	Lineside Huts	grey&black	55-71
R88	Water Crane	black	55-70
R140	Signal Gantry	white	58-78
R142H	Home Junction Signal	white	58-75
R142D	Distant Junction Signal	white	58-75
R145	Modern Signal Box	maroon	61-70
R145	Modern Signal Box	red	71-77
R146	Modern Engine Shed	maroon	61-70
R146	Modern Engine Shed	red	71-77
R180	Viaduct	maroon	61-cur
R188	River Bridge	grey	61-71
R189	Brick Bridge	salmon	61-cur
R264	Suspension Bridge		63-71
R281	Set of 5 Train Figures		61-70
R283	Set of 5 Platform Figures		61-70
R284	Set of 5 Coach Figures		61-70
R299	Trackside Fencing	white	63-87
R413	Set of 2 Loco Crew	black	62-70
R414	Double Curve Level Crossing		65-67
R430	Curved Sidewalls Small Radius	salmon	62-69
R431	Curved Sidewalls Large Radius	salmon	62-69
R453	High Level Pier	salmon	62-72
R456	Straight Sidewalls	salmon	62-69

No.	BUILDING	Colour	Dates
	GENERAL		
R457	Inclined Piers	salmon	62-72
R458	Small Station Set		69-70
R458A	Small Station Set		64-68
R459	Large Station Set		69-70
R459A	Large Station Set		64-68
R460	Straight Platform	grey	62-cur
R461	Platform with Subway	grey	63-69
R462	Curved Platform Large Radius	grey	62-cur
R463	Curved Platform Small Radius	grey	62-cur
R464	Double Curved Platform Ramp	grey	62-cur
R471	Platform Fencing Set		71-72
R472	Island Platform Waiting Room	red	71-72
R474	Upper Floor/Clock Tower	red	71-72
R475	Platform Crane	maroon	63-65
R479	Loading Gauge	white	62-71
R495	Level Crossing Single Track		62-72
R496	Level Crossing Double Track		62-69
R576	Tunnel		64-cur
R582	Ticket Office Platform Unit	maroon	64-70
R582A	Ticket Office Platform Pack	red	71-77
R583	Curved Platform Large Radius Canopy	crm. & grn	64-72
R584	Curved Platform Small Radius Canopy	crm. & grn	64-72
R585	Straight Platform & Canopy	crm. & grn	64-77
R586	Platform Fence/Nameboards	cream	64-67
R587	Trackside Accessory Pack		64-69
R588	Island Platform Set	maroon	64-69
R589	Arkitex Ultra Modern Station Set		66-67
R629	Single Track Level Crossing		70-79
R654	Set of 7 Incline Piers	salmon	70-77
R655	Set of 3 High Level Piers	salmon	70-77
R688	Island Platform Set	maroon	69-70
R689	Mainline Station Set	maroon	70
R2008	Freight Depot	red	71
R5005	2 Road Engine Shed	cream	66-69
R5006	Engine Shed Extension Kit	cream	66-69
R5010	Footbridge	cream	67

No.	BUILDING	Colour	Dates
	GENERAL		
R5015	Girder Bridge	red	66-67
R5020	Goods Depot Kit	cream	66-68
R5030	Island Platform Kit	cream	66-68
R5083	Terminus Through Station Kit	cream	66-68
R5083A	Terminus Through Station Kit	maroon	69-70
R5084	Canopy Extension Kit R5083	cream	66-68
R5084A	Canopy Extension Kit R5083A	maroon	69
R5086	Platform Extension	cream	66-68
R5087	Platform Fence	cream	66-67
R5089	Side Platform Extension R5083	cream	66-68
R5092	Double Track Tunnel	cream	66-67
	FOR USA (all bulk packed)		
R148T	Set 4 Station Accessories		67
R180T	Viaduct		67-68
R188T	River Bridge		67-68
R264T	Suspension Bridge		67-68
R460T	Straight Platform		67,69
R461T	Platform & Subway & Canopy		67
R464T	Platform Ramp		67
R466T	Canopy		69
R469T	Seat Unit & Nameboard		67
R472T	Island Platform Waiting Room		67
R582T	Ticket Office Platform Pack		67
R583T	Large Radius Curved Platform & Canopy		67
R585T	Straight Platform & Canopy		67
T20NP	Platform		68
T26NP	Canopy		68
T39NP	Platform Ramp		68
R834	Station Accessories		68
	FOR CANADA		
R39C	Platform Ramp n.p. (bulk)		67
T20C	Straight Platform n.p. (bulk)		67
T26C	Canopy Unit (bulk)		67
R432	Girder Bridge		68
	FOR AUSTRALIA		
R495	Level Crossing (bulk)		68

DIRECTORY OF BUILDINGS & LINESIDE

FIRST SERIES STATIONS

While the first series station buildings were replaced in the UK around 1962, in Australia and New Zealand production of the original series continued with the tools that they shared between them.

New Zealand

The 1966 New Zealand price list included the following models:

R60	Ticket Office
R61	Signal Box
R63	Central Platform Unit
R65	Platform Short Ramp
R67	Approach Steps
R68	Nameboards (one short and one long)
R80NZ	Complete Station comprising :

1 x R60
2 x R63
2 x R65
1 x R67
2 x R68

These remained available until the end of local production in the early 1970s. All were packaged in locally-made boxes.

Australia

In 1966, Moldex were advertising the following station parts made from the early tools:

R60	Ticket Office
R61	Signal box
R65	Platform Short Ramp
R67	Approach Steps
R68	Station Names Set (box of three sets)

They were packed in locally-made boxes and were not made after 1967.

Large stocks of Hornby-Dublo buildings had to be disposed of and featured heavily in catalogue illustrations of the mid 1960s.

STATION SETS (MAROON)

Small Station Set R458, R458A

The original R458 station set of 1962 had fewer parts and the 'A' suffix was added in 1964 when the set was enlarged. The suffix was retained until 1969 when it was dropped as it no longer served a purpose, all the early sets having been sold a long time before. It remained without the suffix until 1970, the last year it was listed.

R458A and later R458 contained the following:
1 ticket office (maroon)
1 station steps unit (fan type)
2 straight platform units
2 platform end ramps
1 wall unit
1 seat with nameboard
1 platform accessory set (double version)

It is estimated that 55,000 Small Station Sets were dispatched in the Tri-ang Hornby box with the number R458A and a further 20,600 were sold in the last two years with the 'A' suffix dropped. They were sold in the standard red AO size box, now with the Tri-ang Hornby logo.

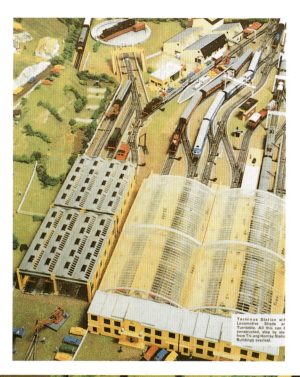

Large Station Set R459, R456A

The Large Station Set was made over the same period as the small one described above and also had an 'A' suffix, from 1964, which was dropped in 1969. The building was maroon and the walls and seat cream. Throughout this period the canopy supports were green. Cream canopy supports belong to an earlier period. The set contained the following:

1 ticket office (maroon)
1 station steps unit (fan type)
3 straight platform units
2 platform end ramps
2 straight canopy units
2 wall units
2 seats with nameboards
1 platform accessory set (double version)

It was not quite so popular as its smaller cousin; about 31,000 being sold in Tri-ang Hornby boxes

before the 'A' suffix was dropped and a further 12,000 afterwards. Despite being deleted in 1971, the Large Station Set has been known to turn up in a Hornby Railways box!

Island Platform Set R588

This was available between 1964 and 1969 and in 1965 was priced 16/11d. About 42,000 were dispatched in Tri-ang Hornby boxes. In 1970 it was replaced by the R688 which had different contents.

The set consisted of :
1 island waiting room (maroon)
1 straight platform unit
1 straight platform with subway entrance
2 platform end ramps
1 straight canopy unit
1 seat (and nameboard until 1966)

Island Platform Set R688

This was an island platform set that was made for 1970 and replaced the R588 set of 1964 as an enlarged and more exciting version. No doubt at the time it was planned, no one foresaw the decision to introduce a completely new range of station sets in 1971. A total of 2,500 sets were made, probably late in 1969, with only 100 being sold that year. The rest were sold in 1970 and there is no record of any being left in the stores at the end of that year.

The R688 set contained:
1 island waiting room (maroon)
2 straight platform units
2 straight canopy units
2 platform end ramps
2 seat units
1 platform accessories set (single)

Maroon station sets and parts typical of the mid 1960s.

Mainline Station Set R689

The R689 station set was also made for one year - 1970. It arrived in the shops in October and all but a handful were disposed of in the pre-Christmas sales. A total of 2,800 were made. The set was not a replacement for a previous one as it filled a gap between the single platform sets and the ex-Hornby-Dublo terminus and through station. It was a two platform set, a forerunner of the R4 Suburban Station of 1971, and contained:

> 1 ticket office (maroon)
> 1 station steps unit (fan type)
> 6 straight platform units
> 4 platform end ramps
> 3 straight canopy units
> 2 seat units
> 2 wall units
> 2 platform accessory packs (single)
> 1 footbridge

It was unusual as a station set in having a footbridge included. It had been intended that this should be the new R71A footbridge but due to problems with the tools for the new bridge the original R71 had to be used instead.

Mainline Station Set R689A

A circular to retailers in February 1971 listed a number of items which had been in the 1971 catalogue but which had subsequently been withdrawn. The list contained the R689A station set and yet I have been unable to find any reference to it in the 1971 catalogue or in the production records. Nor was the plain R689 set of the previous year made again.

I should mention that, while not in the catalogue, R689A was listed in the January 1971 price list and I have seen such a set in a plain red box with a Tri-ang Hornby label printed in red. Sadly I cannot remember in whose collection it was nor what the contents looked like. Logically the building should have been in the new red plastic which would have justified the 'A' suffix.

STATION SETS (RED)

Wayside Halt R1

The model, first introduced in January 1971 when over 13,000 were made, consisted of the following:

R458A Small station set.

R459A Large station set.

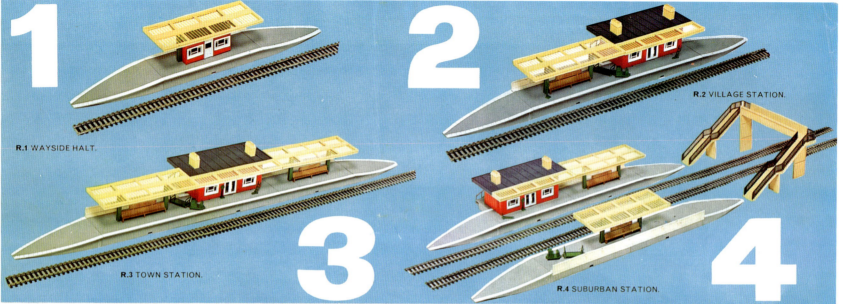

1971 station sets.

1

R.1 WAYSIDE HALT.

2

R.2 VILLAGE STATION.

3

R.3 TOWN STATION.

4

R.4 SUBURBAN STATION.

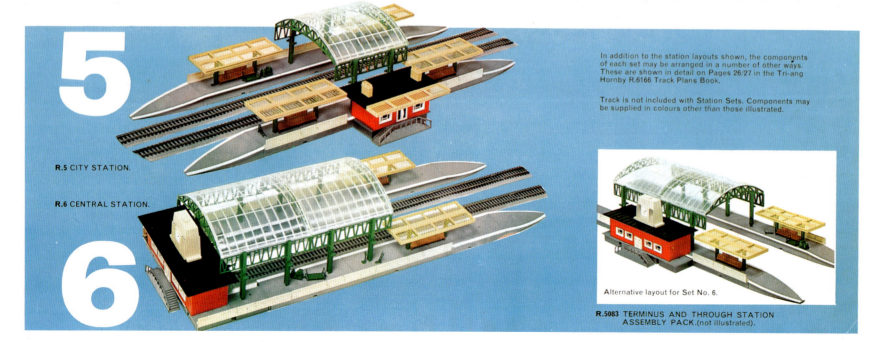

5

R.5 CITY STATION.

R.6 CENTRAL STATION.

6

In addition to the station layouts shown, the components of each set may be arranged in a number of other ways. These are shown in detail on Pages 26/27 in the Tri-ang Hornby R.6166 Track Plans Book.

Track is not included with Station Sets. Components may be supplied in colours other than those illustrated.

Alternative layout for Set No. 6.

R.5083 TERMINUS AND THROUGH STATION ASSEMBLY PACK.(not illustrated).

1 straight platform units
2 platform end ramps
1 island waiting room (red)
1 straight canopy unit

It replaced the R688 station set and was clearly a cost-cutting exercise. It provided the smallest station so far produced and reflected the demand by the public for cheaper models at a time of sudden inflation. It sold for 95p while the R688 had cost £1.25. The model remained available until 1977.

Village Station R2

This was a direct replacement for R458, the differences being the building colour which was now bright red, the wall unit which was off-white instead of cream, an added seat unit and a new steps unit with a handrail. The set contained:

1 ticket office (red)
1 steps unit (narrow railed)
2 straight platform units
2 platform end ramps
1 wall unit
1 seat and nameboard unit
1 platform accessory set
1 straight canopy unit

The set was available between January 1971 and 1977 and about 8,700 were made the first year and sold in Tri-ang Hornby boxes.

Town Station R3

This was also a replacement, in this case for R459. In addition to the change to bright red for the building and off-white for the wall units, it differed from its predecessor in having only one station accessory set instead of two. The contents were:

1 ticket office (red)
1 steps unit (narrow railed)
3 straight platform units
2 platform end ramps
2 straight canopy units
2 wall units

2 seats with nameboards
1 platform accessory set

It was available between January 1971 and the end of 1977. 1,000 of the 6,000 manufactured in 1971 remained unsold at the end of that year.

Suburban Station R4

There was no direct predecessor for this model but it was a shortened version of the R689 Mainline Station Set which was withdrawn at the end of 1970. The set contained:

1 ticket office (red)
1 station steps unit (narrow railed)
4 straight platform units
4 platform end ramps
2 straight canopy units
2 seat units
3 wall units
1 platform accessory pack
1 footbridge (new)

1971 Boxed station sets.

It arrived in the shops in March and a little over 4,800 were made in 1971. Like the others, described above, it remained in the catalogue for seven years.

City Station R5

The two larger station kits added to the range in March 1971 (R5 and R6) were really replacements for the former Hornby-Dublo terminus/through station sets (R5083A) which Rovex had continued to make up until 1969. The Hornby-Dublo station units were made on large heavy tools which were expensive to handle and so Rovex sought a cheaper way of providing a covered station. They therefore

tooled-up their own station roof unit during 1970 and used it in the R5 and R6 station sets.

Except for colours, the R5 was almost identical to the R689 but with a station roof unit added. The contents comprised:

1 ticket office (red)
1 station steps unit (narrow railed)
2 small platform units
6 straight platform units
4 platform end ramps
4 straight canopy units

Above: Miniville station made from Tri-ang TT parts.

R589 Ultra Modern station sets, made for Rovex by Spot-On using their Arkitex tools and showing two different types of box used.

1 station roof unit (with four supporting posts)
4 seat units with nameboards
5 wall units
1 platform accessory pack

In 1971, 4,700 City Station sets were sold in Tri-ang Hornby boxes and although production continued after the change to Hornby Railways, the set was withdrawn in 1974.

Central Station R6

The final station set introduced in 1971 contained two roof units and was intended to be assembled as a terminus station (only two platform end ramps being provided). It was the only set in the 1971 series to feature the new Terminus Building last seen in 1967 when it was the R474 Station Upper Floor with a clock tower. The set contained:

1 station terminus building (red)
1 station steps unit (narrow railed)
4 small platform units
6 straight platform units
2 platform end ramps
2 straight canopy units
2 station roof unit
2 seat units with nameboards
6 wall units
1 platform accessory pack

In 1971, 3,000 Central Station Sets were made. By the end of the year, 1,000 remained in the store. The set was priced at £5.50 but was withdrawn at the end of 1973.

STARTER SET STATION PARTS

Miniville

In 1967 an inexpensive station was required for the Miniville starter set W1 and for this the tools for the TT Island Platform were found and cleaned up.

The station consisted of:

   1 T20 Platform Unit
   1 T26 Platform Canopy Unit
   2 T39 Platform Ramps

The platforms were moulded in vermilion red and the canopy in yellow. As a variation for the W2 Miniville starter set, two of the T24 Nameboard with Seat Units in blue plastic were used instead of the T26. All parts were unprinted.

Canada

As Canada were assembling their own starter sets and required the cheap station, they ordered, and received, the parts for 2,500 stations which were manufactured at Margate in 1967 and bulk packed for dispatch to Canada. Their station was the version with the canopy unit. The parts were listed as T20C, T26C and T39C. It is not known if they were made in the same colours as those used for the British stations.

American Miniville Line

In 1969, ATT of America also required a cheap station for two of their sets to be marketed under the name Miniville Line. So that they could assemble these they were supplied with parts for 60,000 sets. Once again, it was the canopy rather than the name-board/seats unit that was sent. The parts were listed as T20NP, T26NP and T39NP and, again, I do not know what colours were used.

Toy Train Platform R398

This was probably the special platform unit, with built-in steps and a signal, which was planned for 1971 in the Take-a-Ticket set. It eventually arrived in 1972 and will be covered in **Volume 3**.

ARKITEX

Ultra Modern Station Set R589

This was an attempt to sell more Arkitex on the back

Extract from the 1966 catalogue depicting the exciting buildings that could be constructed with the R589 Ultra Modern station set.

R.589 Ultra Modern Station Construction Set. Contains over 550 parts with which you can build this 'over-the-line' Station. Construction is based on real principles, whereby you fit together the 'rolled steel joists' (no adhesives required) and then add the 'cladding'. All parts pre-coloured. Set can be taken down and used to construct other buildings below.

Railway Administration Tower. Another variation that can be built from Set **R.589**.

This Railway Design Office can be built from Set **R.589** above as an alternative to the 'over-the-line' Station.

of the successful railway system but it did not work. About 2,700 sets were made in 1966 and all but a little over 300 sold during the year. In 1967 another 900 sets were made and the residue of these sold the following year. It was not repeated after that. Some were sold in pictured boxes and others in plain red ones, with a Tri-ang Hornby label stuck on the end, but it is not known which type of box was used in which year. The set cost 49/6d which was a lot more than most young railway modellers could afford.

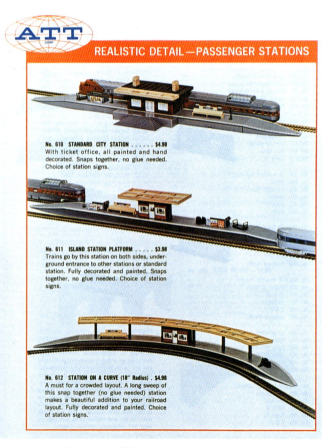

Stations in the ATT catalogue for sale in America.

The 1966 and 1967 catalogues illustrated three examples of buildings that could be made from the construction sets. These were described as a Railway Administration Tower, a Railway Design Office and an 'Over the Line' Station. This latter building was perhaps the most exciting but required a non-standard piece of wood which took the weight of the section over the tracks.

The construction method adopted by Arkitex involved a yellow plastic girder frame onto which the modular wall/window panels were hung and floor and roof panels were placed.

STATION PARTS

Straight Platform R460

The straight platform unit first appeared in 1962 with the introduction of Super 4 track and the new station system designed to go with it. It has remained with us ever since despite major changes to the building system over the years. The platform edges were always painted (later printed) white and the unit was 7 cms wide.

Millions of the unit have been made over the years. During the Tri-ang Hornby period alone there were 138,000 units sold solo and another 600,000 sold in station sets etc. Sold solo they were dispatched to shops six to an AO box. A CX size box may have been used in later years.

For American Station Sets R460T

The American Train & Track Corp. had selected the Tri-ang Hornby second series station units for sale through their own catalogue in the late 1960s and ordered quantities of various units in the range. From these they assembled their own station sets, #610, #611 and #612, which were different to those offered in the Tri-ang Hornby catalogue; indeed, #612 was a completely curved station.

In 1967, ATT ordered 5,000 Straight Platform Units and in 1969 for 30,000 more R460 units. Like the previous batch these were sent out to the States bulk packed and were used in the assembly of the #610 and #611 stations both of which contained three straight station units.

Straight Canopy Set R466

This was not sold as a solo item for general release after 1963 but will be found in **Volume 1**. It was used with the straight platform and seat to make R585 and was also used in many sets.

For American Station Sets R466T

In 1969 ATT of America received a supply of 28,800 R466T Straight Canopies bulk packed for use in the assembly of station sets for the American market. In previous years canopies had been supplied to ATT with platform units (*see* R585T below).

Straight Platform with Canopy R585

Introduced in 1964, the Straight Platform and Canopy outlived the two curved versions. It remained in the catalogue until 1977. In 1965, as a 'Platform Pack' and packaged in specially printed illustrated boxes, it retailed at 5/11d. Over the next seven years 65,000 were made, later ones arriving in the shops in standard 1967 DJ size boxes with printed labels stuck on the ends.

For American Station Sets R585T

5,000 Straight Platforms with Canopies were sent to ATT in America in 1967. These would have been for the #611 Island Station Set advertised in their own catalogue.

Platform with Subway R461

The Platform with Subway was made between 1963 and 1969 and 23,000 were sold during the Tri-ang Hornby era. It sold for 3/6d and like the straight platform unit, they were dispatched in boxes of six

Platform with subway.

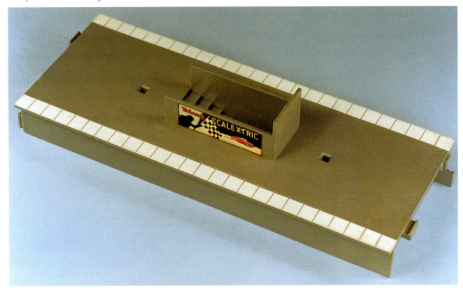

R585 straight platform with canopy and seat.

R584 and R585 platform packs.

Boxes of platform units: R460, R464 and R462.

and split up in the shop. It was also used in some station sets during the period. The advertisement transfers on the raised wall changed over the years. Although dropped in 1970, it returned many years later.

For American Station Sets R461T

The Platform with Subway Entrance was supplied to ATT in 1967 for the #610 and #611 station sets in the ATT catalogue. The former had one R461T and the latter had two. Records indicate that they were supplied with a straight platform canopy but this makes no sense as this would have left ATT with a large surplus of canopies once its quota of station sets had been assembled. 15,000 were supplied that year but none in subsequent years.

Pictures in the ATT catalogue show the unit with advertisements not seen in the UK and it is possible that these were stuck on to the units for the photographs or manufactured specially for the American sets to give a local touch. At a guess, these were printed by ATT rather than Rovex.

Curved Platform Large Radius R462

Originating in 1962, the curved platform units, which are also 7 cms wide, have survived into the 1990s and, at the time of writing, are still made. During the Tri-ang Hornby years, 44,000 of this unit were made. They were sent to shops packed six to a CX box and then sold loose.

The unit was designed to go outside the R486 second radius curve and was the equivalent length of a piece of track. Its own radius was 50.3 cms and it was 22.5° of a circle.

Curved Platform Set Large Radius R467

This was not sold as a solo item after 1963 but will be found in **Volume 1**.

Curved Platform Large Radius with Canopy R583

Made from 1964 the last batch should have been in

1969. Due to public demand, however, another batch was made in 1970 and lasted until 1972. The pack contained the large curved platform and matching canopy with support posts. It was packed in a specially printed illustrated box. Of the 21,000 made, it is probable that only about 3,500 were not dispatched in Tri-ang Hornby packaging.

For American Station Sets R583T
For the 5,000 #612 curved stations ATT intended to make for their own catalogue, they required and ordered 15,000 Large Radius Curved Platforms with Canopies. Records show that only 12,000 were dispatched to the States in 1967.

Curved Platform Small Radius R463
This has had a similar history to the previous item and about 47,000 were made during the period covered by this volume. It was designed to go on the inside of the first radius curve – R484. Its own radius is 30.5 cms. and, as it fits a standard curve, its length is the equivalent of 22.5° of a circle. They were sold six to a BL box which was later changed for a CX.

Curved Canopy Set Small Radius R468
This was not sold as a solo item after 1963 but will be found in **Volume 1**.

Curved Platform Small Radius with Canopy R584
This was available over the same period as the larger one (R583) but cost 4/11d. It is estimated that 16,000 were dispatched in Tri-ang Hornby packaging (pictorial boxes initially) and the last batch was made in May 1970 although it remained in the catalogue until 1972.

Double Curved Platform Ramp R464
Like the straight platform unit, this has been made in vast quantities between 1962 and the present day. About 157,000 were sold solo during the Tri-ang Hornby period and a further 435,000 in station sets.

The unit is 7 cms wide and 12.8 cms long and the radius of the curve was set so that it would fit inside the smallest radius curved track. From 1967, the ramps were sold to retailers in DC size boxes of six and then sold loose in the shop.

For American Station Sets R464T
This was required for all three station sets assembled for the American market by the American Train & Track Corporation. In 1967, 30,000 were dispatched to the States bulk packed. As no other batches were sent, this establishes the maximum number of Tri-ang Hornby station sets that could have been assembled in America as 15,000. Numbers of other parts sent indicate that it was proposed to assemble 5,000 of each of the three station sets.

Orange Ramps – *See* Miniville Station Parts below.

Platform Steps Unit R465
The Platform Steps Unit, which first appeared in 1962 as part of the new station system was not sold as a solo item after 1963 but a large number were made for station sets until it was replaced in 1971 with a new and smaller unit. The R465 was the version that had the steps fanning out in three directions.

Station Steps Unit R67
These were to be made for use in the 1971 series station sets but not initially for solo sales. It was the smallest step unit so far produced and included two narrow flights of steps with handrails. They took their number from the 1952 station steps unit and replaced the R465 fan-shaped steps of 1962 which were not used in station sets after 1970.

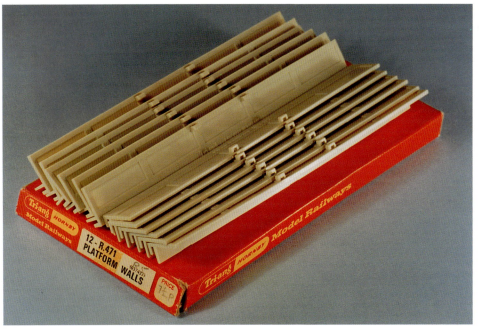

R471 pack of 12 platform walls.

About 48,000 were made in 1971 and, as we have seen, all were used in station sets and were not packaged for solo sales that year. The unit was replaced by yet another design, which had white detachable fencing, in 1978.

Platform Seat Unit R469

This was not sold as a solo item after 1963 but will be found in **Volume 1**. It was used in association with the R585 Straight Platform with Canopy and Seat and included on many of the sets during the Tri-ang Hornby era. It was cream until 1971 when it was finished in the 'teak' coloured plastic used for LNER coaches.

For American Station Sets

10,000 of these went to America in 1967 following an order placed by the American Train & Track Corporation. They were bulk packed for use in station sets for the ATT catalogue. They were fitted with nameboards but it is not known whether they were supplied with station name labels.

Nameboards R470

This was not sold as a solo item after 1963 but will be found in **Volume 1**. It was available, however, as a constituent part of a number of other models and sets including R586 and various station sets.

Platform Fencing Set R471

This, also, was not sold as a solo item after 1963 but will be found in **Volume 1**. Like the nameboards above, It was available as a constituent part of a number of other models and sets including R586 and a number of station sets.

Platform Fence and Nameboards R586

This consisted of two cream or off-white platform 'concrete' wall units, two cream-coloured nameboards and a set of three sets of four self-adhesive station name labels. The station names shown in the

catalogue up until 1965 were:

Bristol	(brown background)
Ramsgate	(green background)
Dundee	(blue background)

From 1966 the names illustrated in the catalogue, all with blue backgrounds, were:

Crewe
Ipswich
York

The pack was available between 1964 and 1967 in a BS box and about 9,000 were released in Tri-ang Hornby packaging. For the next three years there were no nameboards offered. In 1971, they were included in station sets but were now teak coloured and had no name labels.

Ticket Office R473

This was not sold as a solo item after 1963 and is found in **Volume 1**. It was available, however, as a constituent part of a number of other station units. From its introduction in 1962 until 1970 the body of the building was maroon but in 1971, along with the other buildings in the range, it changed to bright red.

The model made through the Tri-ang Hornby period was not fitted with an interior light. It had a black roof, cream canopy and white windows and doors. The maroon version had whited out windows but in the red version the 'glazing' was clear. During this period about 157,000 ticket offices were made in the maroon plastic and about 30,000 in red. Red ticket offices with black canopies were made

R582 ticket office platform unit pack.

during the Hornby Railways period and will appear in **Volume 3**.

Ticket Office and Platform Unit R582

This was an extension pack for station sets and contained:

1 ticket office
1 straight platform
1 platform steps (fan type)
1 double platform accessory set

The retail price was 15/6d and between 1965 and 1970, the last year of production, 36,000 were made.

The illustrated box of the Tri-ang Railways period was reprinted with the new Tri-ang Hornby logo. The picture showed a station scene centred on the elements in the set but seen on a real railway.

For American Station Sets R582T

In 1967, 6,900 Ticket Office Platform Packs were made for ATT against an order for 5,000. Why there was over-production is not known but the packs were required for the assembly of the #610 station set for the ATT catalogue. The buildings were almost certainly in the maroon plastic used for these units at that time. No more were ordered.

Ticket Office Platform Pack R582A

When the station range was revised in 1971 with the introduction of red plastic for the brickwork, the contents of the R582 pack were changed and it was given an 'A' suffix to distinguish it from old stock in the stores. The new pack was available from January and had the new, smaller, steps unit and the reduced platform accessories set, but two small platform units were added. After the revision of contents and colour, a further 4,800 passed through the stores. The price was now £1.

Island Platform Waiting Room R472

Although available as a solo model in 1965, it was withdrawn the following year and used only as part of a station set after this date. Initially it was maroon with white windows and door and a black roof. In 1971, the body of the building became bright red and, once again, it was available as a solo model for two years in an AC box.

The earlier maroon version had been used in two island platform sets – R588 and R688. Besides 4,000 solo models released in 1965, about 45,000 were sold in these sets. In 1971 about 21,000 of the red version were made of which 8,000 were sold as a solo building, the others being used for the R1 station set.

For American Station Sets R472T

This was required by ATT of America for assembly of their #611 and #612 station sets and 10,000 were bulk packed and shipped over to America in 1967. No more were supplied. These would have been in maroon plastic.

LARGE STATION BUILDING

Upper Floor and Clock Tower R474

The upper storey made its appearance in the 1965 catalogue and was maroon to match the other buildings at the time with a cream clock tower. It was designed to sit astride two ticket offices to produce a two-storey building. The model lasted only three years before being withdrawn. Only 4,300 were made in 1965 and this stock in ED boxes lasted two years. Another batch of 1,000 was made in 1967 and packed in DH size boxes after which it was dropped from the catalogue.

Terminus Building Kit R474

The upper floor building was re-released in January 1971 with a selection of other station parts as a conversion kit to extend one of the station sets released that year in the new red livery. Before being re-released it was modified to turn it into a free-standing Terminus Building. Besides giving it clear windows and some doors, the clock tower was now

R472 Island platform waiting room.

moulded in grey instead of cream plastic. It was sold in what remained of the standard DH box printed for the 1967 batch of the Upper Floor and Clock Tower (above).

The R474 kit contained:

1 terminus building
1 straight platform unit
4 short platforms
1 steps unit (narrow railed)
1 platform accessories set

During that first year 3,100 of these kits were made. The Terminus Building was also used in the R6 Central Station Set for which another 3,100 were required that year.

Platform Section R65

These were to be made for use in the 1971 series of

Station Sets but not initially for solo sales. They were almost square in shape and were unprinted. Their main function was to provide a base for the ticket office or terminus building and for this purpose they were clustered together in whatever number were required.

Station Roof Unit R74

As we have seen, the Hornby-Dublo roof unit had been adopted into the Tri-ang Hornby system in 1965 along with the rest of the Hornby-Dublo terminus station. It was made with heavy tools that were unpopular in the factory and so, for the introduction of the new series of station kits in 1971, a new roof unit was designed and tooled up.

The new unit, which was available from March 1971, had three advantages. It was both easier and cheaper to make and it was designed from the start as an extension of the Tri-ang 1962 series station system. Unlike the Hornby-Dublo version it could be used only in conjunction with platform units and so would not look right as a stand-alone train shed.

The two glazed sections for the roof were moulded in clear plastic with unpainted ribs representing glazing bars. The supporting posts (complete with their loudspeakers) and the side and end girders were moulded in green plastic. It is not thought to have been based on any prototype. The unit had a span of 7.5 inches and covered a double track and two half platform widths. It was six inches long, the same as the platform units. The reviewer in *Railway Modeller* saw good potential for the knife men.

During 1971, 7,000 solo units were packaged individually in CX boxes and 4,600 of these were sold during the year. A further 11,000 station roof units were used in the R5 and R6 station sets.

Platform Crane R475

This had been introduced in 1963 and used the crane from the R127 wagon mounted on a short platform section. It was last made in 1965 when 2,650 were produced.

R74 station roof unit introduced in 1971 to replace the former Hornby-Dublo station system.

R474 terminus building in red.

R65 was the only new platform section during the Tri-ang Hornby period.

PLATFORM ACCESSORIES AND FIGURES

Platform Accessories R148

The set contained a Planet platform tractor, a trolley, a sack barrow and a weighing machine. It was first introduced in 1961 and for a brief period it was made in olive green plastic before changing to the brighter SR green. The green grew a little brighter as the years went by. It remained in production for the next 30 or more years. From 1964 it contained two of each item and in the catalogue the following year it was shown as a solo item although not listed in the price list. This suggests that it was a late decision to drop it as a solo model.

From 1965 it was available only in station sets and in 1971 the platform accessory set was reduced to just one of each item. For much of the time the Planet tractor carried the Tri-ang name on the inside of the moulding and it is not known when this was removed.

For America R148T

In 1947 10,000 sets each containing one each of the four pieces were supplied to ATT of America for use with the station sets they were assembling from Tri-ang Hornby parts.

Station Accessories R834

I have no record of this accessory pack having been issued in Britain but it does appear as a product supplied to ATT of America. The number may have been adopted for the 'reduced size' accessory set.

For American Station Sets R834T

The accessory pack was supplied to ATT of America in 1968 when 5,000 were made. The pack appears to have consisted of the standard R148 Platform Accessories Pack (probably the single version) and an R469T Seat Unit and Nameboard although this is by no means certain.

Box of 6 sets of engine crew and a set of coach figures.

Set of 5 Train Figures R281

These came in a small cellophane pack, closed at the top with a folded Tri-ang Hornby card, six to an AA size box. They were ready-painted and consisted of a driver, a fireman, a motorman and two guards. All but the guards were seated. They were priced 2/6d in 1965 and were available between 1961 and 1970. About 32,000 sets were sold during the Tri-ang Hornby period.

Set of 5 Platform Figures R283

The figures included a guard waving a flag, a porter with rolled up sleeves carrying two cases and a man in a suit and hat leaning on an umbrella. The last two figures, a man and a woman, were seated. This set was packaged the same as the last and was the same price. It was available for the same period and it is estimated that about 38,000 were sold under the Tri-ang Hornby label.

Set of 5 Coach Figures R284

Only one of the figures, a waiter with a balanced tray, was standing or for that matter all there! The other four were only half figures cut off at the trunk so that they could be glued onto coach seats and be visible through the coach windows. The passengers consisted of two women and two men. The women clasped their hands in their non-existent laps and the men looked as though they are doing up their flies! Obviously they are just back from the end com-

R360 unpainted railway figures.

partment with the frosted window.

Packaging, availability and price was the same as for the last two sets of figures. It is probable that 28,000 sets were sold during the seven years covered by this volume.

Set of 2 Loco Crew R413

These two gentlemen were first introduced in 1962 and have become a familiar feature of the system. Today it seems that one cannot buy a Hornby Railways steam locomotive without receiving another pair of figures to go in the cab. Initially the 1962 pair were available with certain locomotives only until 1970 after which they had to be bought separately in the same way that the R284 coach figures could be bought to add detail to coaches. It was not until 1980 that they returned as part of the detailing of certain locomotives in the Hornby Railways range and then, unlike the 1960s version, they were unpainted.

The figures cost 1/- in 1965 and traders had to buy a minimum of two AA size boxes of six packets. Between 1965 and 1970 it is estimated that 53,000 packets were sold.

30 Unpainted Railway Figures Set R360

This was planned for release in January 1972 but was (initially anyway) issued in Tri-ang Hornby packaging and so is included here.

The figures were moulded in pink plastic and were made from the same moulds as the other sets listed here. In other words the set contained a mixture of railway personnel, loco crew, standing and sitting passengers and figures for coaches. They were sold unpainted, in polythene bags, still attached to their moulding web. The bag was closed with a Tri-ang Hornby card folded over the top and a printed label indicating the contents.

Retailers bought them in at 70p for six bags (minimum quantity) and retailed then at 21p each. 14,000 were made in 1972 of which 2,300 were carried over as stock into 1973 when a further 11,000 were made. By 1974 the set had been withdrawn.

ENGINE SHEDS

Modern Engine Shed R146

The Modern Engine Shed had replaced the earlier one in 1961. It was sold in a specially printed illus-trated box which showed a similar structure on a real railway with a steam engine pulling out of it amid a cloud of steam and smoke.

Like the signal box, the 1961 engine shed changed to red in 1971. While the roof remained black, the smoke vents were moulded in green plastic. The Modern Engine Shed remained in the range until 1977. About 38,000 of the earlier maroon version and 2,000 of the red left the factory in Tri-ang Hornby packaging.

SIGNAL BOXES

Modern Signal Box R145

The 1961 Modern Signal Box was to remain in production until 1977, but in 1971 the colour changed from maroon to bright red. During the Tri-ang Hornby period about 62,000 maroon models and 9,000 red ones were made. The change of colour was to make the signal box match the new station buildings. In its new colour scheme it had a green roof hatch and carried a transfer inside. It was sold in a DX size box.

Signal Accessory Pack R581

This set was made for two years only – 1964-65. Only about 1,300 were made in 1965 and these were probably packed in Tri-ang Railways boxes. The box was specially printed for the set and, as only about 5,300 sets were made, it is unlikely that there would have been more than one printing of the box. Details of contents are given in **Volume 1**.

SIGNALS

Single Signal R43H, R43D, R43HU, R43DU

The signals were supplied two to an AA box marked R43H or R43D depending on whether it contained two home (red) signals or two distant (yellow) ones. During the Tri-ang Hornby period they were available first with the metal post as introduced in 1955 and from 1968, with plastic posts. This later version remained in production until 1975.

Both types had a metal base to give them stability but from 1968 a plastic clip was used for fixing the base to the track. This led to the signal being given a 'U' suffix to its 'R' number in 1970. Thus the boxes were marked 'R43HU' or 'R43DU' indicating that the model was suitable for both Super 4 and System 6 track. In 1973, in order to simplify the numbering, the Home Signal became R043 and the Distant Signal R143.

During the period 1965 to 1967, 56,000 home and 45,200 distant signals were made, proving, as in the past, that the red signals sold better than the yellow ones. After the change to plastic posts in 1968, 60,000 home and 50,000 distant signals were made before the change to Hornby Railways.

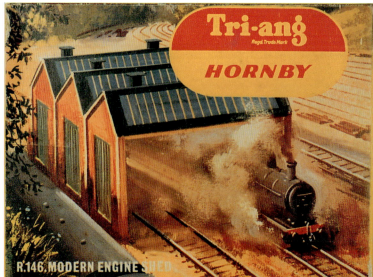

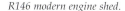

R146 modern engine shed.

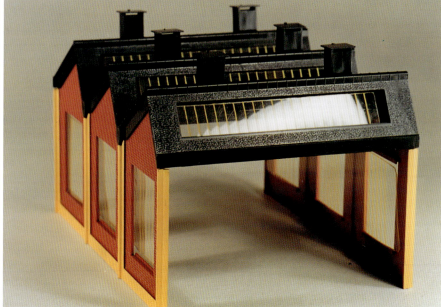

R145 modern signal boxes showing both brown and red versions.

Signal Gantry R140, R140U

The 1958 signal gantry was another model that survived virtually unchanged during the 21 years it was made. It spanned the three periods of production with about 46,000 being sold under the Tri-ang Hornby label. From 1970 until the end of 1972 the model carried the 'U' suffix (R140U) as an indication that it had clips for either Super 4 or System 6 track. Of the 46,000 produced during this period about 1,000 boxes carried the 'U' suffix.

It continued to be sold on an AT size box reprinted with the Tri-ang Hornby logo.

Signals

R.573 COLOUR LIGHT SIGNAL GANTRY 12/15 Volts A.C./D.C. **R.140** SIGNAL GANTRY **R.263L** POINT-OPERATED SIGNAL—LEFT Operation of point changes signal **R.263R** AS ABOVE BUT FOR RIGHT HAND POINT **RT.405** COLOUR LIGHT SIGNAL 12/15 Volts A.C./D.C. **R.142H** JUNCTION SIGNAL—HOME **R.43** HOME SIGNAL or **R.43** DISTANT SIGNAL **R.142D** JUNCTION SIGNAL—DISTANT.

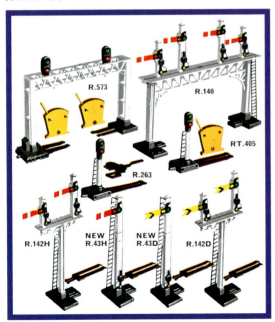

Signals - extract from the 1968 catalogue.

Junction Signal R142H , R142D, R142HU, R142DU

There were both home and distant versions of this double signal. As in the case of the single signals this was indicated on the box with a suffix to the 'R' number. Thus they carried the numbers 'R142H' and 'R142D'. A 'U' suffix was added in 1970 which was dropped in 1973 when, in order to simplify the numbering, the Home Junction Signal became R042 and the Distant Junction Signal R142. It is interesting to note that at this late time, more than a year after the change to Hornby Railways, the signals were still being packed in Tri-ang Hornby AC boxes.

The junction signals had been introduced in 1958 and, unlike the single signals, they had plastic posts from the start. They were to remain in the catalogue until 1975. During the Tri-ang Hornby period about 81,000 home and 58,000 distant signals were made. The peak year for production was 1969.

OTHER LINESIDE EQUIPMENT

Lineside Huts R84

The set of three huts of different sizes had first appeared in the Tri-ang Railways range as early as 1955 and remained in production throughout the Tri-ang Hornby period. Production stopped in 1970 but they were listed for a further year in order to clear the stocks being held in the stores. It is estimated that 44,000 of the sets were sold in Tri-ang Hornby boxes, the AA size being used. The mouldings were made in both black and grey plastic and then mixed so that black buildings had grey roofs and vice-versa. The doors were painted green. The set of cream huts was made much later on in the Hornby Railways period.

Water Crane R88

This had also been introduced in 1955 and survived until 1970 after which production stopped although it remained in the catalogue (not illustrated) one more year in order to clear stocks. During the

Tri-ang Hornby years about 42,000 were sold. Throughout its life it remained black and consisted principally of metal castings. They were sold to retailers in boxes of three.

Cattle Crossing R149

The last batch of these was made in 1965 but the stock was not cleared until 1967. It is not known whether any were sold in Tri-ang Hornby packaging. Late batches would have been in white plastic.

Gradient Posts R173

From 1964 these were included in the R587 Trackside Accessory Pack.

Field Fencing Set R186

During 1965, 4,300 of these field fencing sets, in white plastic, left the Margate stores of which 2,800 had been made that year. The late batches were probably dispatched in Tri-ang Hornby AA boxes.

Pedestrian Crossing Set R187

During 1965, 3,000 sets left the stores and a further 600 in 1966. The remaining 170 were disposed of the following year. This set was also in white plastic. The last ones made were in Tri-ang Hornby AA boxes.

Trackside Fencing R299, R299U

These were sold in a long thin envelope each containing two 1ft long pieces of white plastic fencing and four clips with which to fix it to the track. As it was a fixture, it was one of the range that assumed a 'U' suffix in 1970 to become R299U. It reverted to R299 in 1973.

It retailed at 1/9d a pack in 1965 and dealers had to buy a minimum of six packs. It was available between 1963 and 1987 and about 122,000 packs were sold during the Tri-ang Hornby years.

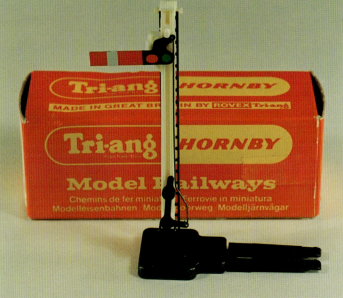

R140U signal gantry and R43HU boxed signal.

R187 crossing and fencing sets were not made after 1965 and yet here is one in a Tri-ang Hornby box.

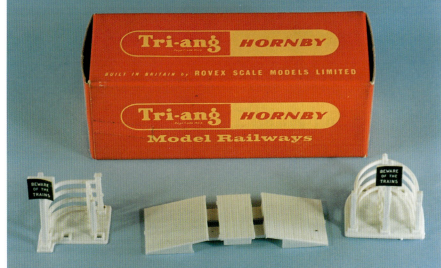

Left: R84 lineside huts etc.

R587 trackside accessory pack.

Telegraph Poles R478

The set of six plastic telegraph poles, sold in an AC box, was made between 1962 and 1965 with 3,000 being made in the final year. These may have been released in Tri-ang Hornby packaging.

Loading Gauge R479

The plastic loading gauge introduced in 1962 was available until 1969. They were packed in boxes of six and retailed at 1/3d each. The foot of the post fitted into a clip with which the loading gauge could be fixed to the track. Over the years covered by this volume it is thought likely that about 40,000 were made.

Trackside Accessory Pack R587

In 1964 a number of small trackside models were brought together and offered as a pack, priced 2/11d. These included five mile posts, four gradient posts and two whistle signs. Although fully illustrated in the 1965 and 1969 catalogues, the catalogue illustrations in the intermediate years persistently missed out the Whistle signs while listing them in the contents. During the Tri-ang Hornby years 14,700 were made and sold in the red and yellow CU boxes. It was last made in 1969 which was also the last year that the pack was listed.

BRIDGES

Footbridge R71

The 1953 model remained in production until 1969 with hardly any alterations in its design. The bridge continued to be sold in standard AT size boxes now carrying the Tri-ang Hornby logo. It was included in a supplementary catalogue page issued to retailers in 1970 designed at clearing outstanding obsolete stock.

Between 1965 and its withdrawal about 66,000 were made. Records indicate a further 10,000 being made in 1970, possibly because the new bridge

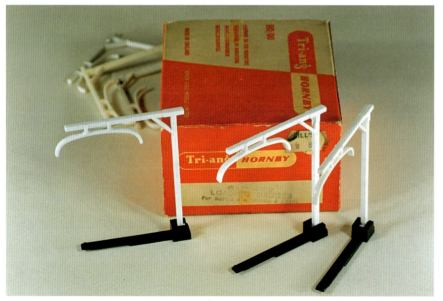

Box of six R479 loading gauges.

R71 footbridge.

(R71A) was not yet in production. Transfers used on late examples included:
Ever Ready Batteries
Bluecol Anti-freeze
KLG Sparking Plugs

New Zealand R71(NZ)
The R71 footbridge continued to be manufactured in New Zealand until the factory was closed down in the early 1970s. New Zealand versions of the bridge are described in **Volume 1**.

Australia R71A
Moldex probably continued to make the footbridge in Melbourne until 1968. Certainly the 1968 price list recorded the number with an 'A' suffix which was an indication of home production.

Double Track Footbridge R71A
The modern style footbridge was due for release in August 1970 but due to problems with the moulds it was not available until the following year, although

R71A double track footbridge as bought and assembled.

New Zealand made R71 footbridge with emergency packaging.

a few were probably made in December. It replaced the 1953 design (R71) and was based on a modern concrete design with concrete slab uprights instead of towers. It looked very much like a footbridge at York which was illustrated in the *Railway Modeller* in February 1964.

R78C girder bridge with plain boss and R78c brown Canadian girder bridge.

It was made in an off-white plastic with black plastic railings clipped into the sides of the steps and the whole came in a simple kit form which required little skill to assemble either with the steps the same side or facing in opposite directions. It was, in my view, one of the ugliest subjects modelled by Rovex!

Initially the model was given an 'A' suffix to distinguish it from the earlier model but this was later dropped. In 1970 and 1971 about 20,000 of the new bridge were made and it remained in the catalogue until 1979 after which it was replaced by the much more attractive R187 GWR Footbridge.

Both R71 footbridges had their height set so that they could stand on the baseboard rather than station platforms. The new bridge was packed in a standard AT size box of 1967 vintage with a printed label stuck on one end panel giving the contents.

R71A was reviewed in the *Model Railway Constructor* in May 1971. This was complementary about the kit both from the point of view of its realism and usefulness to the knife men who would want to convert it. The reviewer did, however, criticise the height of the structure which gave about 7 scale feet clearance above the train, a margin unusual on the real railways. It suggested that if used on a platform, when the clearance would increase to 12 scale feet, the lower section of steps could be left off, leaving a 3'9" clearance which would be fine if catenary was not in use.

Girder Bridge Supports R77

These were the grey girder bridge supports which were sold in pairs, separate from the bridge, until 1968. They had been made since 1954 and from 1965 until the end of 1968 about 20,000 pairs were made. They were packed three sets to a BL size box and were split up in the shop and sold unboxed unless you were lucky enough to buy the last pair.

Girder Bridge R78

This was the 1954 model which was still selling well

and, as we have seen, remained a separate model without the bridge supports until 1968. It is estimated that 21,000 were sold in Tri-ang Hornby BM boxes.

In 1971 4,000 girder bridges were used to make signal gantries for the Tri-ang Big-Big O gauge system. It was probably about this time that the Rovex name was removed from the boss in the centre of each side of the bridge.

Girder Bridge and Supports R78C

The bridge and supports were sold as one in a single box. Some were printed with the contents in red on the end panel and some had a blank panel with a self-adhesive label. During the period 1969 to 1971, 12,400 were made. The bridge was to have been replaced in 1971 by the multipurpose R657 girder bridge but there were delays in its development and so the R78C remained in the range until the end of 1972. To meet demand a further 3,000 were made that year.

Canadian Girder Bridge R78c

Canada was the only market to receive a batch of girder bridges made in maroon instead of grey plastic. A degree of mystery surrounds the release of this model as there is no record of it having been produced. Records suggest that any other girder bridges dispatched to Canada were standard grey issue and only the 1973 Canadian catalogue showed it as maroon. We know that maroon bridges, complete with maroon supports were sold in Canada and several boxed ones have made their way back to the UK but without any record of production we do not know how common it is. It is possible that some, or all, of those standard grey girder bridges made in 1972 were incorrectly recorded. After all the grey and maroon versions had the same 'R' number.

The bridges were sent out ready packaged in the same Tri-ang Hornby boxes as the grey ones sold in the UK. They had 'R.78c GIRDER BRIDGE & SUP-

PORTS' printed in red on the white panels on the box ends. This appears to have been printed on at the same time that the box was printed. An earlier UK sample in the same type of box, but with the High Level & Inclined Track Equipment instruction leaflet HLIT5/33929/1 (which was printed in 1965), had blank panels on the box lid ends to which printed labels, with the same inscription, were attached.

Both examples of the maroon girder bridge in my own collection have a plain boss on their sides. The boxed example, which was sent to me from Canada by the person who had bought it as old shop stock when the local model shop closed down, contains a small assembly instruction slip numbered 04/108/2. The '108' meant that it was the 108th printing order placed since the reformed company came into being and, based on other numbers of known date, this was almost certainly printed in 1973. While the box carried the Tri-ang Hornby trademark, the slip carried the name Hornby Railways.

Viaduct R180

In 1965 the three-arch viaduct was maroon. It had been introduced to the range in 1961 and was to survive well into the 1990s. During the Tri-ang Hornby era about 14,000 were made. The bright red version of the model does not date from 1971 like the station buildings but dates from 1973 and so lies outside the scope of this volume.

After a spell in special illustrated boxes, by 1968 the viaduct was being sold in standard DH size packaging.

American Viaducts R180T

In 1967 5,000 viaducts were bulk packed and sent to the United States to meet an order by American Train & Track Corp. These were boxed up in their own packaging on arrival and sold through the ATT catalogue under the catalogue number '601'. A further 1,000 were sent in 1968. Remnants of these orders were being sold off cheap in America in 1971.

It is assumed that the models supplied were in the standard maroon sold in the UK at that time.

River Bridge R188

Like the Viaduct, the grey River Bridge was added to the range in 1961. It was to survive only until 1971 and about 19,000 would have been dispatched in Tri-ang Hornby boxes. It was sold in the DH size box with early ones being specially printed pictorial boxes.

Versions of the model have been found in three different shades of plastic. These are olive green, light grey and platform grey. The last batch was made in 1970 and it had been planned to withdraw it at the end of the year. However, in order to clear remaining stocks it remained in the catalogue for 1971.

R180 viaduct.

Boxed river bridge made for ATT of America.

American River Bridge R188T

The R188 River Bridge was another model ordered by American Train and Track for sale through their own catalogue in the States. 5,000 in 1967 and a further 1,000 a year later were sent bulk packed for packaging in America. They were allocated the catalogue number '600'. It is assumed that they were made in grey plastic.

Brick Bridge R189

By 1965, this was being made in salmon plastic and this was the colour it was to retain throughout the Tri-ang Hornby period when about 33,000 were made. Early examples were supplied with graffiti transfers that could be placed on the model wherever the customer wished. These seem to have been dropped from 1966 onwards. The Brick Bridge remained in production well into the 1990s.

Victorian Suspension Bridge R264

This dates from 1963 and survived until 1971. It was to return again in the 1980s but that will be covered in **Volume 3**.

The model underwent a number of colour changes but that illustrated in the 1965 catalogue had a platform grey trackbed and salmon cables. In 1966 it was shown with a salmon trackbed and silver cables but by 1968 the colour of the trackbed had

Boxed R265 Victorian suspension bridge with a picture from the 1969 catalogue and an extract from the Hornby-Acho catalogue.

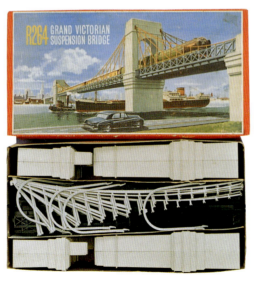

R189 brick bridge with graffiti.

R457 Super 4 inclines and R453 high level piers.

changed again to black. Throughout its life the towers remained off-white.

In 1966/67 4,200 of the bridges were made with the salmon trackbed and between 1968 and 1971 about 7,000 were made in the revised colour scheme. They were sold in a specially printed pictorial box.

There had originally been an extension set (R269) but this was dropped after 1965 and so will not have appeared in any colour scheme other than the first.

American Suspension Bridge R264T

Supplies of the Victorian Suspension Bridge were sent to ATT of America in 1967, 1968 and 1969 in the following quantities respectively: 2,500, 2,800 and 1,500. The 1969 order had been for 2,000 but for some reason 500 were held back and disposed of elsewhere in 1970. The bridge was allocated the ATT catalogue number '602'.

Girder Bridge Set – Super 4 R432

The set which is described in **Volume 1** was available between 1962 and 1965. 3,000 were made in 1965 and it is not known what packaging was used or whether they were made before or after the 1st May.

Canada

In 1968 there was a record of 500 boxed R432 girder bridge sets being sent out to Canada. The fact that they were separately listed in the records indicates that they were assembled specially for the order. Old stock also seems unlikely as only 280 of the 1965 production remained in the stores at the end of the year. It is however possible that the boxes were from unused stock; the contents were all still available as solo models.

Girder Bridge R657

This was planned for the 1971 catalogue but did not materialise until 1973.

INCLINES

Curved Sidewalls Small Radius - Super 4 R430

The salmon coloured incline system had been introduced in 1962 to replace the earlier grey system and to be suitable for use with the new Super 4 track launched that year. The curved walls for inclines came in pairs, with a centre strip to go between the rails similar to that provided with level crossings. They were packed in AE size boxes with three sets per box and retailers had to buy a minimum of four boxes. By 1965 they cost 2/11d a set and they remained available until 1969. During this period, 24,000 sets were sold. A set consisted of a wall for the inside of a curve and one for the outside, together with the centre strip. They had tongues which slipped between the sleepers to hold them in place.

Curved Sidewalls Large Radius – Super 4 R431

These were similar to the above but fitted the larger radius curves. They also survived until 1969 with probably 23,000 sets being sold during the Tri-ang Hornby period.

Straight Sidewalls – Super 4 R456

These matched the curved wall described above in style and retailing and were available for the same period and at the same price. Over the seven years

System 6 incline kit.

of this volume, 36,000 sets were sold. They were dispatched from the factory in boxes of three sets and each set was enough for one ordinary Super 4 Straight rail. The AE size boxes used were the same as those used earlier but the Tri-ang Hornby logo had replaced the earlier Tri-ang Railways one.

Inclined Piers – Super 4 R457

The salmon-coloured set of seven Super 4 incline piers dates from 1962 when it replaced the grey set of six (R79) designed for Standard and Series 3 track. There was a proposal to drop the R457 set at the end of 1970 but it remained available until the end of 1972. Over the years R457 was very popular and 156,000 sets were sold between 1965 and 1971. Later sets were packed in 1967 standard AV size boxes.

High Level Pier – Super 4 R453

These were supplied to retailers six to a DU size box but sold singly at a cost of 1/9d each. They were available between 1962 and 1972 and approximately 215,000 were sold during the Tri-ang Hornby period. It was suggested that they were dropped from the 1971 catalogue but they remained available into

1972. They were designed as an extension to the set of incline piers, being made in the same style and colour but one size larger.

Inclined Piers – System 6 R654

The new set, introduced around April 1970 for the new System 6 track, had seven piers. By May, tool hardening problems were being experienced, but they were back in production again in June.

These were of a completely different design to earlier ones being in kit form with assembly instructions. The kit contained a number of small units that plugged together, each section taking the track a little higher. The shortest pier had no sections at all and the tallest had six units plugged together on each side.

They were made in a pale grey plastic and had no accompanying side walls. It was this lack of walls that made this latest system look inadequate and cheap but despite this they remained the only set of incline piers for many years. The set was priced at 8/6d and during 1970 and 1971, 20,000 sets were made.

3 High Level Piers – System 6 R655

These were released at the same time as the above, in a set of three, priced at 7/6d. When assembled they were one unit taller than the tallest of the incline piers and, like the R654 set, they remained available until 1977. During 1970 and 1971 about 23,000 sets were made and were sold in DC boxes.

LEVEL CROSSINGS

Toy Train Level Crossing R397

This was made in red plastic for use in train sets from 1971. A grey version is also supposed to exist but no record of its production has been found.

Double Curve Level Crossing – Super 4 R414

This feature was released in 1965 and survived only to 1969. The idea was that it could be used on a

corner of a layout, where there is often wasted space, and was designed to fit the first two radii curves of Super 4 track. The largest radius was not yet available. It was priced 7/6d and about 9,000 were made.

It was of a similar design to other level crossings of the time with two grey road sections that carried the gates and clipped into the Super 4 sleepers. There were also matching strips to go between the rails and a section to go between the tracks. The gates were double and each was longer than those used on the straight double-track crossing as the curved track made it necessary to set them further apart in order to clear the trains passing through.

The crossing was packed in a standard DJ box of 1965/66 vintage.

Level Crossing Single Track – Super 4 R495

Priced at 6/11d, the Super 4 Single Track Level Crossing was made between 1962 and 1972. The model had two road sections each carrying a large single gate and posts, and a strip was provided that clipped in between the rails. The hinge post of the gate was tall and provided an anchor point for the straining wire which, on the prototype, took the weight of the gate and helped to prevent it from sagging. About 48,000 were sold in Tri-ang Hornby packaging. The box was initially a DX and the set came with an instruction sheet which it shared with R496 below. A brown base version was made for use in starter sets.

Australia R495A

This was a model that was made in the UK and sent out to Australia bulk packed for use in sets. Moldex then packed surplus level crossings in their own MXT21 type boxes for sale locally.

Blue Black Base

The road section of the Single Track Level Crossing were grey but one sold in Canada in a Tri-ang Hornby box had a blue black base.

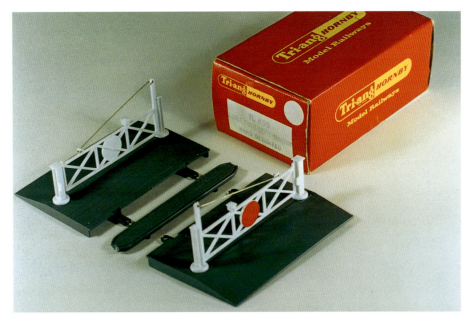

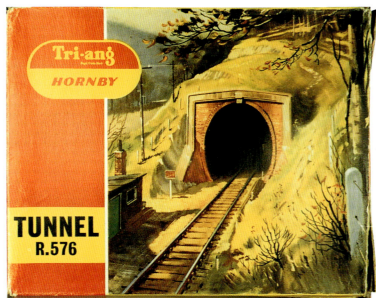

R576 tunnel.

R495 and R496 level crossings.

Level Crossing Double Track – Super 4 R496

The Super 4 Double Track Level Crossing was introduced at the same time as the single-track version described above and was of similar design but with double gates and the extra road section to go between the tracks. The gates, besides being four and individually smaller, did not have straining wires. It retailed at 8/6d in 1965 from which time it is estimated that 35,500 were made. The last stocks ran out at the end of February 1970 after which no more were made.

Initially, the box used was a standard BA as used for points but later a 1967 DJ size box was used.

Single Track Level Crossing – System 6 R629

This System 6 version of the Single Track Level Crossing was released in 1970 and remained in the catalogue until 1979. Records show that the first batch of 1,200 was made in 1969 but were not sold until 1970 when 11,000 were made and a further 11,000 in 1971.

The R629 level crossing was very similar to the Super 4 R495 crossing which remained available for a while. The main change was in the thickness of the road sections and the spacing of the locating clips.

TUNNELS

Tunnel R576

The Tunnel was added to the range in 1964 and has survived to the present day although its appearance has changed from time to time along the way. The plastic used for the moulding was often from reground surplus mouldings of various colours. The first change, to red plastic, occurred in 1972 and is outside the scope of this volume.

The model was made up of three mouldings, two brick portals and the over hill. The shades used for the patchwork of vegetation for the latter varied over the years. An ochre-coloured version belongs to the Thomas series and will be found in **Volume 3**.

The R576 Tunnel sold for 11/6d in 1965 and over the next seven years about 114,000 were made. When originally introduced it had been sold in a DT size Tri-ang Railways box but when this was reprinted for Tri-ang Hornby, it joined the attractive range of red and yellow illustrated boxes, with an illustration of a grassy tunnel mouth on a real railway. The box dates from 1965 or 1966.

TRACK BED

Track Bed R656

In 1970 there was some discussion of producing a moulded track bed but in June it was decided to defer it until 1972.

FORMER MINIC MOTORWAY BUILDING

Freight Depot R2008

This was the former Minic Motorways Freight Depot M2008 which was added to the Tri-ang Hornby range in 1970 when 3,550 were made with

R2008 freight depot building and Minic Car Play version.

Catalogue picture showing Hornby-Dublo buildings absorbed into the Tri-ang Hornby system.

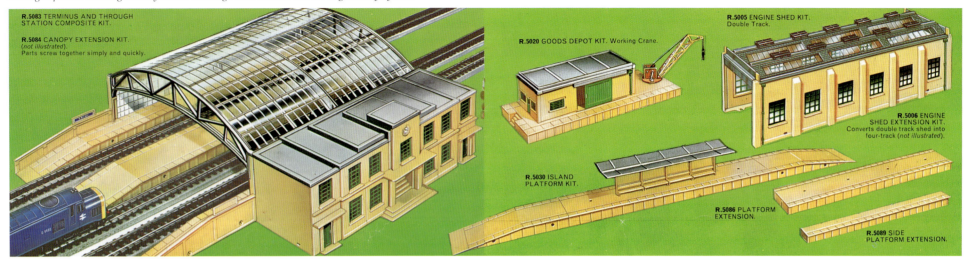

R.5083 TERMINUS AND THROUGH STATION COMPOSITE KIT.

R.5084 CANOPY EXTENSION KIT. *(not illustrated).* Parts screw together simply and quickly.

R.5005 ENGINE SHED KIT. Double Track.

R.5020 GOODS DEPOT KIT. Working Crane.

R.5006 ENGINE SHED EXTENSION KIT. Converts double track shed into four-track *(not illustrated).*

R.5030 ISLAND PLATFORM KIT.

R.5086 PLATFORM EXTENSION.

R.5089 SIDE PLATFORM EXTENSION.

the number 'R2008'. It was described in the catalogue that year as a 'Freight Depot complete with loading bank'. It differed from the Minic model in having a maroon body and black roof. With its grey platform and white windows it matched the colour scheme for the pre-1971 Tri-ang Hornby stations.

In the 1971 catalogue it was shown with red plastic walls and was described as a 'Container Freight Manager's Office and Loading Bay'. A stock of 400 maroon versions was carried over from 1970 and a

further 1,400, now presumably in red, were made. No more were produced after this.

Further information about the model may be found in the section on Minic in the 'Related Series' chapter, which will be found towards the end of this volume.

FORMER HORNBY-DUBLO BUILDINGS

At the time of the Meccano take-over, an effort was made to present it as a merger of Hornby-Dublo and

Tri-ang Railways to avoid alienating Dublo enthusiasts and the model press but it was soon clear that the only items which might be suitable to make at Margate were the buildings considered to be superior to the Tri-ang ones. Consequently they were added to the 1966 Tri-ang Hornby catalogue.

This was fine until the moulds arrived in Margate and were found to be difficult to handle and expensive to operate. No one wanted anything to do with them. As surviving production records are of what

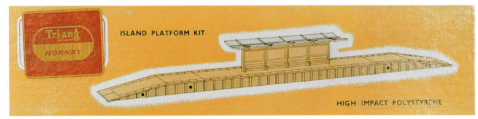

ISLAND PLATFORM KIT

HIGH IMPACT POLYSTYRENE

Hornby-Dublo boxed island platform kit carrying a Tri-ang Hornby sticker and the Hornby-Dublo island platform kit packaged in a Tri-ang Hornby box.

passed through the stores and not what came off the machines it is not certain how many of the former Hornby-Dublo models listed were produced at Margate and how many were transported there from the huge unsold stocks at Meccano Ltd. The only ones we can be sure were made in the Margate factory were the terminus/through stations in maroon plastic.

Two Road Engine Shed R5005

This was a former Hornby-Dublo model of which 2,000 unsold kits were inherited by Rovex in 1965. It was illustrated in the catalogue between 1966 and 1969 and during this period 4,500 left the Margate stores at a rate of about 1,000 a year. The only examples I have seen have been in Hornby-Dublo boxes, with or without a Tri-ang Hornby label attached. To the best of my knowledge all models were in the original cream colour scheme. It was priced at 19/6d.

Engine Shed Extension Kit R5006

About 1,250 of these were issued from Margate between 1966 and 1969, with the exception of 1967 when none were released. The kit cost 14/3d.

Footbridge R5010

In 1967 there were 200 of these former Hornby-Dublo footbridges in the Rovex stores. No other records have been found and it almost certain it was not made there.

Girder Bridge R5015

The only record of the Hornby-Dublo plastic girder bridge going through the Margate stores was in 1966 when just 78 were recorded. Despite this it was illustrated in the Tri-ang Hornby catalogue in 1966 and 1967 and was offered in the price lists at 11/9d. It is known that some went to Australia and that it was not adopted for production at Margate.

Goods Depot Kit R5020

This was one of the former Hornby-Dublo models absorbed into the Tri-ang Hornby range supposedly to be produced at Margate. It appeared in the Tri-ang Hornby catalogue in 1966, 1967 and 1968 and during this period 2,200 passed through the stores. It retailed at £1-9s-0d.

Island Platform Kit R5030

Initially the surplus stock of island platform kits from the Meccano factory at Liverpool were sold to the trade with a Tri-ang Hornby label stuck on the picture label. Later the kit was sold in Tri-ang Hornby boxes with Tri-ang Hornby instructions. It was catalogued between 1966 and 1968 and approximately 4,000 were released during this period. The set consisted of the long platform unit with two end ramps and the island shelter. It was priced at 11/3d.

Terminus Through Station Kit R5083

The ex-Hornby-Dublo Terminus/Through Station was the most famous of the range to be absorbed, as it was the only one to be produced in the colours of the standard Tri-ang range of station buildings. Priced £2-19s-11d, it was illustrated in the Tri-ang Hornby Catalogue between 1966 and 1968 in its original cream colour scheme and in this form 2,360 were released.

The set consisted of:
1 main station building
1 canopy
2 full platform units
1 half width platform unit
4 full ramps
2 half width ramps
4 ramp fences
2 small pieces of platform

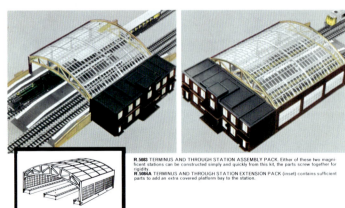

R.5083 TERMINUS AND THROUGH STATION ASSEMBLY PACK. Either of these two magnificent stations can be constructed simply and quickly from this kit, the parts screw together for rigidity.
R.5084A TERMINUS AND THROUGH STATION EXTENSION PACK (inset) contains sufficient parts to add an extra covered platform bay to the station.

Former Hornby-Dublo terminus station made at Margate in Tri-ang Hornby maroon, cream and grey..

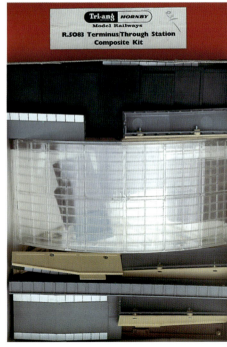

Boxed R5083A terminus through station kit.

The author's much treasured 10 unit maroon station complex.

Maroon Version R5083A

In 1969 and 1970 it appeared in the catalogue in maroon plastic, with white windows and doors, grey platforms and cream roof girders and priced £3-15s-11d. It was further listed, although not illustrated, in the 1971 catalogue in order to clear all remaining stock. Some 3,000 were made in this colour scheme and these were sold in either plain white or plain red boxes with a printed label stuck on the end.

Frog Airfield Buildings

It had been planned in 1970 to add a number of airfield buildings to the Frog model aircraft range which were now being made at Margate. These plans included using the former Hornby-Dublo station roof kit as a hanger (F24) and the station building as a headquarters building (F26). The control tower (F25) was to have been based on the Tri-ang TT railway signal box.

At the time, Rovex Industries were looking to strengthen the Frog range but like many projects explored over the years this one did not end in production.

Canopy Extension Kit for R5083 R5084

While advertised in the 1966 and 1967 Tri-ang Hornby catalogues, none were released until 1968 and then only a little under 1,000 kits. It sold for £1-2s-0d.

Maroon Version R5084A

This was made available in 1969 and 1970 as an extension set for R5083 similar to the ex-Hornby-Dublo R5084 but in the new maroon colour scheme. It contained a canopy, a standard platform unit and two half-width platforms. Only one batch of 1,000 was made in 1969. All but 140 sold that year and the rest had gone by the end of February 1970. The retail price was £1-9s-11d.

Platform Extension R5086

The Platform Extension was a standard platform unit in cream plastic. A batch of 1,600 in 1966 took three years to clear the stores. It retailed at 2/11d a unit and was supplied to shops in boxes of six.

Platform Fence R5087

This was a cream fence unit to go with the standard ex-Hornby-Dublo platform unit described above. It retailed at 1/7d and was supplied to shops in boxes of six. It was available in 1966 and 1967 when about 1,500 sold. It was made in cream plastic to match the platform to which it was bolted with Meccano nuts and bolts.

Side Platform Extension for R5083 R5089

This was the same as R5086 but was half the width. It cost 2/6d and was available between 1966 and 1968, about 2,800 being released in three batches, one each year.

Double Track Tunnel R5092

This was illustrated in the 1966 and 1967 Tri-ang Hornby catalogue after which it was dropped. It complimented the Tri-ang single track tunnel and was priced £1-5s-6d. 2,400 were available in 1966 and a further 500 were added in 1967.

A studio picture showing, inset, three publications. It seems that the Wiring Plans book was abandoned before it reached the printer.

Behind any comprehensive toy system there is bound to be a plethora of associated paraphernalia and ephemera to be collected. These include sundries, spares, sales aids and a considerable quantity of paperwork associated with selling the product. In this chapter we look at these miscellaneous subjects and how they developed over the Tri-ang Hornby period.

For ease of access the chapter has been divided into four subjects:

Sundries
Servicing
Marketing
Packaging

SUNDRIES

Just five of the former Tri-ang Railways range of sundries were adopted into the Tri-ang Hornby range. The five were as follows:

R166   Super 4 Track Plan Book
R207   Packet of Track Pins
RT297   Power Cleaning Brush
RT395   Railway Record
RT521   Capsule of Smoke Oil

Except for the record, these were available throughout the Tri-ang Hornby period. Details of the Track Plan Book are to be found in the chapter on 'Track'.

Two new publications, the R282 Operating Manual and the R701 Tri-ang Hornby Book of Trains, were added in 1969. 1971 saw an expansion of the range of sundries with the introduction of

System 6 track. One of the most common suggestions made in the 1970 customer survey was the need for scale wheels on rolling stock. To allow modellers to convert their rolling stock, packets of wheels were available. The range also included couplings, fishplates and brushes.

The new 16.5mm gauge wheels, which could be bought in packs of four sets for 2/6d from January 1971, had a back-to-back measurement of 14mm and a thickness of 3mm. The flange thickness was 0.75mm and the depth 1mm. Despite their thinness, the wheels still did not strictly conform to the accepted BRMSB or NMRA standards.

CHECK LIST OF SUNDRIES

No.	Item	Dates
R207	Packet of Track Pins	62-
R280	Catalogue	annual
R282	Operating Manual	67-69
RT297	Power Cleaning Brush	64-69
R297U	Power Cleaning Brush	70-72
R298	Home Maintenance Set	63-65
R380	Australian Leaflet	71-72
RT395	Railway 'Sounds' Record	63-65
RT520	Seuthe Smoke Unit	61-65
RT521	Capsule of Smoke Oil	61-68
RT528	6 Charges of Track Fluid	61-65
R611	Pack of 24 flexible track fishplates (Wrenn)	70-73
R696	0.5" Diameter Metal Tyred Wheel Pack	71-

No.	Item	Dates
R697	Pack of 8 Fine Scale Spoked Wagon Wheels	71-72
R698	Pack of 6 NMRA Couplings for Canada	–
R699	Pack of 8 Fine Scale Bogie Freight Stock Wheels	71-72
R700	Pack of 8 Fine Scale Coach Wheels	70-72
R700A	Pack of 8 Fine Scale Coach Wheels with White Rims	70-72
R701	Tri-ang Hornby Book of Trains	68-69
R819	3 Part Coupler Set	68
R835	Pack of 4 Couplings (X171)	71-72
R836	Pack of 4 Brushes (X67)	71-72
R837	Pack of 12 Super 4 Fishplates	71-72

DIRECTORY OF SUNDRIES

Home Catalogue R280   *see* 'Marketing'
Operating Manual R282   *see* 'Servicing'
Australian Leaflet R380   *see* 'Marketing'
Tri-ang Hornby Book of Trains R701   *see* 'Marketing'
Track Plans Book R166   *see* 'Track'.

Power Cleaning Brush RT297
Looking a bit like a toothbrush, the power cleaner was for cleaning locomotive wheels. By passing a current through them to the locomotives motor the wheels turned against the wire brush. The metal bristles at one end of the brush were insulated from those at the other end by a piece of foam. Thus, the two ends of the brush provided a positive and a negative for a 12v DC current.

The bubble-pack card used for the brush when

Power clean brush.

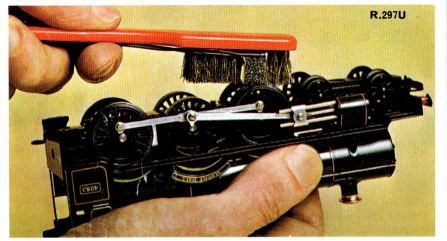

Packets of wheels.

first introduced in 1964, was reprinted with the Tri-ang Hornby logo in 1965 or 1966. It remained in production throughout the Tri-ang Hornby period. 27,000 were made during these years. It had carried the RT297 code since it was introduced in 1964 but in 1970 this was changed to R297U.

Seuthe Smoke Unit RT 520

Despite the change to Synchrosmoke in Tri-ang Hornby locomotives, the Seuthe smoke units were being marketed by Rovex as late as 1965 when 3,600 passed through the Margate stores.

Capsule of Smoke Oil RT521

The RT521 smoke oil remained in production until 1969 and was dropped after 1970. During this time almost 300,000 capsules were made. Retailers bought them in packets of six but sold them individually for 1/3d a capsule.

Railway Sounds Record RT395

The record was last made in 1965 when 5,000 passed

into the stores. It is assumed that it used existing packaging. Over 3,000 were dispatched from the stores during the year.

Home Maintenance Set R298

The last year of production of this set was 1965 and so it just stretched into the Tri-ang Hornby period. It was described in **Volume 1** and in that description reference was made to the set containing the R288 tool kit as part of it. This was in fact untrue as what it contained was a pouch of tools marked RT209. Research has revealed that in 1962 there were plans to release a tool kit which was allocated the number RT209 but none were made. It seems that the pouches had already been printed before the kit was withdrawn and so they were put to good use in the R298 Home Maintenance Set.

I have not seen an example of this set in a Tri-ang Hornby box and assume that Tri-ang Railways boxes were used for the 1965 production run. 2,000 were made that year.

Pack of 4 brushes (X67) R836

This pack appears to have been made for two years. In 1971 11,600 were made up in small cellophane envelopes with a Tri-ang Hornby header card closing the top. The packets were bought by the trade at £1.28 a dozen and sold at 20p each.

0.5" Diameter Metal Tyred Wheel Pack R696

While reference to this wheel pack has been found, I have no evidence that it was made.

Fine Scale Wheels R697, R699, R700, R700A

With the introduction of fine scale System 6 track and fine scale wheels on wagons and coaches it was realised that modellers with older stock would wish to convert it to run on the new track. Consequently, the following four packs were available.

The wheels were mounted on metal pin-point axles and packed four axles to a cellophane packet closed with a red and yellow Tri-ang Hornby folded header card to which a label describing the contents was stuck.

The quantities made quoted in the table below are for those made in 1970 and 1971 and therefore definitely in Tri-ang Hornby packaging.

Pack	Made
8 Fine Scale Spoked Wagon Wheels R697	8,500
8 Fine Scale Bogie Freight Stock Wheels R699	8,400
8 Fine Scale Coach Wheels R700	21,000
8 Fine Scale Coach Wheels with White Rims R700A	12,000

4 Couplings (X171) R835

These were also sold in a cellophane packet with a Tri-ang Hornby header card. The couplings contained in the pack were the IIId type and, in 1971, 23,000 packs were made up.

6 NMRA Couplings for Canada R698

These were advertised in the Canadian catalogue but the factory census sheets show no evidence of

production. The packaging was shown as for the above.

Packet of Track Pins RT207

This was one of the longest running sundries. Introduced in 1962, the R207 pack of track pins is still available at the time of writing (1997). During the Tri-ang Hornby years, 450,000 packs were made.

Track Cleaning Fluid (6) RT528

The fluid was supplied in plastic capsules with a thin neck that could be cut to release the contents. They were packed and sold in white EA boxes of six priced 2/6d a box. 18,000 boxes passed through the stores in 1965, the last year it was made.

Fishplates for Flexible Track R611

These were made by the company's subsidiary, G&R Wrenn Ltd at Basildon, to go with the flexible track also made by that company. The first batch of

3,300 packs went into the stores in 1970 and was followed by a further 38,000 the following year.

Fishplates for Super 4 (12) R837

Introduced in 1971, the fishplates were also packed in a polythene bag with a Tri-ang Hornby header card. 37,000 was packaged thus in 1971.

Re-railer

There was a demand for the reintroduction of the re-railer which had been dropped from the range in 1963 due to problems with multi-wheeled locomotives. Retailers were advised to refer customers to the level crossing which served as a railer.

SERVICING

Rovex Scale Models Ltd, like other members of the Lines Bros. Group, recognised the importance of after-sales service and this service was available

The R282 operating manual.

Packet of track pins.

An extensive range of spares was available to service agents and most were sent out with hand-written labels.

not only for current models but for obsolete models too. For example, production of a number of designs of power unit ceased in 1966 and yet servicing of them continued into the 1970s when spare parts eventually ran out. Rovex also realised early on the importance of providing that service through the hundreds of retailers that were prepared to become service agents not just in Britain but all over the world. The service boxes that were supplied to service agents have, themselves, become collectable. So too have the service sheets, with their exploded drawings of models, that were supplied to agents to assist them in undertaking repairs.

Over the years, Lines Bros. produced a large quantity of printed matter to support the model railway system. This included price lists, maintenance manuals and instruction leaflets, most of which are a rich source of information to collectors and are a specialised collecting area in their own right.

SERVICE AGENTS

1965

By May 1965, soon after Tri-ang Hornby came into being, Rovex had 400 service dealers and sales representatives in the UK and Eire and another 230 around the world. These were stockists whom the company considered qualified and equipped to carry out local servicing and repairs to model railway components.

It is interesting to note the distribution of overseas agents. A third of them (78) were to be found in Australia and another third were split amongst the other three principle Tri-ang buying nations –Canada (39), New Zealand (24) and South Africa (11). The remaining third were spread between 60 countries with an average of one dealer per country. Most interesting amongst the latter were the American and Canadian troops stationed in Germany both of which had their own Tri-ang agent.

1971

By the end of the Tri-ang Hornby era there had been little change. UK and Irish agents for Rovex had risen to 420 while overseas ones remained at around 230. The balance of overseas agents had changed only slightly with 71 in Australia, 49 in Canada, 33 in New Zealand, 11 in South Africa and about 60 in the rest of the world. The latter were somewhat unbalanced. The USA, for example, had only one agent (Amro) and India only two but Barbados with its comparatively small population had four agents.

Service dealers could be identified by a badge that was displayed in the window of their shop. This was originally not dissimilar in design to the Tri-ang Railways lapel badge (R282) but carried the inscription 'Official Service Dealer' in the centre. With the change of name, the window badge was redesigned to incorporate the 'beefburger' logo.

SERVICING SCHEME LEAFLETS

31st Edition

The official Service Dealer's badge also appeared on the cover of the Service Agent Handbook which was given away with every train set or locomotive bought. The book had already run to 30 editions by the 1 May 1965 and the 31st edition, published that summer, was the first to carry the new 'sausage' logo. The Tri-ang Railways and Tri-ang Hornby logos continued side-by-side until the start of 1968 when the Tri-ang Railways one was dropped.

The 31st edition was the first to mention Hornby and Hornby-Dublo indicating that servicing and spares were available from Meccano Ltd. This information continued until early 1968 when the 34th edition of the handbook stated that servicing for

Servicing agents lists, which were packed with each locomotive, ran to many editions during the Tri-ang Hornby years.

Hornby and ex-Hornby-Dublo (2-rail only) models was dealt with by G&R Wrenn Ltd.

If you did not wish to have your model repaired locally, or were not near enough to a service agent, you could post it to one of five factories. These were: Rovex Scale Models Ltd at Margate, UK; Cyclops & Lines Bros. (Aust.) Ltd at William Street, Leichhardt, Sydney, N.S.W, Australia; Moldex Ltd, PO Box 21, Fairfield, Victoria, Australia; Lines Bros. (NZ) Ltd, Tri-ang Works, Panmure, Auckland, New Zealand; or Lines Bros. (SA) (Pty) Ltd, Grimsby Road, Mobeni, Durban, South Africa.

From the advice given in the leaflet, dirty commutators, carpet fluff and over-oiling were clearly common faults.

Logos
The 31st edition of the handbook was the last to carry the Lines Bros. Group logo on the back and the 34th edition was the first to carry the name Rovex Industries Ltd. In mid-1969, the 35th edition was the first to carry the Tri-ang Hornby Minic logo and details of Minic Motorways, production of which had now transferred to Margate. The new shop window badge (with the three names) was carried on the cover of the booklet. (For more information

about logos, *see* 'Marketing' below in this chapter)

It is interesting to note that all the factories listed above were geared up to service Minic Motorways except the one in New Zealand. Also very few of the overseas agents handled Minic and notably none in Australia, Canada or New Zealand. All South African agents, on the other hand, did.

The 35th edition of the Servicing Scheme Handbook was published twice. Once in the autumn of 1968 and again in mid 1969. In the autumn of 1969 the 36th edition was released with the name Rovex Tri-ang replacing Rovex Industries Ltd but we had to wait until the 37th edition, early in 1970, before the address for the Margate factory referred to it as Rovex Tri-ang Limited. Inside the 36th edition the new names for the overseas companies were given.

38th Edition
The 38th edition was published in the autumn of 1970. This differed from the previous one in adding 'Tri-ang Wrenn' models to the ranges serviced by G&R Wrenn Ltd. For years the UK and Irish agents were listed under counties with the exception of Bristol which stood alone. No doubt someone questioned this anomaly and in the 38th edition of the

handbook Bristol appeared under Gloucestershire.

For the final Tri-ang Hornby version of the Servicing Scheme Handbook, the name Minic was dropped and Scalextric was added. The shop window badges had been dropped from the cover and were replaced by an official service dealers ribbon. A number of new agents names appeared in the book as it incorporated those who dealt only with Scalextric. This time all Australian, Canadian and New Zealand agents dealt in both systems. Scalextric was clearly sold far more widely abroad than Minic Motorway had been.

SERVICE BOXES

Large Tri-ang Railways Box
Service dealers each received a wooden service box, two versions of which were described and illustrated in **Volume 1**. Not included in that account was an earlier version.

In contrast to later boxes, the oldest box measured 70.0 x 42.3 x 9.5 cms giving its lid a surface area twice that of the later boxes. It was, however, only half as deep. Internally the box was divided into two. The left-hand side was identical to the pull out tray seen in the left-hand picture on page 374 of **Volume 1** but turned through 90° with the rectangular metal trays to the front but the long compartment on the left side.

The metal dishes were almost certainly made in the Merton factory, the large ones, which measured 16 x 10 cms, being stamped with the Lines Bros. triangular logo. There were 12 of the small metal dishes, painted green like the rectangular ones.

The right-hand side of the box was almost identical to the upper layer of the boxes on page 374 but turned through 90° so that the small compartments, each measuring 10 x 18 cms, were at the back. There were also retaining strips for an ammeter in the large compartment (22 x 33 cms) and a single string supported the lid when open.

Small Tri-ang Railways Boxes

The large service box was superseded by the one illustrated on page 374 of **Volume 1**. No doubt, Rovex received comments about the amount of room the box took up in small shops and so it was transformmed into a two-tier structure measuring 43.2 x 38.0 x 14.0 cms. The lower part was a pullout drawer which was held in place when closed by a sliding wooden panel which may be seen resting in the lid of the box on the right of page 374.

As the illustrations in **Volume 1** show, at some time during the Tri-ang Railways period the lower tray was altered, probably as an economy measure. The drawer of metal trays was replaced by a drawer of wooden compartments which measured 30.0 x 2.0 cms, 33.5 x 7.3 cms and two of 14.5 x 10.0 cms.

The Tri-ang Hornby service box was the same as the mid 60s Tri-ang Railways one. Only the name on the top changed.

There was a place for an ammeter in the upper section of the box which, in the example illustrated in **Volume 1**, was made by Wilkinson.

Tri-ang Hornby Box

All versions of the box up to 1965 had 'Tri-ang Railways' in yellow on the lid as shown in the picture referred to above. From mid 1965 the box, while continuing in the same design, had the Tri-ang Hornby logo in gold on the lid with the word 'Railways' beneath. In contrast to the earlier boxes, the interior of the example I have is a matt red brown instead of red. The Service Manual inside the box now had a plastic cover.

Hornby Railways Box

It is thought that the box remained unchanged until the Hornby Railways period. At some time, however, the construction changed. The new one looked similar but was about 2.5 cms deeper front to back. It was also made up of two identical wooden trays, one inverted over the other, and secured together by a pair of brass hinges at the back.

The box was a slightly darker red but now carried two logos and lettering in gold – the new Hornby Railways decal followed by the words 'Scale Models' and the Scalextric logo.

The interior of the box was quite different. It had a deep lift-out tray divided up into 36 identical compartments each measuring approximately 6 x 6 cms. Under the tray the bottom of the box was divided into six compartments, four measuring 7.0 x 38.5 cms and two 8.0 x 9.5 cms. The interior of my example is painted black and there appears to have been no string to retain the lid. This box has also been seen in blue with just the Scalextric logo on the lid.

SERVICE DEPARTMENT

As we have seen, the public were encouraged to use their local service dealers as a first port of call when their models needed servicing and only if this were not possible to send them back to the factory. Throughout the Tri-ang Hornby days the UK address of the Servicing Department was the factory at Westwood, Margate.

The advice given to customers was:

"Tri-ang Hornby model railways components are tested thoroughly before dispatch from the factory. In the event of a part appearing not to operate correctly, check that the printed instructions have been followed before reporting the matter to the stockist from whom the item was purchased.

"Should a part require attention, it may be taken to an Official Service Dealer, as listed in the Servicing Scheme Booklet. Reasonable charges may be made to cover the cost of servicing and spare parts are charged at authorised prices.

"Carbon brushes, fishplates, mailbags, leads (in various lengths), and similar spare parts may be obtained from Tri-ang Hornby stockists or direct from the factory.

"Complete parts, such as coaches, sections of track, buildings etc., are not supplied direct to the public from the factory but must be purchased from a retail distributor."

With the changes occurring in the company and in order to improve service to retailers a Customer service Department was opened at the Merton factory in November 1970, in order to answer queries from Rovex Tri-ang Stockists. It dealt only with information and did not carry out any servicing and it was run by George Chitty under the control of the Service Director.

SERVICE SHEETS

Service sheets were produced for any item a service dealer might be expected to be asked to repair and they were supplied free of charge to all service dealers. They were updated as necessary and added to as new models were produced. Revised sheets were usually given a letter suffix, thus 5A replaced 5 and

Service sheets were soon appearing with pale green borders.

Service dealers were provided with frequently updated sevice sheet indexes.

Spares price lists were produced for service dealers.

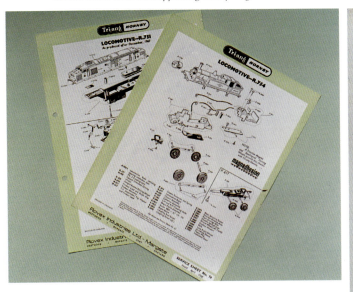

5B replaced 5A. In April 1968 the range of service sheets was revised and the new set had new numbers, starting with 51, and a green border.

Nos. 42-49

The last Tri-ang Railways service sheet to be published was No.41 which provided an exploded view of the X500 motor. This was released in December 1964 on white paper instead of pink but with the same red border as previous ones. After this there was then a gap of two years during which time no further sheets were published.

We were then treated to a batch of eight sheets (Nos. 42-49) in December 1966. These looked similar to the Tri-ang Railways ones using white paper and a red border but with the Tri-ang Hornby logo. Sheets 7 and 8 of November 1955 had been of 'General Spares' and sheets 42 and 43 extended the range covered. No.44 provided a breakdown of the Super 4 'Y' point and No.46 the catenary system.

SERVICE SHEET INDEX.   1969.

Ref. No.	Model	Service Sheet	Power Bogie
*R.52S	Tank 0-6-0	58	
R.53	Green "Princess" 4-6-2 (R.386 A.P. Kit)	59	
R.54	"Hiawatha" 4-6-2	59	
*R.55	Transcontinental Diesel	61	X.3122
*R.150/S.	B.12 Class 4-6-0	67	
*R.152	Diesel Shunter 0-6-0	58	
R.155	Diesel Switcher	61	X.3122
R.157	Diesel Power Car	60	X.3121
*R.159	Transcontinental Double Ended Diesel	61	X.3122
R.251	Class 3F Loco 0-6-0	58	
R.252	Steeple Cab 0-4-0	62	
*R.253	Diesel Dock Shunter 0-4-0	61	X.3166
*R.258	Maroon "Princess" 4-6-2	59	
*R.259/S	"Britannia" 4-6-2	63	
*R.351	EM2 Co-Co Class (R.388 A.P. Kit)	64	X.213
*R.355	Industrial Loco 0-4-0	62	
*R.356/S	"Winston Churchill" 4-6-2	65	
*R.357	A1A-A1A Brush Type 2 Diesel	66	X.337
*R.359	Tank 0-4-0	62	
*R.437	Super 4 Y Point	57	
*R.490	Super 4 Point	56	
*R.491	Super 4 Point	56	
*R.555	"Blue Pullman" Diesel	60	X.3121
R.559	Diesel Shunter 0-4-0	62	
R.651	Stephensons "Rocket" (R.346 Set)	68	
*R.751	Type 3 Co-Co Diesel	69	X.337
*R.753	E.3001 Bo-Bo Class	70	X.3169
*R.754	"M7" Tank 0-4-4	74	
*R.758	Hymek B-B Diesel	75	X.3170
*R.759	"Albert Hall" 4-6-0	73	
*R.850	"Flying Scotsman" 4-6-2	76	
*R.855	S.R. Battle of Britain 4-6-2	65	
*R.869/S	"Blue Flier" (Big Big) Diesel	71	X.3301
*RV.256	Yellow Shunter (Big Big) Diesel 0-4-0	72	
*RV.272	Steam Loco (Big Big)	72	
*RV.276			
*X.04	Mk.IV Motor	51	
*X.500	Motor "Rocket"	52	
*OO	Bogies & Pony Trucks	53	
*OO	Spares & Bogies	54	
*OO	Wagon Chassis & Bogies	55	

Items marked * refer to current models.
Service Sheets are priced at 4d each retail, plus 6d postage.
Servicing sheets numbered 1-50 are no longer available.

The rest of the new sheets depicted new locomotives, one, No.48, the Tri-ang Big Big 'Blue Flyer'.

Nos. 51-75

The next batch of service sheets were those I have already mentioned published in April 1948. There were 25 in all numbered 51-75. Most of these were revisions of earlier sheets but there were three com-

---

pletely new ones. These were No.72 for the Big Big 0-4-0 diesel shunter, No.74 depicting an exploded M7 tank and No.75 showing the Hymek dissected.

Nos. 76-78

Sheet No.76 for the Flying Scotsman chassis detail was released in June 1969 and twelve months later No.77 was the first sheet to carry the Rovex Tri-ang name. This was a revised sheet for the R59 standard tank which had returned to the catalogue.

Since No.51, the sheet border colour had been pale green instead of red but for No.78a/b (Evening Star), published September 1971, the border colour was dropped all together. This was the last service sheet to be issued under the Tri-ang Hornby name.

Service Sheet Indexes

To help Service Agents to find the correct service sheet, they were issued with a typed index which was periodically replaced with an updated one. Items were listed in numerical order together with a description and against each item was the number of the relevant service sheet. In the case of locomotives, a further column supplied the 'X' number of the power bogies they were fitted with.

SPARES LISTS

A list that the public rarely saw was the price list of spare parts which Service Agents received from the factory. In Tri-ang Railways days these were printed in maroon or red ink on thin paper and looked similar to the normal trade list. They were not only a price list but also an order form. By the late 1960s the price list was printed in black and on a better quality paper.

The form listed every part in numerical order starting with the 'X' numbers which were for assemblies such as two wheels on an axle. These were followed by 'complete' units which had 'R' numbers such as tenders or loco bodies (which were listed under the 'R' number but given a '/1' or '/2' suffix to distinguish them from a finished tender). Next came the 'S' numbers which applied to individual spares such as one wheel or one axle.

Beside each number was a description of the item and, in the next column, the numbers of the Service Sheets that referred to the item. Both the trade and

A full list of the Tri-ang Hornby Service Sheets is provided below:

No.	Subject	Date	Border
42	General Spares – 3	Dec 66	red
43	General Spares – 4	Dec 66	red
44	Super 4 Y Point R437	Dec 66	red
45	Locomotive R751 (Co-Co)	Dec 66	red
46	Overhead Power Supply Parts	Dec 66	red
47	Locomotive R753 (E3000)	Dec 66	blue
48	'Blue Flyer' Locomotive RV256	Dec 66	red
49	Locomotive R759 (Hall)	Dec 66	red
51	XO4 Motor	April 68	green
52	X500 Motor	April 68	green
53	Locomotive Bogies & Pony Trucks	April 68	green
54	OO/HO Spares & Bogies	April 68	green
55	OO/HO Chassis & Bogies	April 68	green
56	Super 4 Points R490, R491	April 68	green
57	Super 4 Y Point R437	April 68	green
58	Chassis R52S, R153, R251S	April 68	green
59	Chassis R53, R54, R258	April 68	green
60	Power Bogie R157, R555	April 68	green
61	Power Bogies R55,R155,R159,R253	April 68	green
62	Chassis R252, R355, R359, R559	April 68	green
63	Chassis R259S	April 68	green
64	Locomotive R351 ('Electra')	April 68	green
65	Chassis, Bogies & Pony Truck R356S	April 68	green
66	Locomotive R357 (A1A-A1A)	April 68	green
67	Chassis & Bogie R150S (B12)	April 68	green
68	'Rocket' Locomotive R651	April 68	green
69	Locomotive R751 (Co-Co)	April 68	green
70	Locomotive R753 (E3000)	April 68	green
71	Blue Flyer RV256 (Big Big Hymek)	April 68	blue
72	Locomotive RV272(Big Big 0-4-0DS)	April 68	blue
73	Chassis R759 (Hall)	April 68	green
74	Locomotive R754 (M7)	April 68	green
75	Locomotive R758 (Hymek)	April 68	green
76	Chassis,Bogies&Pony TruckR850,R855	June 69	green
77	Chassis R59S	June 70	green
78a/b	'Evening Star' R861	Sept 71	white

retail price for each spare was quoted and, in some cases, the minimum number that had to be purchased.

In 1969 the form also listed Minic Motorway spares which, incidentally, were given 'Z' (sub-assemblies) and 'Y' (spare parts) numbers.

INSTRUCTION LEAFLETS

Booklet

Rovex were anxious that, having purchased one of their train sets, the customer knew how to use and look after it. Instruction leaflets were supplied with all train sets, locomotives, operating accessories and any other item that required some form of instruction.

The 25th edition of the Instruction and Information Manual was the first to bear the new Tri-ang Hornby name. This was certainly in use by 1 October 1965. Each manual was rubber stamped in the top right hand corner with the 'R' number of the set or loco it accompanied and the date that the model was boxed. The information covered:

power supply and control system
hints on running
home maintenance set R298 and tool kit RT288
track
Super 4 track layouts
locomotives
a lubrication and maintenance chart
smoke units
mark III couplings
availability of servicing and spare parts

By mid 1966, the 26th edition of the Instruction Leaflet was in use. Despite many small changes, the content of this edition was basically the same as the previous one. The most obvious change was the absence of any reference to the home maintenance set and tool kit which had been withdrawn.

Operating Manual

The R282 OO/HO Gauge Instruction and Information Manual was an attempt to upgrade the instruction leaflets normally provided with locomotives and train sets. The cover was of stiffer card printed in three colours so that the Tri-ang Hornby logo could be seen in the correct colours. Inside 20 pages covered the same subjects as the Instruction Leaflets but with several additional pages devoted to the catenary system and the locomotives fitted with pantographs.

The manual was printed three years in a row (1967-69) but all those I have seen have been marked '1st Edition'. The cover carried a picture of a cut-away model of the L1. Over 47,000 were printed for solo sales and it was priced 9d. They were sold to shops in packets of 12. It was announced as out of print and out of stock in August 1970.

Sheet Version

In addition to the Operating Manual, a single double-sided Instruction Leaflet was introduced about the same time to be packed in sets and with locomotives instead of the earlier booklets. It was rightly assumed that not everyone would want to buy the 9d Operating Manual but would still need basic instructions if the Service Department was not to be overloaded with returned models.

The sheet version of the Instruction Leaflet contained most of what was in the original booklets with the exception of details on track and track layouts. There was also an absence of information about the overhead system but readers were advised that:

"Additional information regarding track operated and overhead supply locomotive models is given in the publication R282 'Tri-ang Hornby Model Railways Operating Manual' priced 1/- direct from the factory. Details of track geometry and the elements of layout construction are contained in the booklet R166 'HO/OO Gauge Track Plans' priced 4/- plus 6d postage."

The single-sheet leaflet was in use early in 1967 and a second edition was available by October 1968. This had information on smoke units and track maintenance added.

The edition in use by late 1969 was printed on a much larger sheet of paper to provide room for additional information. This included reference to System 6 track, revised information on power supplies to reflect changes in power unit design, much more information on smoke units and one side devoted to the overhead power supply system.

The Instruction Leaflet in use by late 1970 was greatly revised but contained basically the same information as before. The main omission was the overhead power system. It was a four-page sheet and carried the code number 04/002/-OO/IL/466/1. In August 1971, the 4th edition (04/002/5935/4) of this leaflet was being used and, at the end of Tri-ang Hornby production in December that year, a final version (04/002/4) was being issued with locomotives and sets. This final printing before the change of name had an added section on the steam sound tenders.

DATING TRI-ANG HORNBY PUBLICATIONS

Consecutive numbering of orders continued throughout the Tri-ang Hornby period as did the practice of including the order number in the code at the bottom of each publication. This provides us with an easy way of dating most printed material. The approximate range of order numbers used each year are listed in the following table. Part way into 1970 the 'counter was reset' to nought.

YEAR	ORDER NUMBERS
1965	30951-36500
1966	36501-43750
1967	43751-50500
1968	50501-56500
1969	56501-62000
1970	62001-63016
	1-3000
1971	3001-8500

MARKETING

Getting the message across

While the 1950s had been a period when Tri-ang Railways were very much in demand and advertising was not a high priority, the 1960s was a period of recession in the model railway world and those manufacturers that had survived had to convince the public not only to buy their model railways in preference to those of other manufacturers but also to buy model railways in preference to other toys.

By 1965, Rovex Scale Models had a full page advertisement in each of the three principal model railway magazines. These were *Railway Modeller*, *Model Railway News* and *Model Railway Constructor*. The advertisements were used to announce the release of new products, to encourage people to buy the Tri-ang Hornby catalogue and to emphasise the features that made Tri-ang Hornby models a cut above the rest. The budget was limited and much of the artwork was made up from catalogue pictures reworked. The lack of new products to advertise was also a problem.

There was also a need to find new customers and in the late 1960s more of the marketing budget was directed towards national advertising using the national press and television.

Magazine Advertising

The first advertisements bearing the combined Tri-ang Hornby name appeared in the model railway press in July 1965 and the subject was the TPO. The message was:

"Why not your own mail train? Great, grey thundering ghosts that slip out of their stations every night, unknown to the public and unmarked in timetables. The Travelling Post Offices are miracles of split-second timing and precision between man and machine."

Everything you want to know!

Interested in Model railways or Motor Racing? Then you'll want the fabulous 1969 Tri-ang Hornby Minic Catalogue. **Train enthusiast?** Then you must get the magnificent Tri-ang Hornby Book of Trains.

Real enthusiasts will want both — and they'll be delighted with the presentation. There's a wealth of fascinating top-notch technical information, plus plenty of superb colour illustrations.

You can buy both the Tri-ang Hornby Minic Catalogue and the Tri-ang Hornby Book of Trains **now.** They're both really worth having so get them right away from your usual stockist!

Recommended Retail Price: Tri-ang Hornby Book of Trains 5/-
Tri-ang Hornby Minic Catalogue 1969 1/6

Tri-ang HORNBY MINIC

Rovex Industries Limited · Westwood · Margate · Kent

A typical magazine advertisement of the late 1960s.

The following month the R348 Giraffe Car Set was featured under the title:

"Mind your head. Exactly three things can be said for being a giraffe. 1: no school. 2: the best bits off treetops where other animals can't reach. 3: a view for miles. Giraffes, in fact have it made. But life has trials, too, like being eaten by lions.. . ."

Who wrote these advertisements?!

By September, in case the fact was not apparent to everyone, we were told that Tri-ang Railways and Hornby-Dublo had 'got together' and were introduced to the new converter track and wagon.

December saw a return to an old subject – Synchro-smoke – with smoke generation locos being illustrated. This was an update of the advertisement of a year before.

We had to wait until the following May for a new topic by which time two new locomotives were available. These were the E3000 and the Class 37 or Co-Co English Electric Type 3 Diesel Electric Locomotive to give it its title at the time.

In June 1966 the emphasis switched to point operated colour light signals with a picture of a driver's view from the cab of a diesel. The following month the Big Big train was launched through the full page advertisement and remained its subject for four months.

It was not until December that the new Hall loco was the subject of the Tri-ang Hornby monthly advertisement in the national model railway press, accompanied by that strange catchphrase:

"Where does the 'Albert Hall' hit the track? At your Tri-ang Hornby shop!"

Someone was trying to be 'with it' or 'hip' as I believe it was called in those flower-power days.

In February 1967, the Hall gave way to an illustration of the 1967 catalogue with its Cuneo picture of the M7 tank simmering in an engine yard supposedly somewhere on the Southern Railway. The advert stated that the new catalogue was larger than before with 32 pages. In April the M7 tank itself

became the feature of advertisements with pictures that showed both the opening firebox door and the glow from the fire.

Another new introduction was featured in the May advertisements under the banner:

"Help get liner trains moving here"

This was a promotion for the Freightliner container wagon and was followed in June with a title:

"Great galloping diesels! Tri-ang Hornby have done it again" – which introduced the Hymek.

TO **Tri-ang HORNBY**
CATALOGUE READERS

We would like to know about your particular interests to help us to plan our future models. We hope you will fill in the following questionnaire and send it to us.

1. Name_____

2. Address_____
_____

3. Age *(If over 18 state 'over 18')*_____

4. Do you already own a Tri-ang Hornby Railway?_____
   If the answer is 'No' which items from this catalogue would you like first
   _____
   _____
   If the answer is 'Yes' which item would you add to your system next
   _____

5. Do you prefer Steam or Diesel outline locomotives?
   _____

6. Do you prefer British Rail or Famous Company Liveries?
   _____

7. Which items would you like Tri-ang Hornby to introduce in the future
   _____
   _____

8. What improvements would you like to see made to the Tri-ang Hornby System
   _____
   _____

We will not be able to acknowledge receipt of this card but we do thank you in advance for filling it in and sending it to us.

A readers' survey which went out with the 1970 catalogue.

Rovex never failed to recognise the importance of play value and in July, with the message:

"Tri-ang Hornby trains give you more"

A collage of pictures was used to show the features that other manufacturers did not bother about. These included: Magnadhesion, working catenary, clip-fit accessories, loco crew, isolating points, bell signalling, automatically controlled signals and the new features on the M7 tank.

A new advertisement in October that year introduced the Barclay and Westminster power units that had just arrived in the shops. The power units dominated the monthly advert until after Christmas by which time the products in the 1968 catalogue needed to be pushed.

The January 1968 advertisement featured the R645 'Freightliner' train set, while the new Tri-ang/Hornby/Minic catalogue (now enlarged to 44 pages) was the subject in February with the catch-phrase:

"Who says Inter-city Freightliners and motor racing don't mix?"

In April, another attempt was made at selling the Stevenson's 'Rocket' which had never been a good seller in the UK. The banner read:

"Stephenson's 'Rocket' is a triumph of modern engineering".

It also contained a promotion of the X500, X04 and XT60 motors and the following month these were extracted for a full page advertisement of their own with the headline –

"This is the biggest value you will ever buy!"

The emphasis was, of course, on the tiny X500 motor used in the 'Rocket'.

In June 1968 we were treated to a new advert which played on the transfer of Minic production to Margate, and its promotion alongside Tri-ang Railways, with the message:

"Isn't it time your railway became a complete transport system?"

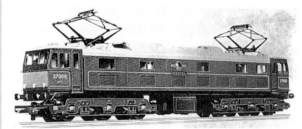

Triang-Hornby

Der größte Spielzeugkonzern der Welt, Lines-Bros aus England, will sich künftig auch auf dem Modellbahngebiet stärker um den deutschen Markt kümmern, wobei das Schwergewicht des Vertriebs auf einer Reihe besonders populärer und gelungener H0-Lokmodelle nach neuen und älteren Vorbildern aus England liegt. Angeboten werden neben dem „Rocket" vier Dampflokomotiven mit Schlepptender, eine Tenderlok, der Hymek-Diesel und die A1 6-Ellok. Von besonderem Interesse scheint uns die Zugpackung mit der Diesellok und zwei „Freightliner"-Behälterwagen mit jeweils drei 20-Fuß-Containern der BR. Die Triang-Hornby-Erzeugnisse sind im Grundmaßstab 1:76 gehalten; das kleinere britische Lichtraumprofil läßt den Unterschied jedoch nicht allzu deutlich werden.

Stolz der British Railways ist der „Liner-Train"-Behälterzug – hier als Triang-H0-Modell

24   moderne eisenbahn 32·68 **Modellbahn**

Above left: In the late 1960s there was quite extensive advertising of Tri-ang Hornby in Holland.

Above: German leaflet.

Left: German review of the liner train.

The picture was the illustration from the box lid of the RMD Moto-Rail set but enhanced with additional background detail. There was a car in exploded form, a hand control, the car-loading ramp and a scene showing road-rail track in front of the freight depot.

In the summer of 1968 the advertisements were reduced to one every other month, no doubt in order to release funds for the increase in national advertising which occurred around this time, and then in October the long awaited 'Flying Scotsman's arrived.

"Now you can have superb models of this famous train, either as an LNER Locomotive and Tender which has Corridor Connections, or BR version both in authentic liveries complete with Magnadhesion, Firebox Glow and Cab Detail, reproduced with the skill of the first name in Model Railways – Tri-ang Hornby."

A revised version of the advertisement appeared in December issues which allowed space for the new *Tri-ang Hornby Book of Trains* to also be advertised. Two months later, this book and the new catalogue were the full feature. The picture of the catalogue showed the mock-up instead of the printed version.

The March 1969 advertisement promoted the new third radius curve and in May an upgraded version of the advert appeared which drew attention to other virtues of the new radius. In fact this was a ploy to strengthen the sales of Super 4 while the new System 6 was being designed. Indeed, the heading of the advertisement '5 Fabulous Features' was, according to Richard Lines in his book *The Art of Hornby*, intended to suggest that this was the new track that it had been rumoured Rovex were working on. It should be noted that System 6 was, originally, to have been called 'Series 5'.

We had to wait until August before the next advertisement and this time it featured the new ISO container liveries with the message that they were in the shops. The picture showed the R634 Containerway and Pickford's combination and the R719 Sainsbury's and Ford.

The following month the Battle of Britain loco in SR livery was in the shops and became the subject of the full page advert. Detail of the crests carried by the locomotive were shown as well as the model itself.

In November 1969 the advertisement was dedicated to the Inter-City coaches with lights which it claimed as:

"Another Tri-ang Hornby first!"

Trix had had illuminated coaches some fifteen years before but this was the first time for Hornby. An early experiment with coach lighting was illustrated on page 205 of **Volume 1** and work done by Wrenn is described elsewhere in this volume.

With its heading:

"Into the 70s with Tri-ang Hornby", the January 1970 advert promoted the 16th edition catalogue. It also drew attention to the new System 6 track including flexible lengths, historical liveries and a modeller's questionnaire. The System 6 track was honoured by a two-page spread for the February advertisement which was headed by:

"This is New System 6 track by Tri-ang Hornby – The inexpensive track that gives you true realism!"

The advert emphasised the realism, positive track joining, clip-fit feature, electrical continuity, Magnadhesion, availability of converter tracks and the new flexible lengths. It declared:

"Tri-ang Hornby – The model railway system designed for the 70s".

There is little doubt that Rovex were feeling the competition from Peco and others.

For the third month in a row the model railway press carried a Tri-ang Hornby advertisement. In March 1970 the subject was the complete Freightliner depot with building, wagons and the Morris crane.

After that Tri-ang Hornby advertising in magazines dried up with the exception of an advertisement for the 1971 catalogue in February that year.

National Advertising

When advertising in the model press Rovex were talking to modellers but they knew that many of their customers and most of their potential customers did not read model railway magazines. In order to convince the man in the street that he should buy his child a train set, and a Tri-ang Hornby train set at that, Rovex had to resort mainly to television advertising which was expensive for such a specialised commodity.

Lines Bros. were one of the few toy manufacturers who recognised the potential of television advertising to sell model railways and there is little doubt that the public's familiarity with the Hornby name greatly increased the impact of those advertisements. A large part of the Rovex advertising budget went on this, with a concentration on the September to December period. Model railways, with their movement and excitement, were a good subject for this medium.

In September 1970, for example, Rovex started their television advertising campaign with a new train set not yet in the catalogue. This was the RS606 set with its reintroduced R150NS B12 locomotive now fitted with the new Steam Sound. It was the Steam Sound that Rovex saw as the big attraction that would sell their train sets that Christmas. The advertisement showed the set laid-out on a printed card mat and the same mat (R707) was made available to stockists for display purposes in their shops. It was intended that the public would associate the train set in the shop with the advertisement on television and so the R707 was available free with every four sets the stockists purchased.

Logos

Logos have long been important marketing tools and it was important that the new Tri-ang Hornby logo should be closely associated with the former Tri-ang Railways one and yet project a new image. The new logo therefore took the lozenge shape that had formed part of the old logo and extended it to contain the whole of the new name. The finished design closely resembled a two-tone medicine capsule. The lozenge became the standard throughout the Tri-ang Hornby era. In the first half it had 'Tri-ang' in yellow on red and in the second half, 'HORNBY' in red on yellow

A variation on this was adopted, from the start, for use in confined spaces such as box ends. This was a 'double decker' version so that the name 'Tri-ang' was placed above that of 'Hornby'. The colour scheme was the same as the standard version but this variation of the logo was short lived.

The new logo appeared on a price list published on 1 May 1965 but it is interesting to note that the Tri-ang Railways logo continued in use in publications in connection with the TT range until the end of 1967.

In New Zealand, while the new logo was immediately adopted, the 'Railways' part of the old logo was retained as an annex to it for the next two years. The New Zealand price list, published in 1972, still carried the company name 'Tri-ang Pedigree (NZ) Ltd' and the product logo which was either two lozenges carrying separately the words 'Hornby' and 'Railways' or just a lozenge carrying the name 'Hornby'.

In Canada, as we have seen, for legal reasons the Canadian company could not initially use the name 'Hornby' and so the products continued to be sold as 'Tri-ang Railways' until 1971 although in 1967 the logo was changed to make it look like that of Tri-ang Hornby. As far as I know, this version of the logo was unique to Canada.

SALES AIDS

To help retailers market Tri-ang Hornby in their shops a range of sales aids was produced. Most of these had to be purchased by the retailer. These aids were sometimes listed at the end of the trade price list but at other times were to be found in special trade leaflets.

The May 1965 trade price list contained the following items that were not for resale:

R168	Set of 10 Train Set Price Tags	9d
RT296	Set of 12 Price Tags with stand	3/-
RT519	Display Stand	4/-
R545	Display Stand	1/-

Sales aids continued to be developed throughout the Tri-ang Hornby period but the best recording of them is to be found in the trade price lists towards the end of the period. Instead of an 'R' number, many of them were coded 'POS' which stood for 'point of sale'. The aids included the following:

Window Pelmet Scheme R721 (POS104)

Printed glass window pelmets could be supplied to model shops and fitted free of charge to the shop window. The pelmet measured 2'6" x 8" but I do not know what happened if the window was a different width. It was later renumbered POS104 and similar ones were available for Scalextric (POS400), Frog Construction Kits (POS200) and Pedigree Dolls and Soft Toys (POS601).

Window Sticker

A new window sticker measuring 7" x 7" was distributed to service dealers in June 1971 with the promise that it would be used on all Tri-ang stands at various exhibitions, toy fairs and trade shows. It was also to be illustrated in future Tri-ang Hornby and Scalextric catalogues with a mention that each Appointed Service Dealer was supplied with a sticker for easy identification in the high street.

Advertisement Block Service

A comprehensive range of single and double column blocks for Rovex Tri-ang products was available to retailers free of charge.

Three Dimensional Action Display R707

This was a printed layout card which cost £2-16s-0d but was given away free to stockists who ordered four RS602 sets. It was available in April 1970 and, as we have seen, featured in the autumn 1970 TV advertisement. It was not available after Christmas.

Show and Sell Pack R706

This was designed to go with the R707 layout card and stood on top of it to display other models. It cost £2-16s-0d and was available in April 1970 for a very short period. It had tunnels beneath it so that the train could run round on the 3-dimensional action display. The display stood 1' high and covered an area 3'4" x 3'0".

Revised Show and Sell Pack R708

This was a revised version of R706 available in September 1970 for a few months only. It provided space for displaying three more sets and, with the sets to be displayed, cost the stockist £16-1s-3d.

In Canada the name Hornby could not be used initially and so the old name was retained but the logo brought up to date.

Clockwork Set Display Stand R709 (POS112)

This was a cardboard display available free to retailers which took three boxed clockwork sets. By January 1972 this had been renumbered POS112.

Tri-ang Hornby Poster R771

A poster was available to stockists, and obtainable from the sales rep, to promote the RS601, RS602, RS604 and RS605 sets for Christmas 1970.

Inter-City Action Display R718 (POS111)

The stand was made specially for the R644A Inter-City train and was finished in British Rail colours. It was motorised with a revolving locomotive pointer that ran on two U2 batteries. The stand was free and was introduced by Rovex Tri-ang in order to raise flagging sales. In 1971 it was renumbered POS111.

Light Box POS105

An illuminated sign bearing the inscription 'Tri-ang Hornby Model Railways' was offered to stockists at the end of 1971. It measured 6" high, 2' wide and 5" deep and was designed to hang by two loops at the top. It was described as being suitable for wall mounting or suspending from the ceiling. The power consumption was 13 watts at 220-240 volts AC and it cost £7. Similar ones were available for Scalextric (POS401), Frog (POS201) and Sindy (POS508).

Perspex Counter Display Cabinet POS106

This was 9" high, 18" wide and 4.5" deep and could display two large locomotives on track. It cost the retailer £3.50 and was available at the end of 1971. A similar one was available for Scalextric (POS406).

DEMONSTRATION LAYOUTS

As we saw in **Volume 1**, over the years Rovex made a large range of layouts that could be purchased by retailers for use in their shops. Some of these con-

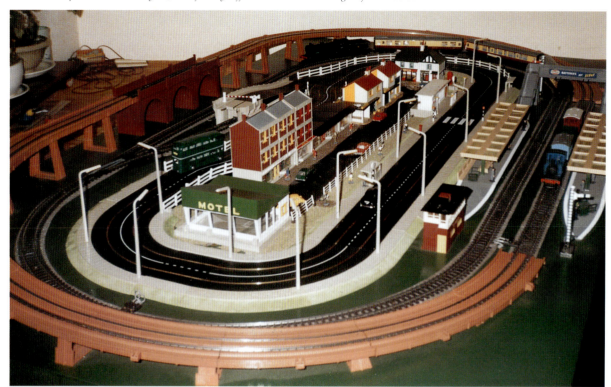

The R600 shop demonstration layout; one of many different ones built at Margate for retailers.

sisted of just model railway items while others combined railway with Minic Motorway. The layouts were built on battened plywood bases for shop display or retail sale.

Railway/Motorway Layout R600

One such was the R600 layout which was made in 1966 and which we know was displayed at the Brighton Toy Fair. The layout was sent out from the factory in a returnable crate on which there was a £25 deposit. This consisted of a double circuit of railway track, one partly elevated, and a circuit of Minic

Motorway in the centre. The buildings that adorned the layout were a mixture of Minic Motorway, Tri-ang Railways and Model-Land.

Other layouts known to have been available during the Tri-ang Hornby period include:

Twin Track Layout and Legs R595 & R515 (POS109 & POS110)

This was a twin track System 6 layout on which two trains could be run independently. From the inner circuit two twin sidings projected into the middle of the table, one pair serving an engine shed and the other pair a container depot with the Morris crane. The layout, which was available from August 1970, could be bought by the retailer for £19-19s-0d who could sell it on to a customer for £37-12s-0d.

If the retailer wished, he could purchase a set of four legs for the table at a price of £2-2s-0d (£3-9s-9d retail). These were numbered R515.

By mid 1971 major display items from all the toy ranges manufactured at Margate had been renumbered under a 'POS' prefix, and the Twin Track Layout and legs became POS109 and POS110 respectively.

Single Track Layout R597 (POS108)

Here was a System 6 layout with a station and single train operation. The station was on a solitary siding which extended into the centre of the oval. The layout was also available from August 1970. The wholesale price was £5-19s-6d and the retail price £11-5s-0d. There were no legs for this layout as the purchaser was expected to provide a table. In 1971, this layout was renumbered POS108.

The above are examples of layouts from early and late in the Tri-ang Hornby period but there were almost certainly others from the period in between. They are excluded here only through lack of information about them.

PUBLICATIONS

There are few model railway systems so well documented through their catalogues, instruction leaflets, price lists, service sheets, books etc. The Tri-ang Railways, Tri-ang Hornby and Hornby Railways catalogues were always amongst the best and the collector can enhance his UK collection of these with those specially designed for various overseas markets. (For servicing and other small leaflets as well as the method of dating Rovex publications, *see* 'Servicing' above)

AMALGAMATION LEAFLET

The 1965 catalogue was the last produced for Tri-ang Railways. In May 1965 the Tri-ang Railways and Hornby-Dublo systems were officially 'merged', and became Tri-ang Hornby.

To mark this occasion and to advise the public of what was going on, an eight-page leaflet was published. The bold red border to the front cover made it quite clear that Tri-ang Railways remained the prominent system and, as we have seen, the amalgamation was in name only.

The leaflet promoted the converter track and wagon (which was shown as green but was in fact black when it arrived) and contained a two-page spread which illustrated the former Hornby-Dublo models that would be incorporated into the combined range. This included seven locomotives but there was no mention that it was old stock that the company was clearing or that there was no intention of manufacturing the models at Margate when stocks ran out. The full list of products offered appears elsewhere in this volume.

Another double-page spread showed a massive layout on which Tri-ang Railways and Hornby-Dublo models feature side by side. The back page featured pre-production models for the proposed Battle Space range.

UK CATALOGUES

Background

Over 3 million general issue catalogues were printed between 1965 and 1971, the greatest number being printed in 1955 when 536,100 passed through the stores. As the following figures show, the number printed dropped off towards the end of the decade:

YEAR	PRINTED
1965	525,853
1966	536,100
1967	492,800
1968	481,215
1969	430,020
1970	317,601
1971	259,421

The figures are from the factory census sheets and judging by the odd numbers quoted, someone must have counted them!

Richard Lines in *The Art of Hornby*, relates that work on the catalogue started the preceding April:

"Alternative layouts, schemes and themes are considered leading up to photography, copy writing and artwork during June-August. By this time in the year it is not always possible to have a clear picture of the current year's successes and failures. The new introductions will only just be finding their way onto the market and whilst orders from retailers may look encouraging it is really only with the onset of the winter season that it can be established which products are moving off retailers shelves. Thus it is too late to react fully to these trends in the next season's catalogue which will already have been printed by the time the trends are clear."

In June 1970 a catalogue supplement was printed to:

"'legalise' certain new items which have been introduced during the year, e.g. RS605A train set, R389S B12 Loco Assembly Pack and others, and at the same time to make it appear that a number of other items which had been inadvertently left out of the main catalogue but of which stocks subsequently appeared, are current products. In fact, they are really in to clear stocks."

The above was an extract from an internal memorandum dated 26th June 1970.

1966 Catalogue

The first Tri-ang Hornby Catalogue was that produced for 1966 and had as its cover illustration the

1969 catalogue.

Terence Cuneo picture that the previous May had been used on the cover of the amalgamation leaflet described above.

Richard Lines recalled the origins of the picture:

"... he (Terence Cuneo) was unimpressed by the electric locomotive that was requested for 1966. His conditions for undertaking this commission were that it must be a night scene in which sparks from the pantograph would reflect off smoke from an adjacent steam locomotive. As we can see it proved to be a most effective cover and, as with his other work, it was carried out extraordinarily quickly. He organised the traffic department at Crewe for the day and had the real locomotives positioned just where they would make a good composition. The train code 1K64 denoted an express passenger train, K meaning destination Crewe, 64 was the year of the painting. The steam locomotive was Jinty No. 47338 at that time allocated to Crewe. The situation was looking down from an over bridge to the eastern end of the depot yard."*

From this account it is interesting to note the date of the painting showing just how far ahead catalogues were planned. The subject was no doubt chosen because the addition of a Class AL1 to the Tri-ang Railways locomotive stud was planned for that year. As it happened the model was not available and it was the Hornby-Dublo version that was featured inside.

The mouse, which is the trademark of Terence Cuneo, was obliterated by the Tri-ang Hornby logo but may be clearly seen seated on the top of a catenary mast in the version of the picture used for the amalgamation document.

1967 Catalogue

In 1967 the cover of the catalogue featured another painting by Cuneo and again it was designed to promote the new locomotive model planned that year. As Richard Lines relates, the painting was in fact executed not on the Southern Region but at Stratford shed in East London where Cuneo discovered an M7 tank (No. 30245) in a state of disrepair. In her book *Terence Cuneo Railway Painter of the Century*, Narisa Chakra tells how the locomotive "had to be ignominiously dragged from its shed by a diesel shunter!" In his painting he gave the locomotive its Southern number '245' and the mouse sat on the window sill. This picture was also used by Arrow Games, another Lines Bros. company, as a subject for a jigsaw.

1968 Catalogue

In 1968 the emphasis was on the amalgamation of Tri-ang Hornby and Minic Motorways and so the 14th Edition Catalogue was a joint one marketing both systems. The cover was one of the least attractive of the period with a picture area divided diagonally to incorporate illustrations of a modern electric train and a motor racing scene. The catalogue was titled: 'Railways & Motor Racing' and carried the new Tri-ang Hornby Minic logo.

The dual theme was repeated in 1969 with an equally uninspiring studio photograph showing a close-up of the AL1 electric locomotive. This catalogue carried the number R855.

1970 Catalogue

1970 saw significant changes. The shape of the catalogue returned to landscape for the first time since 1962 and a Terence Cuneo picture, 'Express Engines at Tyseley', was once again used as the cover illustration. The mouse sat between the tracks. Unlike the last two pictures commissioned by Hornby, this one did not include prototypes modelled by Hornby. This time the stars were GWR 'Clun Castle' and LMS Jubilee Class 'Kolhapur'. At the time, these two locomotives were very much in the railway conservation news and the subject was undoubtedly chosen to tie in with the move into old company liveries.

The message inside was 'New Look for the Seventies' and the new look took in System 6 track, flexible track, pre-nationalisation colours and a public opinion survey. Minic was obviously on its way out being reduced to just six pages at the back. Frog aircraft kits barely mentioned in 1969 now occupied the whole of the outer back cover.

1971 Catalogue

The cover of the 1971 catalogue naturally featured 'Evening Star' which was due in the shops that year. Cuneo again provided the painting and the mouse became an insulator on the telegraph pole. Narisa Chakra, in her book, tells us that the first sketches for the painting were made at Southall Shed.

The 1972 catalogue had already gone to the printers by the time the decision was taken to change the name of the system to Hornby Railways and so when it arrived in January that year it was marked Tri-ang Hornby like those before it. Another Cuneo painting adorned the cover, this time one he had painted for British Rail. It featured the Condor night freight train pulled by a Co-Bo similar to that modelled by Hornby-Dublo several years before.

Catalogue Supplements

From time to time Rovex issued catalogue supplements in order to clear old stock or to draw special attention to particular products.

In 1967 a two-page full-colour supplement (WS/1) was printed to promote the Miniville range. The latter were confined to one side only, the other side being used to advertise Funfair Games which were aimed at a similar market.

The 1968 catalogue supplement (R503) was again a full colour two-sided leaflet, this time just promoting the Tri-ang Hornby clockwork train sets.

A two-colour double-sided supplement published in 1970 (R854) was devoted to old stock. In particular it advertised items for which demand was likely to fall with the introduction of System 6 track.

1966 Australian catalogue supplement.

Supplement published in 1970 to help clear slow moving stocks.

AUSTRALIAN CATALOGUE SUPPLEMENTS

In Australia the 'catalogue' was usually seen as a supplement to the British one as the British range was also being imported. The Australian publication therefore contained only those items made or assembled at the Moldex factory in Melbourne. The first catalogue in 1958 it had been a four-page affair but by the following year it had doubled in size. In 1960 a small, full-colour booklet was available which lasted into 1961 with its own supplement. In 1962 a full colour fold-out leaflet was used which seems to have survived until 1965.

1965 and 1966

In 1965, a four-page full-colour leaflet was printed to the same size as the UK catalogue so that it could be slipped inside as extra pages displaying the Australian range. This contained only six locomo-

tives being made or assembled in Australia. These were:

Jinty
Princess
Sydney Suburban Electric
Double Ended Diesel
Bo-Bo Switcher (assembled only)
A Unit (assembled only)

The last three were shown with the name 'TRANSAUSTRALIAN RAILWAYS' in capitals on their sides, a form which, as far as I know, was not actually used on any Tri-ang models.

A similar leaflet was printed in 1966 but now it bore the name 'Tri-ang Hornby' and made clear its intention with the words:

"The items in this folder are additional to the 12th Edition 'Tri-ang Railways' Catalogue and complete the 1966 range".

The 1966 leaflet added the following:
R358s & R233 'Davy Crockett'
R251 & R33 Deeley/Johnson 3F 0-6-0
R253 Dock Shunter
R353 Yard Switcher
R352 Budd RDC
R355Y 'Connie'
R152 0-6-0 Diesel Shunter
R54 & R32 'Hiawatha'
Some of these had previously been in the 1962 supplement and in the case of some of them evidence of their actually having been assembled in Australia has still to be found.

Later Supplements

In 1967 the same format was used with locomotives on the front page, sets inside and everything else on the back page. The Budd, 'Davy Crockett', 'Connie', 0-6-0 diesel shunter and 0-6-0 3F had gone and the

Spot the difference! 1971 Australian and Canadian catalogues. Note how the models on the studio layout have been changed to match the catalogue contents.

only new locomotive was the top tank locomotive for starter sets.

By 1968, manufacture at Moldex was ending and set assembly was virtually all that remained. This was reflected in the supplement which, with the exception of the R17CA flat wagon with car, the R71 footbridge, lever frames and a page for Tri-ang Big Big, featured only sets. The only new loco was the A1A-A1A in BR blue livery and the whole of the front page was devoted to a picture of the Double Ended Diesel pulling a train.

The 1969/70 supplement also devoted the whole of the first page to an action picture, this time a model scene which included the EM2 which was neither assembled nor used in sets in Australia. Two more individual wagons were offered inside these being the R568A bogie bolster wagon with three cars and the R221A Shell Lubricating Oil Tank Wagon. One page of the supplement was devoted to toy train sets.

1971

Most overseas catalogue supplements were printed locally but in the early 1970s the Australian editions were printed in the UK and given the number R380. Because of this we know that in 1971, 34,800 were printed. This was a 1971/72 edition and carried the 'R' number R380. Like the UK catalogue that accompanied it, it had changed to landscape format.

The cover carried a picture of the same layout used for the Canadian catalogue although from a different angle, with the rolling stock available in Australia. For example the Canadian picture featured Canadian National and CP Rail stock and the former Hornby-Dublo station roof. The Australian version had a mixture of Transcontinental and British stock and the new Tri-ang Hornby station roof due out that year.

Inside the four-page leaflet were the train sets assembled for Australia and on the back page the various power units available there.

1966 New Zealand illustrated price list.

NZ Hornby 1972 North Island illustrated price list.

NEW ZEALAND ILLUSTRATED PRICE LISTS

In New Zealand, illustrated price lists were the norm by 1965. These were printed in two colours and that year they were black and a greeny yellow. The leaflet was approximately A5 size when folded and consisted of six pages. The price lists for 1966 and 1967 were similar, the latter being the first to have decimalised prices.

In June 1968, the shape of the leaflet changed. The new one was square when folded but unfolded to form a long sheet of eight pages. This year there was an attempt at modernising the range of models available and it was noticeable that New Zealand produced items were supported with others imported from the UK to make up more interesting sets.

In June 1969 a new price list was available and again it had eight pages but was of slightly changed shape. This was the first one to carry the company's new name 'Tri-ang Pedigree (NZ) Ltd'.

North Island Price List

An unusual and interesting New Zealand illustrated price list was marked as the '1972 North Island price List'. This was a double-sided sheet that unfolded vertically. Whilst printed in full colour, the colours of the copy I have seen were badly distorted as a result of too little blue. On the reverse it advertised Scalextric. The 'cover' was illustrated with a drawing of a 'Hiawatha' belching black smoke as it hauled its load through wooded country.

CANADIAN CATALOGUES

1965/66 Catalogue

By 1965 Canada had its own full-colour catalogue. As no Tri-ang Hornby models were made in Canada, the catalogue was devoted entirely to Margate-made models and Canadian-assembled sets. The main difference was that most of the models available in Canada were made especially for the North American market. Thus the 'Transcontinental' range was much extended for Canada and by the Tri-ang Hornby period were mostly in Canadian liveries.

The 1965 Canadian catalogue featured a Canadian National A Unit pulling a long train through the Rockies. The message was 'Tri-ang Railways for Canada'. Inside, the train sets carried appropriate names such as 'Le Champlain' and 'The Canadian'. The only Canadian livery at this time was Canadian National but it appeared to have been applied to as many things as possible. The name on the back of the catalogue was Lines Bros. (Canada) Ltd.

The same catalogue was used in 1966 but with it came a supplement of further items by then available. These included power units, presentation sets and a number of additional items from the UK catalogue. The name on the back of the supplement was Meccano – Tri-ang Ltd.

1967 Catalogue

This year the catalogue had one of the most attractive covers of all showing two North American railway scenes. In the top half was the classic scene featuring 'Davy Crockett' as the subject and in the lower half the modern scene with an A Unit pulled passenger train.

Inside, while CN liveries dominated, those of the Canadian Pacific Railroad had begun to appear. The catalogue displayed the new 'Tri-ang Railways' logo used only in Canada and the back page showed us that Meccano – Tri-ang Ltd had now moved to Brown's Line, Toronto.

1969 and 1970 Catalogues

A less inspiring picture adorned the cover of the 1969 Canadian catalogue showing Canadian model trains set out in parallel sidings crossed by a Victorian suspension bridge on which the TC Pacific (CP) pulled a train of pulp wood cars. More Canadian Pacific models had joined the range inside.

A similar picture was used for the 1970 catalogue, this time featuring the container depot crane and Canadian versions of container wagons. The new container range occupied most of two inside pages. CP Rail liveries had now arrived in large quantities eclipsing the short-lived older CP livery.

1971 Catalogue

In 1971, along with the UK catalogue and the Australian supplement the Canadian catalogue was turned 90° to become landscape format. As mentioned elsewhere, the picture was very similar to that used on the Australian supplement that year and on the cover of the new track plan book. All were shot during the same studio session but with models being switched around.

In passing, it is worth mentioning the ghost that appeared on the version used for the Canadian catalogue. The basic layout was unchanged but the new Tri-ang Hornby station roofs and terminus building that appeared in the British and Australian pictures had been replaced by the ex-Hornby-Dublo station roof and building as presumably there was still a good stock of these out in Canada at the time. The river bridge had gone and a viaduct was in its place and lineside huts and yard lamps had disappeared. A nice touch was the signals being turned round to allow right-hand running; a fact drawn to my attention by Peter Duckmanton of Australia who also noticed that, in the Australian catalogue picture, the 'whiskers' on the R159 diesel are upside-down!

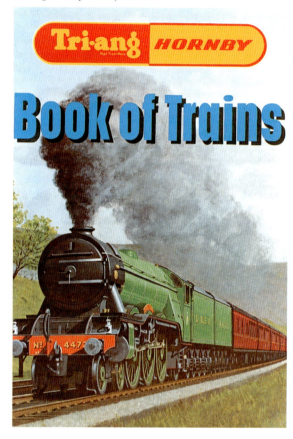

Tri-ang Hornby Book of Trains.

As one might expect, for the Canadian catalogue the rolling stock had changed to items from the Canadian range but the strangest thing of all was the turntable. When the picture was taken it must have still had the Transcontinental A Unit on the turntable, left over from the shot for the Australian supplement, and at a later date it was presumably decided that, as this was not included in the Canadian catalogue, a Canadian loco must be substituted. As it was too late to set up the display again, it was decided to superimposed a picture of a suitable model. The result was a loco and turntable bridge too large for the turntable and the ghost of a lamp post not quite lost during the touching up!

1973 Hornby Railways Catalogue for Louis Marx

It seems that no catalogue was published in 1972 as the Canadian company had disappeared in the break-up of the Lines Bros. empire. Canadian interests had ended up with Louis Marx & Co of Canada Ltd who promptly ordered stocks of the Tri-ang Hornby (now Hornby Railways) Canadian range to sell in Canada. To help sell the models Rovex Ltd printed a catalogue for Louis Marx. 50,800 were printed of which 200 remained in the Margate stores at the year end.

The cover illustration depicted two real Bo-Bo switchers in CP Rail livery pulling a freight train set against a rock wall background. This catalogue was in portrait format but its contents were similar to those that had gone before. There were new train sets and a number of new freight car variations promised. The catalogue also contained a number of locomotives and carriages from the British range and plenty of lineside accessories.

No further editions of this catalogue were published.

THE 'TRI-ANG HORNBY BOOK OF TRAINS'

Before the last war, Meccano Ltd annually published *The Hornby Book of Trains* which was half catalogue and half magazine describing work on the real railways. The *Tri-ang Hornby Book of Trains* was an attempt at reviving the idea but it did not really catch on.

The book was printed in 1968 and 1969 and was reviewed by SW Stevens-Stratten in the November 1968 edition of *Model Railway Constructor*. The book illustrated in colour most of the models made over the years by Rovex but, like its pre-war inspiration, also had a section on prototype operation.

While it is likely that 100,000 were ordered, only 94,000 of them passed through the stores. About 25,000 sold in the first year and 15,000 in the second; then sales went flat. Stockists received their supply together with an eye-catching counter display pack.

PRICE LISTS

Retail Price Lists

The price list has always been a good source of information for researchers as it is an indication of what was made and when it was released. The most common price list one finds is that published in January (or February some years) as this was supplied with the catalogues, most of which were distributed early in the year. In fact, Rovex published price lists at intervals during the year, partly to allow them to revise the list as models became available and partly to take account of major price alterations such as those resulting from tax changes. In the days of Purchase Tax, prices could change every time there was a Budget but since the adoption of Value Added Tax, tax levels have stabilised. It will be seen from the above that price lists issued after January are a guide as to what time in the year a new model was released.

Size of paper and ink colour changed over the years. The early Tri-ang price lists were 20 cm x 13 cm, first as single sheets but, as the range of models increased, they became folded leaflets.

Early price lists were printed in black but those in 1958 were in red and in 1959 they were in blue ink. At this time TT price lists were printed in a variety of colours but on yellow paper. From 1960 black was the standard colour for the 00/H0 lists and between 1963 and 1968 the paper size changed six times before settling down to 285 mm x 213 mm. In June 1965 a special price list was printed on green paper for Rovex Scale Models Limited listing only Hornby-Dublo and indicating that "Items listed may be obtained only from Hornby-Dublo stockists".

The following is a list of 00/H0 retail price lists known to have been published between 1952 and 1971 inclusive. Those in italics were the early illustrated ones printed on art paper:

Year	Cat. No.	Date	Date	Date	Date	Date
1952	-	1.1.52	1.5.52	1.9.52		
1953	-	*1.53*	*3.53*	*10.53*		
1954	-	*3.54*	*9.54*			
1955	1	*1.55*	1.3.55			
1956	2	1.1.56	1.5.56			
1957	3	1.1.57	20.2.57	1.3.57	1.9.57	
1958	4	1.1.58	1.2.58	1.5.58	1.9.58	
1959	5	1.1.59	1.4.59	1.5.59	1.7.59	
1960	6	1.1.60	5.60	1.10.60		
1961	7	1.1.61	21.8.61			
1962	8	1.1.62	5.62	19.6.62	9.62	8.10.62
1963	9	1.1.63	9.4.63	1.11.63		
1964	10	1.1.64	22.6.64	10.9.64		
1965	11	1.1.65	1.5.65	1.8.65	1.9.65	
1966	12	24.1.66	4.5.66	1.8.66		
1967	13	23.1.67	12.4.67	22.6.67	15.9.67	
1968	14	1.1.68	5.4.68	1.9.68		
1969	15	1.1.69	1.5.69	1.8.69	1.10.69	
1970	16	1.1.70	1.2.70	1.6.70	1.9.70	
1971	17	1.1.71	1.4.71			

In 1970, Rovex introduced dual pricing in readiness for the decimalisation of the British currency the following year. Dual pricing was repeated in 1971 but dropped thereafter.

Trade Price Lists

Compared to retail price lists, those for the trade were initially printed on very thin paper approximately 21 x 27 cm in size (sometimes wider). They were usually in red ink with the TT on yellow sheets attached at the back. The printing was larger and by 1958 the price list ran to four sides. From 1966 the trade price lists were printed in black on green paper and that first green edition included an extensive range of Hornby-Dublo stock that the company were trying to clear as well as the Tri-ang Hornby, Model-Land and Tri-ang Railways TT ranges.

There was often a better description of models and minimum quantities that retailers could buy were also quoted. There were two price columns, one for the trade price and the other for the retail price.

Initially, items were dropped from the trade price list when stocks at Margate ran out but sometime in 1970, there was a policy change. Items withdrawn during the year now remained in the wholesale price list but had the word 'withdrawn' printed across the price columns. In the retail price lists the obsolete models remained listed and priced so that stockists could clear their stock.

Overseas Price Lists

Some overseas retail price lists were printed by Lines Bros. and these seem to have the same format as UK lists, each with the name of the country for which it was produced, at the top. Trade price lists for export were also a standard issue from the UK and in the same style as those in use in Britain at the time but printed in a different colour or on a different colour paper. In 1965 they were printed in blue

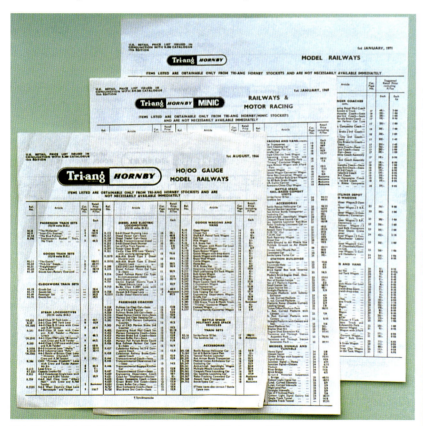

Retail price lists for the 00 range were printed in black on white paper.

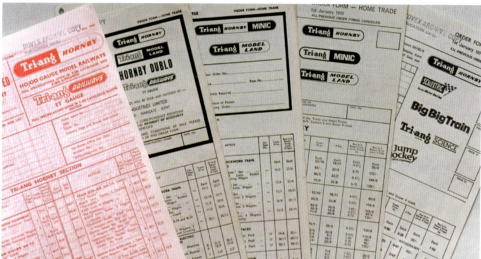

Trade price lists and order forms for 1965, 1967, 1968, 1970 and 1971, and Wholesale Department order forms (pink) for 1968 and 1970 together with one dated 1970 for toy trains (green).

on pale blue paper but from January 1966 the export trade price list was printed in brown on a cream paper.

Overseas, retail price lists were often locally produced but Lines Bros. marketing companies such as Lines Bros. (Holland) NV of Tri-ang House, Amsterdam also printed their own trade price lists for local companies buying through them. In

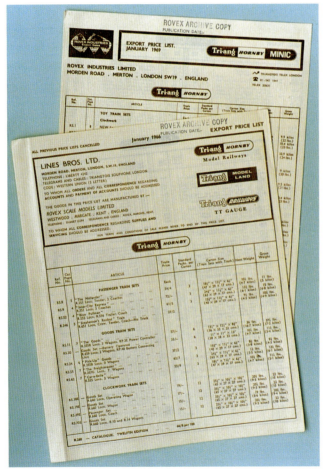

Export price lists were printed on buff coloured paper.

Holland local price lists were available at least from 1957 and throughout the 1960s and changed in shape and character almost every year. In 1961 it was a red sheet, in 1963 it included TT items and by 1970 Hornby Acho had been added.

Lines Bros. Inc. of the United States published their own trade price lists throughout the early and mid 1960s. These were simple, often typewritten, documents. Orders were sent to 236 Fifth Avenue, New York.

Australian Price Lists

In Australia, Cyclops & Lines Bros. (Aust) Ltd published the trade and retail price lists and these had a distinct character of their own.

The trade price list was printed in red on yellow paper and was stapled into a book of approximately quarto size. These are particularly useful for research purposes as the individual models were coded to indicate from where retailers should buy them. Those marked with an 'A' were Australian made or assembled while those marked with an 'E' had to be ordered from Europe (i.e. UK) and those marked with a 'C' were currently available locally but in future would come from the UK. The price lists seem to have covered the same product ranges as the retail price list detailed below.

The early Australian retail price lists were decorated in the margin with drawings of models. From 1962 the list also included Scalextric and Minic Motorway. As the range extended, the drawings disappeared to leave just the NSWR Suburban Electric in the top left hand corner by 1963. In 1965 a supplementary price list was produced listing ex-Hornby-Dublo equipment as 'Tri-ang Hornby' and the new military/space items added to the Tri-ang range that year.

The 1968 price list was headed 'Tri-ang Hornby Minic' in accordance with company policy but also included Scalextric under its own name. In 1969, more trademarks were added to the head of the

paper including 'Tri-ang Power Units', 'Big Big Trains' and 'Jump Jockey'. In 1970 the range was extended further to include Pedigree Dolls, Sindy, Paul, Patch, Pedigree Vinyl Soft Toys, Lifetime Games, Lifetime Teacher Toys and Lifetime Toys. In 1971 the Australian Retail Price List reverted to the products originating from the Margate Factory and/or made in Australia: Tri-ang Hornby, Big Big Trains, Scalextric and Jump Jockey. For 1972 and 1973, the price list contained only items originating in the UK from the Hornby Railways and Scalextric ranges and was headed for the first time with the name 'ROVEX'. It showed Cyclops Industries Pty. Ltd of William Street, Leichhardt, New South Wales 2040 as the 'Australian Agents'.

Canadian Price Lists

As in New Zealand (p435), Canada had illustrated price lists but these were replaced by a catalogue and separate retail price lists in 1965. These and the trade price lists were published by Lines Bros. (Canada) Ltd and were normally limited to railway products although some joint lists were used.

The trade lists were decorated with the Tri-ang Railways logo around the margin and a picture of a three part diesel unit (A and B units) at the head of the paper. The header boasted 'The World's Most Popular Model Railway' and at the bottom of the back page, the name 'Meccano – Tri-ang Ltd' was displayed.

The 1968 trade price list was headed 'Roland G. Hornby (associated with Meccano – Tri-ang)' but by the following year that name had been dropped and the price list was headed 'Tri-ang Railways, Hobby Division, Meccano Tri-ang'. In each case the address given was 95 Brown's Line, Toronto. In 1971 the name 'Tri-ang Hornby' appeared on the price list for the first time.

PACKAGING

It is likely that quite a lot of Tri-ang Railways packaging survived into the Tri-ang Hornby period as stocks were used up. Then, as stocks of individual sizes of box ran out, new printings would have been ordered to previous designs but with the logo changed. While the new logo remained for the whole of the seven years, the detail on the boxes changed to reflect the structural and name changes taking place in the Lines Bros. Group.

As we have seen above, there were two versions of the logo used on packaging: the long lozenge and the two-tiered lozenge. The long lozenge became the standard throughout the Tri-ang Hornby era. The two-tiered lozenge is described on page 429.

Chronology of Change

Initially, therefore, the packaging used was basically the same as that used for Tri-ang Railways products but over the next seven years, the following changes occurred to the printed detail:

1. Previous designs continued with the new logo. Both the long lozenge and the two-tiered lozenge were used in these early days. The boxes carried the words 'BUILT IN BRITAIN by ROVEX SCALE MODELS LIMITED'.

2. Boxes were now marked 'MADE IN ENGLAND ROVEX SCALE MODELS LIMITED'. Some of the boxes were in a darker shade of red and there were some oddities such as converter track in a black and white box, track sold in side flap boxes and the yellow colour missing from the printing for

The last of the Canadian range. Some had a printed label and some were just rubber stamped with the 'R' number.

Pile of Tri-ang Hornby boxes.

certain items. Boxes with red and white backs were still retained for some items such as points.

3. With the reorganisation of the Lines Bros. Group in 1967, 'ROVEX INDUSTRIES LTD' began appearing on packaging instead of 'Rovex Scale Models Ltd'.

4. In the final stage the name 'Rovex Tri-ang Ltd' appeared.

5. Emergency white or brown boxes began to appear.

Choice of box

While good records were kept of models made, no

British-made model sent bulk packed to Australia and packaged on arrival in a locally-made box.

Canadian bubble pack.

records were kept of the types of packaging used. This tended to change according to how stocks were running. For example, the 2-6-2 tank engine has been found in no fewer than four different types of box. When the intended box ran out before new supplies arrived the dispatch room staff would choose the nearest alternative.

This is a subject that would reward further research.

White and Brown Boxes

By 1970 economy measures were beginning to affect packaging. Window boxes were more expensive to buy than the simpler standard end flap types and in the belief that certain wagons were so much in demand that they did not need to be dressed up in a window box, it was decided to sell six of them in the cheaper cartons.

As 1971 progressed, stocks of some printed boxes ran out. With uncertainty about the future of the company, there was naturally a reluctance to order new batches and so unprinted white or brown boxes were used instead with labels stuck on each end to identify the contents. The labels were printed in house as and when required.

During this time of emergencies, it was not always possible to use a box of the correct size and there are a number of examples of a relatively short wagon

being packed in a coach box as there was nothing more suitable. To stop it rattling about yellow tissue paper was stuffed in each end.

Models known to have been packed in white or brown boxes include the following:

No.	Model	Box colour	Notes	Label
R861	'Evening Star'	white	blue interior	TH
R355G	0-4-0 Industrial Tank	brown		TH
R354	'Lord of the Isles'	white	blue interior	TH
R35	B12 Tender	white		TH
R350	L1	white	white interior	TH
R751	Co-Co Diesel	white	yellow interior	TH
R33	3F Tender	white	no box markings	none
R746	LNER Thompson Brake 3rd	white		TH
R747	LMS 1st/3rd Composite	white		TH
R743A	GWR Composite	white		TH
R214	Hopper Wagon	white		TH
R214	Hopper Wagon	brown		TH
R781	NER Coke Wagon	brown		TH
R145	Modern Signal Box	brown		TH
R457	Incline Piers	white		TH
R269	Victorian Suspension Bridge Extension Set	white		TR
R80	Station Set	white		TR
RM921	Car Loading Ramp			Minic
RV258	Open Wagon	white		Big Big
RV277	Freightliner Wagon	white		Big Big

The above list is not comprehensive.

Some models produced for mail order sales were also packaged in white boxes for cost reduction and these should not be confused with the above.

Not all stocks of Tri-ang Hornby boxes were used up at this time. Some survived well beyond the name change and there are examples of silver seal

Emergency packaging used in the closing days of Tri-ang Hornby meant boxes much too large for the contents and, when these ran out, the use of brown or white card boxes bought in as a stop-gap.

models, introduced in 1973, being sold in a Tri-ang Hornby box with the silver seal tabs stuck over the ends.

Sets

Initially the packaging for Tri-ang Hornby sets was the same as before but inscribed 'Tri-ang Hornby' instead of 'Tri-ang Railways'.

After this the style of box was frequently changed. For a while window boxes with vacuum-formed plastic trays were used which avoided the need for expensive artwork. These, however, were weak and did not survive well. In 1970 the sets were sold in boxes with end flaps. These did not survive the constant pulling out and pushing in of the plastic tray and in 1971 Rovex returned to using set boxes with lift-off lids. They also had attractive artwork printed onto a large label that covered most of the top of the box. The artwork depicted scenes of the real railways using the prototypes of the models in the sets. Thus, some of the best set boxes during the Tri-ang Hornby years are to be found at the start and end of the period.

Overseas Production

Australia and New Zealand continued to print their own boxes in the style of those coming out of the factory at Margate. In New Zealand the quality of card used was good but that used by Moldex in Australia was sometimes rather thin. As late as 1968, however, the New Zealand company was sticking Tri-ang Hornby labels on Tri-ang Railways boxes in order to use up old stocks. This came to the notice of the men at Merton and they were instructed to print new boxes.

Locomotives in white boxes.

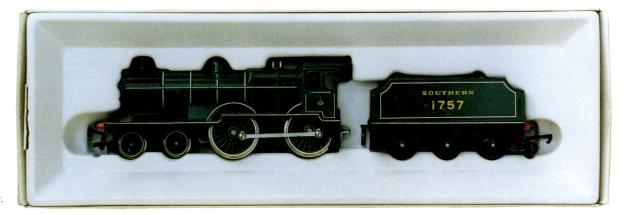

The L1 was not officially released until 1972 and the Hornby Railways era but even so it initially had a white Tri-ang Hornby box.

Set out here in the studio ready for a photograph to be taken for the
1969 Tri-ang Hornby catalogue are the models available at that time.

GUIDE TO RELATED PRODUCTS

In this chapter we look at other products from the Margate factory together with related products from members of the Lines Bros. Group of companies. The following table provides 'at-a-glance' information about many of the associated products covered in this series of books and indicates which volumes include what material and where it may be found:

Product	Detail	Vols.	Chapters
Minic Push-&-Go	Plastic toys of the 1950s of approximately the right scale to complement Tri-ang Railways.	1	'Related'
Tri-ang Railways TT	The TT scale model railway system made at Margate during the late 1950s and early 1960s.	1 & 2	'Related'
Minic Motorway	A slot-car highway and racing system developed and marketed on the back of Tri-ang Railways in the late 1950s and throughout the 1960s. Production was moved from Canterbury to Margate in the late 1960s.	1 & 2	'Related - Minic'
Battle-Space	Part of Tri-ang Hornby range. (Late 1960s)	2	'Wagons'
Hornby-Dublo	Disposal of stock and tools. Locomotives briefly available as Tri-ang Hornby. Buildings absorbed into the Tri-ang Hornby range.	2	'History' 'Locomotives' 'Buildings' 'Related - Wrenn'
Car-Play	A different way of marketing Minic Motorway. (late 1960s)	2	'Related - Minic'
Real Estate/ Model-Land	A series of coloured plastic building kits, originally made by IMA Ltd, transferred to the Rovex factory at Margate in the early 1960s.	1 & 2	'Related'
Minix	This was a Rovex product introduced in the mid 1960s to complement the model railway system.	1 & 2	'Related'
Tri-ang Lionel Science Sets	A series of science kits developed by Lionel of USA and made under licence at Margate. (Early 1960s)	1	'Related'
Miniville	A series of cheap toys developed from existing ranges for sale through small shops. (Late 1960s)	2	'Sets' 'Related - Model-Land'
Tri-ang Ships	A good range of diecast 1:1200 scale ships made by Minic in the early 1960s.	2	'Related - Minic'

Product	Detail	Vols.	Chapters
Minic Ships	A re-introduction of a reduced range of 1:1200 scale diecast ships in the mid 1970s by Hornby Hobbies.	2 & 3	'Related - Minic'
Arkitex	Building construction system developed by Spot-On Ltd which included a station kit for Tri-ang Hornby.	2	'Related - Spot-On'
Tri-ang Wrenn Railways	A re-launch of former Hornby-Dublo models by Tri-ang subsidiary, G&R Wrenn, in the late 1960s.	2	'Related'
Wrenn N Micromodels	Lima N gauge imported by G&R Wrenn in the late 1960s.	2	'Related - N Gauge'
Lone Star	An N gauge model railway system offered to Lines Bros. in the early 1960s	2	'Related - N Gauge'
Hornby Acho	French Hornby remained in production under Tri-ang ownership until 1972.	2	'Related'
Tri-ang Big Big	An O gauge garden toy railway developed at Margate in the late 1960s.	2	'Related'
Frog Kits	Plastic aircraft and other kits, production of which was moved to Margate in the mid 1960s.	2	'Related - Frog/ Pedigree'
Pedigree	Doll and toy production at the sister factory in Canterbury came under Rovex control in the late 1960s.	2 & 3	'Related - Frog/ Pedigree'
Sindy	Sindy was a highly successful product of the Pedigree Division which remained an associate range well into the 1970s.	2 & 3	'Related - Frog/ Pedigree'
Novo Toys	The name given to toys manufactured in Russia from Tri-ang tools in the late 1970s.	2	'Pedigree' + 'Big Big'
Scalextric	Production of Scalextric was moved into the Margate factory at the end of the 1960s.	2 & 3	'Related'
Jump Jockey	A horse racing development of Scalextric made at Margate in the early 1970s	2	'Related - Scalextric'
Scalex Boats	A series of boats transferred to Margate with Scalextric.	3	
Tri-ang Science	A science series from Rovex Tri-ang in the 1960s.	2	'Related - Other'
TMNR	A ride-on garden railway developed by Minic in the early 1960s.	2	'Related'
Ibertren	An associated company in Spain that developed an independent model railway system.	2	'Related'

Product	Detail	Vols.	Chapters
Tank Tactics	A game that used a large scale remote-controlled tank and field gun.	2	'Related - Other'
Arrow Jigsaw Puzzles	Arrow were an associate company whose range of jigsaw puzzles included some based on Tri-ang Hornby catalogue covers.	2	'Related - Other'
Hornby Minitrix	An N gauge model railway system made by Trix in Germany but marketed by Hornby Hobbies under their own name.	3	
Hornby Stamps	Packets of stamps available for a short time in the late 1970s.	3	
Shipwright	A short lived and undeveloped series of ship kits marketed in the late 1970s.	3	
Hornby Steam	The G scale steam model of 'Rocket' and accessories available in the early 1980s.	3	
3-D-S	A short-lived space track series of the early 1980s marketed by Hornby Hobbies.	3	
Crown Railways	A model railway system made in India which was a copy of Hornby Railways.	3	

Product	Detail	Vols.	Chapters
Playtrains/ Early Play	Push-along trains closely related to Hornby Railways.	3	
Hornby Toys	This is a group of various toys sold by Hornby Hobbies in the 1980s including Mites, Pound Puppies, Ani-mates, Gro Toys, Boo Boos and radio controlled vehicles.	3	
Flower Fingers	Fun with artificial flowers in the late 1970s and early 1980s.	3	
H&M Systems	Disposal of the power unit stocks resulting from the take-over of Hammant & Morgan.	3	
System Plus	A 'Lego' like system linked to Hornby Railways and Scalextric. Not developed.	3	
Spacelink	A space series, including Capsela, in the early 1980s.	3	
Flower Fairies	A successful series of doll fairies based on the poems and illustrations by Cicely Mary Barker.	3	
Thomas The Tank Engine & Friends	An extension of the Hornby Railways system but themed on the television films which were themselves based on the books by the Revd. Awdry.	3	

MODEL-LAND & SCENICS

INTRODUCTION

The history of the Model-Land range, with its origins in the Real Estate series made by International Model Aircraft, is described in some detail in **Volume 1** (page 410) and will not be covered again here other than to repeat that production of the kits was transferred from IMA Ltd at Merton to Rovex Scale Models Ltd at Margate in March 1963 and that the name of the kits was changed around this time from 'Real Estate' to 'Model-Land'. The individual models, however, were only touched on briefly in the first volume and, as many of them were exclusive to the Tri-ang Hornby era, the range is described here in greater depth with the help of notes provided by James Day who has made a detailed study of the range.

PACKAGING

The Real Estate/Model-Land series experienced a number of changes in packaging based on two distinct designs – pink boxes and blue boxes.

Pink Boxes

The pink box was actually orange and pink, a psychedelic colour mix which was popular in 1960s Britain. The lift-off lid was made in stiff card. The earliest had a black and white picture. These I have classified as box type A. Later batches had a different and coloured picture and these I have called box type B. Those originally printed for the Real Estate kits made and marketed by IMA (both types A and B) were inscribed with 'Made in England by International Model Aircraft Ltd' on one box end. Following the decision to transfer production of the kits to Rovex at Margate, however, any further

batches of boxes printed at that time had this statement deleted and nothing substituted. These I have called box type B1.

The new Model-Land boxes adopted the same design and colour scheme as the earlier Real Estate ones but with the product name changed and the box end now inscribed 'Built in Britain by Rovex Scale Models Limited, Westwood, Margate, Kent'. These first Model-Land boxes I have called type C. While most of the pink boxes had lift-off lids, some later ones were folded from a single piece of card providing tuck-in flaps at both ends and these I have called type D. The only example of this type of box that I have seen contained a model made in 1965 but containing an instruction leaflet printed early the previous year.

Pink boxes A, B, B1 and C came in four sizes:

Size 1	90 x 85 x 25 mm
Size 2	160 x 135 x 55 mm
Size 3	205 x 155 x 45 mm
Size 4	260 x 175 x 45 mm

The dimensions are given as length x width x height and box D was approximately size 3.

Blue Boxes

The blue box was made like box D with one piece of card that had been folded up to leave a tuck-in lid instead of end flaps. It was initially a window box with an illustrated card behind the window and this I have called type E. Later it was replaced by an almost identical box but which had a non-window printed lid. This I have called box type F.

I have seen only two sizes of blue box:

Size 1	155 x 135 x 40 mm
Size 2	255 x 170 x 40 mm

Type	Box colour	Name	Maker	Picture
A	Pink/Orange	Real Estate	IMA	Black & white, square-on picture, '61
B	Pink/Orange	Real Estate	IMA	Coloured, quarter angle, '62
B1	Pink/Orange	Real Estate	IMA	As B but with the makers inscription removed from the box end, '63
C	Pink/Orange	Model-Land	Rovex	As B but now printed with Model-Land name and new address, Spring '64
D	Pink/Orange	Model-Land	Rovex	Same picture but box folded from a single piece of card, end flaps, '65
E	Blue	Model-Land	Rovex	Coloured picture on card behind a lid window, folded card with top flap lid, '66
F	Blue	Model-Land	Rovex	Same type box and picture which is now printed directly onto the lid, '68

Box Check List

A summary of these box types is given above.

The dates are for the introduction of each box type and are calculated from the limited facts available and not from a written record. We are helped, however, by a practice performed between the transfer of production to Margate in March 1963 and the spring of 1966, of rubber stamping the underside of boxes with the kit number (prefixed with an 'ML' for Model-Land) and the date the kits were packed.

One of the first kits made after the move to Margate was the RML7 Garage which was in production at Margate during late April and early May 1963. Unfortunately the 'ML' rubber date stamp was not ready in time for the start of production and an ordinary date stamp was initially used. The purpose-made stamp came into use sometime between 30 April and 8 May that year.

It will be noted that many type B Real Estate boxes carry an 'ML' date stamp suggesting that large stocks of Real Estate boxes had to be used up by Rovex before they could print their own with the Model-Land name on them. This means that a good number of models assumed to have been made by IMA because they are found in Real Estate boxes, carrying the IMA address, were in fact moulded at Margate by Rovex.

The latest box seen with a rubber date stamp marking was packed on 25 March 1966. This was a Village Inn (RML1) which was presumably made at that time and yet here it was being boxed up in the old Real Estate type B box three years after the name Real Estate had been dropped!

In the check list on page 450, the letter codes have been used to denote in which types of box each kit was sold. This information is based on limited study and should not be considered comprehensive.

The 1967 Model-Land leaflet and the original studio setting for the cover picture.

Box types.   Type A   Type B   Type C

Type B

Type B1

Type C

Type D

Type E   Type F

Kits in blue boxes were typical of the Tri-ang Hornby years.

MINIVILLE

Four of the models were chosen for manufacture in the company's wholesale range in 1968. Each kit also contained parts from which a Minix car (without an interior) could be assembled, and were sold in a polythene bag with an illustrated top card stapled on. They included a printed layout card on which the model could be stood. The buildings themselves were simplified by the exclusion of small parts such as planters or, in one case, the kit's base plate.

DAPOL

The model company Dapol, best known for its acquisition of tools from other manufacturers such as Airfix, Trix and Wrenn, in the late 1980s acquired some of the tools for making the Model-Land building kits. They did not acquire all of the tools and this led to new and interesting variations with some of the buildings.

Dapol adapted the kits according to the demands of the market. For example, with the availability of many excellent bus and coach models in 00 scale from the toy companies, Corgi and Exclusive First Editions, there was a market for a bus garage and the Supermarket kit (RML60) was adapted to this. They also used different names for the shops, including Beatties, and different window displays. On the whole Dapol models were in brighter colours than their Model-Land predecessors but usually

lack the detail such as window boxes and planters and rarely carry painted detail. Some of these variations will be found below under the individual models.

CONSTRUCTION

In true Tri-ang style, the Model-Land kits shared parts. For example, kits RML1, 2 and 14 all had the same roof and wall units with different parts added, and the roof turned up again in RML13. Similarly kits RML4, 5, 7 and 15 had common wall and roof units and were withdrawn at about the same time. A similar arrangement was made for RML18 and 19 and again with RML 10, 11, 12 and 21. The roof pieces from kits RML 4, 5, 7 and 15 turned up again in the later kits RML 38, 44 and 45.

Modular kit assembly was to be used again later with the modern shop and office units in this series. This sharing of parts was one of the reasons for the demise of the series. For example, windows for several models were made on one tool but, if all the models for which that tool made windows were not in production at the same time, there was considerable wastage of mouldings. It also meant that parts had to be sorted out at the moulding machine which was a labour intensive task.

Doors came in a fairly wide range of colours which provided small variations to a given model. The curtains were printed onto the glazing piece but towards the end of production these were dropped in order to keep down the cost of kits.

The Miniville wholesale series made use of some of the Model-Land buildings, renaming them and adding a Minix car.

The observant will notice that the railway in this 1966 catalogue picture is not the Tri-ang Hornby 00 system but Tri-ang TT!

The Miniville models were:

	Miniville name	Date			Made	Model-Land name
W20	Manor House	68	formerly:	RML2	2,000	The Grange
W21	Bahama Villa	68	formerly:	RML12	1,800	Bermuda Bungalow
W22	Ideal Homestead	68	formerly:	RML19	1,400	Parkview
W23	Super Service Station	68	formerly:	RML68	1,700	Garage

No.	Model	Directory section	Dates made	Quantity made	Box
RML1	Village Inn	village buildings	62,63	11,200	A,B
RML2	The Grange	village buildings	62,63	8,100	A,B
RML3	Wood Shed	village buildings	62?,63	6,600	A,B
RML4	Marigold Cottage	village buildings	62?,63	8,500	A,B
RML5	Dove Cottage	village buildings	62,63	7,400	A,B
RML6	Hardware Shop	village buildings	62?,63	6,900	A,B
RML7	Garage	village buildings	62?,63	7,000	A,B
RML8	Accessories Pack	village buildings	62?,63	7,500	A,B
RML9	Oak Tree Cottage	village buildings	62?,63,65	12,600	A,B
RML10	Hollywood Bungalow	modern homes	63,65,68	10,700	B,F
RML11	San Fernando Bungalow	modern homes	63,65,68	11,900	B,F
RML12	Bermuda Bungalow	modern homes	63,65	11,600	B
RML13	Kent Bungalow	modern homes	63,66,68	11,400	A,B,F
RML14	Post Office	village buildings	63,65,68	15,900	A,B,F
RML15	Tea Shoppe	village buildings	63	8,700	A,B1
RML16	Villa Capri	modern homes	63	7,000	B
RML17	Church with Chimes	village buildings	63,65,68,70	15,000	B,C,E
RML18	Heathview	modern homes	64	7,000	C
RML19	Parkview	modern homes	64	7,300	C
RML20	Wishing Well	village buildings	64	6,700	C
RML21	Semi-Built Bermuda	modern homes	65	8,400	D,F
RML22	Lighting Unit	lighting	63-66,68,69	30,000	bag?
RML23	Koepe Pit Head Gear	industrial	not made	-	-
RML25	Winding House	industrial	not made	-	-
RML26	Winding Engine	industrial	not made	-	-
RML27	Boiler House & Stack	industrial	not made	-	-
RML35	Gas Holder	industrial	not made	-	-
RML36	Small Shop Single Storey	shops and offices	66,68	4,400	E
RML37	Medium Shop	shops and offices	66,68	6,600	E,F
RML38	Town House	shops and offices	68	2,600	E
RML44	Small Shop/Office Block	shops and offices	66,68	3,600	E
RML45	Medium Shop/Office Block	shops and offices	66,68	6,100	E,F
RML46	Store/Office Block	shops and offices	66,68	6,300	E,F
RML50	Tree Small	trees	not made	-	-

No.	Model	Directory section	Dates made	Quantity made	Box
RML51	Tree Medium	trees	not made	-	-
RML52	De-Havilland Trident	vehicles	not made	-	-
RML55	Police Station	town centre	not made	-	-
RML56	Town Hall	town centre	not made	-	-
RML57	Georgian House	town centre	not made	-	-
RML58	Village Church without Chimes	village buildings	63-66,68	10,000	B,C
RML59	Factory	industrial	not made	-	-
RML60	Supermarket	shops and offices	66,68	5,000	E
RML61	3 Pylons	industrial	66-68	13,000	E
RML61T	12 Transmission Towers	industrial	68-69	11,000	?
RML62	Light Green Scatter	scenery	64	10,200	bag
RML63	Dark Green Scatter	scenery	64	10,900	bag
RML64	Brown Scatter	scenery	64	9,700	bag
RML65	Factory Power House	industrial	68	4,000	E
RML66	Factory Loading Bay	industrial	68	3,000	E
RML67	Factory Twin Prod Bays	industrial	68	3,800	E
RML68	Garage Single Storey	shops and offices	66	2,000	E
RML69	Vac Formed Scenery	scenery	not made	-	-
RML70	Figures Set 1	figures	64,65,67,70	12,000	blister
RML71	Figures Set 2	figures	64,65,67,70	12,000	blister
RML72	Figures Set 3	figures	64,65,67,70	12,000	blister
RML73	Figures Set 4	figures	64,65,67,70	13,000	blister
RML74	Figures Set 5	figures	65,67,68,70	14,000	blister
RML75	Figures Set 6	figures	65,67,70	12,000	blister
RML80	British Taxi	vehicles	not made	-	-
RML81	BR Van	vehicles	not made	-	-
RML82	GPO Van	vehicles	not made	-	-
RML100	Cement	cement	64-70	640,000	loose
RV60	Ballast Grey	scenery	65-72	31,000	bag
RV61	Cornish Yellow	scenery	65-70	18,000	bag
RV62	Spring Green	scenery	65-72	76,000	bag
RV63	Autumn Green	scenery	65-71	45,000	bag
RV64	Earth Brown	scenery	65-72	39,000	bag
RV65	Moorland Mixture	scenery	65-68,70-72	31,000	bag
RV66	Poppy Red	scenery	65-67	7,000	bag
RV67	Lichen	scenery	65-70	19,000	bag
RV68	Cork Bark	scenery	65,67-70	18,000	bag
RV69	Mod-Roc	scenery	68	4,800	bag?

Base Plates

All of the series up to RML21 had base plates which were usually a fawn or grey moulding; in the case of the wishing well and wood shed, however, the bases were green. Early bases had some of the features picked out in colour, eg white doorsteps and drain surrounds with black drains. An early version of the RML10 Hollywood Bungalow came with a black driveway. These painted details were lost in an early cost cutting exercise. Later models in the series were issued without bases.

The base provided in the church kit was vacuum formed and covered in green flock. This piece is often missing in kits that have survived. Earlier examples of the base had a raised section marking the boundary of the building over which the church fitted but in later kits the base was flat.

The base plates provided with early kits contained one or more circular holes into which the lighting unit (RML22) could be inserted.

CHECK LIST OF MODEL-LAND & SCENICS

On page 450 is a table of all the Model-Land kits and associated scenic materials, planned and/or made, cross-referenced to the directory below. The dates the model was known to be in production are given with the total number made and the type of packaging used.

DIRECTORY OF MODEL-LAND & SCENICS

VILLAGE BUILDINGS

Village Inn RML1

The building had an exposed wood frame but with parts of the upper walls tiled. The walls of this model and RML 2 and 14, with which it shared parts, came in two variations. Early versions had black mouldings with the panels between these spray painted white. Later kits had white wall mouldings with the timbers coloured black. The windows were printed with a diamond pattern to give a leaded appearance. With the building came an inn sign on a post with its own base. There were also window boxes, planters and a seat. It was sold in a size 2 box and had a brown roof whose design is shared by several other models in the series.

Dapol re-released this model as C52 and called it the 'Village Local' or 'Village Inn' but theirs had the flower planter and seat missing. It also lacked the timber cladding on the right-hand end wall and featured a brown inn sign and wall tiles.

The Grange RML2

Sold in a size 2 box, this model was the same shape as the last one and shared a roof, a base and some walls. Also like RML1 it had a Tudor look with white walls, external timber-work (note the two moulding colours referred to above) and leaded windows. However, its 'between-the-wars' suburban roof suggested that it was mock-Tudor. The kit was supplied with lichen that could be used to create garden plants on the base plate.

The model was one of the four released in 1968 by the Wholesale Department in their Miniville series. As such it was the Manor House W20. Dapol later re-released the model as C50 'The Mount' with a black roof.

Wood Shed RML3

More suited as an accessory to one of the other buildings, this was released as a model in its own right. Due to its small stature it came in a size 1 box. The 'wooden' walls were dark brown and the roof red. Attachments, finished in aluminium colour, included a bent chimney, a bath to hang on the wall, a drain pipe and a water butt. As previously mentioned, the model had a green base.

Marigold Cottage RML4

This was made up from Dove Cottage (RML5) and the Wood Shed (RML3) but had a grey base plate of its own with moulded crazy paving picked out with colour. The walls were white and the roof red; the roof of the timber lean-to shed being grey. The doors and window frames were red as were the guttering and down pipes. The kit was packed in a size 2 box.

The model was reissued by Dapol as C76 in their Dapoland series under the name of 'Buttercup Cottage' with a grey roof.

Dove Cottage RML5

This was by far the smallest of the cottages and, indeed, was so small that it came in a size 1 box. It was made in 1962 by IMA and a second batch was made by Rovex in September the following year.

It had white walls and a simple pitched grey slate roof. The doors, window frames, gutters and chimney pot were red and the porch roof black. The base plate was pale brown with door steps picked out in white. The down pipes were grey and probably moulded with the roof. Lichen provided in the kit could be used to create climbing plants on the walls.

It was re-released as Dapoland model C66 'Robin's Nest'.

Hardware Shop RML6

Packed in a size 2 box, the Hardware Shop had cream walls and a grey roof and base. The most striking features were its bright green doors, window frames, guttering, down pipes, name board and shop front. The roof was double pitched and there was a small lean-to on one end of the building. The base was large enough for a parking place or garden beside the shop and this was provided with tall, closed, board double gates (also in green) and a brick wall and fence. Three pollarded trees were provided to go inside this area.

This was proposed in the Dapoland range, called 'Ironmonger's Store' and numbered C69 but at the time of writing it had not been released.

Garage RML7

Named 'Tri-ang Garage', this was a very attractive model which could be complemented by the addition of Minix cars. It consisted of two white walled buildings joined together, the smaller residential unit being RML5 Dove Cottage. The roof was either all red or part red and part grey. It also had either a red or yellow matching gate and front door. The larger roof was the same as that used for the RML18 and RML19 kits.

The gate referred to above was set in a stone wall surrounding the front garden. In reality this would have been sacrificed for a vehicle parking area as was so often the case with village garages. With the kit came a petrol pump and an open/closed sign in red.

The model was first made by IMA probably in 1962 and sold in size 2, type A, boxes. As we have seen, another batch was made at Margate during late April and early May 1963, the new stamp had arrived and was used on the rest of the batch. These were sold in a type B, size 2, boxes.

Dapol have indicated their intention to re-release the model as C53 'Country Garage'.

Accessories Pack RML8

This contained four small models with parts each moulded in three different colours so that they could be mixed when boxed up. The models were:
1. Stile – a base with a short three-bar fence and stile, finger direction sign on a post and a notice board on a pole.
2. Memorial – a single moulding of a needle-style stone memorial with three steps at its base.
3. Stocks – a base with both leg stocks and a head stock and seat.
4. Trough & Inn Sign – a base with a stone water trough and inn sign.

Dapol have re-released these as individual models in grey plastic.

Oak Tree Cottage RML9

Oak Tree Cottage was the only thatched building and came in a size 2 box. The thatch and the base were pale brown with paving slabs on the base picked out in white. The walls were white and the single door and window frames were red.

A slightly deeper red was used for the small tiled parts of the fire place/chimney protrusion on one end of the building and these matched the roof of the lean-to wooden shed at the other end. The latter was immediately recognisable as the Wood Shed RML3. The model was finished with a white paling gate and posts and lichen with which to make a hedge around the property.

Post Office RML14

This was a third version of the model that had already appeared as RML1 and RML2 (note the two moulding colours referred to for RML1 above). The facade had been changed by moulding a new panel which gave the building two shop fronts and a more modern appearance. This was always moulded in white. The larger of the shops was shown in the leaflet as a greengrocer's and the smaller one, which protruded slightly, was a post office with a glued on post box. The leaded front windows were replaced with sash ones which were later used in the brick town house RML38 and its derivatives.

A late version of the post office had a cream base instead of the normal brown one. The model also normally had a brown roof but a late example has been found with a red one. There was also a version with a cream base, bright blue roof and green shop windows. This is believed to date from 1968.

An early batch in type A boxes was made in 1963 and a second batch in type B boxes in June that year. Both batches were sold in a size 2 pink box and presumably the final batch which was made in 1968

was sold in size 1 blue boxes.

A post office van was proposed as RML82 in the Model-Land series but did not go into production. This would have complemented the Post office building which has been listed as C51 in the Dapoland range, although, at the time of writing, this is still awaited.

Tea Shoppe RML15

A small single-storey building with the wood shed attached at one end and a smaller lean-to at the other end. All were white while the doors were red. The model had a brown base and a free-standing 'inn' sign. The kit also contained a white gate, benches and some lichen with which to make creepers and a hedge.

The kits were sold in size 2 boxes, the second batch in B1 type boxes being assembled in July 1963. The model was given the number C70 in the Dapoland range and renamed the 'Coffee Shop'.

Village Church with Chimes RML17

This must be the best known of the Model-Land series both because of the typically Tri-ang idea of providing it with bells and because of its reintroduction to the Hornby Railways range in the 1980s, this time without 'bells'.

Through necessity the church was somewhat under scale but, even so, was an impressive building with immediate appeal. It was based on a stone prototype with a red tile roof which always looked wrong. It had a square tower with a clock on three faces and a flagpole plus a St George's flag. The doors of the church changed colours over the years. While most seem to be buff or light grey they ended up brown.

A small sheet of clear plastic streaked with various coloured paints was provided to cut up and stick behind the windows giving the appearance of stained glass and the base board was vacuum formed and coated in a very fine green flock to represent the grass of the churchyard. Yellow had been sprayed outside the church doors to represent gravel. The base, which is usually missing from second-hand models, contained a hole for a lighting unit. (Note the comments on variations in the section on bases above).

The 'bells' mechanism was a small unit from a music box made by Matthey of Switzerland, which when wound up with the fixed key played a peal of bells. The mechanism, which came wrapped in a piece of waxed paper and a roll of corrugated card, was fixed to three prongs on the inside wall of the tower attached to the church and the key protruded through the wall into the main body of the church. This meant that the musical unit could not be wound up without lifting up the church. On the other hand the concealed key did not spoil the appearance of the model.

The kit was sold in a size 4 pink box and later in a size 2 blue one. It was one of only two kits in the series to be revived in the Hornby range in the 1980s as R599; the other being the set of electricity pylons.

Village Church without Chimes RML58

This was the same as RML17 but did not include the musical unit. It was sold in the same pink box but with a small printed label inscribed '58 Village Church without Chimes' stuck on each end and a larger one beside the picture on the top. The blue box, which I have not seen, probably had a differently printed card behind the window.

The Church kit (R599), reintroduced by Hornby Hobbies in 1986, was without chimes. It is different from the original. The stained glass windows were much improved and the clock, which had originally had a black face, was now white. The roof was red.

Wishing Well RML20

This had a green base, dark brown timber work and a stone-effect well surround. A spindle, winding handle and bucket were also provided but these

are often missing from second-hand assembled models.

The kit was made in March 1964 and was sold in a size 1 pink box.

MODERN HOMES

Hollywood Bungalow RML10

Four of the Model-Land kits formed a group of luxury bungalows; one of them (RML21), still only in the construction stage. The other three, RML10-12, shared a common roof.

The Hollywood Bungalow and its detached garage had green pantile roofs. The walls were white and the window frames brown. Both the bungalow and the garage had light oak double doors with large wrought iron hinges and coach lamps (which glowed if a lighting unit was fitted inside the house) on wrought-iron brackets on either side. The back door on my model is blue. The external chimney, the front wall and gateposts were of stone and lichen could be used to provide plants along the top of the hollow garden wall. The driveway gates and the gate between the bungalow and the garage were black wrought iron to match the other iron features. The large grey base plate was moulded to represent crazy paving except for the driveway to the garage. Early examples exist where the driveway has been painted to resemble tarmacadam.

It was originally sold in a size 3 pink box and later in a size 2 blue box. Dapol acquired the tools for this model in the late 1980s and reintroduced it as C65 'Executive Bungalow' but it was a shadow of its former self. They had not acquired the tooling for the garage and so it was given the smaller one from the Villa Capri.

San Fernando Bungalow RML11

This was a cheaper version of the Hollywood Bungalow. It used a smaller base, still grey and with a crazy paving effect, and the garage, front wall and

gates had been dispensed with. In all other respects it appeared to be the same including the glowing front door lamps. The kit sold in a size 3 pink box and a size 2 blue box. In the Dapoland range it was C63 'Holiday Bungalow' and had a grey roof instead of the familiar green one.

Bermuda Bungalow RML12

This had the same roof as the last two models, but in blue, and the same base as the last kit but used the opposite way round and minus the stone exterior chimney. One corner of the building was recessed to give a covered area outside the front door, this corner being supported by a brick pillar. The walls were again white but the window frames and doors were green. The covered area by the front door was painted red. A bird box on a post was provided for the garden and lichen formed bushes.

The kit was sold in a size 3 pink box. Later kits may have been released in D type boxes but I have not seen any.

The model was one of those adopted by the Wholesale Department in 1968 for their Miniville series in which it was W21 Bahama Villa supplied without a base plate. Dapol later offered it as C64 'Retirement Bungalow' with a bright red roof.

Kent Bungalow RML13

This was the smallest of the bungalows and used the roof and the base from RML1 the Village Inn. A rarer version of the model has a red roof but blue was more normal. Although shown pink in the 1964 Tri-ang Railways catalogue, the walls were cream and the chimney was white. The window frames and guttering were dark brown and the doors yellow. The building had a timbered (painted on) gable front similar to the Post Office (RML14). Lichen was provided for garden plants and the base was buff coloured.

The kit was sold in size 2 pink boxes and size 1 blue boxes. It was not added to the Dapoland range.

Villa Capri RML16

This was principally a single-storey building but with a room in the roof complete with dormer window. At one end of the house was an external chimney stack and at the other a flat roof garage built onto the house. The front window had carved shutters and these together with garage doors, window frames, gutters and down pipes were blue. The front door was yellow and the walls white. There was a balustrade around the top of the garage making a small roof garden and the garden proper had a large white gate inscribed 'Villa Capri' and a brick wall.

The kit will only be found in a size 2 pink box as no more were made after 1963. Dapol bought the tools and reintroduced the model as C68 'Seaside Mews' with a bright red instead of maroon roof and no flower boxes.

Heathview RML18

Heathview and Parkview were very similar, with many common parts, and could be displayed either way round making four different modern houses of the type one might expect to find on a private housing development.

The principal difference between the two buildings was that Heathview had a built-on garage on its left and Parkview had a built on covered car port on the right, both of which had a flat roofs. The house roofs were red and both buildings had a yellow match board timber panel down each end of the house. The doors were blue and the window frames were yellow. The guttering and down pipes were grey and both kits included yellow planters and a piece of lichen.

Heathview had a partly recessed front panel which gave the house more character. To this, was fixed a blue panel and a yellow timber panel. A larger yellow timber panel fitted on the rear of the house covering one of the upstairs windows.

The kit was made in April 1964 and sold in a size 2 pink box. It had not, at the time of writing, been offered as part of the Dapoland series.

Parkview RML19

Besides the car port, the main difference when compared with Heathview, described above, was that while the front and rear walls of the house were the same, a blue match-board panel was stuck over one of the windows at the back in the case of the Parkview model.

Parkview was sold in a size 2 pink box and was made in March 1964. It was chosen as a Miniville model in 1968. As such was the Ideal Homestead W22. At the time of writing it has not appeared as a Dapoland building.

Semi-Built Bermuda RML21

This was the last model to be introduced to the range with a base plate, which was the same as that used for RML11 and RML12. The bungalow had its walls, a stone-built fireplace and chimney stack and the rest was open timber frame. All the mouldings used in the assembly of this kit were purpose made for the model. This included the internal walls, chimney stack and timber frames. The fine timberwork was excellent and included the frames for internal walls, woodwork, ladders, and workman's trestles with planks across them.

While records show that it was made only in 1965, it has been found in both size 3 pink (type D) and size 2 blue (type F) boxes. The first batch had richer colour woodwork.

It seems that the tools were not amongst those acquired by Dapol in the 1980s.

LIGHTING UNIT

This was RML22 in the series and consisted of a yellow three-prong moulding which supported an LES bulb holder. It was designed to fit snugly into the hole in the base of the early buildings (no hole was provided in later models). In the case of the Supermarket, nipples were provided in the roof to take the unit.

INDUSTRIAL BUILDINGS

The following industrial buildings were planned and it is claimed that some were not made due to insufficient orders. It has been observed, however, that even as late as September 1964 no price was given in the trade price list and so they would have been very difficult to order. It seems more likely that with so many models planned, the more complicated ones were left to the end when it was decided that they would be too expensive to make as they shared few or no parts. They were illustrated in the 1964 Tri-ang Railways catalogue and only the parts for the factory and the pylons reached production.

Koepe Pit Head Gear RML23

This was illustrated as a grey and orange structure on a brown base plate and 1,270 were ordered before it was decided not to proceed with the model.

Winding House RML25

This would have been a large square, red brick building with two opposite sides glazed to maximise daylight. It was designed to be used in conjunction with the RLM23 Pit Head Gear but was not made. 1,160 were ordered.

Winding Engine RML26

This was the engine to drive the Pit Head Gear (RML23) and was designed to go inside the Winding House (RML25). It was intended to be a working model fitted with a 9/12v DC electric motor but was not made. Orders received for it in 1964 totalled 1,140.

Boiler House and Stack RML27

This was again a large, square red brick building with a brick chimney at one end. Only 1,140 were ordered before it was dropped.

Gas Holder RML35

This would have been a particularly useful building although somewhat out of scale but as orders for only 1,300 were received it was not proceeded with.

Factory RML59

Had this been made, it would have been a large building consisting of six factory units joined together. For some reason, no orders were recorded and it was not made in this form. Instead, in May 1964, it was offered as three units and orders amounted to about 1,200 per unit but again none were made at that time. One batch of each was made, however, in 1968 and these remained in the stores until finally disposed of in 1972. All had cream walls and black roofs. For further details, *see below*.

Factory Power House RML65

This model shared nothing but window frames, doors and shutters with the other factory units. It was a much taller building with huge glazing panels in the roof. The chimney, attached to the outside of the building was the prefabricated type typical of modern factories. The shutters, doors, guttering and down pipes were black and the window frames green. The shutters were made to be raised and so were held in place by a retaining plate. Approximately 4,000 were made in 1968 and were sold in the size 1 blue boxes.

Factory Loading Bay RML66

In its complete form, this is thought to be one of the rarest Model-Land buildings. This is largely due to the delicate nature of the model which, while having the same basic body shell and roof as one of the RML67 units, had in addition a gantry crane one end of which was normally glued to the roof of the building and was easy to break. One batch of 3,000 was made in 1968 and sold in size 1 blue window boxes.

It also came with a small stand-alone gate keeper's hut which very rarely survives. This shed shared a roof with the RML3 Woodshed but its other components were different. It also had no base.

This and the two units that made up RML67 had uneven pitched roofs, said to make the most efficient use of winter light and common to many factories at the time, including the Tri-ang factory at Margate. The steeply pitched section was usually glazed but on RML66 it was made of solid black plastic. The other pitch of the roof was grey and looked like corrugated asbestos sheets. The gantry was also grey while the doors, ventilators, shutters, guttering and down pipes were black and the window frames green.

Factory Twin Production Bays RML67

The basic buildings were similar to RML66 but the

The electricity pylon which, to this day, stands outside the office windows of the factory and was used as a prototype for the Model-Land kit, shown here with one of the made-up models.

steep slopes of the roofs were glazed and the other slopes grey 'asbestos'. All other fixtures were black. At first sight it would appear that the kit made up into two identical units but in fact there was insufficient shutter door equipment in the kit to make them both the same. 3,800 sets were made in 1968, making 7,600 of these units in all, and the kits were sold in size 1 blue window boxes.

3 Pylons RML61

This was reviewed in the November 1966 edition of *Railway Modeller* with the following remarks:

"The Tri-ang Model-Land pylon kit is, we feel, one of the best scenic accessories we have yet met. In fact, the only point at issue is whether you feel that a power line is an improvement on the scenery or not!

"On a purely pedantic note, the model is not provided with the festoons of barbed wire around the lower legs to

inhibit would-be climbers. Apart from this, it is an accurate 4mm scale model of a pylon carrying two 3-phase supplies."

The model was easy to make and the insulators even had holes in them to take 'wires'. Each box contained enough parts for three pylons and the price was 7/7d.

The first batch of 5,900 kits was made in 1966 with 2,000 following in 1967 and 5,500 the year after. By now sales had dropped off and the last remnants of the final batch went out of the stores in 1971. In 1986 the tool was found in the tool store and cleaned up for reuse and the set of three pylons remained in the catalogue for several years selling as R530. These kits were made with a lighter grey plastic and had darker brown feet and insulators.

12 Transmission Towers RML61T

The pylons were amongst the models ordered by American Train and Track in 1968 and 1969. 5,000 sets of 12 were sent out to America for packaging by ATT during the first year and 6,000 the next.

SHOPS AND OFFICES

These depended largely on modular construction including a shop unit, residential unit and an office unit which could be combined to make a variety of buildings.

There was both a single and double shop front unit each with the shop nameboard attached. The residential unit was in brick and was used as a house in RML38 or as flats above a shop in RML44 and 45. Originally (1966), the brick panels were moulded in maroon plastic but the later (1968) batch was in the lighter salmon plastic. The office was a concrete panel type construction typical of the 1960s and was used as the upper floor of RML46.

Small Shop Single Storey RML36

This consisted of a single-storey, single shop unit

with a modern black roof. The choice of printed paper shop names to attach to the nameboard over the front of the shop was: 'Tri-ang Toys', 'Home Deco' and 'Confectioners 'Le Bon Bon' Tobacconist' and these came with appropriate paper window displays. It was first made in 1966 when about 2,650 passed through the stores, all being sold before the end of the year. Only one other batch was made in 1968 when 1,750 were supplied to retailers. It was sold in size 1 blue boxes and was later listed in the Dapoland series as C74 'Single Fronted Shop'.

Medium Shop RML37

This was effectively a double version of RML36 (above) consisting of the double shop front unit above which were fitted the black modern roof units. The kit came with a choice of three names: 'Boots - Dispensing Chemist', 'F.W.Woolworth' and 'Co-operative' and a choice of six printed paper window displays. Almost 3,000 were made in 1966 and a further 3,670 in 1968. The last of these left the stores at Margate, in their size 1 blue boxes, in 1970. This was also made by Dapol but as C75 'Double Fronted Shop' with browner walls.

Garage Single Storey RML68

This first appeared in 1966 for one year only. The model was effectively the RML37 Medium Shop – Single Storey, but with a different coloured roof unit, new stickers, two Minix cars, a set of three pumps, an oil cabinet, a tyre rack and a lot of self-adhesive trade stickers and notices including 'Energol', 'BP', 'We Give Stamps' and 'Retreads'. It was a BP station, carried the name 'Westwood Garage' or 'Parade Car Sales' and sold for 9/-.

Only 1,900 kits were made and it is not easy to find, especially complete. It is consequently, possibly, the rarest of the Model-Land buildings.

The model was to be released as a Miniville kit in 1968 when 1,700 were made. This was sold in a polythene bag as the Super Service Station W23.

Town House RML38

The back cover of the 1966 catalogue showed amongst the models on the layout, a row of Victorian houses, or more correctly the rear facades as no doors were shown. In actual fact they had no doors the other side either as they were assembled from the upper sections of the shops RML44 and RML45. The picture on page 19 of the same catalogue clearly showed what was meant to be the Town House but this must have been a mock-up as we know that, although the kit was planned for 1966, none reached the stores until 1968 and the plastic colours used for most of the parts were different to those shown.

Brickwork mouldings used in this kit were common to other kits in the series but, being a late model, was a deep salmon rather than maroon. It was packed in a size 1 blue window box and a total of 2,565 were made in 1968. The kit had mid-green window frames, blue doors and a grey door surround. The porches (common to at least three other kits) over the front and rear doors were of different designs, one being flat and the other pitched. This in effect made the kit reversible. Furthermore the door and porch could be placed over either of the downstairs windows in each side producing a left-handed or right-handed door.

Small Shop/Office Block RML44

The upper part of this building was a brick structure. Approximately 2,800 kits were released in 1966 which had maroon brickwork, red windows and bright green back doors. A further 840 kits were made in 1968 and these had salmon brickwork, green windows and blue back doors. Both batches were supplied with yellow front doors. The model had a standard shop front and the same selection of names and window displays as RML36.

Medium Shop/Office Block RML45

This kit was packed in a size 1 blue window box and labelled 'Medium Shop and Older-Style Office

Dapoland kits, sold in the 1990s, made from the Model-Land tools.

Block'. Again, the 2,650 kits made in 1966 had maroon brickwork but with yellow windows and a yellow back door. The ground floor windows were green or yellow. In the case of the 3,440 kits made in 1968, the brickwork was salmon, the upper windows green, the lower windows could be yellow and the back door either blue or green. In both batches the front doors were red or yellow and there was the same choice of shop names and window displays as provided with RML37.

Store/Office Block RML46

The illustration of this model in the 1964 Tri-ang Railways catalogue showed a four-storey building which was four shops wide. In fact, the kit that eventually arrived was the same as the last model but had a modern, flat-roofed, office block over the shop made up of two identical single units suggesting scope for a single fronted version at a later date. The upper storey was designed in such a way that several units could be stacked on top of each other to produce a much taller city centre type building.

The roof panels were not plastic but a thin black fibreboard which had a tendency to warp. The modern windows of the office were blue and the blue panels below them were paper which had to be glued on by the model builder. The lower window frames and door at the rear were green while the front doors were yellow. This also had the same shop papers as supplied in RML45. The kit was packed in a size 1 blue box.

In 1966 2,350 kits were dispatched and no more were made until 1968 when a further 3,900 were produced. It took until 1971 before the last of them cleared the stores.

Supermarket RML60

This appeared late in 1966 and carried the name 'Tri-ang Supermarket'. It had a Lines Bros. Group flag which is normally missing on surviving examples found today. The price of this kit, which came in a size 2 blue box, was 13/8d and it assembled into a model 8" x 5.5". The interior unit was either white or black and the windows were green or brown.

No doubt this proved to be a little too large for many layouts but it made a useful addition to a good-sized town scene. As the *Model Railway Constructor* pointed out, the model had the advantage that it could be converted into a cinema or warehouse without much difficulty. Dapol were later to list the kit as C71 and, as we saw earlier, also produced a bus garage version!

About 4,600 were made. These were made up of 2,800 in 1966 and a further 2,200 late in 1968. The last stocks left the stores in 1970.

TREES

These were not made and I have seen no illustrations of what were intended. They were not even included on the factory census list in 1964 although the other models, that did not reach the production stage, were listed. The intended models were:

Tree Small RML50
Tree Medium RML51

TOWN CENTRE BUILDINGS

These were illustrated in the 1964 Tri-ang Railways catalogue and so we can see what was intended. The buildings were all to be large but there was little evidence of shared parts seen elsewhere. None of these models was made and it was claimed that orders were insufficient. As we have seen, however, even as late as September that year they were listed without prices suggesting that the models had not yet been developed. They did not appear in the Model-Land list the following year.

Police Station RML55

This was to have been a red brick building in fine town-centre style. Probably of Victorian design, it was to have had a flight of steps up to the entrance over which would have hung the famous 'blue lamp'. A little over 1,100 kits were ordered in 1964.

Town Hall RML56

This was to have been a red brick building with portico, steps and a Union Jack flying from a flag pole on top of a clock tower on the roof. The roof was to have been double-pitched. 1,150 kits were ordered in 1964 but none were made.

Georgian House RML57

This would have been a three-storey stone building in typical Georgian style. The wooden door was shown with a half light over it and on the storey above a single arched window. The lower windows would have had window boxes. It attracted 1,170 orders.

MODEL-LAND FIGURES

Model-Land figures came in six sets, each of six pieces. Sets RML70-73 were available in 1964 and RML 74 and 75 were added the following year.

The figures were moulded in different coloured plastics according to the principal colour required for the figures. For example, the children were in white and the industrial workers in blue. The shades of plastic for a given set could vary; with evidence of more than one batch of each having been made. The painting was kept to set colours but again shades varied.

They had large circular grey painted bases, similar to those found on later Merit figures, an important feature when identifying unpackaged ones.

They were sold in a blister pack with the figures in a line, each with its own blister. The order of the individual figures in packs of the same set seem to remain the same and where it differs there is usually evidence that the pack has been opened and re-stapled.

The figure sets were deleted in 1971.

Figures Set 1 RML70

This was the 'Pedestrian Figures' set and actually had seven figures, as one consisted of a woman and young girl moulded together on an oval base. There was one other woman and the rest were men; one of them running. Flesh-coloured plastic was used.

Figures Set 2 RML71

This was the 'Workmen' set and was moulded in

white plastic. It consisted of five men and a woman. The postman was possibly chosen to go with the Model-Land Post Office (RML14) and proposed (but not made) post van (RML82).

Figures Set 3 RML72

This was the 'Child Figures' set and the contents were amongst the most recognisable in the whole range. In particular, the girl running with a hoop and the girl skipping were most distinctive. A third girl looked as though she was in her Sunday best. The other three figures were boys. One was flying a model aircraft and the other two appeared to be playing a game possibly of leapfrog. The figures were moulded in white plastic.

Figures Set 4 RML73

This was the 'Urban Figures' set which contained two policemen; one on traffic duty and the other giving chase. There was a robber running away and no doubt this and the chasing policeman were made for the proposed police station kit (RML55). Unfortunately the latter was not made.

The other three figures were a window cleaner in a khaki boiler suite, a lollipop man holding up a 'Stop – Children Crossing' sign and a man in a black uniform carrying a long case. With a bit of imagination one could see the latter being a taxi driver (to go with the proposed RML80) helping his customer with luggage!

Black plastic was used for this set.

Figures Set 5 RML74

The 'Industrial Workers' set contained the only figure in the series not to have a base as he was seated. Two of the set were dressed as miners with white hard hats fitted with lamps, possibly with a view to their being used with the mine buildings planned in the Model-Land range but not made. The seated figure may have been intended as a driver for the vehicles that also were not made. The other three included

a man with a shovel, one with a clipboard and a third with a spanner and tool box.

The figures were moulded in mid-blue plastic for Model-Land but also in white as a set of figures for Minic Motorway race track where they were sold as M1709 Pit Crew.

Figures Set 6 RML75

This was called the 'Road Workmen' set and was moulded in grey. It contained a dustman with a silver bin on his shoulder, a road sweeper pushing a broom, a sweep carrying a sack of soot in one hand and his rods and brush over his shoulder, a street vendor with a tray of things for sale and a workman with a silver pneumatic drill.

RML74 'Industrial Workers' set of Model-Land figures, in a modified paint scheme, was also used as M1709 'Pit Crew' for Minic Motorway. Pictures of other figures will be found in Volume 1, page 416.

VEHICLES

Sadly none of the Transport models were made. They were illustrated in the 1964 Tri-ang Railways catalogue and listed (unpriced) in the trade price lists that year. Clearly they had not been developed by the end of the year and were dropped from the list in 1965 supposedly because the company was by then developing a range of model road vehicles called Minix.

De-Havilland Trident RML52

It was intended that this would be a model fitted with remote control taxiing and steering operating off 9/12 volts DC. It was designed by John Hefford and a patent No. 29153/63, dated 23 July 1963, was applied for under the title 'Improvements to model aeroplanes and vehicles, tricycle drive'. Only 600 orders were received and the model was not made.

British Taxi RML80

This looked as though it was to have been a black Austin London Taxi. Orders for the model reached 1,050 by the end of 1964.

BR Van RML81

This was to have been a yellow Scammel Scarab with a van body, typical of many seen at the time collecting parcels from the station parcels office. 1,085 had been ordered by the end of the year.

GPO Van RML82

A red delivery van of unknown origin would have carried 'Royal Mail' and 'ERII' on its sides. 1,035 were ordered.

CEMENT

Cement RML100

A soft plastic tube of polystyrene cement was originally supplied in each kit with a small instruction sheet warning of its dangers. This was later left out and the instruction sheet was over printed 'not now supplied'. A further slip advised the use of Frog polystyrene cement.

The tubes of polystyrene cement were packed in boxes of 12 and the minimum order for retailers was three packs. It was by far the most popular item in the Model-Land range achieving sales in excess of 640,000 by the end of 1971.

SCENERY

Model-Land Scatter

Bags of scatter were made under the Model-Land label for 1964 only and were replaced in 1965 by Tri-ang King Size Colourings. The scatter was packaged in polythene bags with a printed card stapled at the top. The card was printed in red and yellow with the contents rubber stamped in red on a white panel. It was described on the card as 'Landscape Material'.

The range was limited to just three colours:

Light Green RML62	10,200 made
Dark Green RML63	10,900 made
Brown RML64	9,700 made

Vacuum-Formed Scenery RML69

This was planned but was not made although a three-piece vacuum-formed layout was made and sold in later years under the Hornby Railways label.

King Size Colourings

These replaced the earlier Model-Land scatter and came in a much greater selection of colours as follows:

Ballast Grey RV60	22,400 made
Cornish Yellow RV61	12,500 made
Spring Green RV62	46,600 made
Autumn Green RV63	28,900 made
Earth Brown RV64	21,800 made
Moorland Mixture RV65	12,600 made
Poppy Red RV66	7,500 made

A packet of rare Model-Land scatter and packets of the more common Tri-ang Hornby scenics.

Other aids available in this series were:

Lichen RV67	10,300 made
Cork Bark RV68	13,200 made
Mod-Roc RV69	4,200 made

Scenic Materials

In 1968 it was decided to change the printing, dropping the name 'King Size'. The new head card was to be red and carry the Tri-ang Hornby logo and the words 'Scenic Materials'. The proposed packaging appeared in the 1969 catalogue on pages 28 and 29 but before the cards could be printed, it seems that there was a change of heart. Instead of all red, the cards were printed half red and half yellow as illustrated in the catalogue the following year. The RV number and colour were rubber stamped on a small white panel in the background printing.

The above are assumptions based on the fact that the red and yellow cards have 'Rovex Industries Ltd' on the reverse. Had they been printed in time for 1970 instead of 1969 they would have carried 'Rovex Tri-ang'.

The range included the following:

Ballast Grey RV60	8,700 made
Cornish Yellow RV61	6,300 made
Spring Green RV62	29,700 made
Autumn Green RV63	16,900 made
Earth Brown RV64	17,800 made
Moorland Mixture RV65	18,900 made
Lichen RV67	9,200 made
Cork Bark RV68	5,300 made

MINIX CARS

INTRODUCTION

Original Series

Over the years between 1965 and 1972 some 10 million Minix models were made of which a third to a half were sold as wagon loads in the Tri-ang Hornby system. The cars were marketed as 'The Greatest Little Cars in the World' and, unlike many other model road vehicles manufactured in Britain at that time, they were, with one exception, consistent in scale which was of course, 4mm. The one exception was the bus which was approximately 1:135. They were excellent models.

Coded 'RC' for 'Rovex Cars' Minix models were first introduced by Rovex in 1964 to complement their railway system and Model-Land buildings. They then realised that they had a toy range which could stand alone in its own right. The name Minix, incidentally, was chosen to link it to Minic in people's minds but the 'X' on the end indicated no mechanism!

Richard Lines, remembers well their introduction:

"We wanted to sell little cars to go with the trains at 1/- retail. To get the price down we designed and built an amazing automatic wheeling machine. You fed rolls of aluminium coils one side and rolls of wire in at the other.

The machine impact extruded the wheels and forced them onto the axles completely automatically at a rate of about 30 per minute. The machine had a large red light on top which came on if anything went wrong. It was visible all over the factory.

I took the first samples of the cars to Woolworth's buying office in Marylebone Road with a view to getting an order for millions. Unfortunately Corgi had been in the day before with their Huskys which had diecast bodies and Woolworth's preferred them."

The first batches came off the production line towards the end of 1964 and sold for 1/- each (the bus was 1/6d). The price had been deliberately targeted to undercut Matchbox cars which sold at 1/6d each. Only numbers RC1-RC3 and RC5 were produced that year when a total of just under 20,000 Minix cars were made.

Colours

Different body colours were not applied individually but whatever model was being made at the time

The range of models made.

was made in the colours in use. Thus, each model appears in a wide range of colours, those cars made for the shortest period having the smallest colour range. Evidence suggests that the colour range was limited until 1967.

In all twelve colours were used for the bodies: vermilion (called orange by some collectors), red, deep yellow, pale yellow, pale green, emerald green, dark green, blue, white, black, lime green and maroon. Of these, the last three are the hardest to find and the lime green and maroon ones are rare. I have heard of eight of the models turning up in black, six of them in maroon but only Nos. 1, 7 & 16 have, to my knowledge, been seen in lime green.

Each car had a chromed base with bumpers and grill and carried the model name and the number in the series, on its underside. An insert placed during assembly gave the models windows.

From No.9 onwards, the cars had interiors. These varied in colour between pale blue, beige, white, red, medium blue and pale green. Pale blue and beige were the most common colours.

Following a suggestion from a new Marketing Manager, towing hooks were added to eight of the cars. These were RC4, RC5, RC6, RC8, RC9, RC11, RC12 and RC15.

The wheels came in two sizes, the larger size appearing on just three models – the Ford Thames Van, Vauxhall Cresta Estate and the Rambler.

Packaging

The individual models were packed in small open top boxes designed as display packs with a cellophane wrapping to complete the job. The box had been deliberately designed to look like the packaging used for sparking plugs. The flat blanks were folded up and the car inserted fully automatically. The number and name were rubber stamped on the end of the box. The reverse of the box illustrated six Minix models. They each carried a cut-out gold token which could be saved up for gifts.

Retailers were supplied with attractive display

The common colour range and scarcer models including black and maroon cars and the only American car – the Rambler. Nos. 1, 2 & 3 each appeared in 11 different colours.

chests crammed with the little boxes.

A more conventional blister pack was used for exported models. Sales of the model in the United States was handled by FJ Strauss Co. Inc. of New York. That company had their models supplied in a bubble pack printed specially for them.

PRODUCTION

1965-66

In 1965 production reached 1.74 million and included only RC1- RC7, RC9 and RC10. A further 0.5 million were used as loads in the recently released

Minix packaged for Strauss of America.

Car-a-belle car carrier of which 50,000 were sold in sets in 1965 and a further 24,000 as solo models.

The same range was available the next year with the addition of the Ford Van (RC13) and Caravan (RC16) both of which were important as wagon loads. The first of the twin packs also started to appear in the shops and there was a wholesale box which contained 72 cars. That year, 1.6 million cars were made for solo sale and a further 0.5 million as wagon loads. Of the solo sales, a third were sold in the wholesale boxes. There were now five wagons with a car load. In just two years, in excess of 4 million Minix cars had been made.

In 1966 it was planned to add a series of lorries based on a Ford type D chassis but these did not reach production. Also planned were a Hillman Minx, Vauxhall Victor 101, Austin 1800, Vauxhall Cresta Estate Car, Rambler Classic and a Pacesetter Bus none of which arrived that year but were subsequently made.

1967

The Vauxhall Victor (RC11), Austin 1800 (RC12),

Vauxhall Cresta Estate Car (RC15) and Rambler Classic (RC17) all arrived in 1967 but there were large stocks to clear from the previous year. Approximately 120,000 of the new models were made for solo sales and 330,000 cars were produced for use in Twin Packs. A further 460,000 were used as wagon loads.

1967 was a year marked with export successes for the range. 87,000 cars went to Holland and 9,000 to Australia. About 15,000 went to France as loads for the R342F Car Transporter which was sold with Hornby Acho couplings. The largest success however was the sale of almost 1 million cars to the United States. Total production of Minix models at Margate in 1967 was approximately 1.9 million.

Also planned for 1967 was a Ford Van magnet pack and a tractor, neither of which appear to have been made.

1968

1968 saw the first deletions from the range. The RC15 Vauxhall Cresta Estate which had been introduced only the year before was one of the casualties

possibly due to damage to the tools. Others were the Vauxhall Viva (RC3), Ford Anglia (RC1) and the Austin Cambridge (RC5). There were however two new additions. The most welcome was the Strachan Bus (RC14) which offered hope of other commercial vehicles and the long-awaited Hillman Minx (RC8) for which a space had been kept open in the number sequence since 1965.

1968 also brought two new twin packs (RC36-37) and plans for both 4-car (RC29-31) and 7-car packs (RC40-41) none of which are thought to have materialised. A special series of Playpacks were, however, made for Woolworths. These were numbered RC50-55 and are described below. 44,700 of these packs were made during the year. There was also a special pack of 36 cars made for a customer in Austria who purchased 96 packs and an Australian company was supplied with 13,172 of each of the six twin packs which accounted for over 158,000 cars. Why such unusual quantities were chosen and how the packaging varied from the usual is not known.

In addition to the Woolworths, Austrian and Australian specials, sales of the solo models stood at 284,000 this year, with a further 152,000 models sold in multiple packs. Not taking account of Minix models used for wagon loads, it seems probable that about 687,000 models were made during the year. If wagon loads were added, this figure would more than double.

France would also have received a further 15,000 boxed cars when a second batch of Car-a-belle wagons was sent. Some of these were to be sold as loads for the Hornby Acho wagon porte voitures and as late as 1972 this wagon was being sold with recently supplied cars (with white bases and black wheels).

Wholesale

With the setting up of a Wholesale Department within the company to sell to small retail outlets, Minix cars were amongst the products chosen for special packaging for this service. These were given a 'W' prefix to the number and items chosen in 1968 were WRC2, WRC4, WRC6-14, WRC16-17, WRC25-32, WRC36-37 and WRC40-41. RC1-17 were also available from the Wholesale Department in packs of 72 of each model priced £2 a pack (except the RC14 Pacesetter Bus which was £3 a pack). Retailers were also offered an assorted pack of 72 models for £2.

The Wholesale Department also sold a range of Model-Land building kits each of which contained a Minix car. Details of these will be found in the Model-Land section of this volume.

1969
During 1969 there were no further changes in the models available. All local sales appear to have been through the Wholesale Department.

Solo sales were down to 58,600 and twin packs to 41,000. 360 of the 72 car wholesale packs were also made and sold and 450 of the 48 car packs. ATT of America ordered 30,000 vans which were not delivered but Australia received a further 75,260 twin packs and Austria another 230 of the 36 car packs.

A further 22,000 playpacks were made for Woolworth's while Miner Industries purchased

Minix cars used in a Penguin building set. These may well have been part of the batch of cars sent to the Belfast factory in 1969.

A Minix order form.

ALL PREVIOUS PRICE LISTS CANCELLED   1st January, 1968   ORDER FORM — HOME TRADE

ROVEX INDUSTRIES LTD.
WESTWOOD, MARGATE, KENT
TELEPHONE: THANET 22294.   TELEGRAMS: Rovex, Margate, Kent

MINIX REGD.
THE GREATEST little CARS IN THE WORLD

INVOICE TO:
(If different from despatch address)   A/c No.

NAME
ADDRESS

DESPATCH TO: A/c No.

NAME
ADDRESS

Your Order No.
Date.............   Reps. No.
Delivery Required
Signature of Person
Placing Order

Qty. Singles	Order in Multiples of	Ref. No.	ARTICLE	Trade Price	Rec'm'd Retail Price (incl. P.Tax)	Qty. Singles	Order in Multiples of	Ref. No.	ARTICLE	Trade Price	Rec'd Retail Price (incl. P.Tax)
				Doz.	Each					Doz.	Each
	36	RC. 2	Morris 1100	6/8	1/-		12	RC.25	2 Car Pack	CAR PACKS 13/4	2/-
	36	RC. 4	Triumph 2000	6/8	1/-		12	RC.26	2 Car Pack	13/4	2/-
	36	RC. 6	Ford Corsair	6/8	1/-		12	RC.27	2 Car Pack	13/4	2/-
	36	RC. 7	Sunbeam Alpine	6/8	1/-		12	RC.28	2 Car Pack	13/4	2/-
	36	RC. 8	Hillman Minx	6/8	1/-		12	RC.36	2 Car Pack	13/4	2/-
	36	RC. 9	Simca 1300	6/8	1/-		12	RC.37	2 Car Pack	13/4	2/-
	36	RC. 10	Hillman Imp	6/8	1/-		6	RC.29	4 Car Pack	26/2	3/11
	36	RC. 11	Vauxhall Victor 101	6/8	1/-		6	RC.30	4 Car Pack	26/2	3/11
	36	RC. 12	Austin 1800	6/8	1/-		6	RC.31	4 Car Pack	26/2	3/11
	36	RC.13	Ford 15cwt Van	6/8	1/-		6	RC.40	7 Vehicle Pack	52/11	7/11
	36	RC.14	Bus	10/-	1/6		6	RC.41	7 Vehicle Pack	52/11	7/11
	36	RC.16	4/5 Berth Caravan	6/8	1/-				SALES AIDS		
	36	RC.17	Rambler Classic	6/8	1/-		—		"Many-Way" Display Stand Wall Hanging Display Card	Free Free	

TERMS AND CONDITIONS OF SALE:

components for the assembly of around 24,000 of each of three play packs of their own design.

With the introduction of the Lowmac with a car load, G&R Wrenn ordered and received about 5,500 Anglia cars which were supplied bulk packed, and a supply of nearly 83,000 bulk packed cars to the Group's Belfast factory also took place in 1969.

Of the 1.24 million Minix models made during the year, approximately half a million were used as wagon loads.

1970

This appears to have been the last year that Minix cars were offered for sale as a model range in their own right. After this they were used only as wagon loads.

46,000 solo-packed Minix models were in the stores at Margate during the year of which only a little over 14,000 were sold. The fate of the rest is not known. They may have been unwrapped for use as wagon loads or may have been sold off abroad as a job lot. During the year there were also 11,000 twin packs in the stores of which only 850 sold.

ATT again ordered 18,000 vans which for some reason were not supplied although they were in the stores. There were also 107,000 Triumph 2000s made for an American customer of which only 40,000 were sent. The French company at Calais ordered over 36,000 cars, presumably for the Hornby-Acho car transporter, the New Zealand company received 1,000 assorted cars as wagon loads and Australia received 500 for the same purpose.

Thus a little over 55,000 Minix were sold in 1970 but a further 500,000 were used as wagon loads in the UK. 116,000 of the latter were used in the new Cartic car transporter that was added to the Tri-ang Hornby range in 1970.

1971

Car requirements for wagon loads were down to 250,000 in 1971; half that of the previous year and as there were no recorded separate sales this represents the annual output of Minix that year.

1972

Production fell further to a little over 10,000 in 1972 all of which were again used as wagon loads.

HORNBY RAILWAYS

White Bases

After 1972 the models were far less attractive with their plain non-chromed white (from 1974) base and black (or sometimes grey) wheels (from 1972) and it appears that only numbers RC1, 2, 4, 6, 8 and 12 were available. Black or grey based models were much later. They were no longer referred to by their individual numbers but were listed as R160 which ever version they were. They were now being made just as wagon loads or for use as fillers in starter sets and so, when the wheeling machine broke down, there was not sufficient justification to invest in new plant. The black plastic wheel mouldings were a cheaper alternative.

The plastic was now softer and the colours used were white, grey, yellow, dark blue (two shades), dark green, medium green, red, gold and silver. None had seats. To start with the new plastic colours appeared on silver bases until these were used up and models of the chromed chassis period may sometimes be found with black wheels. These were probably made between 1972 and 1974.

In the early to mid 1970s car carriers had a mixture of white-based cars but when the R126 car transporter was introduced late in 1977, it came with three Triumph 2000s or Hillman Minx's with white bases and black wheels. In 1982, the wagon was 'improved', becoming R124, and the Triumphs were issued in new colours including gold and with black bases and wheels. One could also buy a single Triumph on a new short flat wagon (R005).

Ford Sierra

The 1989 catalogue showed that the Cartic was to be reintroduced but in Silcock's livery and a Ford Sierra was to be modelled as a load. Unfortunately there was some delay in developing the new car and for 1990 the model was issued with an updated Sunbeam Alpine instead. This had a black base and wheels and was available in red, yellow, blue and green. There is a theory that it was chosen from a number of samples of the earlier Minix models that were run off for the purpose, using the original tools. This is supported by the existence of a Vauxhall Victor 101 in a soft blue plastic and the newer black wheels and base. The latter, however, was fitted with a Simca glazing unit!

The Ford Sierra was the only new car model to be made after 1968 and it seems likely that it was eventually available in 1990. Besides being used as a wagon load, it was sold in bubble packs of three.

CHECK LIST OF MINIX

No.	Model	Years	Notes
	INDIVIDUAL VEHICLES		
RC1	Ford Anglia	64-67, 69	
RC2	Morris 1100	64-70	
RC3	Vauxhall Viva	64-67	
RC4	Triumph 2000	65-70	
RC5	Austin A60	64-67	
RC6	Ford Corsair	65-70	3 versions
RC7	Sunbeam Alpine	65-70	
RC8	Hillman Minx V	68-70	
RC9	Simca 1300	65-70	
RC10	Hillman Imp	65-70	
RC11	Vauxhall Victor 101	67-70	
RC12	Austin 1800	67-69	
RC13	Ford 15 cwt. Thames van	66-71	
RC14	AEC Reliance/Strachans bus	68-70	
RC15	Vauxhall Cresta Estate	67	
RC16	4/5 berth caravan	65-72	
RC17	Rambler Classic 770	67-69	

No.	Model	Years	Notes
	TRAILERS		
	trailer boat	68-69	in sets only
	high sided trailer & 2 ladders	68-69	in sets only
	vehicle trailer	68-69	in sets only
	FORD COMMERCIALS		
RC18	Ford pantechnicon (type D)	not made	
RC19	Ford open truck	not made	
RC20	Ford high capacity tipper (type D)	not made	
RC21	Ford tanker (type D)	not made	
RC22	Ford platform wagon	not made	
	TWIN PACKS		
RC25	twin pack	66-70	
RC26	twin pack	66-70	
RC27	twin pack	66-70	
RC28	twin pack	66-70	
RC29	4 car pack	not made	
RC30	4 car pack	not made	
RC31	4 car pack	not made	
RC32	72 car wholesale pack	66-69	
	MISCELLANEOUS		
RC34	Ford vans/magnet pack	not made	
RC35	tractor	not made	
	TWIN PACKS		
RC36	twin pack	67-70	
RC37	twin pack	67-70	
	MULTI-PACKS		
RC40	7 vehicle pack	not made	
RC41	7 vehicle pack	not made	
RC42	36 vehicle pack 6x6	not made	
RC43	48 car pack	69	
	PLAYPACKS		
RC50	Woolworth's	68-69	
RC51	Woolworth's	68-69	
RC52	Woolworth's	68-69	
RC53	Woolworth's	68-69	
RC54	Woolworth's	68-69	
RC55	Woolworth's	68-69	
	DISPLAYS		
	many-way display stand		
	wall hanging display		

COMPONENTS LIST

All components used in the assembly of models at Margate carried individual numbers. These were usually prefixed with an 'S' but in the case of the Minix cars they were prefixed by 'SC'. The following list, prepared by Bernard Taylor of London, appeared in *Model Collector* in October 1995:

No.	Car	Part
SC1011	1	body
SC1012	1	floor
SC1013	1	glazing
SC1014	2	body
SC1015	2	floor
SC1016	2	glazing
SC1017	3	body
SC1018	3	floor
SC1019	3	glazing
SC102	4	body
SC1021	4	floor
SC1022	4	glazing
SC1023	5	body
SC1024	5	floor
SC1025	5	glazing
SC1026	not	used
SC1027	not	used
SC1028	not	used
SC1029	6	body
SC1030	6	floor
SC1031	6	glazing
SC1044	15	body
SC1045	15	floor
SC1046	15	glazing
SC1050	8	body
SC1051	8	floor
SC1052	8	glazing
SC1056	7	body
SC1057	7	floor
SC1058	7	glazing

No.	Car	Part
SC1059	13	body
SC1060	13	floor
SC1061	13	glazing
SC1062	10	body
SC1063	10	floor
SC1064	10	glazing
SC1065	9	body
SC1066	9	floor
SC1067	9	glazing
SC1068	9	seating
SC1069	12	body
SC1070	12	floor
SC1071	12	glazing
SC1072	12	seating
SC1073	10	seating
SC1074	13	seating
SC1075	11	body
SC1076	11	floor
SC1077	11	glazing
SC1078	11	seating
SC1079	16	body
SC1080	16	floor
SC1081	16	glazing
SC1082	16	seating
SC1083	15	seating
SC1085	14	body
SC1086	14	floor
SC1087	14	seating
SC1088	14	axle ret.
SC1089	14	glazing
SC1091	17	body
SC1092	17	floor
SC1093	17	glazing
SC1094	17	seating

Numbers SC1026 to SC1928 are thought to have been reserved for the proposed No.6 Vauxhall Victor (FB).

DIRECTORY OF MINIX

Most of the prototypes chosen by Rovex for the Minix series had very distinctive shapes that made them easily recognisable; an important point when modelling in a small scale.

RC1 Ford Anglia

This model first appeared in 1964 and had the distinctive feature of a rear window that sloped inwards. Over 20,000 were sold as solo models and it was selected by G&R Wrenn in 1969 to be the load on their Lowmac wagon and were supplied bulk packed. It is one of the Minix models that was made in black plastic and lime green as well as all the nine common colours. The model had no interior and no towing hook. Production survived beyond 1972 and so may be found with a white base and black wheels. Indeed, the 1973 catalogue showed it as the only car used as the load for both the R342 car transporter and the R666 Cartic.

RC2 Morris 1100

The Morris 1100 was one of a number of BMC cars that shared a common body shell. This model may also be found in the full basic colour range and black and maroon. It too was one of the four models that arrived in 1964 when about 6,000 were made and did not have an interior or towing hook. 22,000 were made in 1965 and a further batch of 11,000 for solo sales in 1968. It was made after 1972 with a white base and black wheels.

RC3 Vauxhall Viva

This was a model of the Vauxhall Viva HA. It arrived just before the end of 1964 with no more than 250 being made that year. The rest were made in 1965. For some reason it was dropped after 1967 and yet was later seen in the 1977 catalogue with a white chassis and black wheels. A total of 22,000 were made for solo sale and it can be found in any of the common colours and black and maroon. It was not given an interior or towing hook. The later Vivas had bent sills and a large hump in the roof.

RC4 Triumph 2000

The Triumph 2000 arrived in 1965 when over 24,000 were made for solo sale. The last of these sold in 1971. Over 100,000 of the model were made for export to America in 1970 but it appears that only 40,000 went.

It survived into the 1970s to become the sole wagon load at the end of the decade. For this the name 'Minix' was removed from the base of the car, possibly in 1977 when the model was needed for the R126 Car Transporter. This was one of the models that carried a hook at the rear but it was not fitted with an interior unit.

RC5 Austin A60

Nearly 11,000 of the model of the Farina-styled Austin A60 (Cambridge) were made in 1964 making it the most numerous in that first year. Another 22,000 were made the following year for solo sales but these were the last to be sold in this form. This stock eventually ran out in 1968. In 1966, '67 and '68 batches of the A60 were made for the RC25 twin pack which would have used 56,000 of the model.

The early version of the model had no towing hook but as it was to be one of the cars to be used with a caravan as a wagon load for the R713 and R878 wagons a hook was added for later batches. As no more A60s were individually packaged after 1965 none of the hooked ones would have been sold in solo packets in the UK. It was not given an interior unit. A strange feature of this model is the scarcity of colour variation. Over many years observing the models I have seen only deep yellow, red, blue, emerald and dark green ones. I have not seen a post 1972 version.

RC6 Ford Corsair

The reverse of the packaging in which individual models were sold illustrated a Vauxhall Victor FB which was to have been RC6 in the range but this was not made. Instead the Vauxhall Victor 101 was made later as RC11 and the number RC6 was given to the Ford Corsair.

The first batch of models was made in 1965 when over 25,000 went for solo sales. Another batch for individual packing was made in 1968; this time the model had been up dated by the addition of a V-4 logo on the base and the badge on the front of the bonnet. This later model was to be further adapted with the addition of a towing hook sometime later. None of the Ford Corsairs had interiors.

The model was a popular one for multiple packs. It appeared in at least one of the Woolworth's Playpacks and was supplied to Miner Industries for their No.501 pack. It was also used for at least two of the twin packs. Multiple packing accounted for a further 180,000 models of which 51,000 went to Australia. Many more were made for export for wagon loads of which we have no figures.

The Ford Corsair was made in all the basic colours and black and maroon and survived into the 1972 appearing with a white base and black wheels.

RC7 Sunbeam Alpine

The only sports car in the Minix series, the Sunbeam Alpine arrived in 1965 when nearly 129,000 were individually boxed for solo sales. Another small batch was made the following year and probably in 1968 and 1969 bringing the total for solo sale to about 143,000. In addition to these it is known that 7,000 went to Holland and that probably 162,000 were used in twin packs made between 1966 and 1969 of which nearly 51,000 went to Australia. The model also appeared in the Woolworth's Playpack RC55 of which 1,000 were made.

At least one batch of the model was made in a strange lime green plastic, a variation worth looking out for. It may also be found in black and none had towing hooks or interiors.

The model may have been dropped from the series before 1972 but it was used again later when a grey based version with tinted windows appeared on the R666 Silcock Express Cartic car transporter. These had red, blue, green or yellow bodies and may be seen in the 1990 catalogue. In this form the model was listed for one year only, being replaced by the Ford Sierra in 1991, but stocks of the Alpine must have remained at the stores for a batch was released to retailers in 1995. That year, Hatton's of Liverpool were offering the cars at five for £3 or 20 for £10. They were also giving them away free with other orders.

RC8 Hillman Minx

The Rootes Hillman Minx Series VI, with its restrained lines, was one of the least distinctive designs chosen but its very simplicity probably lay behind its choice as a wagon load in the late 1970s.

It is not known why this model was delayed. Although listed from the start it did not reach production until 1967 and then only 6,800 were recorded and all of them for a Dutch customer. The main production was in 1968 and 1969 when a total of 13,700 were made for solo sales. We do not know how many were used in packs or as wagon loads. In addition, it was one of the models chosen by Meccano France in 1970 but there is a question mark over whether it was delivered.

The model had a towing hook but no interior moulding. It was made in the common colours but also in maroon.

RC9 Simca 1300

This was the only Continental car in the series and as such was amongst those chosen for France but may not have been delivered. It was also the first in the range to be fitted with an interior seating unit and it had a towing hook. 7,000 went to Holland and no doubt other countries received it as a matter of course. The first batch of solo-wrapped models went into the stores in 1965 and a total of 22,000 were sold in this form. Another 14,000 were sent to the Belfast

factory to be used for something they were making. Only silver base models have been seen by the writer. It seems to have been available in almost all the common colours.

RC10 Hillman Imp

Nearly 15,000 were sold as solo models from 1965 onwards. A further 15,000 were used in the Belfast factory and we know that 1,000 more were used in Woolworth's Playpacks. 7,000 also went to Holland. All had towing hooks and interiors although the latter appears to have been an after-thought if the sequence of the component numbering is anything to go by. This too seems to have been available in all nine common colours although I have yet to see pale green.

RC11 Vauxhall Victor 101

As we have seen, a Vauxhall Victor FB had been planned as No.6 but was dropped. This is possibly due to a change in the prototype which led to Rovex adopting the Vauxhall Victor 101 instead. The prototype always looked as though its body style was the work of Frank Hampton who drew some interesting futuristic cars in the Dan Dare cartoon strip for Eagle comic. All had towing hooks and interiors.

The first batch of 26,000 solo-packed models passed through the stores at Margate in 1967. It appears that further batches were made in 1968 and 1969 making the total solo sales production 47,000.

In addition to these 6,500 were sold to Holland, at least 2,000 were used in Woolworth's packs, 24,000 were sold to Miners Industries for their No.502 set and 17,500 were used at the Belfast factory. An unknown quantity would have been used in multiple packs and as wagon loads. The range of colours seen has been rather limited but includes black.

RC12 Austin 1800

This seems to have made its first appearance in 1967 when nearly 38,000 were made for solo sale of which very few had sold by the end of the year. This suggests that production started late in the year. More were made the following year after which production for solo sales ceased. These batches accounted for a total of 48,000 models.

In addition 24,000 were sold to Miner Industries, and 16,000 to the Belfast factory. The Dutch customer received another 6500. The rest are unrecorded.

A black-bodied version exists and the model was fitted with a towing hook and an interior.

RC13 Ford 13 Cwt Thames Van

This was one of the most important models in the Minix series, featuring as a load for a number of wagons. Indeed, only about 20,000 were sold as solo models. This compares with 177,000 sold with the R563 bogie bolster wagon and 411,000 with the R888 integral flat wagon; the latter in starter sets. An additional 24,000 were sold to Miner Industries for their No.503 set and 46,000 were used in the RC28 twin pack.

In catalogues, the vans were normally shown in red, white and blue but the model may be found in all the other common colours and a rare one is maroon. It is unlikely that it will be found other than with a silver base and wheels, although a silver-base version has been found with a body in a post 1972 shade of blue. The model had larger wheels but no towing hook. It had an interior unit but the component numbering suggests that this was an afterthought.

RC14 AEC Reliance/Strachans Bus

This model went under a number of names in the literature produced by Rovex. It was referred to as a 'Continental Style Luxury Long Distance Coach' in their 1967 leaflet but also as the 'Pacesetter'. For the leaflet illustration, the negative was reversed to show the doors on the offside of the vehicle. This would suit its description as a 'Continental' coach. The actual model was typically British.

As we have seen, the model, unlike the rest of the Minix series, was to 1:135 scale. This allowed it to fit into the standard packaging for solo sales. 38,375 were made for this purpose in 1968 and most remained unsold by the end of 1970. 1,000 were used in the RC55 Woolworth's Playpack and 24,000 were sold to Miners Industries for their No.501 set. Most of the common colours have been seen as well as maroon which is a very unusual colour for the Minix range. It had an interior unit.

RC15 Vauxhall Cresta Estate
This is one of the scarcer models and did not appear until 1967, although the components carried numbers that suggest that it was originally to have followed No 6, the Ford Corsair of 1965. It was made for one year only and just 24,500 were produced for solo sales. Despite this, it is known to have been made in at least six of the common colours and black. The model was fitted with a towing hook and interior (another afterthought) and also had larger wheels than most of the other cars.

RC16 4/5 Berth Caravan
Unlike the rest of the range, this had a grey base. The earliest recorded production was in 1966 when 6,400 were made for solo sales. Two further batches of 13,000 and 10,000 were made in 1967 and 1968 respectively. At least 58,000 were made for use in the RC27 twin pack and 24,000 were sold to Miner Industries for their No.502 set. A further 1,000 were used in one of the Woolworth's sets. It had an interior and at least six of the common body colours have been seen as well as lime green.

The caravan was also used, with a car, as a load on two wagons. These were either the R713 or R878 used in the Clockwork starter set RB3 in 1970 of which 24,000 were made and the R713CN integral wagon of which nearly 2,500 were made in 1969 and sent to Canada for production of starter sets.

Trailer Boat
This is thought to have been available only in Playpacks in 1968 and 1969. It was based on the caravan chassis and is quite scarce. It is also known that 24,000 boat trailers were sold to Miner Industries for their No.502 set in 1969.

High-Sided Trailer and Ladders
This also was based on the caravan chassis and came in Playpacks in 1968-69. The vehicle had high sides and a bar with slide stops on which the two ladders could lean. It was 55mm long and is today another scarce model. Miner Industries in 1969 received 24,000 trailers and 48,000 ladders.

Vehicle Trailer
This came in one of the Playpacks and was clearly intended to be a trailer for transporting a car. The example seen was blue.

RC17 Rambler Classic 770
The Rambler Classic was a late addition arriving in 1967. It was originally thought to have been made for two years only during which time 36,000 models were boxed up for individual sales, but an example has turned up which has a white base and black wheels. The silver-base version has been found in at least six of the common colours. It had the larger wheels, an interior but no towing hook.

FORD LORRIES

The following Ford commercial vehicles were planned but not made although a mock-up of an open truck has been found at the factory which could have been for RC19. It is now in a private collection. The following is the range of lorries that was proposed:

RC18 Ford Pantechnicon (Type D)
RC19 Ford Open Truck
RC20 Ford High Capacity Tipper (Type D)
RC21 Ford Tanker (Type D)
RC22 Ford Platform Wagon

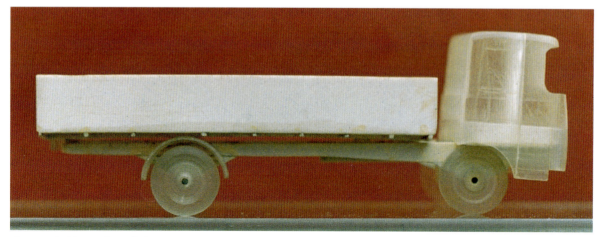

A pre-production mock-up of a lorry.

DISPLAY MATERIAL

Many-Way Display Stand

This was a large square cardboard sleeve from which two square trays were pulled out. The sleeve was then stood on end and one tray plugged into the base of the sleeve and the other sat in the top. Both were angled to face the customer and both contained two layers of boxed models; 18 to a layer. Each tray contained 36 of the same model and a printed label stuck on the end of the tray stated the contents e.g. '36 RC6 Ford Corsair'. Thus, a retailer had to buy several trays and mix up the contents to offer the customer choice.

The box was orange and bore the inscription 'Minix – The Greatest Little Cars in the World' and '1/-'. It also advertised the gold token scheme.

Wall Hanging Display Card

Details of this are not available.

TWIN PACKS

There were six of these, numbered RC25-RC28 and RC36-RC37. Each contained two models in a red window box with a yellow inner card former. On the reverse of the box were illustrations of numbers RC1 to RC13 and RC16. The actual contents for each set seem to have varied but the following were originally announced:

RC25	Austin A60 & Sunbeam Alpine
RC26	Simca & Sunbeam Alpine
RC27	Ford Corsair & Caravan
RC28	Ford Corsair (or A60) & Ford Van
RC36	Vauxhall Victor & Caravan
RC37	Ford Corsair & Ford Anglia

The packs were offered to retailers in packs of 144 assorted boxes at a wholesale price of 13/4d per dozen. They were then sold in shops at 2/- each.

A Minix shop display.

A Woolworth's play pack and a Minix twin pack.

PLAYPACKS

Woolworth's Playpacks

These were bubble packs 5"x 6¾", priced 2/9d each, and made for sale in Woolworth's. They did not carry the Tri-ang or Hornby name but were numbered RC50-RC55. Each contained a mixture of vehicles and figures including in some cases a trailer made from the chassis of the caravan but given a different body as described above.

The packs were made in 1968 and 1969 and in excess of 96,000 were supplied. RC50 and RC51 were made in greater quantity than the rest; 23,000 and 26,000 respectively. Just over 10,000 were made of each of the other four packs.

The contents are known of four of the packs and these were as follows:

RC?	RC6 Ford Corsair, RC11 Vauxhall Victor 101, RC16 4/5 Berth Caravan, two yellow petrol pumps on a base and a Model-Land figure.
RC?	RC10 Hillman Imp, RC11 Vauxhall Victor 101, Vehicle Trailer, two yellow petrol pumps on a base and a Model-Land figure.
RC?	Two vehicles, a boat trailer, an orange plastic tent and a Model-Land figure.
RC55	RC7 Sunbeam Alpine, RC14 AEC Reliance/ Strachans Bus, orange red bus shelter and a Model-Land figure.

They were blister packs with a background card with a picture of cross roads, a zebra crossing and a pond.

Miners Industries Play Packs

In 1969, it seems that Miner Industries expressed an interest in assembling their own Play Packs and a supply of parts for 24,000 of each of three sets went through the stores. The sets were numbered 501, 502 and 503 and contained the following:

501	Bus Station Play Pack	bus shelter, RC14 bus and an RC6 Ford Corsair
502	Camping Set Play Pack	tent, RC16 caravan, RC11 Vauxhall Victor and the trailer boat
503	Utility Repair Set Play Pack	RC13 van, RC12 Austin 1800, petrol pumps, high sided trailer and two ladders

RC3X Series

No details are available of this series.

OVERSEAS SALES

In 1967 87,000 cars were sold to a customer in Holland. These were probably in special packaging as the model numbers were given an 'H' suffix. These were RC1H-RC12H and the following numbers were sent:

No.	Model	Sent
RC1H	Ford Anglia	6,500
RC2H	Morris 1100	6,500
RC3H	Vauxhall Viva	6,500
RC4H	Triumph 2000	6,500
RC5H	Austin A60	6,500
RC6H	Ford Corsair	6,966
RC7H	Sunbeam Alpine	7,000
RC8H	Hillman Minx V	6,500
RC9H	Simca 1300	7,000
RC10H	Hillman Imp	7,000
RC11H	Vauxhall Victor 101	6,500
RC12H	Austin 1800	6,500

That year also saw bulk sales going to FJ Strauss of New York. Some were pre-priced and others unpriced. It is thought that these were batches of 600 and 1,000. As we have seen, the models were dispatched in a blister pack pre-printed with the customer's name. There were 1,415 batches, all in the one year, and if the above assumption is correct these contained a total of 905,800 cars.

ATT of America ordered 30,000 vans in 1969 but for some reason the delivery was not made although the vans went into stores. A further order of 18,000 vans the following year also remained in the Margate, although there was a sale to the States of 67,000 Triumph 2000s against a larger order. This may have also been ATT.

An Austrian customer requested a pack of 36 models in 1969 and this was duly made with a total of 326 packs (11,736 cars) being sent and 270 remaining in store and possibly broken up for solo sales or wagon loads at a later date.

MINIX FOR OTHER LINES COMPANIES

Besides supplying major outside customers, as we have seen, Margate received orders from other companies in the Lines Bros. Group. These would have been for cars that could be used in the assembly of toys at their own factories.

In 1970 an order for 36,000 bulk-packed cars came from the Meccano factory at Calais but although the order was put together there is no record of it having been delivered. The proposed use of the cars is not certain but may have been the Hornby-Acho Car Transporter (No.7292 Wagon Porte Voiture) which was illustrated with five Minix cars in the 1970 Hornby-Acho catalogue. The models chosen were the Morris 1100, Ford Corsair, Sunbeam Alpine, Hillman Minx and naturally the Simca 1300. They were sold, loose packed with the wagon. I understand that the cars were silver based.

The Belfast factory also put in an order in 1969 for the Simca 1300, Hillman Imp, Vauxhall Victor and Austin 1800, all to be bulk packed. This order of 63,000 cars was delivered that year.

Australia also received all six twin packs and the batches went out in 1967, 1968 and 1969 accounting for a total of almost 200,000 packs or 400,000 cars.

The New Zealand factory received 'assorted cars' in 1970, 71 and 72, presumably for wagon loads. The total supplied was just 3,300.

One of the better known customers was G&R Wrenn which came under the wing of Rovex Industries Ltd. They had relaunched the former Hornby-Dublo Lowmac wagon with a car load and in 1969 ordered 5,447 Ford Anglia cars from Margate. There is no record of further batches.

TRI-ANG MINIC NARROW-GAUGE RAILWAY (TMNR)

INTRODUCTION

In **Volume 1** in the section on Company History, the Tri-ang Minic Narrow-gauge Railway (TMNR) was mentioned together with a photograph of the train set-up outside the Canterbury factory. At that time, however, I had little information about the system and so it was not expanded upon.

It is believed to have crossed the time zone between Tri-ang Railways and Tri-ang Hornby and as I have had several enquiries for further information, it was decided to expand upon the subject a little further in this volume.

The system was made by Minic Limited at their Canterbury factory from 1963.

Choice

The gauge of the track was 10.25" but train and track for a gauge of 9.5" could be supplied. The system was designed as a garden 'ride-on' railway and we know that some were bought by commercial operators including Butlin's. Prospective buyers were encouraged to visit the factory to discuss their requirements and the coloured leaflet contained a detachable form for booking a visit:

"The Directors of Minic Limited extend a welcome to any customer wishing to consider further the purchase of a TMNR layout and it is hoped that the children of customers will also come to the factory to drive by themselves these wonderful brand new Golden Arrow Locomotives.

All individual layout problems can be studied at such a meeting and the visit places the customer under no obligation whatsoever should the train prove unsuitable for any reason."

Hire purchase terms were available and the equipment was delivered to the customer in a special BR container direct from the factory.

The Loco

The locomotive was based on the E5000 class electric locomotive of the Southern Region but where the pantograph should have been there was a large hole in which the driver (one adult or two children) sat. This choice no doubt was influenced by the Minic factory being sited in Kent where E5000's could be seen hauling the prestigious Golden Arrow expresses.

The motor was located under the driver's seat and it had three forward and a reverse gear. The model was available with one (TMNR1) or two (TMNR3) 0.75 horse power GEC electric traction motors fitted and the double motor version was available also with external controls (TMNR4) if required. Butlin's ordered the twin-engine version with an up-rated power supply with built in speed and direction controller, so that Redcoats could terminate the ride.

Power pickup was from the track and all four wheels on the bogie were driven. The motors ran on a 40 volt supply and the TMNR172 mains power and control unit gave a continuous 40 amp rating. There was also a mains power unit available that

Above: The TMNR leaflet.
Left: TMNR locomotive and coach.

provided a 20 amp rating (TMNR171), and it was claimed that the train could run up a gradient of 1 in 25 with a light load.

The dashboard in the locomotive had two levers. One controlled direction of travel and the other the speed. As optional extras there was also a lamp, which lit to indicate that the power was on, a horn button and a switch for lights. Another extra was brakes which were only considered necessary where long steep gradients were used in the track layout.

The body of the loco was a plastic moulding mounted on a metal baseplate which floated on independently sprung bogies. It was 60.75" long, 17.75" wide and 20.75" high. The distance between bogie centres was 32.5". With the driver and two adults in the Pullman car it could travel at about 8 mph. To break forward movement, the engine was put into reverse; cable-operated brakes were an extra.

The single-motor version had cost £104 in 1964

while the two-motor version with internal controls was £138.5.0 and with external controls £132. The TMNR171 power unit cost £27 and the TMNR172 £64. Covers for the locomotive and coaches cost 17/- each. In all around 85 locos were made of which about 25 are known to have survived.

Pullman Coach

This was TMNR11 and had a plastic body riding on sprung bogies and was finished in normal Pullman livery. The name chosen for the leaflet was 'Emerald'. There were two seats each capable of taking an adult or two children. The coach was 56.5" long and 17.75" wide and cost £47-12s-6d.

Open-Sided Coach

This was available by the start of 1964 but has not been seen by the writer. It contained three toast-rack seats and presumably did not resemble a coach. It cost £34-17s-6d and was numbered TMNR13.

Shunting Locomotive and Trucks

It had been planned to add an 0-4-0 tank locomotive to the stud but it is thought that this did not reach production. This was to have been powered by a 12 volt GEC traction motor and accumulator carried on the engine with a built in 1.5 amp charger.

The loco would have looked very much like 'Nellie' of Tri-ang Railways fame and the child driving it would have sat in the coal bunker with his feet in the cab. It was to have had a whistle as an optional extra. Scale trucks were also proposed.

Literature had claimed that it was due in the autumn of 1963 but by the time the January 1964 price list was published both the loco (TMNR2) and the four wheel non-passenger carrying truck (TMNR12) were still unavailable although the plastic outdoor covers for them were!

The Track

This was made from galvanised steel and it was claimed that it was specially designed to create the safest possible conditions. The rails were screwed to hardwood sleepers and the track sections bolted together.

Sections available included a 6' straight (TMNR101) priced £2-0s-6d, a large radius curve (TMNR110) which cost £2-6s-0d, a small radius curve (TMNR111) priced £2-2s-0d and buffer stop (TMNR141) at £3-3s-6d. Points were also available. The large radius curve required sixteen units to make a 36' circle while the small radius curve needed twelve pieces for a 24' circle. Each piece of track came with fishplates, nuts and bolts.

The system had taken a year to develop during which time it was stringently tested for safety.

Tri-ang Ships shop display including the rare rubber cliff sections made by Young & Fogg Ltd.

MINIC

INTRODUCTION

The production of Minic Motorway was transferred to Margate in 1967 in order to make room at the Canterbury Factory for doll production which was being moved from Merton. Minic had been a trademark of the Lines Group since before the war when production was based in a unit in the Merton factory.

When the success of Tri-ang Railways became apparent, Richard Lines had moved to the Minic factory in 1957 to sort it out. He had found it in a sorry state largely due to the transfer of many of the most experienced staff to overseas factories to help Tri-ang production. His cousin Tony Edmondson worked with him and helped dispose of all the wartime tools used for munitions production. The unit had expanded onto the staff football field to provide storage space for cartons.

In 1958, Minic became a company in its own right and in 1960 was moved into a purpose-built factory in Canterbury. Richard Lines and Tony Edmondson moved with the company but with the death of John Doyle, the Managing Director at Rovex, Richard was soon moved to Margate to take charge of Rovex Scale Models Ltd.

TRI-ANG SHIPS BY MINIC

Introduction

The company had previously made tin-plate, and later, plastic toys but in 1959 launched Tri-ang Ships (more commonly known today as Minic Ships) which were 1:1200 scale diecast waterline models of merchant and fighting ships. These were excellent models available either in sets or singly. There was also a wide range of harbour parts which allowed the collector to build an extensive display laid out on a very realistic plastic sea.

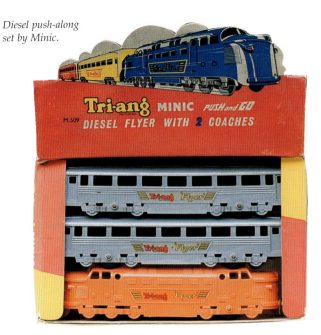

Diesel push-along set by Minic.

Despite the considerable difference in scale, some of the ships were included in the large model road/railway exhibition displays built by Rovex to promote Tri-ang products in the early 1960s.

Merchant Fleet

Taking pride of place amongst the large ships were the RMS 'Queen Elizabeth' and the RMS 'Queen Mary' as well as more modern liners such as RMS 'Canberra' and SS 'United States'. The range included many less famous and smaller liners such as SS 'Flandre', RMS 'Saxonia' and MS 'Port Brisbane'.

There were ferry boats, a pilot ship, life boat, oil tankers, light ships, tugs and HM Yacht 'Britannia'. There were even a whale chaser and a factory ship which came complete with a plastic whale.

Naval Craft

Battleships, aircraft carriers, cruisers, fleet escorts, destroyers, frigates and mine sweepers made up the bulk of the fighting ships but there were also submarines and naval tugs. The smaller ships were available with an assortment of names on their base even if they did not look very different above the waterline. This added to the fun of collecting them.

Accessories

The dock units included breakwaters, quays, docks, sheds, cranes, storage tanks, lighthouses and beacons, bridges, piers, lifeboat shed, ocean terminal buildings, factories, oil terminal and the Statue of Liberty!

The ships were displayed at the New York International Fair in 1960 and were sold through numerous outlets in the UK. After three years, sales had not reached expectations and production ceased. Shops were disposing of their stocks of Tri-ang Ships in the late 1960s and early 1970s.

One of the most interesting items to find is a shop display unit measuring approximately 3'6" x 2'6". This contained a large range of harbour parts and ships fixed down to a board covered with a sheet of plastic sea. Behind the harbour was a section of cliffs moulded in rubber and almost certainly made for Minic Ltd by sister company Young & Fogg Ltd, the makers of the rubber Countryside Series and embankments for Tri-ang Railways.

Hornby Minic Ships

A limited range of the ships and harbour parts was to return in 1976 under the name 'Minic Ships'. There were two gift sets, two harbour sets, four major liners and six large warships including four foreign ones not previously released. The ships all had maroon plastic bases which distinguish them from earlier models. By 1978 they had been dropped from the price list.

Some of the models also appeared in the catalogues of the Spanish manufacturer Anguplus of Barcelona under the name 'Mini Ships'.

Plastic Sea

An attractive feature of the Minic Ship series was the plastic 'sea' supplied with each set and available separately. It was a turquoise semi-opaque plastic sheet embossed to suggest waves. The result was very effective. Two sizes were available: M857 was 26" square and M858 was 52" square.

This plastic sea was also available in the Minic Motorway system for the 'river' in the ferry set M1578. Here it was numbered M857.

MINIC MOTORWAY & CAR PLAY

Introduction

Minic Motorway was launched in 1960. It had been the idea of a man called Rogers, a buyer for Curry's who had seen an American slot-car highway system and had suggested that Lines should make one as an extension of their successful model railways. It was developed at Merton but production was to be carried out in a new factory in Canterbury. From the start it was designed to ride on the back of Tri-ang Railways and so was made compatible, although showing little sense of scale. The system was briefly covered in **Volume 1** and really deserves a book of its own. It will not be covered in any detail here except where it relates to the railways.

Tri-ang Hornby Years

By 1965, the Minic Motorway system had reached its peak of development and a slow decline lay ahead. The first car to go that year was the Austin A40. Despite this, 1966 saw new editions although the emphasis was moving more towards motor racing and away from model highways.

1967 saw two more new cars for the race track and the fire station transformed into a heliport designed to take the Tri-ang Hornby helicopter on a pad on the roof. We also saw the conversion of the bus station to a motorway patrol police station. Track rationalisation meant the loss of the roundabout which

The cover of the 1968 joint Tri-ang Hornby Minic catalogue.

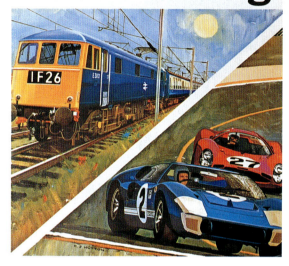

The cover of the 1968 joint Tri-ang Hornby Minic catalogue.

A page from the 1968 Tri-ang Hornby Minic catalogue.

had been such an attractive feature especially when assembled with the complementary centre and verge units.

Move to Margate

In 1968, following the move of production from Canterbury to Margate and the incorporation of the Minic Motorways system into the Tri-ang Railways catalogue, the final new racing cars were introduced. In passing it should be noted that some Minic production, excluding the Motorway range, was moved to the Belfast factory in 1968.

The name 'Minic Motorways' was dropped and the railway related items hived off into a separate part of the price list under the name 'Tri-ang Hornby Minic'. The rest stayed under the name 'Tri-ang Minic'.

Moto-Rail Set

The system was to survive for two more years but these were marked by further rationalisation. This included selling crossings already fitted with gates or barriers and car loading ramps with a car transporter; it also resulted in renumbering of some parts and the marketing of a new combined road and rail set.

The latter contained the dock shunter and the rail car carrier together with the loading ramp and enough track for an oval of road and a circle of rail joined by a common siding. This 1968 issue and some special issue road/rail sets are described in the chapter on 'Sets' earlier in this volume.

Road & Rail Link

Throughout the mid-60s, the joint road and rail equipment had stayed intact. It had always been kept together in the Minic catalogue and from 1964 occupied the back cover. The term Road & Rail Link dates from 1966. The equipment had 'RM' rather than 'M' numbers, and consisted of four subjects:

Level Crossings
Car Transporter and Ramp
Road/Rail Equipment
Freight Depot

Double Track Level Crossing RM905

Three Motorway items, from the start, were advertised in the Tri-ang Railways and Tri-ang Hornby price lists as, for some reason, they were to have been made at Margate rather than Canterbury. These were the RM905 Double Track Level Crossing, RM907 Set of Two Booms (for RM905) and RM908 Automatic Crossing.

RM907 Set of Two Booms did not materialise and the leaflet issued with the Double Track Level Crossing indicated that the "RM904 set of two lifting booms and pillars made by Minic Ltd in the Minic Motorways range" should be used instead. The leaflet incidentally, a first edition, was printed in mid-1964.

The RM905 Double Track Level Crossing had been planned for 1963 and the pre-production model that would serve as the basis for this and the RM908 Automatic Crossing has survived, having been kept by the model-maker when he left the company. It was carved out of a block of clear plastic. The roadway was the equivalent of a double straight and the tracks were single straight rails.

Almost 800 Double Track Level Crossings had been ordered by retailers in 1963 but production did not start until the following year. The 1,460 made in 1964 were sold in red EC size Tri-ang Railways boxes and only 700 were dispatched during the year. A further 3,500 were made over the next three years, 1967 being the last year it was available in this form.

With the transfer of Minic production to Margate, the double track crossing was absorbed into the Minic range. From 1968, it came with booms and was renumbered RM928. In this form it lasted a further two years during which time 2,550 were made. Throughout its life the model was black, and late examples had black track insulation rubber instead of the usual orange ones.

Automatic Crossing RM908
The RM908 Automatic Crossing was not made. A rough mock-up for the Automatic Crossing was seen in a photograph used in the 1962 Gamages catalogue and appears to have made use of the Tri-ang Railways R170 gated electric crossing as a basis for the model. As we have seen, in actual fact it was planned to use the same basic unit as the RM905. The following year the Tri-ang Railways 9th Edition catalogue illustrated both a manual and automatic version of the double track crossing on page 30. Both were in grey plastic but only the automatic version had barriers in place.

The operating mechanism was shown on a protruding plate and the inscription explained that it could be covered by the R145 Modern Signal Box.

Unlike the prototype shown in the Gamages catalogue, this appeared to be operated by a solenoid. The base of the Double Track Level Crossing was designed to also be used by the automatic version, fitted with contact strips which would cut the power to the roadway. There were also to have been breaks in the roadway conductors to allow for isolated sections. Evidence that the RM905 moulding was to have been used for the automatic version, may be seen from the moulding marked on the underside 'RM905-8'. There was also plenty of clearance underneath for the operating rods and closer inspection reveals that anchor points for pivots had been moulded in.

Single Track Crossing RM901
While the Double Track Level Crossing was made at Margate as part of the Tri-ang Railways system, the RM901 Single Track Crossing, based on a straight road unit and a straight rail track, was initially produced in Canterbury as part of the Motorway system. This had pavement sections in each corner into which gate or barrier posts could be plugged. The pair of gates (RM903) and pair of booms (RM904) were available separately. The booms illustrated in the 1962 Minic Motorways catalogue look very much like the ones used on the R171 Transcontinental Level Crossing but those actually released were quite different consisting of a red and white pole. These and the gates were sold separately until 1967.

From 1968, for the final two years, the Single Track Crossing was sold with either gates (RM926) or with booms (RM927) fitted and during this short time, 2,750 of the former version and 1,800 of the latter were made.

While early Single Track Crossings had a grey road section, all those made during the Tri-ang Hornby years should have been black, the colour change having occurred in 1963. The rubber used for

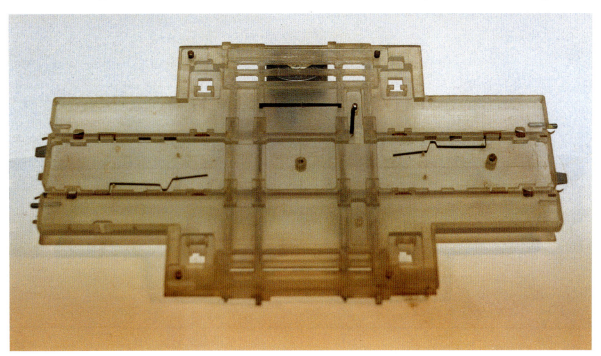

The pre-production sample of the double road/rail crossing.

the insulation in the grooves of the road sections was orange until the late 1960s when it changed to black.

Car Transporter and Ramp RM930

This car loading equipment is described in **Volume 1** and remained in production well into the Tri-ang Hornby period, being finally dropped from the catalogue in 1970. The Car Loading Ramp was RM921 and the Railway Car Transporter was RM922 until production moved to Margate when the two were sold together as RM930 Car Loading Ramp and Transporter. This joint issue first appeared in the 1968 combined catalogue. 2,660 sets of RM930 were made during the two years it was listed.

As we saw in **Volume 1**, the wagon was made in a variety of finishes over the years and it is not known when changes occurred. The only further information I can add to this is that the later wagons have been found without a running number and, at some stage, some ramps were issued in emergency packaging.

Road/Rail Track

In 1961, British Rail had experimented with a system developed on the Chesapeake and Ohio Railroad. The prototype did not have the opportunity to reach revenue earning capacity on BR, being replaced instead with liner trains and container depots.

The road rail track and mechanical horse are described in some detail in **Volume 1**. Like all the road rail components, the track was made to Super 4 standards. The most difficult part to design was that where the rails of the railway track had to cross the slots in the roadway. This had to be done in such a way as to guarantee the smooth running of both the road vehicles and the trains. It was achieved with sprung pads which were depressed as the wheel flanges of rail vehicles ran over them and sprang back into position once the wheel had passed. This was not completely successful and the

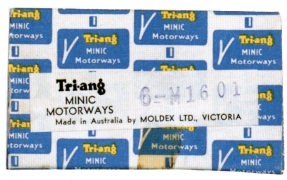

Above: Australian packaging for Minic Motorway track.

Left: One of the last Minic sets to be made at Margate, whose cars are amongst the more sought after in the Minic Motorway range.

track/road units could not manage locomotives larger than 0-4-0s without some difficulty.

Although initially issued in grey plastic, most road/track sections are black and this was certainly so by 1965.

As we saw in **Volume 1**, there were just five sections of road/rail track – a straight, two junctions and two stops. With the incorporation of the system into the Tri-ang Hornby railway system in 1968, it was planned to drop the two stop sections and offer the two junctions (left and right) together as a pair under the number RM929. While the junctions and straight section (RM910) were illustrated in the 1968 catalogue, none was made at Margate and they were not to be seen again.

Mechanical Horse RM925

It had been planned to reintroduce the mechanical horse at Margate as a single unit numbered RM 925 but, while it appeared on the January 1968 price list as being available later, by September it had been

removed. It is interesting to note that it was not even illustrated in the catalogue that year.

Freight Depot M2008, R2008

This first appeared as M1808 and was called a Railway Goods Shed. It had salmon coloured walls and the grey roof of the bus depot. The model had been designed to be extendible both lengthways and sideways and was of a slot-together construction. Following the move of production to Margate, in 1967 it changed colour to red and blue (although still salmon in the catalogue), was renamed Freight Depot and given the number M2008. Only 850 were made in 1968 to orders of just 760. Things did not improve in 1969 with only 210 ordered. A blue roofed building was not what people wanted.

In 1970 the Freight Depot ceased to be a Minic building and was absorbed into the model railway system as R2008. It appeared in the catalogue alongside the container depot crane and was given a new colour scheme – the maroon walls and black roof of the Tri-ang Railways station system. Orders immediately jumped to 3,500 and over 3,550 were made (*see* 'Buildings').

The following year, with the station buildings changing to bright red, it was also illustrated in this colour scheme and 1,400 were made. Although it could be seen in one of the pictures in the 1972 catalogue, the model had been dropped from the price list and no more were made.

Car Play

The name Car Play seems to have appeared first in 1966 and referred to the clip-together buildings that were so much a feature of Minic Motorways. These were models previously available under a different number and some in a different colour scheme. They included:

Car Play number	Model	1966 price	Original number
M2001	Filling Station	9/11	M1801
M2002	Service Station	13/6	M1802
M2003	Drive Through Building	17/6	M1804
M2004	Bungalow with garage	15/11	M1806
M2005	Double Chalet with Garages	13/6	M1814
M2006	Transport Depot	14/6	M1803
M2007	Racing Pit	9/1	M1809
M2008	Railway Goods Shed	11/6	M1808

The R1811 Heliport was later added to the range. Car Play eventually had its own packaging without reference to Minic or Tri-ang Hornby.

French Minic Motorway

Minic Motorway was made in the Calais factory for the European market under the name 'Tri-ang Autoroutes Minic'. At least six sets were available.

The equipment also included the rail/motorway level crossing (Passage à Niveau) RM901 and lifting booms (Barrier de Passage à Niveau) RM904 which were made originally at the Margate factory. Calais

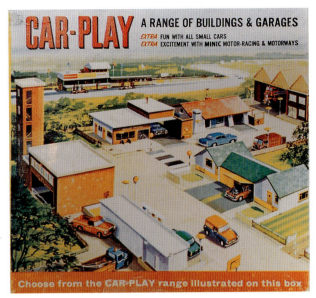

Packaging for Minic Car Play.

provided the two combined in a set with two pieces of Acho adapter track. This combination was numbered RM901/904A.

The End of Minic Motorway

While the public liked the Minic Motorway system, it was not appreciated by retailers who were reluctant to stock the large range of parts available, fearful that if demand fell they would be left with a lot of unsaleable stock on their hands. Production had limped along and in February 1970 the decision was taken not to purchase any further components or packaging. It was decided to limit production to replenishing finished goods stocks.

Thus passed Britain's only comprehensive, railway compatible, slot car system. The fact that such a large range was available and that so much has survived to enhance our Tri-ang Railway and Tri-ang Hornby collections is a blessing.

SPOT-ON LTD

THE COMPANY

Three Products

When Spot-On Ltd was established in a factory built at the back of the large Lines Bros. factory at Castlereagh in Belfast, it was planned to make three new ranges of products. These were Spot-On diecast cars, Spot-on dolls' house furniture and Arkitex construction kits. They all reached production together in 1959.

The cars were an attempt to get into the diecast market which, at that time, was dominated by Dinky, Corgi and Matchbox. They were not very successful despite the fact that unlike other manufacturers Spot-On kept to a single scale (1:42) for all their models. They were also the only range to offer a comprehensive roadway system in pressed metal and plastic allowing the purchaser to make up realistic street scenes.

The End of Spot-On Ltd

With the purchase of Meccano Ltd, there was little future for the Spot-On diecast range and production was terminated in 1967. That year Spot-On Ltd became part of Rovex Industries Ltd, the Model Division within the Lines Bros. Group. Some of the tools were sent out to the Lines Group factory in New Zealand where they were put to work for a time with a range of 12 models which were released by the middle of 1968. With the exception of a Land-Rover, all were cars. In 1969, the remainder of the Spot-On tools were moved out of the Belfast factory, effectively marking the end of the company as a recognisable unit.

The dolls' furniture was also unsuccessful and it is thought that the tools went to Kent where Rovex put some of them to immediate use for the Jenny's Home range which they were marketing by 1970. It is not known whether these were made in the Margate or Canterbury factories.

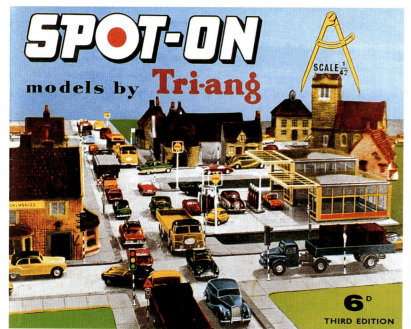

The three product ranges of Spot-On Ltd were diecast cars, dolls' house furniture and Arkitex building sets.

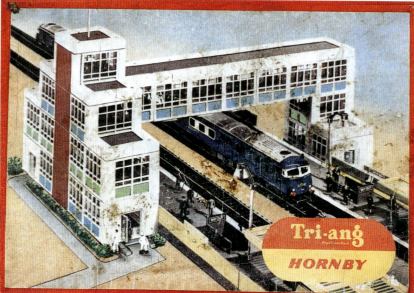

A boxed Tri-ang Hornby ultra modern station set made by Spot-On Ltd.

ARKITEX

Of the company's three products, only Arkitex bore any relationship to Tri-ang Hornby.

The company's own description of the product sums up the system:

"Tri-ang Arkitex modern building equipment is a true to scale modern building construction which will appeal to all ages. Simple self-locking parts, moulded in high impact Polystyrene – the girder framework is erected on a base and the floors are positioned on each storey. Window panels, door panel, plain panels and vertical corners are then dropped into place, followed by the roof edging.

"Each kit is complete in itself, and the basic kit enables the builder to complete simple modern buildings. Step by step instructions are given in the manual and details of expansion to each kit from the wide range of spares and accessories is illustrated and explained – allowing for the construction of unlimited variety of buildings from single office blocks to towering skyscrapers."

Arkitex was available in two sizes: 1:42 scale (for use with Spot-On Cars) and what was claimed to be '00' scale. They were released at the same time. As the '00' sets are far more common today than the larger ones, they must have been more popular. Unfortunately I do not have the production figures with which to make comparisons.

During the decade that Arkitex was available, five main sets and four junior sets were made. In addition to these there was a lighting set, a stairway set and supplementary packs. Just as the trains had an 'R' (for Rovex) as their code prefix, all the Arkitex parts were prefixed with an 'A'. In the case of the small scale series, this was followed by '00' and then the number of the part. In the case of the sets, the suffix was a letter rather than a number.

Set E seems to have been a late introduction and is the only one to have brick panels instead of the usual concrete ones. It allowed modellers to introduce residential buildings amongst their towering office blocks.

The larger scale series had no junior sets and had '42' following the 'A' prefix. The sets were given a number rather than a letter. The 42nd scale series was more intricate and had a greater variety of parts. It also had a different design for panel fixing.

Ultra Modern Station Construction Set R589

To give the system a helping hand, it was decided to market a set under the Tri-ang Hornby label and this was featured in both the 1966 and 1967 catalogues as R589 Ultra Modern Station Construction Set. In order to get that splendid span across the railway tracks a special timber floor unit had to be made for the set, which also contained the Minic Motorway flag and flagpole. The instruction leaflet in the box was published by Rovex.

3,600 of these sets were sold and further details will be found in the chapter on 'Buildings'.

Arkitex was a sophisticated building system but it had to compete with the unsophisticated and much cheaper Chad Valley girder building series. Arkitex was totally unsuccessful despite attempts to link the '00' series to Tri-ang Railways.

TRI-ANG RAILWAYS TT

TT remained available well into the Tri-ang Hornby period although no new models were introduced. A fairly detailed account of the system up to its demise is provided in **Volume 1** and so little further information is given here.

1967 Limited Runs

Special production runs of six models to meet demand were advertised through the *Model Railway Constructor* in 1967. These were:

T91   Castle and Tender (T92)
T97   Britannia and Tender (T98)
T182 Composite Coach in BR brown and cream livery
T183 Brake/2nd Coach in BR brown and cream livery
T184 Restaurant Car in BR brown and cream livery
T370 GWR Toad Brake Van

In May 1967 all but the Castle were being advertised as available by Eames of Reading.

A total of 495 Britannia locomotives were made

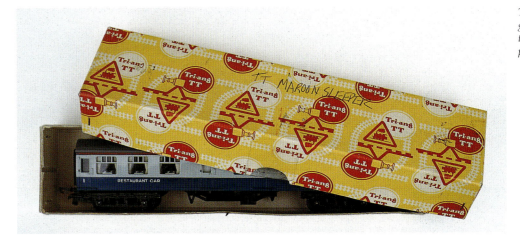

The late blue and grey BR coach in the final style of packaging.

that year together with 566 tenders (T98). The coaches were made in batches of between 500 and 600 and there were about 800 GWR brake vans produced at that time.

There were also 700 tenders (T92) made for the T91 Castle Class loco but the loco itself was crossed out on the census sheet indicating that while production of it had been intended, none were made. Possibly the tools for some part of the loco were no longer available and this was not discovered until the tenders had been made. On the other hand, by June that year Lou Nadin of Warrington were advertising the range of 'Limited Runs' including the Castle. Another possibility is that there was already a stock of tender-less Castle locomotives to dispose of and so it was decided to pair them off with the new tenders and to not make any more.

It is interesting to note that Eames were no longer advertising any of these models by June but, in November Lou Nadin advertised the 'Limited Run' Castle, Britannia and two WR coaches at reduced prices. They were also offering the chassis and tenders of both locos separately.

There were TT models that did not reach production stage and the pre-production models of two of these are in my possession. The first of these is a 2-6-0 loco minus tender. It is finished in black and numbered in yellow '1757' which is strange as the body is Great Western but the GWR did not have a mogul carrying that number. Nor indeed were there any Stanier 2-6-0s with such a number. The other model is of a Great Western Siphon in early BR red.

Wrenn Take-Over

At the 1968 Toy Fair we learnt that Wrenn were to take over the distribution of TT and that blue and grey coaches were planned. Only 1,200 of these were made which being very rare today are exchanged at high prices. Even rarer is the blue version of the diesel A1A-A1A of which little is known.

The *Model Railway Constructor* for September 1970 carried a list of Tri-ang TT spares available from the Service Department at Margate.

TRI-ANG WRENN

INTRODUCTION

G&R Wrenn had three units of a four unit block on an industrial estate at Bowlers Croft, Basildon in Essex. The remaining unit lay between theirs and they could not persuade the occupant to move. Wrenn made motors and assembled models but all casting and moulding work was put out to contract. The casting was done in North London.

The Wrenn Motor

One year, at the Brighton Toy Fair, George Wrenn was seen demonstrating a clever motor he had developed which started when it was spun. He was marketing it in a tug boat called Master Mariner which floated in an inflatable pool which he had on his stand. Lines Bros. liked the motor's potential and so they bought a 66% holding in his company.

Master Mariner was not very successful due to the public's reluctance to allow their children play with water around the house but Lines subsequently used the motor for the Magicars made by Spot-on. When you pushed them they went and when you stopped them the motor stopped. It was also probably a development of this motor that was used to turn the propellers of a Spitfire in the Dinky Toys range and, in 1971, a series of Frog aircraft kits called 'Spin-a-prop'.

New Products

After the take-over of the company, G&R Wrenn was placed under the protective wing of Rovex Ltd.

The main task was to broaden the range of prod-

ucts handled by the company and it quickly became a clearing house for other companies in the Lines Group. Frog Fix N' Fly aircraft, formerly sold by AA Hales Ltd, had been passed onto Rovex to make, but they did not fit in at Margate. These were vacuum-formed and included a Spitfire, Brigand, Buccaneer, Bantam and a Tadpole. Wrenn took over their manufacture but they were not very successful and production ceased in 1969. George Wrenn was looking for something else to do and so, when Rovex rejected Lima's offer to market their N gauge trains in Britain, he took it on.

He also took in some of the old Hornby-Dublo stock from the Meccano factory and Tri-ang TT and sold it through his price lists. He advertised two battery powered (6V) motor boats possibly fitted with the revolutionary engine and from the Pedigree factory in Belfast he took in Doodle Discs and Pull Back 'n Go Mini Cars, as well as the Pedigree Baby Alarm developed in the Tri-onic factory in Hampshire. I acquired one of these second-hand many years ago and we had it working in our house when the children were young. It worked very well and I still have it.

The Hornby-Dublo Tools

For years, Wrenn had been making model railway track but this had become a very competitive field as standards required by a discerning public were steadily rising. The company needed a purpose and they were still worried that they had no major products to sell. George Wrenn was visiting the Meccano factory one day looking for things his company could make when he spied the Hornby-Dublo tools. What followed was something like this:

"There are a lot of tools there not being used", he told Richard Lines. *"I am sure I could do something with them."*

Richard Lines then contacted Joe Fallmann at Meccano and negotiated to obtain two or three sets of tools.

Early temporary packaging for Cardiff Castle.

He probably suggested the 2-6-4T, the 8F and the Castle. Joe Fallmann wanted in the region of £80,000 for the tools so Richard went to Graham Lines and came to an agreement. Wrenn then probably tried some to see if they could use them.

TEMPORARY
PACKING

G.W.R. "Castle" Class

The Tri-ang Wrenn page in the 1969 Tri-ang Hornby catalogue.

Heavy Diecast Metal Locomotives
Built for Tri-ang *HORNBY* **by WRENN**

Early G&R Wrenn price list for obsolete Hornby-Dublo.

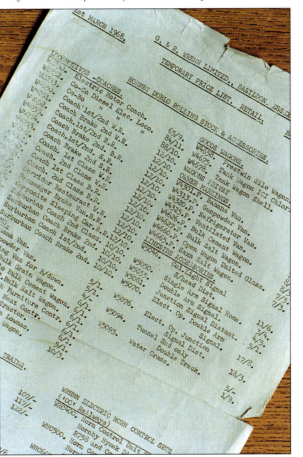

Trademarks

Initially, the models that came out of the factory bore the name 'Wrenn Railways' and no clear logo was used. By August 1969 the Tri-ang Wrenn logo appeared. This was identical to that used for Tri-ang Hornby only the word 'Hornby' had been replaced by 'Wrenn' and a grey and yellow colour scheme had been adopted. This was clearly intended to indicate to the public the close association of the two companies and the compatibility of the products.

With the separation of the companies in 1972, the 'Tri-ang' half of the logo was removed.

The First Locomotives

In December 1966 Wrenn advertised 'The Cardiff Castle' as the first of three:

"fine new Wrenn OO/HO locomotives available in limited supply commencing December price 5.5 guineas."

The advertisement told us that the other two locomotives were to be the 2-8-0 8F priced £5-19s-6d and the 2-6-4 Tank Locomotive priced 5 guineas, both to be delivered in January 1967. The advertisement claimed that:

"Wrenn Locomotives are engineered to the highest standard of quality with finely detailed die castings and powerful DC motors. Each is equipped with a choice of couplings, Tri-ang as fitted, with an exchange clip-on Hornby-Dublo type included."

The advert also stated:

"It is anticipated that Wrenn OO/HO locomotives will become the most sought after locomotives on the market".

Thirty years later, when the company ceased production, they were!

At the 1967 Toy Fair the news was given that initially it was intended to reissue several of the old Hornby-Dublo range, but these were to be improved, and the Hornby trademark would no longer be used on these items. In March that year the first pictures of the 'Cardiff Castle' and 2-8-0 8F were seen in the model railway press for immediate delivery. By December all three locomotives were being advertised.

1967 was also the year in which Wrenn started their other major venture, news of which was released at the toy fair at the start of the year. This was the agreement to sell Lima N gauge under the Wrenn name. More of this is told on page 490.

1968 started with an advertisement for three 00 train sets without track which made use of some of the surplus Hornby-Dublo stock but used the three locomotives so far reintroduced by Wrenn. By now Wrenn were also offering other Hornby-Dublo surpluses including coaches, wagons and signals. They were also marketing the remainder of the

Tri-ang TT stock under the name Wrenn Table Top Railways.

The Wrenn Diesel Horn

In February the now much sought after horn sets appeared. These combined a Tri-ang Hornby loco with a Wrenn horn unit in a Wrenn box. It was made clear, however, that the sets were made for Tri-ang Hornby whose logo they carried.

At the London Toy Fair the horn unit was also demonstrated in use with a Minic Motorways police car and it was not long before two Minic vehicles were offered with the unit, the other being a fire engine.

A spin-off from this was to have been a coach lighting kit but it seems that this did not get beyond the experimental stage unless Rovex adopted the system for their Inter-City coaches.

Private Owner Wagons

By October, Wrenn were advertising eight former Hornby-Dublo wagons in private owner liveries as being available for November release. Again the

models would be fitted with tension-lock couplings but with Hornby-Dublo type couplings provided in the box for those who wished to convert them.

These, when they arrived, were a strange mix. The Hornby-Dublo moulding tool had been used, as had the detailing tool so the wagons carried BR numbers and codes. On top of what would otherwise be a standard BR wagon a modern product logo had been placed. With a few exceptions, the wagons looked anything but authentic but they came at a time when the public were becoming interested in private owner liveries. In the early batches, only two carried their original Hornby-Dublo finish. These were the Saxa Salt wagon and the UD six-wheeled tanker.

Range Expansion

The 0-6-0 tank was the next to be offered early in 1969 and this was quickly followed by the West Country, both in BR livery. By the summer the A4, City and N2 tank were going into production and again BR livery was chosen with the names previously used on the Hornby-Dublo versions. The number of 'private owner' wagons reached 16 by the end of the year.

The supply of Pullman cars for the WP100 set must have been drying up as an alternative set was offered with three BR coaches instead.

There was no mention of the train sets on the 1970 price lists suggesting that stocks of these had cleared. Six of the existing Wrenn locomotives were introduced in new liveries or, in the case of the West Country, a new name. Flexible Tri-ang Hornby System 6 track was also now being marketed by Wrenn and the stocks of the Tri-ang TT appeared to have been cleared.

In 1971 there were no new locomotives but five additional 'private owner' wagons.

Separation

In 1972 G&R Wrenn Ltd ceased to be associated with Rovex (and so will not appear in **Volume 3**). Following the break-up of the Lines Group, Wrenn became an independent company again. It is interesting to note, however, that the Hornby Railways stand at the Brighton Toy Fair in 1972 included the Tri-ang Wrenn range suggesting that at that stage (late January) Wrenn had not completely separated from Rovex although Rovex was now owned by DCM.

The answer lay in a report in the *Railway Modeller* which said that although the company had separated, Rovex would continue to list Wrenn for a year at least. This was presumably to give Wrenn the time to set up their own marketing organisation having relied, until now, on the vast Tri-ang network.

Wrenn continued to make a range of locomotives and rolling stock using the old Hornby-Dublo tools until October 1992 when it closed down. Assets of the company were purchased by Dapol with a view to bringing Wrenn products back into the shops.

Many have asked why the Dublo name could not have been used by Wrenn in later years. The answer is that the names 'Hornby' and 'Dublo' remain the property of Hornby Hobbies Ltd.

CHECK LIST FOR TRI-ANG WRENN

No.	SETS	Dates
WP100	'Cardiff Castle' & 3 Pullman Cars	68-69
WE100	'Cardiff Castle' & 3 BR Coaches	69
WF200	8F & 8 wagons	68-69
WRG300	2-6-4 Tank & 2 Suburbans & 4 Wagons	68-69

No.	HORN SYSTEM	Dates
WHC400	Hymek & Horn	68-70
WHC500	Type 3 Co-Co & Horn	68-70
WHC600	Minic Fire Engine & Horn	68?
WHC700	Minic Police Car & Horn	68?
WHC800	Control Unit Kit	68-70
WHC2	Horn Condenser Unit	69-70
WHC10	Control Box and Leads	69-70

No.	LOCOMOTIVES	Colour	Dates
W2206B*	R1 0-6-0 Tank BR	black	68 -
W2206C	R1 0-6-0 Tank (chassis)		69 -
W2206G*	R1 0-6-0 Tank BR	green	69 -
W2211	A4 'Mallard' BR	green	67 -
W2212	A4 'Sir Nigel Gresley'	blue	70 -
W2216	N2 Tank BR	black	69 -
W2217	N2 Tank LNER	green	70 -
W2218	2-6-4 Tank BR	black	67 -
W2221	'Cardiff Castle' BR	green	67 -
W2221K	'Cardiff Castle' Kit	green	69-70
W2222	'Devizes Castle' GWR	green	70 -
W2224	2-8-0 8F BR	black	67 -
W2225	2-8-0 8F LMS	black	70 -
W2226	'City of London' BR	maroon	69 -
W2227	'City of Stoke-on-Trent' LMS	black	70 -
W2235	'Barnstaple' BR	green	68 -
W2236	'Dorchester' BR	green	70 -

* The suffixes were only to be found on the original price list and did not appear on the boxes.

No.	WAGON	Colour	Dates
W4300	Blue Spot Van - Findus Frozen Foods	white	69-74
W4301	Banana - Fyffes	brown	68 -
W4305	Passenger Fruit Van - Babycham	maroon	70-74
W4311	Guards Van - LMS	brown	69-83
W4313	Gunpowder Van - Standard Fireworks	brown	68-70
W4315	Horse Box - Roydon Stables	green	70 -
W4317	Ventilated Van-Walls	red	68-71
W4318	Ventilated Van - Peek Freans	grey	68-74
W4320	Refrigerator Van - Eskimo	white	68-71
W4323	Utility Van - SR	green	70-74
W4325	Ventilated Van - OXO	white	68-71
W4600	Ore Wagon - Clay Cross	grey	69-71
W4626	Bulk Cement - Blue Circle	grey	68-71
W4627	Bulk Salt - Cerebos	blue	68 -
W4635	Coal Wagon - Higgs	grey	69-76
W4652	Lowmac - Auto Distributors	brown	69-72
W4657	6 Wheeled Tank - UD	white	70-76
W4658	Prestwin - Fisons Fertiliser	red	71-77
W4660	Open Wagon - Twining	brown	68-71
W4665	Salt Wagon - Saxa	yellow	68 -
W4666	Salt Wagon - Sifta	blue	71-74

NB. A date followed by a hyphen means that the model remained in production for a long time.

DIRECTORY OF TRI-ANG WRENN

SETS OF EX-HORNBY DUBLO STOCK

Early sets were made up using some Hornby-Dublo unsold stock. These sets may easily be identified as they were sold in the first batch of boxes printed with the wrong spelling of Basildon.

WP100 'Cardiff Castle' & three Pullman coaches

This set was offered at the start of 1968 and called the 'Pullman Train'. The loco was a Wrenn reintroduction but the Pullman cars were from surplus Hornby-Dublo stock. The price was £7-7s-6d.

The set with this number may be found with other contents. As models from the old Hornby-Dublo stock got used up, alternatives were used. Thus, the WP100 set may be found with the Co-Bo or 'Crepello' with three maroon coaches.

WE100 'Cardiff Castle' & three BR coaches

An advertisement in October 1968 offered this variation on the WP100. Again the locomotive was one of the new batch made by Wrenn but the coaches were old stock Hornby-Dublo and could be either maroon or chocolate and cream. The set was priced £7-12s-6d.

WF200 8F 2-8-0 & eight assorted wagons

This was called the 'Freight Train' and was priced £7-12s-6d. It was available from the start of 1968 and contained old Hornby-Dublo wagons from the Meccano factory but with the recently reintroduced 8F loco now made by Wrenn.

WPG300 2-6-4 Tank, two Suburban coaches & four assorted wagons

At the start of 1968, the set, consisting of Wrenn loco and Hornby-Dublo rolling stock, was sold under the name 'Passenger/Goods Train' and was priced £6-17s-6d. Either maroon or green suburban coaches were used and the wagons varied according to what was available that fitted the box.

HORN SETS & EQUIPMENT

Wrenn had developed a diesel horn unit which was sold either separately, or in a set with a Tri-ang Hornby locomotive or a Minic Motorway vehicle. The two-tone horn fitted inside the locomotive or road vehicle and was remotely controlled using the WHC10 control box.

The horn control box was wired between the power unit and the track with a second 12 DC (uncontrolled) power supply to it from the power unit. This proved to be a poor arrangement because the power unit took both of its 12 DC supplies from the same winding and this had the effect of slowing down the train each time the horn was sounded. It used a 2.5 micro farad 63 volt non-polarised electrolytic capacitor in series with the speaker unit.

The sets contained a general instruction sheet (WHC1167) while the WHC800 contained a supplementary sheet (WHC1167B).

The following sets and separate equipment were available:

WHC400 Loco Set R758 Hymek, WHC10 horn control unit & 2 control leads

The horn control unit was made by Wrenn but the R758 Hymek was the standard Tri-ang Hornby one supplied by Rovex. It sold for £7-9s-3d and was first advertised in February 1968. It came in a grey box of Wrenn design with a cross in a square at each end to indicate which loco the box contained. Although the box lid illustration was of a green liveried Hymek, the one in my box was blue which is what one would expect bearing in mind the date of its release.

WHC500 Loco Set R751 Co-Co Type 3, WHC10 horn control unit & 2 control leads

This was similar to the WHC 400 set but made use of the standard R751 Tri-ang Hornby Co-Co diesels, which was generally blue but could be green. It was also first advertised in February 1968.

WHC600 Car Set No.1550 Fire Engine & WHC10 horn control unit

Offered in 1968 only, priced £5-19s-6d and, if made, would have had the plastic chassis fire engine.

WHC700 Car Set No.1552 Police Car & WHC10 horn control unit

Offered in 1968 only, priced £5-19s-6d and, if made, would have had the white police car with the plastic chassis. It seems that it was intended that the blue lamp on the roof of the police car or fire engine would flash only when the two-tone horn was sounding.

WHC800 Control Unit, Leads and Horn

Available in 1968 priced £5-5s-0d. This was intended for those who wanted to convert their own locomotive, and consisted of a control box, leads and the horn/condenser unit. It came with fitting instructions in a small square brown box.

WHC2 Horn/Condenser Unit

Available in 1969 priced 12/6d. These could be added to any loco or road vehicle providing you already had the control box.

WHC10 Control Box & Leads

Available in 1969, priced £3-19s-11d, and sold in the same small brown box as WHC800 with a tick box to indicate contents.

COACH LIGHTING UNIT

This had been planned as an extension of the horn system in 1968 but I have no evidence that it was made. It is thought likely that Tri-ang Hornby coaches were to be used. It is possible that the experimental coach lighting pre-production model illustrated on page 205 of **Volume 1** was the one developed by Wrenn or that the work done by Wrenn was used in the lighting system adopted for the Tri-ang

Hornby Mk2 coaches in 1969. The following were planned:

WLC800	Control Unit with Lights
WLC801	Composite Coach with lights
WLC802	Brake 2nd with lights
WLC803	Buffet Car with Lights

LOCOMOTIVES

Initially Wrenn had old Hornby-Dublo stock to dispose of and this was sold under a Wrenn number, consisting of the original Hornby-Dublo number prefixed with a 'W'. This stock was included on a price list issued in March 1968 and comprised three locomotives:

W2250	Electric Motor Coach
W2232	Co-Co Diesel Electric
W2233	Co-Bo Diesel Electric

The following locomotives are new production commissioned by Wrenn. All tender locomotives were sold with their tender:

CASTLE CLASS

'Cardiff Castle' W2221

The model was available by the spring of 1967 and was reviewed in the April editions of *Model Railway Constructor* and *Model Railway News*.

Looking at the picture on the page it was hard not to make comparisons with the recently released Tri-ang Hornby Hall. The first obvious criticisms were that the motor not only filled the cab but protruded out almost to the tender. As the model depended on a diecast body it lacked the fine detail possible with plastic moulding; for example, the rivets were over scale and the lining was too heavy. The good points were the wire hand rails and, what everyone mentions, the weight of the engine which gave you a feeling of 'quality'.

The driving wheels had been plated but not those of the front bogie or tender, and the tender chassis had been altered to take the tension-lock coupling.

A black fibre insert filled the hole in the tender front through which pickup wires had passed on the Hornby-Dublo model. The castings carried the inscription 'G&R Wrenn Ltd. Made in England'.

It is surprising that Wrenn chose to reintroduce the model with the same name as the original model as surely with a different name it would have been easier to sell to Hornby-Dublo operators.

Such was the pressure to get the model on the market that it was initially sent out before the boxes designed for the new series had arrived. Consequently it was dispatched in a plain white window box marked 'Temporary Packaging'.

The *MRC* review referred to the bad flashing left on the sample they had received and the poor detail on the safety valve. It also pointed out that there was very little difference between this and the Hornby-Dublo model and yet the price had increased by 25%. The maximum speed when running light was 130 scale mph but this dropped to 100 mph with five Hornby-Dublo coaches attached and 110 mph with five Tri-ang Hornby coaches. In the *MRN* track test, the loco managed a 1 in 30 incline with six coaches without trouble.

'Cardiff Castle' Kit W2221K

This was listed for production in 1969 priced at £5-17s-6d. It came in a long box with the parts sorted out into compartments. It was described as a:

"part assembled kit, designed for easy construction with no painting or soldering required".

It is understood that only about 50 were made and it was found that some kits had wrong parts and so it was withdrawn from sale.

The kits were offered by Hattons of Liverpool with GWR enamelled and engraved brass alternative name plates. The following names were chosen: 'Caerphilly Castle', 'Windsor Castle', 'Harlech Castle' and 'Clun Castle'.

GWR 'Devizes Castle' W2222

Released in the early months of 1970, the model was priced £7-9s-6d. The loco was the same as 'Cardiff Castle' but for its number and the 'G (crest) W' on the tender sides. The lining was that of a BR locomotive. 'Cardiff Castle' remained available although not illustrated in the Tri-ang Hornby catalogue that year.

2-8-0 8F

BR 48158 W2224

This appears to have been available in February 1967 and was illustrated in *Model Railway News* that month as one of three former Hornby-Dublo locomotives to be reissued by Wrenn. No reviews were given in the popular model railway press nor track tests carried out. It was fitted with Tri-ang tension-lock couplings fore and aft and spare Hornby-Dublo couplings were provided in the box. The model had plated driving wheels.

LMS 8042 W2225

This was new in the spring of 1970 when it was priced £7-9s-6d. It was identical to the previous model, which remained in the list, but it had a yellow 'LMS' on the tender sides and a yellow '8F' and '8042' on the cab sides.

2-6-4 STANDARD TANK

BR 80054 W2218

The actual date of introduction is not known but the advertisement in *Model Railway News* in March 1967 said that it was 'available in limited supply'. In December 1967 it was the subject of a *Model Railway Constructor* track test.

It was basically the Hornby-Dublo model which was already commendably accurate as far as dimensions were concerned, but the reviewer was critical of the heavy paint spraying that had swamped much of the casting detail and the lining that was still too heavy. The compromise for track versatility had meant that the rear bogie was slightly too short and the three sets of footsteps had been omitted. The driving wheels were plated but not those of the forward pony truck or the trailing bogie. The centre wheels were flangeless and the back-to-back measurement was 14.5 mm.

The locomotive was driven by the traditional Hornby-Dublo 3-pole vertical motor through worm and gear onto the rear driving axle. The gear ratio was 16:1. The Hornby-Dublo motor had a screw adjustment on the motor which was accessible through the rear of the bunker but this feature was missing from the Wrenn model although the access hole remained.

The adaptation to the Tri-ang tension-lock coupling had not been totally successful as the rear coupling fouled on the buffer beam on tight curves. The bogie also had a tendency to come off the rails during the trial.

The model was finished in BR lined black with the later (BRc) logo and carried the number '80033', the same as the last Hornby-Dublo version. Alternative Hornby-Dublo couplings were provided with the model which clipped onto rivets on the bogies after the tension lock couplings had been removed.

The loco reached 140 mph on the test track but slipping occurred with seven Hornby-Dublo coaches. The speed was smooth as low as 5 mph.

R1 0-6-0 TANK

BR Black W2206B

The model was available in small quantities in December 1968 priced £2-17s-6d. It looked very much like the original model. It had plated wheels and Tri-ang couplings fitted with Dublo couplings provided in the box.

BR Green W2206G

Available in 1969 priced £2-17s-6d. This also looked like the Hornby-Dublo original with its malachite green body and the conversions indicated above.

0-6-0 Chassis with Motor W2206C

Available in 1969 priced £2-9s-6d.

WEST COUNTRY 7P5F CLASS

BR 'Barnstaple' W2235

This was released in time for Christmas 1968 but only in small quantities and priced £8-9s-11d. It looked very much like the Hornby-Dublo model but came fitted with a Tri-ang coupling on the tender and a spare HD coupling in the box. The model was in standard BR green and had plated drivers. At the time Wrenn were criticized for once again not taking the opportunity to produce the model with a different name and Hattons sold their supply of the model with a selection of alternative names on brass plates. The names chosen were 'Exeter', 'Tavistock', 'Ilfracombe' and 'Dorchester'.

BR 'Dorchester' W2236

With the exception of the nameplate and number, this model was identical to 'Barnstaple'. It was available from early 1970 and was priced £7-19s-6d.

N2 0-6-2 TANK

BR Black W2216

This tank was available in the autumn of 1969 and priced £4-5s-0d. The Wrenn leaflet published in August 1969 showed the tank in BR black with the early emblem (BRb) but the model in fact had the later one (BRc) as illustrated in the 1970 Tri-ang Hornby catalogue.

LNER Green W2217

The LNER N2 tank was probably available by the spring of 1970. It was priced £4-9s-6d. The metal body had been sprayed in a bright green to resemble LNER apple green. The LNER lettering, in gold shaded red, was carried on its tank sides and beneath it, in the same colours, was the number '9522'.

DUCHESS/CITY 8P CLASS

BR 'City of London' W2226

Available from the autumn of 1969 priced £7-9s-11d, this was in BR maroon livery but with yellow lining on the footplate, cylinders, cabside and smokebox only. It was the version with a modified front end as last made by Hornby-Dublo and looked very like the Hornby-Dublo model but with a Tri-ang tension-lock coupling on the tender. It had plated driving wheels. The model had a plastic tender on a metal chassis with the late BRc logo, although the November 1969 advertisement in *Railway Modeller* showed it with the old tender markings and BRb logo.

LMS 'City of Stoke-on-Trent' W2227

This was another loco released in the early months of 1970. It was priced £7-19s-6d and was produced in post war LMS black livery lined out in maroon and straw.

CLASS A4

BR 'Mallard' W2211

It must have been a late decision to include this model as the May 1969 leaflet did not show it while the August one did, stating it would be available in November or December. The price was £7-19s-6d. The loco was BR green with the late emblem (BRc) and had plated drivers and a tension-lock coupling on the tender, which was plastic and supplied by Rovex from the A3 production line at Margate. In Wrenn's hands, the A4 was substantially modified.

LNER 'Sir Nigel Gresley' W2212

This was listed as being available in the spring of 1970 priced £7-19s-6d. It was in the LNER blue of the post-war Hornby-Dublo model but had a plastic tender supplied by Rovex. It carried a '7' on its cab sides.

WAGONS

In their first sales list Wrenn offered old stock Hornby-Dublo wagons and also listed eight they intended to make in private owner liveries.

The old stock was given the following numbers:

W4301	Banana Van
W4313	Gunpowder Van
W4318	Packing Van
W4625	Bulk Grain Wagon
W4626	Prestflo Bulk Cement Wagon
W4627	ICI Bulk Salt Wagon
W4647	Wagon with Furniture Container
W4648	Wagon with Meat Container
W4649	Wagon with Tractor
W4652	Lowmac Machine Wagon
W4657	UD Milk Tank Wagon
W4658	Prestwin Silo Wagon
W4675	Tank Wagon ICI Chlorine
W4678	Tank Wagon Shell

The new private owner wagons, which were initially given a 'P' suffix when they were listed, were available from late in 1968. Some renumbering took place in 1970 to eliminate suffixes. The dates given when the model was dropped is not an indication that the model was not later reinstated. As these new batches fell outside the Tri-ang Hornby period

and therefore after G&R Wrenn became an independent company, they are not included. Later, some wagons were available in pre-nationalisation and some in standard BR liveries. Few were authentic and all had tension-lock couplings.

BANANA VAN

Fyffes W4301

Available in the autumn of 1968, priced 7/6d, it survived in the range for many years. The van was initially brown but later in other colours including yellow. It resembled the Hornby-Dublo banana van but had a large blue and white oval transfer in the middle of each side which said 'Fyffes Blue Label Brand'.

GUNPOWDER VAN

Standard Fireworks W4313

The model which may be found in a number of different coloured plastics including brown and dark green, was programmed for release in the autumn of 1968, priced 7/6d. It survived only until 1970. The model was the Standard gunpowder van but with 'Standard' written across the van at an angle in red and white and below it 'Fireworks' in white.

REFRIGERATOR VAN

Eskimo W4320

This used the WR meat van with its white body and grey roof. The name 'Eskimo' was reversed out of a large black panel with a stylised picture of an Eskimo. It was programmed for release in November 1968, priced 7/6d, and survived until 1971.

VENTILATED VAN

Oxo W4325

The ventilated van had a white body and the name 'OXO' in large letters transferred across each side. It was released in November 1968, priced 7/6d, and was available until 1971.

Walls W4317

Originally catalogued W4318P, this was the ventilated van in red plastic with a large 'Walls' logo in yellow in the middle of each side. These may be found with the name applied thickly or in a very thin paint. It was released late in 1968, priced 7/6d, but survived only a short time.

Peek Freans W4318

This model was originally catalogued as W4318P/A. It had a grey body although brown ones may be found. 'Peek Freans' was written in white in the centre of each side with a red and white chef logo to the left. The van was available late in 1968, priced 7/6d, and survived in the range until 1974.

BULK CEMENT WAGON

Blue Circle W4626

This dates from 1968, priced 8/6d, and remained until 1971. This was the Prestflo wagon now in grey plastic with a rectangular board on each side to carry a yellow and blue name and trademark. The wagon was also listed at one time as W4326.

A Tri-ang Wrenn Sir Nigel Gresley and box.

BULK SALT WAGON

Cerebos W4627

This was one of the eight wagons to be available from November 1968, priced 8/6d. It remained in the price list through 1969 and 1970 although the Tri-ang Hornby catalogue for 1971 shows it as a new wagon to be available from mid 1971. This was an error in the catalogue as there were only two new wagons that year.

The wagon had a green/blue body and the same nameboards as the last model but this time they carried the words 'Cerebos Salt' on a blue background.

OPEN WAGON

Twining W4660

This was initially listed as United Glass Wagon but when it arrived it was a brown 5-plank wagon with 'Twining' in large white letters shaded in black and 'Bristol' in smaller white shaded black. It arrived in 1968, priced 7/6d, and remained until 1971. Over the years it was made in three colours – orange, brown and very dark brown.

Higgs W4635

This was a 12-ton coal wagon and had a black 'coal' in-fill. The wagon was grey but may be found in various other colours. It carried the name 'Higgs' in

A boxed Tri-ang Wrenn Higgs open wagon.

large white letters outlined in red, and the name 'London'. It was first listed late in 1969, priced 9/11d, and remained until 1976. It was to return to the catalogue in 1979.

SALT WAGON

Saxa W4665

This was the only one of the eight original private owner wagons to appear in its original Hornby-Dublo livery. It had a yellow body with the roof painted grey and 'Saxa Salt' in red in a curve across each side. It arrived in November 1968, priced 7/6d, and remained for many years with successive batches differing in the shade of plastic used.

Sifta Table Salt W4666

The wagon, which was expected in the shops in the summer of 1971, was priced 12/-. It had a blue body with 'Sifta Table Salt' on the left-hand side and the sailor logo on the right. It survived until 1974.

ORE WAGON

Clay Cross W4600

This used the lower part of the Prestflo wagon with the nameboards on the sides. The wagon had a grey body and a red board. The writing, 'Clay Cross Ltd Limestone', was in white. The wagon was first listed late in 1969, priced 9/9d, and survived until 1971.

LOWMAC MACHINE WAGON

Auto Distributors W4652

This was the brown Lowmac wagon with advertising boards mounted on posts on each side bearing the name 'Auto Distributors Ltd Coventry' on a pale blue background. Between the boards sat a Minix Ford Anglia and caravan. We know that 5,447 Minix Ford Anglias were supplied to Wrenn for this purpose which suggests that the wagons were being

made in batches of about 5,000. The model was first listed in 1969 and remained until 1972.

BLUE SPOT VAN

Findus Frozen Foods W4300

This was the white Blue Spot van printed with 'Findus' reversed out of a red background and 'Frozen Foods' reversed out of blue. It was introduced at the end of 1969, priced 9/11d, and remained until 1974. It was then reintroduced to the catalogue in 1979.

MIDLAND GUARDS VAN

LMS W4311

This was identical to the original BR(LMR) version but with 'LMS' in large white letters on its sides. It was first listed late in 1969, priced 8/11d, and survived until 1983.

6-WHEELED MILK WAGON

United Dairies W4657

This was identical to the original Hornby-Dublo version. It was first listed in November 1969 but probably did not arrive until the spring of 1970. It was priced 13/6d and remained in the range until 1976. Further batches were made in 1978 and 1982.

PASSENGER FRUIT VAN

Babycham W4305

This was available in the latter half of 1970, priced 15/-. It remained in the range until 1974. The model was the standard maroon BR passenger fruit van but with a single advertisement panel on each side promoting Babycham.

HORSE BOX

Royden Stables W4315

The standard BR(SR) green horse box with opening doors now carried a poster on the drop-down door advertising the Foxhunter Championships between October 6-11 at Wembley. Under the windows in white is written 'Roydon Stables Brighton'. The model was probably in the shops in the second half of 1970. It was originally priced 18/- and remained in the range well into the 1980s. Later horse boxes had the date deleted.

UTILITY VAN

SR W4323

The green BR(SR) four-wheeled utility van had 'Southern Railway' printed in yellow under the roof line. The model, which was originally priced 18/-, had opening doors and was to be available from the second half of 1970. It remained in the shops until 1974 but another batch was made in 1977.

PRESTWIN

Fisons Fertiliser W4658

This was available between 1971 and 1977. Further batches were made in 1979 and 1983. It was priced 15/- when it first appeared. The model had a red body with standard Prestwin printing but with 'Fisons Fertiliser' in black on a white band around the top.

COACHES

Wrenn did not make any coaches during the Tri-ang Hornby period but the following former Hornby-Dublo stock was sold under a Wrenn number:

W4050	BR(WR) Composite	W4053	BR Brake 2nd	W4070	BR(WR) Restaurant Car
W4051	BR(WR) Brake 2nd	W4054	BR(SR) Composite	W4075	BR Passenger Brake
W4052	BR Composite	W4055	BR(SR) Brake 2nd	W4078	BR Sleeping Car
		W4060	BR(WR) 1st	W4081	Suburban 2nd
		W4061	BR(WR) 2nd	W4082	Suburban Brake 2nd
		W4062	BR 1st	W4083	Suburban Composite
		W4063	BR 2nd	W4084	Suburban Brake 2nd

Breidt uw modelbaan uit met een locomotief van Tri-ang Hornby

CASTLE CLASS LOC EN TENDER „CARDIFF CASTLE"

Sneltreinloc voor personenvervoer ontworpen door C. B. Collett. Eerste produktieserie augustus 1923. In juni 1932 bereikte de locomotiefnummer 5006 over de afstand Swindon-London, 62 km, een gemiddelde snelheid van 144 km/u. Meer dan 160 exemplaren van dit type loc's werden gebouwd. De „Caerphilly Castle" met het serienummer 4073 staat tegenwoordig in het Kensington Museum te London tentoongesteld. Lengte 19,88 m, gewicht 126,5 ton. Trekkracht 14,358 kg.

2221. Loc met tender . . . . *f* 69.50

HALL CLASS LOC EN TENDER „ALBERT HALL"

Deze locomotief voor personen- en goederenvervoer werd voor Great Western Railways door C. B. Collett ontworpen en na 1926 in de locomotieven fabriek Swindon gebouwd. Enkele van de 250 exemplaren uit deze bouwserie deden meer dan 40 jaar dienst. Lengte van loc met tender 19,2 m, totaalgewicht 122,5 ton. Trekkracht 12,380 kg.

R 759. Loc met tender en personeel . . . . *f* 38.50

EEN TRI-ANG PRODUKT EEN KWALITEITSPRODUKT

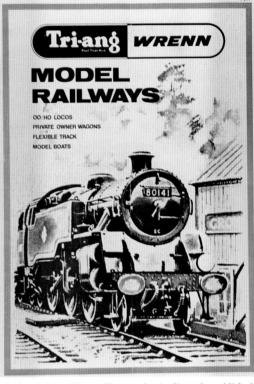

The last Tri-ang Wrenn illustrated price list to be published before the demise of Lines Bros.

Abroad, Tri-ang Hornby and Tri-ang Wrenn were sometimes sold side by side.

LINESIDE ACCESSORIES

The March 1968 list issued by G&R Wrenn also included the following and are believed to be old Hornby-Dublo stock they were clearing:

W5046	Colour Light Signal
W5050	Single Arm Signal Home
W5055	Double Arm Signal
W5061	Junction Signal Distant
W5070	Electrically Operated Double Arm Signal
W5076	Electrically Operated Junction Signal Distant
W5094	Tunnel End only – Double Track
W5095	Water Crane

POWER UNITS

Wrenn produced the WN7 and WN8 units in 1969, almost certainly for use in their own train sets. The residue were sent to Margate in 1971 for disposal, probably by sale overseas.

Power Controller WN7

This was a Wrenn controller that had the same body moulding as the Tri-ang PR15. The control knob looked slightly different and it had metal instead of plastic terminal nuts. There was a single controlled output of 12v DC 0.25 amps (3VA). It was first advertised in the model railway press in September 1969 and cost £2-2s-6d. 338 units passed through the Margate stores in 1971.

Circuit Control Unit WN8

This Wrenn unit used the mouldings of the Tri-ang P42 circuit controller. According to the press advertisement it:

"connects to the 12 volt DC uncontrolled output of type RP14 and the older RP4.5 and RP5.5 power controllers of Tri-ang to provide resistance and reverse polarity control."

It could also be used with the RP 40 or any other suitable 12v DC supply. It was priced £1-4s-6d and the remaining 489 unsold units went to Margate for disposal.

WRENN N GAUGE

INTRODUCTION

Despite the fact that scratch-built models in N gauge, including a brass-built diesel chassis and a large white plasticard warehouse, have been found in the factory suggesting that Rovex were experimenting with the smaller gauge, they did not actually want to make N gauge. They did, however wish to keep their finger on the pulse and in 1967 they did a deal with Lima of Italy under which they would sell Lima N gauge through the Rovex subsidiary, G&R Wrenn, and Lima would be allowed to make Tri-ang Big Big trains under licence.

LONE STAR TRAINS

Rumours about Tri-ang Hornby entering the N Gauge field had been rife for sometime when the 1967 Toy Fair opened and the first models were seen. As early as the summer of 1964 they had been approached by Stan Perrin of Lone Star to ask if they would take the Treble-O-Lectric system off their hands.

Rovex liked the simple rubber-band drive which DCMT used on their Lone Star locomotives. It was a system used by some American manufacturers and Rovex had experimented with the idea themselves but none of their designs reached production stage. DCMT had a patent for this and the way that it retained the wheels in place.

DCMT's problem was distribution. They did not have a salesmen network like Lines Bros. but instead sold their toys through wholesalers. This did not work with a system toy where the customer liked to come back to the supplier for advice when things went wrong.

DCMT had also found that they could not get their prices sufficiently low to make Treble-O-Lectric attractive enough for people to make a change. Their small and rather crude diesels sold for 45/- while a purchaser paid only 49/6d for the A1A-A1A OO model in the Tri-ang range.

Perrin thought that Rovex would be able to make his trains cheaper than he could but the men at Rovex did not agree. The offer came at the time they were realising that their own TT system was not going to be profitable and when they were considering how they could merge the Tri-ang and Hornby-Dublo systems. On 14 August 1964, Richard Lines wrote to Stan Perrin diplomatically turning down the offer.

WRENN N GAUGE MICROMODELS

The Lima offer was different. From the start it was clear that they were to be made by Lima and just marketed under the name of G&R Wrenn Ltd. The news was that a good range of track, locomotives and rolling stock would be available by June 1967 together with two boxed sets which would include British outline models. The first of the British locomotives was to be the AL1 type Bo-Bo 3001.

A full page advertisement for the 'new' system was to be found in July model railway magazines but all the models illustrated with drawings were of Continental outline. The caption read:

"4 into 1 will go – With WRENN N gauge micromodels you can build true replicas of OO railways in ¼ the space. Junctions on window sills and sidings on shelves with over 50 different, perfectly detailed locos and rolling stock items to run on them!"

A fully illustrated coloured brochure and price list were available.

The following month the half-page advert contained photographs of three wagons. These were 405 Mineral Wagon, 406 brake van and 454 BP tank wagon. The first two were clearly models of British prototypes, the standard BR brake van having been adopted for the model, while the tank wagon was equally clearly an existing continental model produced in a familiar livery.

By November two sets were offered for the Christmas trade. Set No1 had the E3001 and 2 coaches and set No2 had the same loco but with four wagons. At the same time the loco was available as a solo purchase.

It was not until the summer of 1968 that we heard any more about new British outline models and then it was the availability of the two coaches in maroon livery and a Buffet car in both maroon and blue/grey. Then in November the Wrenn full page advertisement showed the British and Swedish 2-6-4 tanks, a parlour car Wagon Lits and one of the BR Mk1 coaches in blue and grey livery.

Packaging

The models carried the name 'LIMA' on their undersides and the grey and yellow packaging was inscribed "Made in Italy by Lima for G&R Wrenn Limited, Basildon, Essex, England."

The sets, locos and rolling-stock came in window boxes but other models had simple end flap boxes. All had the catalogue number and contents roughly printed onto a white panel on the box ends.

Later, the livery of the packaging changed to black and orange and all reference to Lima was dropped. By now, the locomotives and rolling stock were sold in clear plastic boxes which had proved popular on the Continent for the smaller scale models.

HORNBY MINITRIX

While Rovex continued to sit on the fence with regard to N gauge, Graham Farish seized the oppor-

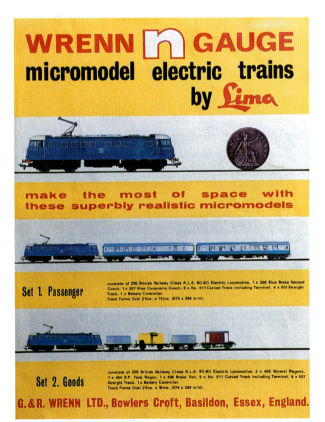

Rovex considered developing N gauge but in the end left it to their associate company, G&R Wrenn, to market the Lima system.

tunity to become Britain's leading developer of the system. The association with Lima probably ended with the collapse of Lines Bros. and the separation from Rovex of G&R Wrenn as an independent company. This, and probably the immediate success of Graham Farish, seems to have given Rovex the opportunity to review its N gauge policy. Still unconvinced that it would prove a good market, they turned instead to a tie up with the German

Minitrix system and its marketing in Britain under the name of Hornby Minitrix. Details of this arrangement will be found in **Volume 3**.

HORNBY-ACHO

Hornby-Acho was made throughout the Tri-ang Hornby period in the Calais factory of Lines Bros. As this company did not become part of Rovex Industries when the model manufacturing companies were put together as a group within the Tri-ang empire and, as it has been been well covered in Michael Foster's book on Hornby-Dublo, I have limited my coverage to the Tri-ang Hornby models marketed by Meccano Tri-ang Lines Freres SA. These will be found in the main body of the book under the standard models. Further reference to these will be found under Company History.

TRI-ANG BIG BIG

INTRODUCTION

Origins
Big Big was the idea of W. Moray Lines, the then Chairman of Lines Bros. Ltd, and was developed at the Tri-ang Research Centre at Canterbury and then handed to Rovex for tooling and production.

The idea was to involve children in copious play participation which could involve the outdoors as well as the house. The plan was that the track could be laid from the house into the garden and a child could send trains outside while staying indoors himself even if it was raining. With this in mind, the system was designed to withstand all weathers. This principle was illustrated in the leaflet (below) produced to launch the system which shows children playing with a set in this way.

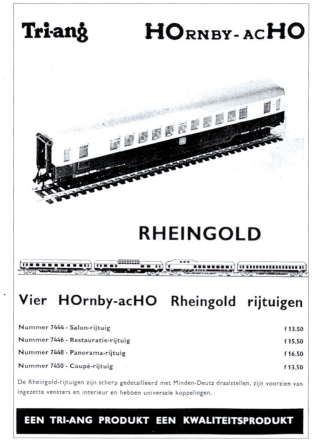

A Dutch advertisement for Hornby-Acho.

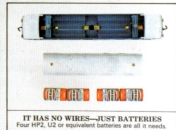

The 1966 Interview

In an interview for the February 1966 edition of *Model Railway Constructor*, Richard Lines was asked about the company's proposals for an O gauge system. He said:

"Our new O Gauge train set is coming along very well and will be on the market very shortly. It is not, however, intended as a model railway so much as a toy train which has the great merit of being able to operate indoors and outdoors. We thoroughly enjoyed ourselves recently testing it running under a shower! We do intend to add some further items to the range in due course; we are just hoping that we do not get too much criticism from serious modellers with regard to the wheel flanges."

In July 1966 the model railway press carried advertisements for Big Big under the heading 'Model or Toy'. They extolled its virtues referring to safety, authentic detail and suitability for outdoor use. Apparently its pliable polypropylene track took the ups and downs of a garden layout in its stride. The trip switch was illustrated as were the two sets available at that time; one had the Blue Flyer and

two trucks and the other the Blue Flyer and four trucks. The new models were reviewed in the March *Railway Modeller* with the words:

"A new 0-gauge product is rare enough; an 0-gauge 'toy' which happens also to be a scale model is an event!"

The reviewer was astounded by the price of the loco – just 39/11d! In attempting to appeal to children, the bright colours were not to everyone's taste but the serious modeller could always paint them.

Couplings

The models were fitted with a coupling similar to the Peco type used on the Hornby-Dublo locos and rolling stock. They were at first attached with a brass screw into a type of heli-coil fixed to the chassis. In 1968 the type of fixing changed to the more familiar brass eyelet.

Packaging

Big Big solo models were packed in yellow cardboard boxes. Initially, they were individually printed with a picture of the contents and the name 'Big

Tri-ang Big Big catalogues.

Big Trains' between two heavy black lines. Later the boxes depicted the contents only on the end flaps and the logo took on a different style with the heavy lines being dropped.

With an uncertain future, in 1971 a number of models were sold in other yellow boxes with a label stuck over the printing or in temporary white boxes when stocks of the normal packaging ran out. These had a printed label stuck on the box ends describing the contents; the labels being produced on a small printing machine in the factory.

Imitations

While the system had a reasonable reception in the UK it sold well on the Continent. In 1967 the concept was sold to Lima in exchange for a deal to supply Wrenn with N Gauge trains. Lima tooled up their own version of Big Big under the name Jumbo. Later they went on to a more realistic 0-gauge model railway system using power from the track.

Big Big was also sold to the American Machine & Foundry Co. (AMF) who tooled up their own version. This was where the idea of the caboose came from.

Yet another version was made in Hong Kong for an Australian store but this was a pirate.

Advertising left you in no doubt that Tri-ang Big Big was for little children and, also, that it was for use in the garden as well as the house.

Tensator Motor

In February 1970 consideration was being given to using the Tensator constant tongue spring clockwork motor in the Big Big locomotives. Mr Fisher of Tensator Ltd UK was to be contacted. It was found that, allowing for a write-off of tools over a 50,000 production run the production cost of a model would be £1-2s-7d. This was quite expensive and would have been a good reason why it should not go ahead.

End of the System

Big Big was discontinued in 1972. In the final years various new models were considered using parts for which tools already existed. One of these was a low steel wagon with two cable drums and another was a steel-covered wagon. Both these are described in the wagon section below.

While a nice product, Big Big was not really profitable due to the heavy tooling costs. Although large quantities were exported, these had a low profit margin – a common problem with exporting.

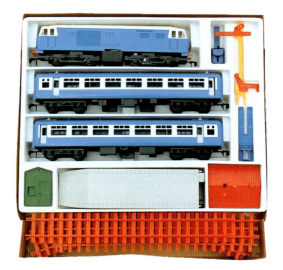

The Novo set from Russia.

Eventually the tools were sold to Russia and used to make the Novo train sets that were available in the UK in the late 1970s. These Novo sets frequently turn up and so, before leaving the subject, it is worth explaining their origin and mentioning the pirate set referred to above.

Novo

Dunbee-Combex-Marx, who had acquired Rovex Ltd at the break-up of the Lines Group in 1971, entered into an agreement with the Soviet Ministry of Light Industries in 1975 under which various factories in the Soviet Union would manufacture kits and toys using tools and materials supplied by DCM. The finished products would be packaged under the Novo name and be delivered to DCM without charge in payment for the tools. Amongst the tools cleaned up and test run for this deal were Frog aircraft kits, Tri-ang TT and Tri-ang Big Big. I understand that linked to this, Russia also supplied stamps which were offered in thematic packs by Hornby Hobbies in 1976.

Novo Toys Ltd was set up in 1975 by DCM as a British company, based in Peterborough, with the sole purpose of handling this business. Once the agreed number of Novo kits, toys etc. had been delivered the tools became the property of the Russians and no further toys were supplied.

With the demise of Dunbee-Combex-Marx in 1980, Novo Toys Ltd went first into receivership and then was wound up when a buyer for the company could not be found.

Mighty 'Red Rocket'

The 'Red Rocket Set' was made in Hong Kong possibly in the early 1980s. When I first saw it I was convinced that it was made from the Big Big tools but Tim Curd, whose knowledge of Big Big I have fed upon in writing this chapter, has convinced me that it was an excellent copy.

The set included the Hymek in red with a white roof and raised white letters giving the name 'Red

The Mighty Red Rocket set from Hong Kong.

Rocket' in the same style as the original 'Blue Flier'. The code on the front of the loco was '4K-86' and the cab windows, unlike the Tri-ang and Novo versions, were glazed.

The other contents were two yellow mineral wagons with black chassis, two green side-tipping wagons, two black switches and an oval of track in dark blue plastic instead of red. The couplings were the Peco type and the only inscription carried on the parts was 'Made in Hong Kong'. The mineral wagons differed from the Rovex ones in having domed buffers.

The box was white with the train flying out of a scenic picture circled in track. The set was called the 'Mighty Red Rocket' and the box carried a sticker with the name 'Artin' on it and a racing car logo.

I do not know the history of this set but it is thought to have been manufactured in Hong Kong for an Australian store. The likeness to the Big Big models is remarkable and a credit to the pirate tool makers!

As they were not Rovex models, neither the Novo set nor the Mighty Red Rocket are dealt with any further here.

CHECK LIST OF TRI-ANG BIG BIG

No.	SETS	Dates
RV266	Train Set No1 (2 trucks)	66
RV267	Train Set No2 (4 trucks)	67?
RV278	Yellow Shunter Train Set	67-68
RV279	Blue Flier Train Set No.3 (freight)	67-68
RV280	Big Big Passenger Set	68-70
RV281	Mixed Traffic Train Set	68-70
RV282	Goods Train Set	68-70
RV320	Express Train Set	70-71
RV321	Lumber Camp Train Set	70-71
RV322	Goods Train Set	70
RV323	Mining Depot Train Set	70-71
RV324	Freight Yard Train Set	70-71
RV325	Zoo Train Set	71
RV326	Local Goods Train Set	71

No.	COACHES	Colour	Dates
RV257	Mk2 Coach	blue	68-71
RV274	Continental Coach	yellow	68-71

No.	LOCOMOTIVES	Colour	Dates introduced
RV256	Hymek	blue	67?-69
RV256A	Hymek	yellow	70-71
RV262	0-6-0T	green	71
RV272	0-4-0DS	yellow	67-69
RV272A	0-4-0DS	blue	70-71
RV276	0-4-0T	red, yellow	68-69
RV276A	0-4-0T	blue	70-71

No.	WAGONS	Colour	Dates
RV258	Open Wagon	red, green	67?-71
RV259	Gondola	yellow, blue	67-71
RV273	Side Tipping Wagon	red, green	67?-71
RV275	Steam Roller Wagon	various	67,68,71
RV277	4-section Change-a-Truck	various	70-71
RV283	Zoo Wagon	yellow + red	71
RV293	Crane Truck	-	not made
RV297	3-Section Change-a-Truck	-	sets only
RV298	Caboose	green	70-71
RV330	Message Carrier	green	72?
?	Cable Drum Wagon	-	not made
?	Covered Steel Wagon	-	not made

No.	TRACK	Dates
RV260	Straight Track	67?-71
RV261	Curved Track	67?-71
RV264	Y Point	67-70
RV269	Half Curve	67-71
RV270	Uncoupler	not made
RV304	Buffer Stop	71
TRACK PACKS		
RV260P	4 Straight Track	70-71
RV261P	4 Curved Track	70-71
RV264P	Y Point and 2 Half Curves	70-71

No.	MISCELLANEOUS	Dates
RV265	Reversing/Stop Switch	67?-71
RV268	Display Unit	67
RV285	Level Crossing	69-71
RV286	Barrel Loader	69-71
RV289	Station	70-71
RV290	Catalogue	70
RV296	Signal	70
RV299	Level Crossing without Track	not made
RV308	Girder Bridge	71

DIRECTORY OF TRI-ANG BIG BIG

SETS

RV266 Blue Hymek, mineral wagon, side tipping wagon, eight curves, two straights

This set sold for 69/6d and was advertised in 1966 but I have no record of the number made. It was referred to as the No1 Train Set and there was sufficient track for a 15ft run. It was the only set made that did not have one or more trackside switches.

RV267 Blue Hymek, two mineral wagons, two side tipper trucks, two switches, eight curves, four straights

This auto-reversing, or No2 Train Set, sold for 99/6d and was available by 1967 and seems to have remained in production until 1970. The track provided with it was sufficient for an 18ft run. Almost 40,000 of these sets were made.

RV278 Yellow 0-4-0DS, steam roller truck, two switches, eight curves, two straights

This was called the Yellow Shunter Train Set and was made for two years. In 1967, 21,000 were produced and a further 3,000 the following year. The catalogue that year indicated that it had only one

The Zoo Set.

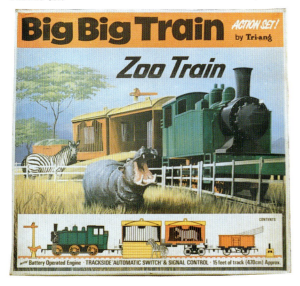

switch in 1967 and two in 1968.

In 1967, supplies sufficient to assemble 5,000 of these sets was sent out to Canada. The Canadian sets would have had only one reversing switch in them.

RV279   Blue Hymek, mineral wagon, side tipping wagon, switch, eight curves, two straights

In 1967, the Blue Flier Train Set No.3 replaced the ill-fated RV266 Blue Flier Set planned for 1966, the difference being that the set now had a reversing switch. It was made for two years during which over 26,000 left the factory.

Canada also received parts for the assembly of 10,000 of this set in 1967.

RV280   Blue Hymek, two blue and white coaches, two switches, eight curves, six straights

The Big Big Passenger Set was introduced in 1968 and remained in production for three years. It made use of the new Mk2 passenger coaches introduced that year and a total of 35,000 were made.

RV281   Red 0-4-0T, blue gondola, yellow and red Continental coach, two switches, eight curves, two straights

Nearly 30,000 of the Red Puffer Sets were made between 1968 and 1970. It was later renamed Mixed Traffic Train Set.

RV282   Yellow 0-4-0DS, mineral wagon, side tipping wagon, two switches, eight curves, two straights

The Yellow Shunter or Goods Train Set was also made between 1968 and 1970 and over 65,000 left the factory during these years.

RV320   Blue Hymek, two blue coaches, switch, signal, eight curves, six straights

The Express Line Train Set was made in 1970 and 1972 and almost 32,000 sets passed through the stores.

RV321   Yellow 0-4-0T, six section change-a-truck, caboose, switch, signal, eight curves, two straights

This was the Lumber Camp Train Set of 1970 which was also available in 1971 and 1972. During the three years, almost 30,000 were released.

RV322   Blue 0-4-0T, mineral wagon, side tipping wagon, switch, signal, eight curves, two straights

The Goods Train Set introduced the 0-4-0 tank engine in a new colour. It was made in 1970 only when over 64,000 were produced.

RV323   Blue 0-4-0DS, mineral wagon, gondola, barrel loader, switch, signal, eight curves, two straights

19,000 of the Mining Depot Train Set were made spread over three years – 1970, 1971 and 1972.

RV324   Yellow Hymek, four section change-a-truck, caboose, switch, signal, level crossing, eight curves, four straights

This was the Freight Yard Train Set of 1970, 1971 and 1972. During these years over 11,600 were made.

RV325   Continental 0-6-0 tank engine, mineral wagon, two cages, bogie section from the change-a truck, a zebra, a hippo, switch, signal, 15 ft oval of track and six Minic Motorway M1707 fence sections

Zoo Train Set was made in 1971 and 1972 and over 24,000 were produced.

RV326   Continental 0-6-0 tank engine, mineral wagon, tipping wagon, switch, signal and 15 ft oval of track

Local Goods Train Set of 1971 and 1972 when a little over 42,000 passed through the stores.

LOCOMOTIVES

HYMEK

Blue Flyer RV256

The *Model Railway Constructor* introduced a review of this model in July 1966 by saying:

"Visitors to the recent Railway Club Exhibition at Central Hall, Westminster, may have noticed two large blue diesel locomotives shuttling backwards and forwards on the Tri-ang Hornby stand. At first glance this appeared as little more than a child's toy railway not for consideration of serious modellers, and it must be admitted that the name ('The Big Big Train'), the plastic track and wagons, in various bright colours, tended to prompt this reaction. When the first locomotive arrived in our offices the other day opinions were altered very quickly indeed. It is an almost exactly scaled replica of a Beyer Peacock Diesel Hydraulic currently in service on the Western Region, the leading dimensions being absolutely correct."

An annoying feature was the name 'Blue Flier' moulded into the locomotive's sides and no doubt many had this ground off and the body re-sprayed. In early examples of the model the words were picked out in white but on later ones they were not painted. Later still the name was removed altogether and a blank rectangular plate left in their place. The body was electric blue and the roof and window frames white and carried the headcode 3D95. Originally the bogies and couplings were secured with a screw/heli-coil but on later ones it was a brass eyelet.

It was driven by a powerful and reliable 3 volt electric motor (X3304) made in Hong Kong and powered by four U2 1.5 volt batteries that were said to last two hours if standard or eight hours if of the long-life type. Fitting the batteries required the removal of the roof which was held on by two screws on early models and a single screw and a lug on the last ones to be made.

Drive was on one axle only and this was gear driven. This axle had toughened rubber wheels for very good adhesion while all other wheels were plastic. The wheel standards were: back to back – 28mm; wheel width – 5mm; flange thickness – 1mm; flange depth – 2mm.

The review described how, when tested, the loco could pull five fully-detailed Exley coaches at a scale speed of 60 mph. On a 1 in 30 bank the speed was reduced to 30 mph and then rose to 85 mph down the other side. This was from a locomotive that weighed only 26 ounces! A review in the *Model Railway News* claimed that it could comfortably pull 10 to 12 wagons and 20 with a bit of help.

The loco had a switch on either side. One marked 'R' was to reverse the loco when it struck a trip on the lineside and the other marked 'S' was for starting and stopping it. The latter could also be operated by a lineside trip.

As the locomotive carried its batteries within the body casing it could be run on the floor without track. For this purpose the bogies were linked by a

The Hymek in its yellow livery and the diesel shunter.

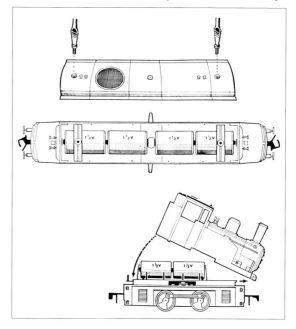

Fitting batteries. Rovex always provided good instructions with their models; a standard that Hornby Hobbies maintains today.

rubber band which prevented them from swinging independently. This had to be removed before the loco was returned to the track. Batteries left in locos were a common cause of corrosion.

It is not known exactly how many were manufactured by Rovex but certainly in excess of 150,000. Almost all were sold in train sets. Only a little under 6,500 were sold as solo models between 1967 and 1971. 10,000 were sent out to Canada in 1967, bulk packed, for use in assembling the RV279 Big Big set over there.

For those interested in converting one of these locomotives to a scale model, there was a lengthy article in the September edition of the *Model Railway News* describing blow-by-blow how it could be done. The writer declared that:

"For many years 00 modellers have benefited from a huge toy market which has provided a wide range of articles that, with a minimum of time and skill, can be converted into very acceptable scale models. Now, for the first time since 1939, an O gauge locomotive has appeared on the market which, whilst unashamedly a toy for younger children, can be transformed into a decent model by even the most ham-handed kit-basher!"

Another article on the subject was published in *Model Railway News* in March 1970.

Yellow & Red Version RV256A

In 1970 the Hymek was introduced in a second colour scheme. It had a yellow body, black chassis and roof and red upper cabs and buffers. The bogies and couplings were fitted with brass eyelets and the roof with a single screw. There was no name on the side of the model. In this form it featured in the RV324 Freight Yard Train Set. A total of 16,500 were made in this livery, 4,700 of them being sold as solo models.

RUSTON 0-4-0 DIESEL SHUNTER

Yellow Version RV272

The model, which was priced 33/2d, was a narrow gauge Ruston shunter but designed to run on O gauge track. It had been designed by Raphael Lipkin, another member of the Lines Group but, rather than produce two similar systems in competition with one another, Raphael Lipkin withdrew and Rovex took over the work already done. The model was first seen by the public at the 1967 Trade Fair.

The loco had a yellow high-impact polystyrene body, black cab roof, running plate and below and red coupling rods. Early examples had glazed cab windows. Levers projected from either side of the body to engage the trackside reversing and stop switches. A small screw between the outside frames held the wheel assembly, which also carried the

very neat and powerful little motor. This was the X3304 Hong Kong motor also used in the Hymek which ran off four 1.5 V batteries. The overall length of the model, excluding couplings, was 6.85" and its height was 4.25".

It is very hard to find a diesel shunter body which has not been split or had its fixing lugs broken off. This was due to poor design in not leaving enough room for the batteries.

It first appeared in 1967 and about 90,000 were sold in train sets and a further 8,300 were sold solo. Another 5,000 were sent to Canada for assembly of the RV278 set over there.

Blue Version RV272A

This first appeared in the 1970 catalogue. It had a blue body, black chassis and red coupling rods. A little over 22,000 of the blue version were made of which about 3,700 were sold solo.

0-4-0 TANK ENGINE

Red Version RV276

The model which first appeared in the 'Red Puffer' set of 1968 had a red body, black chassis (from the 0-4-0DS) and yellow chimney, dome, coupling rods and cab roof. This model often suffers the same problem of body splitting as the diesel shunter. About 32,400 were made in this colour of which a little over 2,600 were sold as solo models.

Yellow Version

This was shown in the 1970 catalogue as the power unit for the lumber camp Train Set. As far as we know about 30,000 were sold; all of them in sets.

Blue Version RV276A

This first appeared in 1970 and had a blue body and yellow chimney, dome and cab roof. The coupling rods were red and the chassis black. 4,000 were sold as solo models and a further 56,000 in sets.

0-6-0 TANK CONTINENTAL ENGINE

Green RV262

In 1970 there were three locos in production and thought was being given to introducing a fourth for 1971. The choice was between a Continental 0-6-0 tank engine and a tender-powered steam outline loco. The former could be manufactured for 13/10d while the latter would have cost 16/10d. The tank engine won and it was tooled up at a cost of about £7,000.

It was reviewed in the *Model Railway Constructor* in August 1971. They found it had less pulling power than the 0-4-0 diesel shunter and could manage only two coaches. They also suggested that like the diesel shunter, the 0-6-0 tank engine was nearer the scale of gauge 1 than 0 gauge.

It appears to have been based on an American tank brought over to Britain during the war to help with munitions trains and afterwards purchased by the Southern Railway. This was a good choice as not only was it recognised as a prototype to be seen on British rails but it had international appeal. The model had a green body and red wheels and the same trackside controls as the other locomotives in this series. It was made only in 1971 and 1972 when a little over 5,300 passed through the stores as solo sales and 66,000 were used in sets.

The model had plenty of room for four HP11 batteries but, because of its fragile chassis construction and the weight of the batteries it was easily damaged by dropping. Finding one today with a good chassis is difficult.

The 0-6-0 tank.

COACHES

Slide switches on the roofs of both coaches allowed the operator to open the doors. Both coaches were made between 1968 and 1972. Sometime in 1968 the inscription on the bottom of the body moulding was blanked out suggesting that the tool was used overseas at this time.

BR Mark 2A Coach RV257

The Mk2a coach arrived early in 1968 and was reviewed by the *Model Railway Constructor* in their March edition. It was thought to be based on an open second with one compartment and the centre door missing, thus reducing the length to a scale 57′. The height and width, however, were virtually to scale.

Plastic moulding was used throughout and the body sides, ends and roof were moulded in blue as one unit. The window panels were sprayed white and the doors were sliding. They were operated by moving a sliding ventilator on the roof. Buffers were omitted but the underframe and floor were well moulded in black and secured by clips and two screws.

The bogies were very good reproductions of the B4 type and were accurate enough to be of interest to those wanting them for other uses. The plastic wheels were a little over scale but they ran well.

It is a fairly common model and invariably turns up with the Blue Flier loco. 133,000 were sold in sets and a further 24,000 were sold solo. Despite being a toy rather than a model, it was a good representation of a BR Mk2a coach and no doubt many were 'improved' by scale O gauge railway modellers; indeed, an article in the June 1968 edition of *Model Railway News* showed how this could be done.

Continental Coach RV274

This used the same body tool as the Mk2a coach but the side splits of the mould were different giving a corrugated effect presumably with the overseas market in mind. It was made in yellow plastic with red window panels, doors and roof switches.

The model was released at the same time as the blue and white Mk2a and 30,000 were sold in sets with a further 14,000 as solo models.

WAGONS

Mineral Wagon RV258

This was released in 1966 and was one of the first two wagons available in this series. It sold for 7/11d and was described in *Railway Modeller* as:

"an accurate model of a BR open mineral".

The subject obviously was chosen because it lent itself to plastic moulding. It was based on a 10' wheelbase, vacuum-braked prototype. Colours at the start were red and green but it was later available in blue. There were four versions of the chassis:

1. Couplings fixed with a screw/heli-coil system and carrying the words 'TRI-ANG MADE IN ENGLAND' slightly off set from the centre of the underside of the chassis. Body clipped to chassis with three lugs and the disc wheels have three holes in them.

2. As 1 but the word 'TRI-ANG' blanked out.

3. Couplings secured with a brass eyelet and the body glued to the chassis in the absence of lugs. The body had four small extensions added to its sides for use in the change-a-truck system and the disc wheels had no holes.

4. As 3 but the words 'TRI-ANG' and 'MADE IN ENGLAND' had gone and 'MADE IN BRITAIN' had been added in small letters under the opposite axle.

About 300,000 were made; 10,000 of them being sent to Canada bulk packed in 1967 for use in assembling RV279 sets.

Side-Tipping Wagon RV273

This had been designed by Raphael Lipkin to go with their Ruston diesel but the Managing Director

of Lines Bros. Ltd decided that Rovex would take over production and merge it with their proposed Big Big system.

It was produced at the start of the series in 1966 and was not prototypical for standard gauge for the reasons given above. It sold for 6/11d and no doubt appealed to those modelling in 58 inch scale. This was made in a choice of red or green plastic. Over 274,000 were made of which 10,000 went to Canada for the RV279 sets assembled over there.

Gondola RV259

This appears to have been styled on an American vehicle which would explain the lack of buffers. The bogies had a scale 6' wheelbase. Colours found include blue and yellow but early on a batch of red ones was made and these seem to be quite rare. The floor had a wood grain finish while the sides were made to look like pressed-steel plates. The bogie frames were excellent plastic mouldings of diamond pattern which were common at the time. The length of the body was 11.25" and it was priced 12/6d.

The model first appeared in the 1967 catalogue with a blue body. Around 1968 the inscription on the bottom of the body moulding was blanked out. This suggests that the tool was used overseas at this time. The gondola was unusual in carrying its 'RV' number embossed on the underside. About 74,000 were made. In June 1968, the *Model Railway News* contained an article which showed how the wagon could be rebuilt and detailed as a Great Central bogie coal wagon.

Steam Roller Wagon RV275

This consisted of a red flat wagon with a black mineral wagon chassis and a royal blue Minic plastic steamroller with grey wheels and roller as a load. The plastic body of the wagon contained three depressions to accommodate the wheels of the steamroller. The steamroller may be found in other colours including pink with a gold smoke stack.

The cost of tooling was a constant worry and the

fact that the tools for the load were already in existence made this an inexpensive wagon to produce. It arrived in the shops in 1967 and over 37,500 were made; 6,500 of them going to Canada for set production. It was priced at 10/3d.

4-Section Change-a-Truck RV277

This was an ingenious gimmick wagon that could be assembled in a variety of forms based on two, three or four units with bogies on the outer units. Besides the four chassis units, two of which carried the bogies, there was a mineral wagon body, a low-sided wagon body, a tipper body that could also be inverted to look like a tank wagon top and also two catwalks which could also be used to turn the wagon into a double-deck car carrier. The low-sided wagon body added to the mineral wagon body turned it into a container. The wagon was released in June or July 1970 and featured in the RV324 Freight Yard Train Set. 30,000 were sold in the set and a further 7,000 as solo models.

Zoo wagon RV283

This was a late wagon arriving as it did in 1971. It consisted of the animal cage, containing one animal from the Zoo Train Set, glued onto the mineral wagon chassis. The elephant, lion, zebra, hippo and polar bear used with this wagon were supplied by Scale Figures Ltd, a recently acquired subsidiary of Lines Bros. Ltd. It is not known how many were made but it is thought to be about 6,000.

Bogie Zoo wagon

This was found only in the R325 Zoo Set of 1971 and consisted of two cages mounted on a bogie chassis from the 4-change wagon. One cage contained a zebra and the other a hippopotamus. About 24,000 were made.

Crane Truck RV293

The model was not made but it had been intended that the wagon would carry the Dinky Toys crane.

3-Section Change-a-Truck RV297

The log truck, as it was also referred to in the factory, was due out in June/July 1970 but was not available as a solo model. It did however appear in the RV321 Lumber Camp Set of 1970. Over 11,000 were made, all of them for use in sets.

It consisted of six parts. These were two end chassis sections each carrying a bogie, a centre chassis section, 10 logs (two lots of five wooden dowels) and a unit that was a tipper wagon one way up and an oil tank unit when inverted. The latter came with two cat walks which could also be used to support an upper deck if required or serve as cradles for the tipper unit.

Caboose RV298

This was a green bodied wagon, released around June or July 1970, which used the mineral wagon chassis. It was not unlike the Old Time Caboose made in the mid 60s for the 00 system. It is thought that about 50,000 were made. The model was later used for the message carrier (see below).

Message Carrier RV330

It was also planned to produce an operating mail van. The mail coach was to allow children to post letters in through a slot in the roof which would be ejected at a predetermined point on the track. It was dropped from the 1971 programme for a short while because of some technical trouble in ejecting the pieces of paper but was reinstated in May 1970 after the design had changed so that the messages were put in a cylinder which was inserted into the roof of the caboose and ejected at the appropriate point along the track. The model eventually used was the green caboose and then not until 1972. It is consequently a rare item. No solo models were sold.

Cable Drum Wagon

This model did not reach production stage but a mock-up was produced. It had a blue 'steel' body consisting of open wagon unit, from the Change-a-Truck model, mounted on the mineral wagon chassis and two Minic cable drums (as used on the Transcontinental cable drum wagon) glued inside.

Steel Covered Wagon

A mock-up for this wagon was also produced at the factory. It consisted of a mineral wagon with the Change-a-Truck open wagon unit inverted on top of it making it look a bit like a Presflo wagon. This also failed to reach production.

TRACK

The track was all moulded in bright red flexible, weatherproof, plastic and had a simple integral clip-on fishplate to join sections together. It was designed to be easy for children to use.

Beatties of London made points to go with the Big Big track which they advertised as early as mid-1966. I have not seen any and assume that they were made by cutting up the red plastic track. These points sold for 15/- each.

Straights RV260

Released in 1967 and priced 1/11d, over 1.5 million were made. 30,000 were bulk packed to Canada in 1967 for use in assembling sets over there. Solo sales in the UK accounted for 430,000 and a further 190,000 were sold in packs of four which were numbered RV260P. The rest were used to make up sets. The straight track was 17 inches long.

Curves RV261

Released in 1967 and priced 1/11d, about 4 million were made of which most were used in sets; each set requiring eight curves which formed a 51 inch diameter circle. Canada received almost 120,000 bulk packed for set assembly over there and 106,000 were sold in the UK in packs of four between 1970 and 1972. These packs were given the number RV261P. Solo sales in the UK accounted for nearly 200,000 pieces.

Y Point RV264

This first appeared in 1967 and was hand operated. About 115,000 were made of which 35,000 were sold in the RV264P pack with two half curves. On its own it was priced 6/1d and was 10" long. The switch rail was separately moulded and fastened with a brass rivet which acted as a pivot. It was operated by a simple spring switch and was half the length of a standard curve. There were holes in the sleepers in three places so that it could be fixed down.

Half Curve RV269

This also appeared in 1967 and about 215,000 were made. Almost 70,000 of these were sold in the RV264P track pack referred to above; the rest being sold solo. This was priced 1/3d, was 9.5" long and could be used with the Y point to either complete a standard curve or to pull back the line of the track to its original direction. Two half curves could therefore be used to create two parallel tracks away from the point.

Uncoupler RV270

Proposed but not made.

Buffer Stop RV304

This hydraulic buffer stop did not arrive until 1971. As a gimmick the buffers rang a bell when compressed by a train running into them. As far as we know, only 5,300 were made.

MISCELLANEOUS

Reverse/Stop Switch RV265

This was released with the first models and sets in 1966 priced 4/11d. Referred to in publicity material as a 'boomerang' switch, it was clipped to the track side and set so that it would catch a lever on the side of the loco causing it to change direction. It could be set to just stop trains or to let trains through in one direction but not in the other.

The colour of the reversing switch was changed to

yellow in 1968 and by 1970 it was blue. In all, a little under 835,000 switches were made of which some 60,000 were sold solo and 15,000 supplied bulk packed to Canada. The rest were used in sets over here.

Display Unit RV268

This was the only item released in 1966 and was available to retailers that year. 1,160 were made.

Level Crossing RV285

This had a green base, red posts and white booms with yellow and black warning discs on them. It came with two pieces of track and was added to the catalogue in 1968. The booms were dropped manu-

ally. 25,000 were made of which 11,600 were used in the RV324 set.

Barrel Loader RV286

This was green and red and used the same base as the level crossing. The overhead loader released a barrel automatically as a wagon passed underneath. It was supplied with six barrels. The model arrived in the shops in 1968. It was also referred to as a 'mining depot' and featured in the 1970 RV323 Mining Depot Train Set for which 19,000 were made. A further 12,000 were sold as solo models.

Station RV289

This clip together unit was introduced in May 1970. It consisted of two platform sections and two track

clips. On the platforms sat a small green building with a red roof. When a button on the roof was pressed a bell rang. This is not a common model, only 8,970 being made for solo sales.

Catalogue RV290/1

This was renewed each year and the 'RV' number varied between RV290 and RV291. In 1970 37,000 copies of the English version were printed but it was also available in five other versions as follows:

Export	16,000
Swedish	10,000
Norwegian	9,800
Dutch	10,000
German/French	10,300

Boxed action accessories.

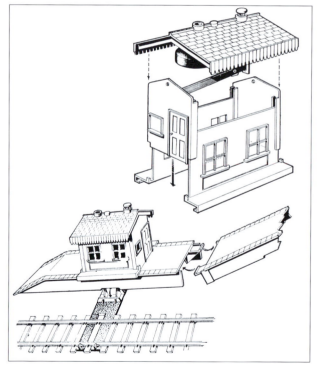

From the instruction sheet on assembling the station.

Signal RV296

This upper quadrant signal first made its appearance in 1970 and was used in several train sets. It had a yellow post and blue or red arm and lever and clipped onto the track. When set to 'danger', the signal lever acted as a stop switch thus halting the loco at the signal.

To connect it to the track it had a blue clip, the same as that used on the station. 'Made in England' or 'Made in Great Britain' were the normal markings but at least one of these clips has been found inscribed 'Tri-ang Hornby'. Over 173,000 signals were made of which a little over 25,000 were sold solo.

Level Crossing without Track RV299

Planned for 1971, this would have consisted of a standard level crossing without the two pieces of track supplied with RV285 but was not made.

Girder Bridge Signal Gantry RV308

This was introduced in 1971 and made use of the old Tri-ang Railways girder bridge tools but had two signals which clipped onto the girder base sections. It came with two pieces of track. Unlike the early bridges this did not have the Lines Bros. logo the boss in the middle of each side. It is an uncommon item, only a little over 4,000 being made.

FROG KITS AND PEDIGREE DOLLS

INTRODUCTION

Aircraft

The International Model Aircraft Company's Frog aircraft kit production was transferred to Margate in 1966; the Real Estate building kits, also developed and made by IMA, having been transferred there in March 1963.

The International Model Aircraft Ltd had been formed in 1931 specifically to make model aircraft. It joined Lines Bros. Ltd the following year to

Frog plastic kit production was transferred to Rovex at Margate in 1966 where the range underwent extensive development.

assist its marketing potential. It took the frog as its trademark and production was moved to a newly built unit at the Merton factory in 1934.

The company's most successful products were its petrol, diesel and hot-point motors for model aircraft. Plastic aircraft kits were a later development and with them IMA became the centre for plastic moulding in the Lines Group.

Pedigree Dolls

The Pedigree trade mark came into being in 1936 and in 1938, Pedigree Soft Toys Ltd was established at Merton to make dolls and other soft toys. Originally the dolls were made of china clay, saw-dust and resin but in 1948 the company went over to plastic doll production. As IMA had the moulding machinery, and it was not being used to capacity, the plastic doll mouldings were made there and sold to Pedigree Soft Toys, in another part of the factory, for assembly and sale. This led to an amalgamation of the two so that IMA made Pedigree Dolls.

The job of stuffing sawdust into soft toys was unpopular with the women at Merton and so it was decided to split production with soft toys going to the Belfast factory in 1959. The Soft Toys became a separate unit in the Belfast factory and in 1967 this formed part of Select Nursery Goods (SNG) and, the following year, Pedigree Westline Ltd.

Separation and Sindy

Meanwhile plastic doll production continued at IMA but there was confusion in the factory with workers not knowing whether they were making aircraft kits or dolls. In 1963 Sindy was launched and became an instant success but by now the technology involved in doll production was different to that used for aircraft kits and in January 1964 Frog and Pedigree Doll production at IMA parted company.

The popularity of Sindy meant that doll production required more room and with Minic production in the doldrums it was decided to move Pedigree Dolls to Canterbury and Minic Motorway production went to Margate where it complemented the railways. (See The History of Sindy by Colette Mansell).

French Production

The aircraft kits had been amongst the products chosen for manufacture in the Calais factory when it opened and kit production started to tick over in France in 1962, reaching full production in 1963. It later came under the control of Meccano (France) Ltd. Feeling that the name 'Frog' might cause offence, the kits were sold under the Tri-ang name on the Continent.

The Move to Margate

With heavy competition in the world of plastic kits, a market dominated increasingly by the Far East and in this country by Airfix Ltd, plastic kit production at Merton shrank to such a low level that Frog aircraft, ship and car kit production was gradually moved from Merton to Margate where advantage could be taken of the more modern moulding machines and Rovex had the benefit of added turnover.

The IMA unit at Merton was closed down around 1967 and the space taken by Leeds Ltd, a subsidiary making cheap photocopiers.

The IMA staff did not move to Margate and this meant that Rovex had to start from scratch with designing, making and marketing Frog kits. A group of kit designers was formed from Rovex's own staff and new models quickly came into production. Technical aeronautical advice was supplied by a company called Avrecon.

Collaboration

In 1967, a deal was worked out with AMT of America who borrowed Frog moulds to produce aircraft kits for the American market while Rovex sold AMT/Frog car kits in Europe. Frog car kits, in

1967, were transferred to Minic Ltd as Rovex were finding it hard to cope with a range of 90 different kits! Another overseas deal was struck with an associate company of AMT, Hasegawa of Japan, who were major players in the kit market. This lead to an exchange of kits between the companies for sale under the recipients' names.

Some of the aircraft kits were also used by the Wholesale Department when Rovex Industries recognised the need to get their products into every corner shop instead of selling them just through their established dealers.

Novo Toys Ltd

The kit range continued to expand at Margate well into the 1970s, the Frog trademark being taken over by DCM on purchase of Rovex Tri-ang Ltd. in 1972.

In 1974, DCM made the controversial decision to sell almost all its Frog moulds to Russia, payment being made in finished kits which were sold under the Novo trademark in Britain by Novo Toys Ltd, established by DCM for this purpose. A few moulds including some not previously put into production were sold to Revell. With the transfer of the tools to Russia, Frog production came to an end in mid 1976.

Pedigree Dolls remained at Canterbury where they were joined by Pedigree Soft Toys when the Belfast factory closed thus completing the circle. The grouping of companies in 1966 had placed the Canterbury factory in the 'Model Division' which was firstly called Rovex Industries Ltd and later Rovex Tri-ang Ltd. As such, as we have seen, they were taken over by Dunbee-Combex-Marx Ltd in January 1972 and continued to be manufactured by Rovex Ltd at Canterbury. In 1980, with the break-up of DCM, the soft toys and dolls parted company with Rovex, being acquired first by Tamwade Ltd and in 1986 by Hasbro, the current makers of Sindy.

Four of these kits were made during the Tri-ang Hornby period and two date from the mid 1970s.

SCALEXTRIC

INTRODUCTION

Minimodels Ltd

In 1970, production of Scalextric was moved to the Margate factory and today forms a major part of the production there.

Minimodels Ltd had been founded by a Mr B Francis in 1947 at Tennyson Road, London, NW6, for the manufacture of tin-plate clockwork toys. These include a motor racing system called Scalex which he launched in 1952. The metal sports or racing cars had a novel type of motor. By pressing the car down while pulling it backwards a fifth wheel engaged the floor and wound up the clockwork spring. When released, the cars raced forwards.

In 1954, they moved to New Lane, Havant and, with sales falling off, looked for a way of renewing public interest in their product. Mr Francis had seen a demonstration of cars on a track operated by electricity. In 1956 he started electrifying his models and using the same basic techniques employed today. This lead to the production of the company's first slot-car set which was given the name Scalextric.

The sets were launched at the January 1957 Harrogate Toy Fair and production commenced in May that year. By mid-1958 Mr Francis realised he needed more money than he had to develop his system and he looked around for an investor.

Tri-ang Take-Over

Lines Bros. Ltd, seeing the potential of this product, purchased the company in November 1958 and applied their knowledge in toy production and factory economics to Minimodels and developing the slot-car system. The situation was almost a repeat of that which had resulted in the establishment of Tri-ang Railways seven years earlier. H K Badcock from Merton was put in charge of the development with the help of Walter Graeme Lines.

Scalextric production was moved into the Rovex factory in Margate in 1970.

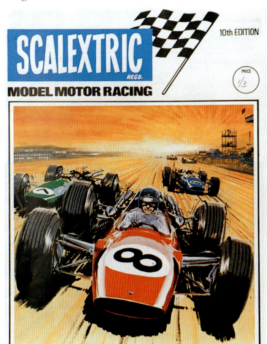

The system was quickly expanded and the first four plastic-bodied cars with Tri-ang RX motors were introduced in 1960. Buildings were based on the Goodwood circuit which was nearby and easy for designers from the factory to study.

As the company expanded, they needed new premises and so a new factory, covering 60,000 sq ft, was built for them on a five-acre site in Fulflood Road, Leigh Park, Havant. The building was to the same design as those at Margate and Canterbury and it opened in May 1961. Their old factory in Havant became the home of Lines Bros. Ltd Electronics Department which was moved there

Production of Jump Jockey transferred to Margate along with Scalextric.

from Merton. This became Trionics Ltd and made, amongst other things, the Trionic construction sets.

Overseas Production

With the completion of the factory in Calais, the doorway into the European Common Market was well and truly wedged open and Scalextric was one of the products chosen for manufacture there.

A Scalextric production unit was also set up in the Moldex factory in Melbourne, Australia and the slot-car system was eventually to be made in America by Lionel under licence. The New Zealand company had a production line and it was also produced in Spain.

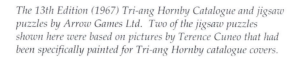

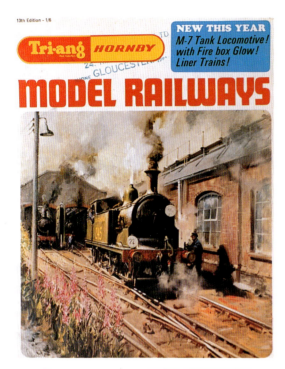

The 13th Edition (1967) Tri-ang Hornby Catalogue and jigsaw puzzles by Arrow Games Ltd. Two of the jigsaw puzzles shown here were based on pictures by Terence Cuneo that had been specifically painted for Tri-ang Hornby catalogue covers.

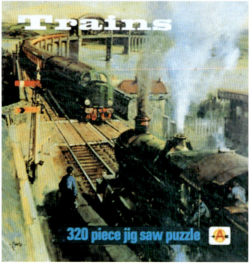

Scalextric clubs formed all over the country to cater for slot-car racing which had become a craze by the early 60s. The National Scalextric Championships were held in Gamages Store and the system was displayed at the New York International Fair in 1960 as well as exhibitions in Stockholm, Brussels and Nuremberg. It was also used as a gimmick at dinners and banquets as something different with which to entertain the guests. A replica Goodwood circuit was constructed for the Monte Carlo Rally Ball in London. The Duke of Richmond had a circuit installed at Goodwood House.

Move to Margate
Plastic track replaced rubber in 1963 making it possible to make it in the factory. Sprung rear suspension, steerable front wheels, blow-out simulation, realistic noises, Race Tuned cars, film related models, the 1:24 series, Power Sledge chassis and 'You Steer' control all followed. Some were good developments but a few were expensive mistakes.

Demand fell off and it was realised that Scalextric no longer needed a factory all to itself. With production of Tri-ang Hornby reduced, it was decided to move slot-car production to Margate. In 1970, Minimodels Ltd ceased to exist and the Havant factory was closed.

The first Scalextric sets began to leave the Margate factory in 1971 and this was the beginning of Scalextric's big fight back.

Jump Jockey
This was a sister product of Scalextric which used the same principle but developed it as steeplechasing.

It was first advertised as part of the Rovex range of products in January 1971 when four sets and a number of accessories were available. All had a 'JJ' prefix to their catalogue numbers.

IBERTREN

INTRODUCTION

In 1960, Spain was closed to imports. Exclusivas Industriales SA (EXIN) wrote to Lines to tell them that they had registered the Scalextric trademark in Spain and wished to start manufacturing it under a licence agreement. They proved to be excellent partners and eventually Lines bought a holding in the

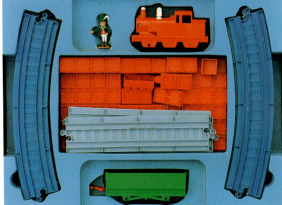

Penny Brix.

company and it changed its name to EXIN-LINES BROS SA.

The owner was Mr Jose Arnau and in 1967 he decided he wanted to go into trains. What he had in mind was a 3 rail N gauge system based on the Marklin design. He went ahead despite advice from Lines Bros. to go for 2 rail. The system was successfully launched at the 1970 Valencia Toy Fair. To handle the train business, a wholly-owned subsidiary called Model Iber SA was set up but this was later absorbed into the parent company. The first products probably started to arrive in the shops in September 1970.

A 2 rail N gauge system was also introduced but their 3 rail survived into the mid 1980s. 2 rail was still available in the early 1990s.

Ibertren also made HO trains which first appeared in the UK in the early 1980s. Looking at the HO range in 1985 it was very difficult to see any Rovex influence. It is true that the motors used in many of the models had a close similarity to the Tri-ang XO4 and that the oil for the smoke generator came in packaging identical to that used by Hornby except for the name that it carried but these were very minor points.

OTHER RELATED PRODUCTS

Arrow Jigsaw Puzzles

Arrow Games, who were another company in the Lines Bros. Group, produced a series of jigsaw puzzles called 'Trains' some of which were pictures that had been used on Tri-ang model railway catalogue covers.

Pennybrix

This was a plastic plug-together building brick system, dating from around 1965-67, probably developed in the Canterbury factory, which also incorporated a push-along railway to give them more 'action'. The track for the latter was similar to that used

Hot Shot. By pressing down on the head of the player, one of its legs kicked out at the ball.

for the Playcraft push-along trains but was blue and had larger sections. The four-wheeled tank engine, which had been a solo toy, the Veteran Puffer (M516) in the Minic range for several years, had the body of an 0-6-0 and there was a goods truck. The railway sets were marketed as 'Playtime Railways' in the late 1960s.

Hot Shot

This was a goal shooting game that was made in the Margate factory around 1970/71. It contained two footballer figures that stood 7" high, two footballs and two goals. The latter came complete with nets but had a card fitted into its front onto which holes had been punched. The idea was to kick the ball through the target holes. This was done by pushing down on the footballer which caused his movable leg to flick out. This may have been based on an idea developed by another Lines Bros. company – Scale Figures Ltd, who made Subbuteo.

Chase Games

Bandit Chase (CG501) was a slot-car chase game developed by AMF of America and made under licence by Spot-On at Belfast, around 1967. Race 'N Chase (CG503) was a larger version dating from 1969. With the closure of the Belfast plant and the winding up of Spot-on Ltd, sales of the chase games

Funfair Games were made by Rovex in the Margate factory for the Tri-ang Wholesale Department.

were handled from Margate.

Rota Race (RR1) was another slot car game, available in 1970, but much smaller and was simply a non-electrified race track ready assembled in a box.

Toys

In 1968 and 1969, a small number of the RV114 Two Telephone Set and the RV115 Two Telephone/Exchange Set were ordered. They did not work very well and there is no evidence of stock passing through the stores to meet these orders, in either year. On the other hand, about 350 of RV253 Table Top Cricket were made and sold during these two years.

Tank Tactix

This was a large scale remote control battle toy using a tank and a field gun that fired at each other until one disabled the other. It had been developed at the Minic factory in Canterbury and production moved to Margate in 1967, along with other Minic production. Originally called 'Shoot or Surrender' by 1969 it was sold as Tank Tactix and about 800 M310 sets were made in 1970.

The only parts that were common to the model railway system were the six red rockets fired by the

field gun. These were the same as the rockets used by the R216 Rocket Launching Wagon. In June 1970 retailers were told that no more could be made as they had run out of boxes and some parts. The set was sold in a display window box with an outer plain cardboard sleeve. The M302 Tank was available separately in 1969. The set had been used for mail order.

Tri-ang Science Sets

Walter Lines had been very interested in the educational value of play. When Minimodels Ltd, the makers of Scalextric, were moved into their large new premises at Fullford Road, Havant, in 1961, it left a vacant factory. Tri-ang, believing that the relatively new electronic age would create a demand for science toys, established production of a range of science sets, called Tri-onics, at the old Minimodels factory at New Lane. Tri-onics Ltd went on to become a company in its own right within the Lines Bros. Group.

Rovex, having undertaken to make the Tri-ang Lionel inventors and laboratory sets at Margate, strengthened their range with an agreement to also produce the German Kosmos sets under licence. This led to the decision to transfer Tri-onic produc-

Tank Tactics used the rocket from the R216 Rocket Launching Wagon.

tion to Margate so that all the science sets could be manufactured under one roof.

I have very limited knowledge of these sets and what follows has largely been gleaned from factory census sheets. In particular, it is not known which were made before 1965 and the years quoted below are those when stock of the products was known to be in the Margate stores.

Tri-onic Communication Equipment

RE102	Baby Alarm	67-69
RE103	Intercom	67-69
RE104	Extension lead	68-69

The baby alarm had a small speaker for the cot and a large one for the parents. The latter had an on/off switch, volume control and 'press-to-talk' switch. The intercom, on the other hand, had two large units, one with a volume control and both with a 'press-to-talk' switch.

Made originally at Havant, in 1967 production of the Baby Alarm and Intercom was transferred to Rovex Scale Models Ltd at Margate where they continued to be made under the Tri-onic name. In 1971, G&R Wrenn took over the RE102 and RE104 and renumbered them WE102 and WE104. At that time WE105 Microphone Unit was added.

Biology Slides

| RE112 | Wasp Microslides | 65 |
| RE113 | Fly Microslides | 65 |

Radio and Electronic Construction Kits

RE202	Single Receiver Kit	66
RE203	Single Amplifier Kit	66
RE205	Earphone for RE202	66

Carpentry Sets

RE500	Carpentry set	66
RE501	Carpentry set	66
RE502	Carpentry set	66
RE550	Junior Chippy	66
RE551	Senior Chippy	66

Tri-onic science sets were made by Minimodels at Havant.

The Tri-onic communications equipment was made by Rovex for a while.

These were probably initially made at Havant and moved to Margate. It seems that very few of the carpentry sets were handled by Rovex and none of the Chippy sets. They were dropped as a result of low orders and the cost of the tools that had to be bought in to make up the sets.

Tri-ang Kosmos Sets
(renamed Tri-ang Science in 1967)

RE508	Junior Technician	65
RE508	The Senior Technician	66-71
RE509	Junior Optician	65
RE509	The Senior Photographer	66-71
RE510	Junior Microscopist	65
RE510	The Senior Microbiologist	67-71
RE511	Junior Chemist	65
RE511	The Senior Chemist	67-71
RE512	Junior Electrician	65
RE512	The Senior Electrical Engineer	67-71
RE513	Radio Technician	65
RE514	Kinderlabor	65
RE514	The Young Technician	66-68

RE515	Elektrofilius	65
RE515	The Young Electrician	66-67
RE516	The Young Weatherman	67-68
RE530	The Weatherman	67-71
RE531	The Electronic Scientist	67-71
RE560	Uncle Intercom	67

These sets were made under licence. The copyright, which dated from 1957, was held by Franckh'sche Verlagshandlung W Keller & Co of Stuttgart. Some of the parts were marked 'Kosmos' suggesting that these were bought in from the originator. Sets I have seen came with a booklet, carrying the name Rovex Scale Models Ltd, each of which contained a large number of experiments by Dr.W.Fröhlich.

Young Constructor Sets

| RE600 | Electric Motor | 65-68 |
| RE602 | Electric Bell | 65-68 |

RE603	Pendulum Clock	65-68
RE604	Padlock	65-65
RE605	Bird Clock	65-67

These were originally made by Tri-onics Ltd at Havant where the kits were numbered CK/600 etc. When transferred to Margate, some of the sets were released in the old packaging but with instruction booklets/sheets which carried the Margate address. The earliest booklet I have found was printed in 1963. The series originally included a water pump kit.

Chemistry and Physics Sets

RE700	Magnetic Power	65
RE701	Chemical Changes	65
RE703	Chemical Moon Rocket	65
RE800	Electricity	65
RE801	Magnetism	65
RE803	Light Waves	65-67

| RE804 | Optical Illusions | 65-65 |
| RE900 | Senior Electro-Magnetic Set | 65 |

Tri-ang Lionel Science and Inventor Sets

The remnants of the Tri-ang Lionel Science and Inventor series were being run down and were listed in the RE3000 series. The last of these were finally cleared out of the Margate stores by the end of 1966. They are well covered in **Volume 1**.

Tri-ang Experimenter Fun

In 1966, a series of mini sets was marketed. The series consisted initially of 10 but was later extended with two composition sets. Each set was contained in a small box which was fixed to a stiff card. This was a little over A4 in size and carried the details, on both sides, of the experiments that could be done with the contents of the box.

Tri-ang Kosmos sets were made under licence from a German company.

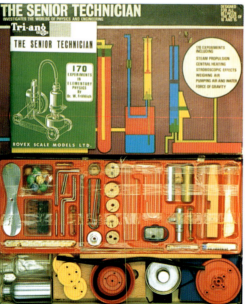

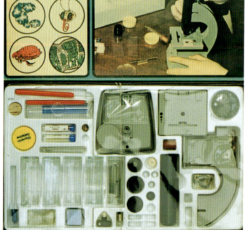

That concludes our study of the Tri-ang Hornby range and its associate products. Of course the story did not end here as the factory at Westwood, Margate, remains in production today. Only the name of the product has changed. We will therefore continue the story in **Volume 3** when we follow the development of Hornby Railways through the 1970s, 1980s and 1990s.

I hope that you have enjoyed this book and that it has brought back many happy memories for you. It has been a great privilege to write. If you have any further information to add to the story I should be pleased to receive it. Any correspondence should be sent through New Cavendish Books.

Tri-ang Experimenter Fun sets were made at Margate from 1966 until 1970.

Left: Young Constructor Sets were made at Margate between 1965 and 1968.

The last of the Tri-ang Lionel sets were cleared from the Margate stores during 1966.

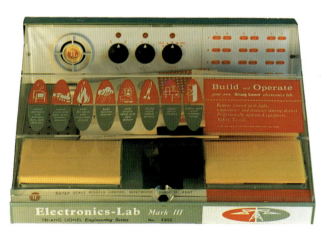

Subjects in the following index have been cross-referenced for ease of access and individual subjects printed bold. Wider areas of study are listed in capitals and illustration page numbers are shown in italics.